Aquatic Oligochaete Biology V

Developments in Hydrobiology 95

Series editor
H. J. Dumont

Aquatic Oligochaete Biology V

Proceedings of the 5th Oligochaete Symposium,
held in Tallinn, Estonia, 1991

Edited by

T. B. Reynoldson & K. A. Coates

Reprinted from Hydrobiologia, vol. 278 (1994)

Springer Science+Business Media, B.V.

Library of Congress Cataloging-in-Publication Data

Aquatic oligochaete biology V : proceedings of the 5th symposium, held in Tallinn, Estonia, 1991 / edited by T.B. Reynoldson and K.A. Coates.
p. cm. -- (Developments in hydrobiology ; 95)
ISBN 978-0-7923-2686-1 ISBN 978-94-011-0842-3 (eBook)
DOI 10.1007/978-94-011-0842-3
1. Oligochaeta--Congresses. 2. Aquatic invertebrates--Congresses. I. Reynoldson, T. B. (Thomas Braban) II. Coates, K. A. III. Title: Aquatic oligochaete biology 5. IV. Series.
QL391.A6A67 1994
595.1'46'0916--dc20 93-47117

ISBN 978-0-7923-2686-1

Printed on acid-free paper

Contents

Hydrobiologia **278,** 1994.
T.B. Reynoldson & K.A. Coates (eds), Aquatic Oligochaete Biology V.

Preface

The Fifth International Symposium on Aquatic Oligochaeta was organized by Drs E. Parmasto, O.A. Scarlato, T. Timm and H. Timm and colleagues at the Estonian Academy of Sciences (Institute of Zoology and Botany, Tartu) and the Võrtsjärv Limnological Station. The meetings were held 15–21 September, 1991 at the Olympia/Sport Hotel in Tallinn. Although the showers only sporadically ran hot, this was not reflected in the consistent, intense exchange of ideas and opinions of the attendees. Over 50 scientists attended, a truly international body with representation from France, Italy, The Netherlands, Canada, USA, Sweden, Finland, Spain, Russia, Armenia, Germany, Estonia, Yugoslavia, Denmark, Poland, Ireland, Belgium, Ukraine, Hungary, Norway and Argentina (poster presented *in absentia*). Included in these Proceedings are 36 of the papers presented at the meetings, covering the many, diverse aspects of oligochaete biology from ecological through evolutionary studies.

The meeting in Tallinn was originally planned as an opportunity to increase contact between Soviet and western researchers in this field. Aquatic oligochaete science has been well-developed in the USSR by such workers as Lastochkin, Svetlov, Malevich, Chekanovskaya, Isossimov and Sokolskaya, among others. Six All-Union symposia on the aquatic oligochaetes had been held in the 20 years 1967 to 1987 but their proceedings were only partially published and then only in Russian. The collection of papers from the first of these meetings was finally translated into English and published as *Aquatic Oligochaete Worms*, Amerind Publishing Company, New Delhi, 1980. Further participation of the Soviet's in scientific interactions among aquatic oligochaete biologists was obstructed by political, financial and language barriers.

Unfortunately, in the period just prior to the meeting in Tallinn the economic circumstances of the USSR and the whole Socialist system reached a point where funds for travel, even within the Union, were not available. Considerably fewer Soviet and Eastern European scientists than planned came to Tallinn. Nonetheless, they were better represented than at any previous meetings and this volume represents the single largest contribution of soviet oligochaete biology presented in English since the 1980 publication. Nine of the 36 papers in this collection are written by authors of the former USSR.

Just a month before the meeting, the USSR collapsed. Thus, the Symposium also became a small, international, celebration of the peaceful restoration of an independent Estonian Republic.

As in earlier Proceedings of the Symposia on Aquatic Oligochaete Biology all manuscripts were rigorously peer reviewed. This process was, at the very least, time-consuming. We thank Barbarann Ruddell of the Department of Invertebrate Zoology, ROM, who prepared the manuscripts for review and later revised the manuscripts subsequent to the authors changes. She also kept track of all the authors and of about 50 reviewers who were contacted by KAC. All the reviewers are thanked for their valuable comments and criticisms. The logistic and financial support of the ROM and Environment Canada were critical to the completion of this project.

We thank also our very gracious hosts in Estonia, especially Drs Vivii and Henn Timm and our banquet hosts at the Võrtsjärv Limnological Station.

K.A. Coates
T.B. Reynoldson
T. Timm

Hydrobiologia **278**: 1–6, 1994.
T.B. Reynoldson & K.A. Coates (eds), Aquatic Oligochaete Biology V.

Some characteristics of genetic variants of *Tubifex tubifex* (Müller, 1774) (Oligochaeta: Tubificidae) in laboratory cultures

Andreas Anlauf
Zoologisches Institut der Universität Köln, Lehrstuhl für Physiologische Ökologie, Weyertal 119, D-5000 Köln 41, Germany
Current address: Bundesanstalt für Gewässerkunde, Schnellerstrasse 140, 12439, Berlin, Germany

Key words: oligochaetes, *Tubifex tubifex*, enzyme electrophoresis, cultures

Abstract

Genetic variants of the oligochaete *Tubifex tubifex* were identified with enzyme electrophoresis and subsequently reared in laboratory cultures. Three types are abundant in field populations. Individuals that show homozygotic bands of glucosephosphate-isomerase (GPI) 22 together with isocitrate-dehydrogenase (IDH) 35 were labeled Type A. Type B is characterized by GPI 23 together with IDH 11 and Type C is characterized by GPI 11 with either IDH 34 or IDH 33. Initial results on freshweights of adults and cocoon production revealed differences between the two main types, A and B. In the same period, Type B reached higher weights and produced five times as many cocoons as Type A, whereas number of eggs per cocoon were not different between these Types. Type B also had the lowest mortality in 16-week experiments with changing temperatures.

Introduction

The oligochaete *Tubifex tubifex* (Müller, 1774) has been observed to have great variability in life cycle and habitat distribution. Laboratory cultures studied by various research groups indicate that growth rate and time until reproduction can vary over a wide range (Bonacina *et al.*, 1987, 1989; Paoletti, 1989; Poddubnaya, 1980; Timm, 1974 cited in Poddubnaya, 1980). Data obtained from field observations showed different peaks of abundance in eutrophic as well as in oligotrophic freshwater habitats (Milbrink, 1973, 1978, 1983; Reynoldson, 1983; Saether, 1980; Särkkä & Aho, 1980; Särkkä, 1987). Some authors, therefore, suggested the species consisted of different ecological races (Korn, 1963; Poddubnaya, 1980; Milbrink, 1983). This suggestion was supported by results of investigations on genetic variability performed with enzyme electrophoresis (Anlauf, 1989). Subsequent to the genetic analysis of field populations (Anlauf, 1989, unpublished data), I established laboratory culture stocks of different genetic variants. The worms of the cultures were collected from field populations and identified using enzyme electrophoresis. They could be used after enzyme analysis for laboratory cultures because they regenerated the posterior part which was used for the electrophoretic analysis. The present paper gives some data that were obtained from these cultures in the first year of observation.

Materials and methods

The worms were collected from profundal populations of the lakes Kaarster See and Broichsee, two gravel-pit lakes in the lower Rhine valley with depths of 17 m and 14 m, respectively. Each worm was cut transversely with a scalpel behind the clitellum. The postclitellar part was used for electrophoresis and the anterior part was set into culture vessels corresponding to the electrophoretic results. All the worms in the cultures regenerated their posterior part completely. The plastic culture vessels (7 by 7 by 5 cm) contained groups of 20 animals in a layer of 1–2 cm silica sand as substrate. Brewed Urtica powder (Synopharm GmbH, Postfach 1205, W-2000 Barsbüttel, Germany) was added as food once every week.

Cultures in long term experiments were kept under constant temperature (5 °C, 15 °C, 20 °C) and light dark cycle (L:D 16:8) for several months except in experiments that were performed to check the constancy of electrophoretic banding patterns of individual worms under varying conditions.

In the short term experiments with variable conditions each worm was reared in a small cylindrical plastic tube (2.5 cm diameter, 5.5 cm height). The bottom of each tube was a net with 100 μm mesh size, which prevented escape by the worms. The tubes were filled with 1–2 cm silica sand, supplied with food and put into temperature controlled water baths with aerated tap water. In the course of each experiment electrophoretically identified specimens were reared for a period of eight weeks. This time is sufficient to guarantee stable physiological conditions in the worms (see Hipp *et al.*, 1984; Hoffmann *et al.*, 1986; Precht *et al.*, 1973; Seuss, 1981). After this time, the temperature was raised from 5 °C to 17 °C (experiment 1) and from 17 °C to 28 °C (experiment 2), respectively. A first cohort (expt.1, $n = 32$, expt.2, $n = 33$) of the experimental group was checked electrophoretically four weeks later. A second cohort (expt. 1, $n = 48$, expt. 2, $n = 33$) was checked eight weeks after the change of temperature. Control worms were kept at the same temperatures for all 16 weeks.

For electrophoretic analysis the posterior part of the worm was homogenized in cold buffer, soaked on one or two filter strips and placed in the starch gel (SIGMA starch, 12% w/v) for horizontal electrophoresis. Details of the electrophoretic method are described by Anlauf (1989) and follow the schemes of Wolf (1982) and Shaw & Prasad (1970). After the electrophomorphs were visualized it could be seen that *T. tubifex* is a diploid organism and I named the genotypes with two digit numbers. Different numbers represent heterozygotes corresponding to different bands and equal numbers represent homozygotes visible as one band.

The populations of *T. tubifex* include several distinct genotypes (Anlauf, unpublished data). Individuals that showed homozygotic bands of glucosephosphate isomerase GPI 22 together with isocitrate-dehydrogenase IDH 35 were labelled Type A. Type B is characterized by GPI 23 together with IDH 11. A third type, C, is characterized by GPI 11 with either IDH 34 or IDH 33. Details of relative proportions in the populations and genetic peculiarities are reported by Anlauf (unpublished data – in prep.) and Anlauf & Neumann (unpublished data – in prep.).

The fresh weight of the worms was measured with an analytical balance (Sartorius M1702P ± 0.1 mg). The worms were taken from the water, placed on a wet filter paper to soak up the outer water and then placed on a tray which was covered with vaseline which prevents the worms from sticking. Stable measurements were obtained after a few seconds.

Statistical analysis comparing mean values of weights was performed according to Sachs (1978). Because the overall numbers of Type C that I found were relatively low, statistically significant results have not been obtained in long term laboratory experiments.

Results

Short term experiments

Individual worms in the laboratory experiments showed the same genetic alleles after the change of temperature as before for both four and eight week periods. However, the mortality of the three genotypes tested during these two experiments

Table 1. Mortality in changing temperature conditions, weight of adults and cocoon production of different genotypes of the oligochaete *Tubifex tubifex* reared in laboratory cultures. Explanations of genotype labels and experimental conditions are given in the text. Numbers in brackets give the total number of individuals introduced in the experiments.

Genotype label	A	B	C
Genotype of GPI	22	23	11
Genotype of IDH	35	11	34/33
Relative mortality in experiments with changing temperatures			
5 → 17 °C	28.6% (35)	10.4% (48)	62.5% (8)
17 → 28 °C	26.9% (48)	5.0% (57)	69.2% (13)
Fresh weight of adults (mg) mean ± s.d.			
5 °C	2.3 ± 0.8 (8)	2.84 ± 1.1 (16) (n.s.)	
15 °C	1.73 ± 0.7 (35)	3.15 ± 1.8 (72) (***)	
20 °C	2.3 ± 0.8 (16)	3.16 ± 1.3 (34) (*)	
No; of cocoons per ind. and week			
Mean ± s.d.	0.1 ± 0.05 (59)	0.5 ± 0.19 (60)	
Range	0.03–0.17	0.23–0.75	
Eggs per cocoon	2.84	2.96	

(Weights: T-Test $p<0.05$*, $p<0.001$***)

was different. Type B was the form most resistant to changing temperature, showing the lowest mortality (Table 1). Type A was more sensitive to temperature changes than Type B but the highest mortality, in both experiments, was observed in Type C (Table 1).

Long term experiments

The worms of Type A and B grown in the laboratory cultures reached significantly different weights. After growing for 3 months, Type B reached a fresh weight heavier than Type A at temperatures of 15 °C and 20 °C (Table 1). There was no significant difference in the fresh weights obtained at a temperature of 5 °C (Table 1).

All genotypes of *T. tubifex* produced eggs/cocoons in winter, spring and early summer with a decline of production beginning in July/August and an intermission during November/December. By counting the production of cocoons per individual per week in early summer, I found Type B worms to produce five times as many cocoons as Type A although the number of eggs per cocoon did not differ between these types (Table 1).

Discussion

Electrophoretic analysis of polymorphic enzymes in populations of the oligochaete *T. tubifex* supports the suggestion of intraspecific races in the species (Anlauf, 1989). Distinct, dominant genotypes (Anlauf, unpublished data) are characterized by distinct allozymes (see Table 1). Anlauf & Neumann, (unpublished data) show a habitat specifity of one of these genotypes, Type C, which was primarily collected from shallow water habitats. The other dominant Types, A and B, were found in all types of field locations, including in deep-water (Anlauf & Neumann, unpublished data), in nearly constant portions. In some of the lakes that were sampled (Anlauf & Neumann, unpublished data) anoxic conditions occurred during summer stratification. In the populations living under these anoxic conditions Type B temporarily increased in abundance whereas the frequency/abundance of Type A decreased (Anlauf, 1989). Hence, I suggested that Type B has a lower mortality during anoxia.

Differences between the main genotypes of *T. tubifex* observed in laboratory experiments can now be added to the field observations of Anlauf (1989) and Anlauf & Neumann (unpublished data). Laboratory cultures were established based on the genotypes that were found and subsequently reared in genetically homogeneous groups. Although Type B worms reached the highest weights in these cultures, their fresh weights were lower than other authors have reported from cultures of *T. tubifex*. For example, Bonomi & Dicola (1980) fed mature specimens with lettuce and found the worms obtained a weight of 4.6 ± 1.08 mg/I ($n = 22$). Marian & Pandian (1984) observed *T. tubifex* with a maximum weight of 5.6 ± 0.4 mg in a sandy sediment with 75% cow dung by weight (28 °C, $n = 30$). Since the fresh weights obtained in the present study were below the values obtained by these authors, one may assume that culture conditions were not as good. It is possible that the food, Urtica powder, is not the most favourable nutrient for culturing all genotypes of *T. tubifex*. However, temperatures in the laboratory may have been too low for high weights, and densities of worms in the vials may have had a negative influence on growth. Paoletti (1989) demonstrated negative correlations of reproduction and fecundity with density in *Tubifex* cultures, but the densities of my cultures were much below the injuring values indicated by Poddubnaya (1980) and Paoletti (1989) and therefore they may be of secondary importance.

Food quality is a major factor influencing growth and fecundity of *Tubifex* in laboratory cultures (Marian & Pandian, 1984; Bonacina *et al.*, 1987). From personal observations Urtica powder is inferior to field sediments. Nevertheless, Urtica powder is useful for culturing the worms because (1) its homogenous structure and dark green colour facilitate the sorting out of cocoons and juveniles from the vessels, (2) genotypes are reared under controlled conditions, and (3) differences in weight and fecundity were obtained.

Temperature is one of the non-tested factors in this study but it is one of the main factors influencing growth and reproduction of *Tubifex* (Poddubnaya, 1980; Reynoldson, 1987). From the results I assume Type B grows more efficiently, at least at 15 °C. To date the attempts to culture genotypes at different temperatures, different densities or with different food supply have not been undertaken. These variables should be investigated further.

The data of the present study do show differences between the genotypes with Type B worms having a higher fecundity and adult fresh weight. Unfortunately, from the fresh weight data it is not clear that the genotypes have different growth rates. Reynoldson (1987) and Bonacina *et al.* (1989) showed that worms reaching higher weights had a higher growth rate and that larger worms also produced more eggs. These results coincide with my data on production of cocoons and fresh weights which show that these differences are genotype dependent. Type B genotypes reached the highest weights and produced one cocoon per individual during a two week period whereas Type A needed 10 weeks. Hence, Type B produced a mean of 0.42 eggs per day and reached similar values of egg production as were described by Bonacina *et al.* (1989), 0.33–0.51 eggs per day, 20 °C, but lower than that of Kosiorek (1974), 0.61 eggs/day, 24 °C, Wisniewski (1979), 0.4–0.9 eggs/day in the field, 5 °C, and Finogenova & Lobasheva (1987), 2 eggs/day.

Due to its higher weight and greater production of eggs one may assume that Type B is more efficient than Type A in gaining energy from the food. If true then this may also explain why Type B can survive better in the profundal of gravel-pit lakes in the lower Rhine area (Anlauf, 1989). These lakes have organic carbon levels in the sediments (3–5% w/w) (Anlauf, 1989; Berndt, 1988) much lower than other eutrophic lakes (Rybak, 1969; Hakanson, 1984) and oligochaete populations do not reach the densities found in highly eutrophic lakes (e.g. Wagner, 1968) and rivers (e.g. Lazim & Learner, 1987; Lazim *et al.*, 1989).

Following are two possible explanations for the lower densities of *Tubifex*. One is that the food provides different organic compounds which are selectively used by the genotypes and are more favourable to Type B worms. A differential assimilation of bacterial compounds from organic nutrition was only observed, to date, from different species (McMurtry *et al.*, 1983) although it has not been analyzed in the same species. Further investigations on the genotypes should take this into account. Another explanation may be a metabolic adaptation of Type B. If Type B has a greater assimilation efficiency, then how can we look for metabolic adaptations? Anlauf (1990) reported a lowered metabolic activity in encysted specimens of *T. tubifex*. Since all encysted worms that have been genetically analyzed, belong to Type B, I believe this genotype is able to reduce its routine metabolic activity. If true then it might have more efficient growth and reproduction when food is limited.

A correlation of growth rate and metabolic rate with genetic phenotype, especially the rate of heterozygosity, was reported for some species of marine mussels (*Mulinia lateralis*: Garton, *et al.* 1985; *Mytilus edulis*: Koehn & Gaffney, 1984). Garton *et al.* (1985) observed correlations of routine metabolic rate with rates of heterozygosity (Gaffney, 1990; Koehn & Gaffney, 1984; Koehn & Shumway, 1982) and correlations of growth and fecundity with lowered routine metabolism. Some authors observed specimens with the highest rates of heterozygosity to grow faster and reach higher weights and fecundity and then proposed that the routine metabolism was different in the genotypes. Further physiological investigations of the different *Tubifex* genotypes should include studies of anaerobic metabolism.

The present data support the suggestion of different ecological races in *T. tubifex*. The genetic variability of other laboratory cultures (Bonacina *et al.*, 1987; Paoletti, 1989; Timm, pers. com.) should be checked for genetic characteristics.

Acknowledgements

I am grateful to Prof. Dr D. Neumann for his support on this study. Further thanks are due to M. Volpers, D. Simons, R. Mittmann, and M. Ryl for technical support in the field and laboratory.

References

Anlauf, A., 1989. Die Charakterisierung von Populationen des Schlammröhrenwurms *Tubifex tubifex* (Müller) mit Hilfe von enzymelektrophoretischen, populationsgenetischen und ökologischen Methoden. Ph. D. Thesis, Universität Köln, Germany, 132 pp.

Anlauf, A., 1990. Cyst formation of *Tubifex tubifex* (Müller) – an adaptation to survive food deficiency and drought. Hydrobiologia 190: 79–82.

Berndt, J., 1988. Die ökologische Bewertung von niederrheinischen Baggerseen mit Hilfe von Makrozoobenthosarten als Bioindikatoren. Ph.D. Thesis, Universität Köln, Germany, pp.

Bonacina, C., G. Bonomi & C. Monti, 1987. Progress in cohort cultures of aquatic oligochaeta. Hydrobiologia 155: 163–171.

Bonacina, C., G. Bonomi & C. Monti, 1989. Population analysis in mass cultures of *Tubifex tubifex*. Hydrobiologia 180: 127–134.

Bonomi, G. & G. Dicola, 1980. Population dynamics of *Tubifex tubifex*, studied by means of a new model. In R.O. Brinkhurst & D.G. Cook (eds), Aquatic Oligochaete Biology. Plenum Press, New York: 185–203.

Finogenova, N.P. & T.M. Lobasheva, 1987. Growth of *Tubifex tubifex* Müller (Oligochaeta, Tubificidae) under various trophic conditions. Int. Revue ges. Hydrobiol. 72: 709–726.

Gaffney, P.M., 1990. Enzyme heterozygosity, growth rate, and viability in *Mytilus edulis*: another look. Evolution 44: 204–210.

Garton, D.W., R.K. Koehn & T.M. Scott, 1985. The physiological energetics of growth in the clam, *Mulinia lateralis*: an explanation for the relationship between growth rate and individual heterozygosity. In P.E. Gibbs (ed.), Proc. 19th Europ. Mar. Biol. Symp. Cambridge University Press, Cambridge, England: 455–463.

Hakanson, L., 1984. On the relationship between lake trophic level and lake sediments. Wat. Res. 18: 303–314.

Hipp, E., U.A. Sedlmaier & K.H. Hoffmann, 1984. Aerobic metabolic trends after anoxia in the freshwater oligochaete *Tubifex*. Comp. Biochem. Physiol. 78B: 125–129.

Hoffmann, K.H., J. Seuss, E. Hipp & U.A. Sedlmaier, 1986. Aerobic and anaerobic metabolism in *Tubifex*, a freshwater oligochaete. Zool. Beitr. 30: 153–170.

Koehn, R.K., & P.M. Gaffney, 1984. Genetic heterozygosity and growth rate in *Mytilus edulis*. Mar. Biol. 82: 1–7.

Koehn, R.K., & S.E. Sumway, 1982. A genetic/physiological explanation for differential growth rate among individuals of the american oyster *Crassostrea virginica* (Gmelin). Mar. Biol. Letters 3: 35–42.

Korn, H., 1963. Studien zur Ökologie der Oligochaeten in der oberen Donau unter Berücksichtigung der Abwassereinwirkungen. Arch. Hydrobiol. Suppl. 27. 181 pp.

Kosiorek, D., 1974. Development cycle of *Tubifex tubifex* (Müll.) in experimental culture. Pol. Arch. Hydrobiol. 21: 411–422.

Lazim, M.N., & M.A. Learner, 1987. The influence of sediment composition and leaf litter on the distribution of tubificid worms (Oligochaeta). Oecologia 72: 131–136.

Lazim, M.N., M.A. Learner & S. Cooper, 1989. The importance of worm identity and life history in determining the vertical distribution of tubificids (Oligochaeta) in a riverine mud. Hydrobiologia 178: 81–92.

Marian, M.T. & T.J. Pandian, 1984. Culture and harvesting techniques for *Tubifex tubifex*. Aquaculture 42: 303–315.

McMurtry, M.J., D.J. Rapport & K.E. Chua, 1983. Substrate selection by tubificid oligochaetes. Can. J. Fish. aquat. Sci. 40: 1639–1646.

Milbrink, G., 1973. On the Use of Indicator Communities of Tubificidae and some Lumbriculidae in the assessment of water pollution in swedish lakes. Zoon 1: 125–139.

Milbrink, G., 1978. Indicator communities of oligochaetes in scandinavian lakes. Verh. int. Ver. Limnol. 20: 2406–2411.
Milbrink, G., 1983. An improved environmental index based on the relative abundance of oligochaete species. Hydrobiologia 102: 89–97.
Paoletti, A., 1989. Cohort cultures of *Tubifex tubifex* forms. Hydrobiologia 180: 143–150.
Poddubnaya, T., 1980. Life cycles of mass species of Tubificidae. In R.O. Brinkhurst & D.G. Cook (eds), Aquatic Oligochaete Biology. Plenum Press, New York: 175–184.
Precht, H., J. Christophersen, H. Hessel & W. Lascher, 1973. Temperature and Life, Springer, Berlin.
Reynoldson, T.B., 1987. The role of environmental factors in the ecology of tubificid oligochaetes: an experimental study. Holarct. Ecol. 10: 241–248.
Reynoldson, T.B., 1983. The population biology of the aquatic Oligochaeta in the english Lake district. Ph.D. Thesis, University of Lancaster, England. 245 pp.
Rybak, J.L., 1969. Bottom sediments of the lakes of various trophic type. Ekol. pol. 17: 661–662.
Sachs, L., 1978. Angewandte Statistik. Springer, Berlin. 552 pp.
Saether, O.A., 1980. The influence of eutrophication on deep lake invertebrate communities. Prog. Wat. Tech. 12: 161–180.
Särkkä, J., 1987. The occurrence of oligochaetes in lake chains receiving pulp mill waste and their relation to eutrophication on the trophic scale. Hydrobiologia 155: 259–267.
Särkkä, J. & J. Aho, 1980. Distribution of aquatic oligochaetes in the finnish Lake District. Freshwat. Biol. 10: 197–206.
Seuss, R.J., 1981. Anaerobiosestoffwechsel und Gluconeogenese bei *Tubifex* sp. (Annelida, Oligochaeta). Ph.D Thesis, Universität Ulm, Germany, 159 pp.
Shaw, C.R. & R. Prasad, 1970. Starch gel electrophoresis of enzymes – a compilation of recipes. Biochem. Genetics 4: 297–320.
Timm, T., 1974. On the life cycles of the aquatic oligochaeta in aquaria. Acad. Sci. Estonian SSR Inst. Zool. Bot. Hydrobiol. Res. 6: 97–118.
Wagner, G., 1968. Zur Beziehung zwischen der Besiedlungsdichte von Tubificiden und dem Nahrungsangebot im Sediment. Int. Revue ges. Hydrobiol. 53: 715–721.
Wisniewski, R.J., 1979. Investigations into the reproduction and mortality of Tubificidae in lakes. Ecol. Pol. 27: 463–479.
Wolf, H.G., 1982. A comparison of different electrophoretic techniques for the detection of isoenzymes in single daphnids. Arch. Hydrobiol. 95: 521–531.

Hydrobiologia **278**: 7–15, 1994.
T.B. Reynoldson & K.A. Coates (eds), Aquatic Oligochaete Biology V.

Morphology of some species of the genus *Aulodrilus* Bretscher

Nonna P. Finogenova[1] & Nina R. Arkhipova[2]
[1] *Zoological Institute, Academy of Sciences, Universitetskaya 1, St. Petersburg, Russia;* [2] *Institute of Biology of Inland Waters, Academy of Sciences, Borok, Russia*

Key words: Oligochaeta, Tubificidae, *Aulodrilus,* morphology, taxonomy

Abstract

The external and internal morphology of *Aulodrilus limnobius* Bretscher, 1899, *Aulodrilus pluriseta* (Piguet, 1906) and *Aulodrilus japonicus* Yamaguchi, 1953, and morphology of the setal apparatus of *Aulodrilus pigueti* were studied on new material from Russia and compared with various literature data. Besides the peculiar "wings" on bifid setae, dominance of asexual reproduction is regarded as a primary synapomorphy of the genus and is accompanied by an anterior shift of the whole reproductive system and a tendency to doubling the gonads. Naidid-like cocoons, with a single egg, are another element of this reproductive mode. Similar characters seem to have arisen independently in several groups of Oligochaeta.

Introduction

The genus *Aulodrilus* (Tubificidae) deserves special attention because its morphology combines a number of primitive and derived features characteristic of tubificids and even naidids. Its taxonomy has not been clarified completely in spite of the vast literature. This has been due to a rarity of mature individuals and inadequacy of light microscopic techniques for study of the setae. It was indicated quite recently that the prostate gland, at least in some species, opens into the atrium by a wide base, not by a narrow stalk, and that the species have true penes. Hrabě (1981) gave a fairly complete description of *Aulodrilus pluriseta* (Piguet, 1906), *Aulodrilus japonicus* Yamaguchi, 1963, *Aulodrilus limnobius* Bretscher, 1899 and *Aulodrilus pigueti* Kowalevski, 1914. Giani *et al.* (1984) in a revision of *Aulodrilus* did not refer to the work of Hrabě (1981) although the descriptions and detailed drawings of the male ducts of *A. pluriseta* and *A. limnobius* confirmed Hrabě's observations. It appeared interesting to us to continue the study of this genus. We give descriptions of the external and internal morphology of *A. limnobius, A. japonicus, A. pluriseta* and describe peculiarities of the setal apparatus of *A. pigueti* from our own material and then discuss characters of this genus.

Materials and methods

The following material was used for species descriptions. *A. limnobius*: the River Oka, a tributary of the Volga, 24 m, with plants, collected 2 July 1953–1 mature specimen (spm) (fragment); the River Usa, a tributary of the Pechora, grey silt, collected August 1956 – juvenile specimens (spms). *A. pluriseta*: culture from a cocoon found in Rybinsk Reservoir, basin of the Volga, collected 15 June 1983–3 mature spms; Lake Kelziaty, basin of the Usa, silt, 25 July 1956 – several mature spms. *A. japonicus:* Neva estuary, 5 m depth, grey silt, clay, sand, collected 24 August 1982–1 partially mature spm (fragment; spermathecae not fully developed), 5 juvenile spms; 2.5 m depth, grey silt, sand, collected 15 August 1990–1 semimature spm (fragment) apparently damaged by a parasite (only with gonads, spermathecal pores and compact mass of sperm in IX). *A. pigueti*: Rybinsk Reservoir, grey silt, collected August 1990 – several juvenile spms.

Setal apparati of *A. limnobius* (from the Usa), *A. pluriseta* (from culture) and *A. japonicus* were studied by light microscopy with a resolution of 0.43 μm and scanning electron microscopy on a Hitachi S-570, and the setae of *A. pigueti* on a JSM-25S. For anatomical studies a series of paraffin sections 6–10 μm thick (1 series of *A. limnobius*, 3 series of *A. pluriseta*, 2 series of *A. japonicus*) were made.

Description

Aulodrilus limnobius Bretscher, 1899 (Figs 1, 2, 4, 5 & 19)

Prostomium small, triangular. Pharynx with a complex, folded configuration. Ventral and dorsal setae only bifid, with upper tooth shorter and thinner than lower (Figs 1, 2, 4 & 5). Setae expanded distally, in a plane perpendicular to the plane of the primary setal teeth, oar-shaped (Figs 1 & 5). Several setae with a distal, lateral cavity (Fig. 4). Setae 4–7 per bundle in preclitellar and mid-body segments, 2–4 in posterior ones. No modified genital setae. Ventral bundles near the male openings with 3 normal setae. Clitellum over $^{1}/_{2}$VI–VIII. Spermathecal pores in front of the dorsal setae of VI. Male openings on VII, lateral to the ventral setae.

Two pairs of testes in V–VI, those in V smaller. Ovaries in VII. One pair of male funnels on septum 6/7. Vasa deferentia 12–15 μm wide (n = 10; several measurements from the single mature individual) and 350 μm long, entering the atria apically. Atria with fusiform ampullae 50 μm wide, lined with glandular epithelium; efferent ducts 25 μm wide. A massive, lobate, stalked prostate gland entering the central part of atrial ampulla. Atrial duct ending with a voluminous, muscular, rounded penis 70 μm wide at its base and 40 μm long, located in a deep, folded muscular chamber about 130 μm wide and high (Fig. 19). Length of atrium and penis together equal to the length of vas deferens. Sperm sac and egg sac unpaired, the first from 6/7, the second from 7/8. Spermathecae in VI, with long ampullae and comparatively short ducts. Ampullae with sperm in loose masses. Ventral nerve cord with glands in V–VI like those of *Limnodriloides agnes* Hrabě, 1967.

Remarks. Brinkhurst (1966) first saw distal expansions on the setae of this species and named them lateral "wings". The overall form of the gonoduct in our material is similar to the descriptions made by Hrabě (1981) and Giani *et al.* (1984).

Aulodrilus japonicus Yamaguchi, 1953 (Figs 3, 6–10 & 18)

Prostomium short, bluntly conical. Pharynx with complex, folded configuration. Dorsal bundles with short hairs and bifid nodulate setae with upper teeth shorter and thinner than lower, sometimes with an intermediate tooth; ventral bundles with bifid setae similar to dorsal ones. In SEM, bifid setae wider in plane perpendicular to primary setal teeth than in plane of teeth and expanded distally, with one to several lateral teeth on both sides of a seta (Figs 3, 6–10). When lateral teeth are numerous, bifid setae appear as if they are broken (Figs 3, 6, 8, 9). In anterior bundles additional lateral teeth fewer than posteriorly, hardly visible in lateral view (Fig. 10). Dorsal bundles of II with 3 hairs and 4 bifid setae, those of the other segments with 4–6 hairs and 6–8 bifid setae, several segments anterior to py-

Figs 1, 2, 4 & 5. Setae of *Aulodrilus limnobius.*
Figs 3, 6 & 9. Ventral setae of *Aulodrilus japonicus.*
Figs 7 & 8. Dorsal setae of *Aulodrilus japonicus.* Scale bars equal 8.5 μm for Figs 1 & 2; 12 μm for Fig. 3; 5 μm for Figs 4–6; 13.5 μm for Figs 7 & 8; 10 μm for Fig. 9.

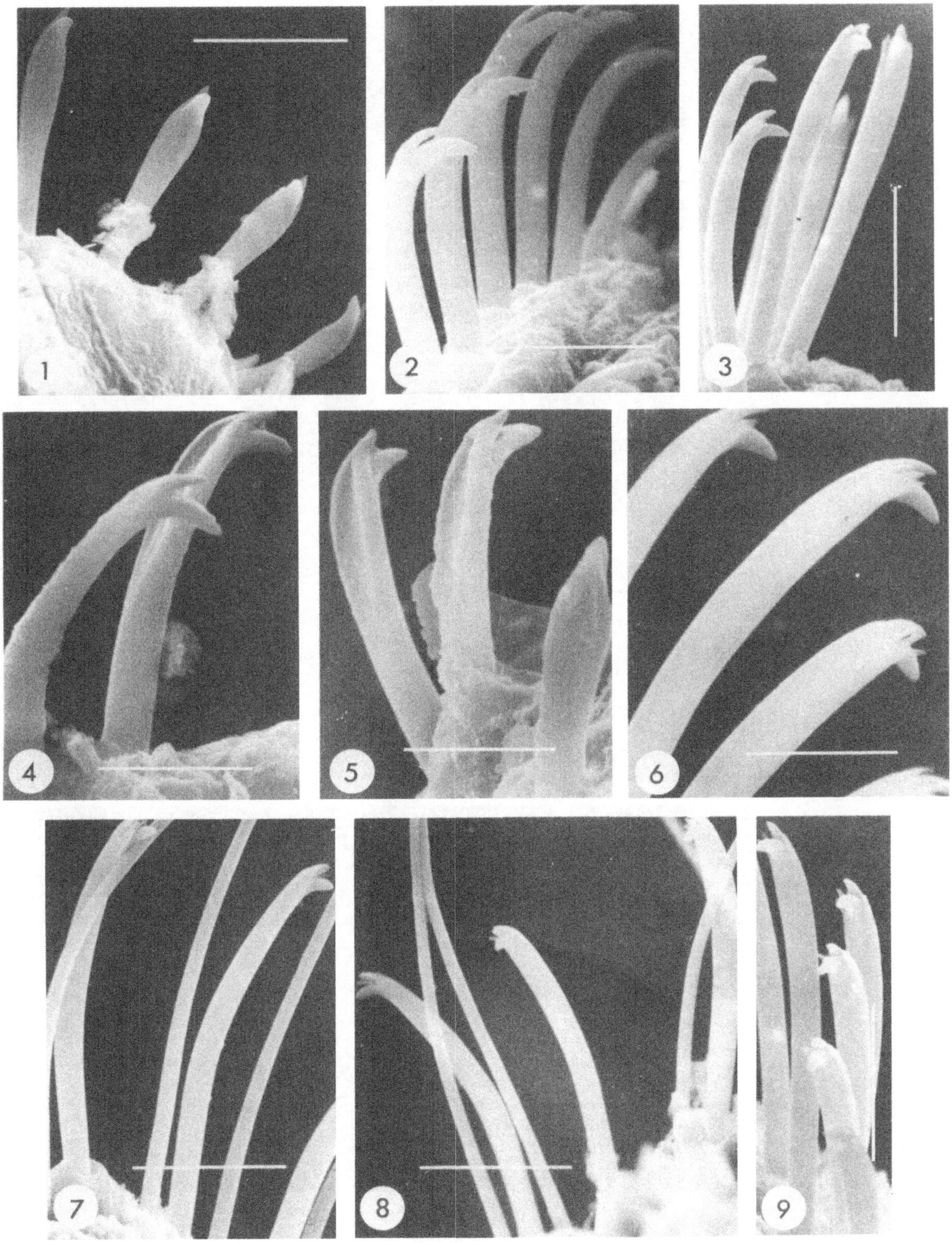
1
2
3
4
5
6
7
8
9

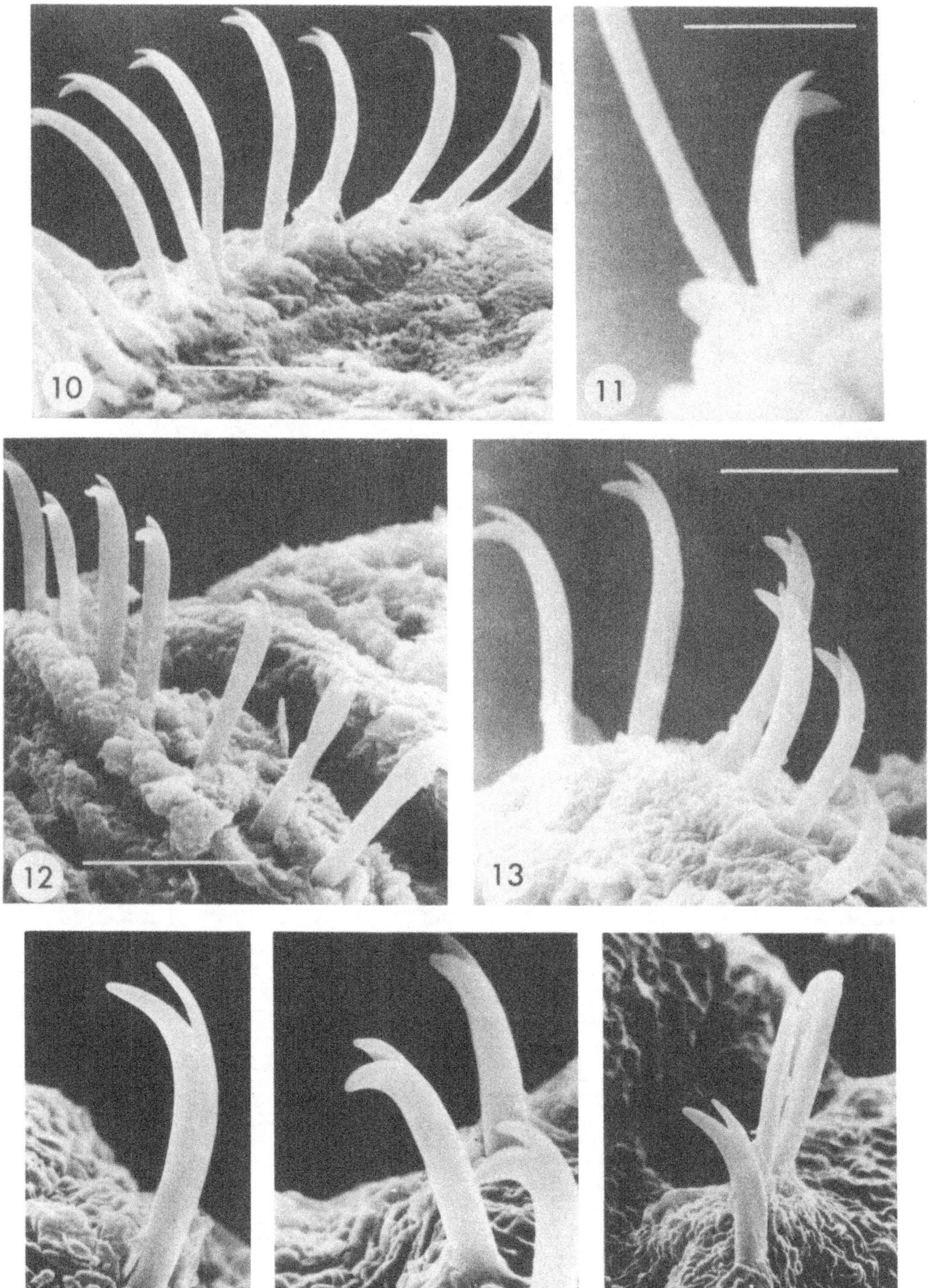
10
11
12
13
14
15
16

gidium with 2 hairs and 6 bifid setae. Ventral bundles of II with 6 setae, those of III–IV with 7–8, in most segments 9–11, in segments near pygidium 5–9 setae. No modified genital setae. Ventral bundles near the male openings with 5 normal setae. Hair setae 2–2.5 times longer than bifid ones. Clitellum over $^{1}/_{2}$IX–XI. Spermathecal pores in the lateral line of VIII and IX about halfway between setal bundles and anterior septum. Male openings in X lateral to the ventral setae.

Two pairs of testes in VIII–IX. One pair of ovaries in X and sometimes another small pair of ovaries in XI. Male funnels on septum 9/10. Slightly looped vasa deferentia about 300 μm long, maximum 25 μm wide, entering atria apically. Atria with spindle-shaped ampullae up to 50 μm wide, and 150 μm long; muscular efferent ducts 35 μm wide, about 140 μm long. A massive, lobate stalked prostate gland entering central part of atrium. Atrium ending with a conical penis 40 μm wide and long, located in small penial chamber (Fig. 18). Sperm sac unpaired, large, in X–XII. Ovisac in XI–XII. Spermathecae poorly developed in only two specimens. Glands on the ventral nerve cord in III–IX.

Remarks. Our description of the gonoduct of *A. japonicus*, from a single mature specimen, conforms to that given by Hrabě (1981). Slight differences apply to size but not the proportions of the atrium (in Hrabě, *op. cit.*: ampulla 55–70 μm wide and 160–170 μm long, penis about 50 μm long). Hrabě noted aberrant additional testes on 6/7. He also indicated a somewhat higher number of setae in the bundles (12–14 in ventral bundles, 5–8 hairs and 5–10 bifid setae in dorsal bundles) but only 2–4 setae in bundles of the 9 anteriormost segments of budding individuals (in the following segments, 10–14 setae per bundle as usual). According to his data, 9 head segments regenerate after amputation of 12–20 anterior segments. Descriptions of setae given by Hrabě (*op. cit.*) correspond to what can be seen under our light microscope.

Brinkhurst (1971) proposed that *A. japonicus* was a synonym of *A. pluriseta*. In his earlier works (Brinkhurst, 1965, 1966) he noted for *A. pluriseta* an extra tooth in the dorsal bifid setae and male openings on VII or IX. Probably two species were confused here: *A. pluriseta* with male openings in VII and *A. japonicus* with male openings in X (it is possible they were erroneously indicated in IX as the male openings of this species are very small and difficult to see) and lateral teeth on its setae. *Aulodrilus japonicus* should be considered distinct from *A. pluriseta*. It differs from *A. pluriseta* by the setal apparatus, the male gonoduct, and the location of the male openings. Apparently, *A. japonicus* like other species of *Aulodrilus* has a wide geographic distribution.

Aulodrilus pluriseta (Piguet, 1906) (Figs 11–13 & 20–22)

Prostomium very short, rounded. Pharynx with complex, folded configuration. Length of body 14–17 mm ($n = 4$), 107–112 segments ($n = 4$). Posteriorly, diameter of body decreasing but pygidial zone wider than the immediate prepygidial part. Dorsal bundles with short hair setae and bifid setae with upper tooth shorter and thinner than lower (Fig. 11). Ventral setae bifid, similar to dorsal bifid setae (Figs 12 & 13). Dorsal bundles with 6–8 hairs and 6–8 bifids in precliteral region, 4–5 hairs and 5–6 bifids on clitellum and in the midbody, posteriorly setae per bundle decreasing to 3–4 hairs and bifids, then to 1–2 hairs and 2–3 bifids; about 20 segments anterior to pygidial zone with only 2–3 bifids per bundle. Hair setae 2–2.5 times longer than bifids. Ventral bundles of II

Fig. 10. Ventral setae of *Aulodrilus japonicus*.
Fig. 11. Dorsal setae of *Aulodrilus pluriseta*.
Figs 12 & 13. Ventral setae of *Aulodrilus pluriseta*.
Fig. 14. Ventral seta of segment II of *Aulodrilus pigueti*.
Fig. 15. Ventral setae of segment XIII of *Aulodrilus pigueti*.
Fig. 16. Dorsal setae of segment IV of *Aulodrilus pigueti*. Scale bars equal 17.5 μm for Fig. 10; 7.5 μm for Fig. 11; 15 μm for Fig. 12; 12 μm for Fig. 13; 5 μm for Figs 14–16.

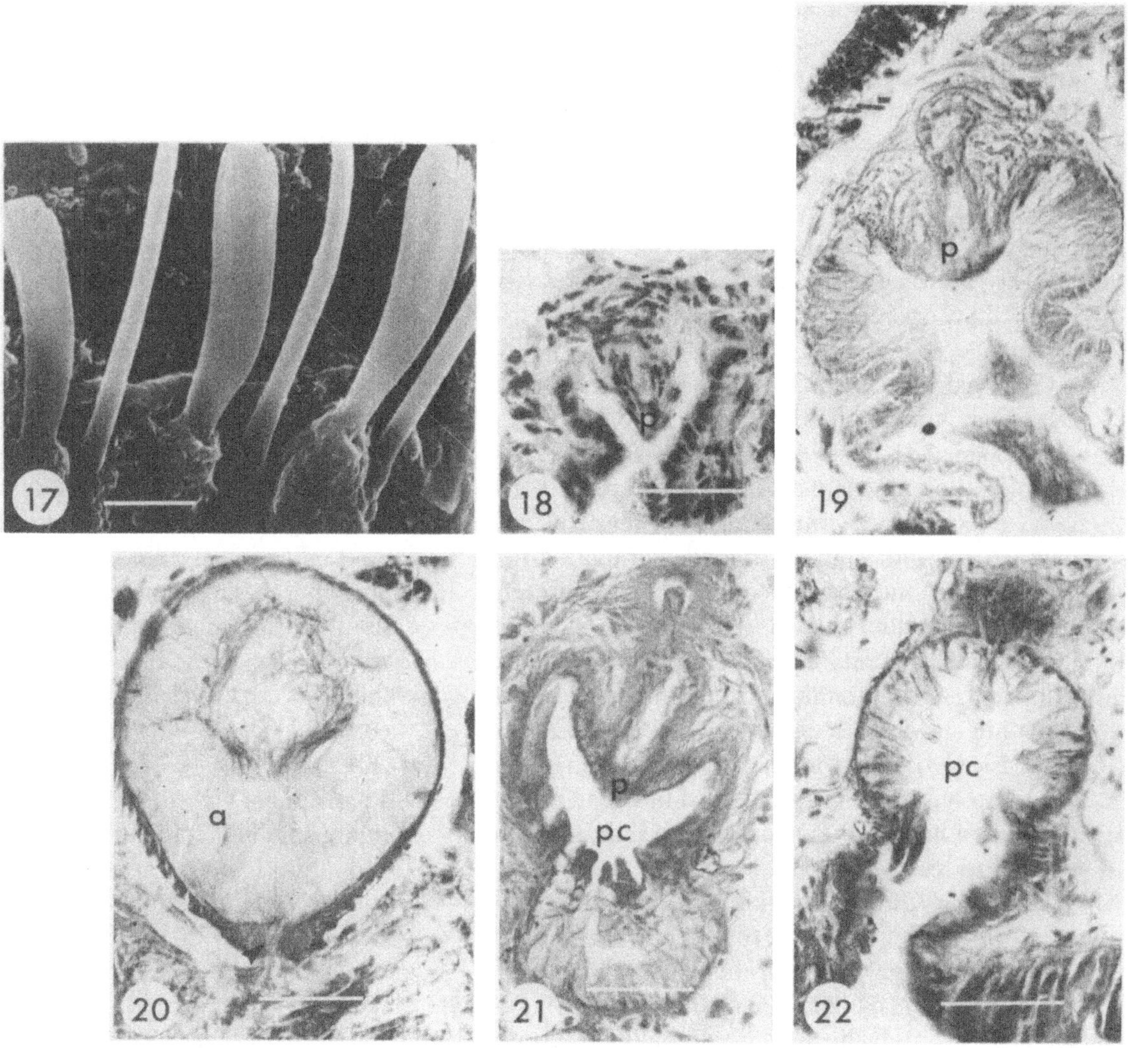

Fig. 17. Dorsal setae of segment XIV of *Aulodrilus pigueti.*
Fig. 18. Penis of *Aulodrilus japonicus*. p, penis.
Fig. 19. Penis and penial chamber of *Aulodrilus limnobius*. p, penis.
Fig. 20. Atrium of *Aulodrilus pluriseta* at union with prostate gland (arrow). a, atrium.
Fig. 21. Penis and upper part of penial chamber of *Aulodrilus pluriseta*. p, penis; pc, penial chamber.
Fig. 22. Lower part of penial chamber of *Aulodrilus pluriseta*. pc, penial chamber.

with 8 setae, III–IV with 9–12 setae, bundles of VII, near male openings, with 4 ordinary setae, posteriorly 7–9 setae, in last segments 1–4 setae per bundle. Clitellum over $^1/_2$VI–VIII. Spermathecal pores in lateral line, anterior to setal bundles of VI. Male openings lateral to ventral setal bundles of VII.

Two pairs of testes in V–VI. One pair of ovaries in VII. One pair of male funnels on 6/7. Slightly looped vasa deferentia 10–15 μm wide ($n = 3$), 215 μm long, entering atria apically. Atrium with oval ampulla and short muscular efferent duct joining conical muscular penis; penis 65 μm wide ($n = 3$) at base and 50 μm long ($n = 3$), in a deep folded penial chamber, length about 160–200 μm ($n = 3$) (Figs 21 & 22). Large, lobate, stalked prostate gland opening into the proximal part of atrial ampulla (Fig. 20). Length of atrium plus penis about equal to length of vas deferens. Sperm sac and ovisac unpaired. Spermathecae with large saccular ampullae and short muscular ducts. One of ampullae with sperm in loose masses. Ventral nerve cord with glands in II–VI and VIII–X, most frequent in III–VI.

Reproduction. The progenitor worm hatched 11 July 1983 from a cocoon found in the Rybinsk Reservoir. In slightly more than a month the culture contained four juvenile individuals dividing rapidly to form a large population. The culture was kept for 1.5 years at a temperature of 15–20 °C. The worms remained sexually immature under these conditions. In October 1984 the aquarium was placed in a refrigerator with a temperature of 2.4–7 °C. In March 1985 there were 15 juvenile worms. In May the aquarium was transferred into a basement with the temperature 17–20 °C. In July it contained 30 mature individuals and more than 50 inactive cocoons. In mid-November dead embryos were found inside the cocoons.

Later, cocoons were obtained again from the same population. A culture with cocoons was kept for a few months (from February until May) in a refrigerator and then was placed in a basement. Worms hatched from cocoons 1.5 months later. We did not succeed in obtaining any mature specimens that had not previously gone through the stage of architomic reproduction. Each cocoon always contained one egg occupying the whole volume of the cocoon. The cocoons of this species resemble those of naidids (Arkhipova, 1989). Thus the wider tip of the cocoons of *A. pluriseta* as well as the cocoon of *Ripistes parasita* (Schmidt, 1847) is closed both by a plug and operculum which opens at the moment of hatching of a worm. An operculum is not characteristic of tubificids.

Remarks. In experiments on regeneration, six anterior segments were restored in *A. pluriseta* and *A. pigueti* (Hrabě, 1937). According to the data of Hrabě (1981), *A. pluriseta* has 3 pairs of testes in IV–VI. Of these the first pair is poorly developed. Giani *et al.* (1984) indicate only 2 pairs of testes in V–VI. The distribution of the gonoducts of our specimens corresponds to descriptions of Giani *et al.* (*op. cit.*). According to Giani *et al.* the penial chamber is somewhat larger (length 200–250 μm) than in our specimens and the vas deferens is slightly longer (approximately equal to the length of atrium + penial chamber).

Aulodrilus pigueti Kowalevski, 1914 (Figs 14–17)

Dorsal bundles of II–III with 2–4 bifid setae with upper tooth shorter and thinner than lower, in IV–VI(VII) with 1–2 short hairs and 2–3 bifid setae with upper tooth shorter and thinner than lower, from VII–VIII with 1–3 hairs and 2–3 oar-shaped setae (Fig. 17), expanded in the plane of the primary teeth. Distal part of bifid setae sometimes with lateral cavities, one on each side of a seta (Fig. 16). Ventral bundles with 3–6 bifid setae with upper tooth thinner and shorter than lower (Figs 14 & 15), distal expansion of setae apparently absent.

Discussion

As mentioned previously (Cekanovskaya, 1962; Brinkhurst, 1971), other species of the genus, *Aulodrilus remex* Stephenson, 1921 and *Aulodrilus kashi* Mehra, 1922 should be regarded as synonyms of *A. pigueti,* and *Aulodrilus trivandranus* Aiyer, 1915 as a synonym of *A. pluriseta. Aulodrilus prothecatus* Chen, 1940 is close to *A. pigueti* but so far is considered to be a separate species because its spermathecal pores are close to septum 5/6. We regard *Aulodrilus stephensoni* Mehra, 1922, similar in many traits to *A. japonicus,* as a separate species since it has penial setae and its spermathecal pores are located slightly in front of the ventral setae. *Aulodrilus pectinatus* Aiyer, 1928 requires additional study as it is the only species in this genus having one pair of testes, in VI (reduction or plesiomorphy?). *Aulodrilus paucichaeta* Brinkhurst & Barbour, 1985 probably does not belong to *Aulodrilus* since it does not display the characteristic anterior shift of the sexual apparatus and does not have setae with either anterior-posterior expansions or lateral wings. *Aulodrilus americanus* Brinkhurst & Cook, 1966 and the *Aulodrilus* sp. treated by Brinkhurst *et al.* (1990) belong to *Aulodrilus* but many characteristic features of this genus are not mentioned in their descriptions.

The genus *Aulodrilus* is distinctly separate from other genera of Tubificidae and is remarkable for a number of features. Foremost is a shift of sexual apparatus and multiplication of gonads which may be related to architomy. These phenomena may also accompany parthenogenesis or regeneration as a result of damage and are known among tubificids as facultative aberrations in some individuals. Shifting of the sexual apparatus and doubling of spermathecae occur more frequently than doubling of the testes (in *Bothrioneurum* Stolc, *Svetlovia* Cekanovskaja, *Rhyacodrilus* Bretscher, *Aktedrilus* Knöllner, *Bathydrilus* Cook). For most species of *Aulodrilus* two pairs of testes and one pair of ovaries is usual. An additional, third pair of testes (*A. pluriseta, A. japonicus*) or second pair of ovaries (*A. japonicus*) occurs only rarely. Both the usual pairs of testes are functional as the body cavity of segment V of *A. limnobius* and segments V–VI of *A. pluriseta* were found to contain sperm. Among microdriles such series of gonads are also found in lumbriculids. Nonethless, the gonad series of *Aulodrilus* may be a synapomorphy. It is possible that in the lineage leading to the recent lumbriculids, multiple pairs of testes and ovaries could have arisen from two pairs of gonads (GII, testes -GIII, ovaries) which is characteristic of tubificids, naidids and other microdrile worm families.

In the life history of *Aulodrilus* architomy is the basic reproductive strategy in well-studied habitats, accounting for the rarity of mature specimens in nature. Our experience of keeping *A. pluriseta* in culture shows that worms do not mature sexually without passing through the architomic stage. It is possible that the shift of the sexual apparatus is "fixed" in accordance with the number of regenerated ("larval") segments of the anterior part of the body. As a rule the testes of architomic species are placed in the last newly developed segments. Thus in *A. pluriseta,* in which six anteriormost segments regenerate, the testes are placed in V–VI and in *A. japonicus,* having 9 regenerated segments, in VIII–IX. As mature individuals of these species originate from architomic individuals the normal position of the sexual apparatus for tubificids is not observed. Anterior displacement of the gonads is not fixed genetically (Hrabě, 1937, 1981) for *Bothrioneurum vejdovskyanum* Stolc, 1888, *Lamprodrilus mrazeki* Hrabě, 1928, and *Lumbriculus variegatus* (Müller, 1774).

Asexual reproduction is undoubtedly significant to the dispersion of these species. A number of other freshwater oligochaetes with cosmopolitan distributions also can reproduce asexually (Brinkhurst, 1971; Christensen, 1984; Timm, 1987) via parthenogenesis, architomy or paratomy.

Similarity in the structure of cocoons of *Aulodrilus* and naidids (operculum) is probably convergent as is the presence of one egg in a cocoon. Like many tubificids, *Aulodrilus* has sperm in loose masses, presumably a plesiomorphic state. The position of spermathecae in this genus is variable: median (*A. pigueti*), in front of the setal bundle (*A. limobius, A. pluriseta*), between the setal bundle and anterior septum (*A. japonicus*) or close to the anterior septum (*A. prothecatus*). One could propose that the plesiomorphic state of the spermathecal pore position is seen in *A. prothecatus*. Possibly an unpaired, posterior sperm sac is also plesiomorphic. The shape of the denticulate, bifid setae is apomorphic as widening of the somatic setae in a plane perpendicular to the plane of the primary teeth is apparently unknown for other oligochaetes. The presence of penial setae in some species of *Aulodrilus* is possibly a plesiomorphic trait, a state that is shared both with some other Tubificidae and some Naididae.

Study of this unusual genus is far from completion. It is necessary to investigate additional new material of *A. pectinatus*, *A. pigueti*, *A. americanus*, and *A.* sp. from China.

References

Aiyer, K.G.P., 1928. On a new species of the Oligochaeta genus *Aulodrilus* Bretscher. Rec. Indian Mus. 30: 345–352.

Arkhipova, N.V., 1989. Cocoons of some oligochaetes. Trudy Instituta biologii vnutrennihk vod (USSR Academy of Sciences) 56: 115–132. [in Russian]

Brinkhurst, R.O., 1965. Studies on the North American aquatic Oligochaeta II: Tubificidae. Proc. Acad. Nat. Sci. Philadelphia 117: 117–172.

Brinkhurst, R.O., 1966. A contribution towards a revision of the aquatic Oligochaeta of Africa. Zoologica Africana 2: 131–166.

Brinkhurst, R.O., 1986. Guide to the freshwater aquatic microdrile oligochaetes of North America. Canadian Special Publication of Fisheries and Aquatic Sciences 84. Department of Fisheries and Oceans, Ottawa, 259 pp.

Brinkhurst, R.O. & M.T. Barbour, 1985. A new species of *Aulodrilus* Bretscher (Oligochaeta: Tubificidae) from North America. Proc. Biol. Soc. Wash. 98: 931–934.

Brinkhurst, R.O. & D.G. Cook, 1966. Studies on the North American aquatic Oligochaeta. III. Lumbriculidae and additional notes and records of other families. Proc. Acad. nat. Sci. Philad. 118: 1–33.

Brinkhurst, R.O. & B.G.M. Jamieson, 1971. Aquatic Oligochaeta of the World. Oliver & Boyd, Edinburgh: 444–625.

Brinkhurst, R.O., Qi Sang & Liang Yanling, 1990. The aquatic Oligochaeta from the People's Republic of China. Can. J. Zool. 68: 901–916.

Cekanovskaya, O.V., 1962. The aquatic Oligochaeta of the USSR. Opred. Faune SSSR 78 1: 1–441. [in Russian]

Chen, V., 1940. Taxonomy and faunal relations of the limnitic Oligochaeta of China. Contr. biol. Lab. Sci. Soc. China Zool. 14: 1–131.

Christensen, B., 1984. Asexual propagation and reproductive strategies of aquatic Oligochaeta. Hydrobiologia 115: 91–95.

Erséus, C., 1987. Phylogenetic analysis of the aquatic Oligochaeta under the principle of parsimony. Hydrobiologia 155: 75–89.

Erséus, C., 1990. Cladistic analysis of the subfamilies within the Tubificidae (Oligochaeta). Zool. Scr. 19: 57–63.

Giani, N., E. Martinez-Ansemil & R.O. Brinkhurst, 1984. Revision du statut taxonomique des Aulodrilinae (Tubificidae, Oligochaeta). Bull. Soc. Hist. Nat. Toulouse 120: 17–22.

Hrabě, S., 1937. Prispevek k poznâni zízâlice *Lamprodrilus mrazeki* a nítenek rodu *Aulodrilus*. Sb. Klubu přirodověd. v Brne za r. 1936, 19: 1–7.

Hrabě, S., 1967. Two new species of the family Tubificidae from the Black Sea, with remarks about various species of the subfamily Tubificinae. Spisy přir. Fac. Univ. Brne 485: 331–356.

Hrabě, S., 1981. Vodní málostětinatci (Oligochaeta) Československa. Acta Univ. Carolinae, 1979, Biologica 1–2: 1–167.

Mehra, H.R., 1922. Two new Indian species of the little known genus *Aulodrilus* Bretscher. Proc. zool. Soc. Lond. 59: 943–954.

Timm, T., 1987. Oligochaeta of waterbasins of North-west USSR. Tallinn, 299 pp.

Hydrobiologia **278**: 17–26, 1994.
T.B. Reynoldson & K.A. Coates (eds), Aquatic Oligochaete Biology V.

A description of spermatozoan ultrastructure and some anatomical characters in *Branchiobdella kozarovi* Subchev, 1978 (Annelida: Clitellata), and review of the spermatozoan morphology within the genus

S.R. Gelder[1], M. Ferraguti[2] & M.A. Subchev[3]
[1]*Department of Biology, University of Maine at Presque Isle, Maine 04769, USA;* [2]*Dipartimento di Biologia, Università degli Studi di Milano, 20133 Milano, Italy;* [3]*Institute of Zoology, Bulgarian Academy of Sciences, Sofia 1000, Bulgaria*

Key words: Annelida, Clitellata, *Branchiobdella kozarovi*, taxonomy, anatomy, spermatozoan ultrastructure

Abstract

The use of spermatozoan ultrastructural morphology to produce a meaningful phylogeny requires a critical number of descriptions from representative species in a taxon. *Branchiobdella* spermatozoan morphology is considered to be aberrant within the Clitellata. The description of the spermatozoon of *B. kozarovi* has extended the information on interspecific variation in this genus. The spermatozoa of this species have an acrosome (16 μm long), nucleus (4.5 μm long), midpiece (1.5 μm long), and tail (158 μm long), with a total length of about 180 μm. The acrosome is a thick helix with about 15 gyres, contains an acrosome tube and acrosome vesicle which has branches penetrating into the central subacrosomal space. The acrosomal "button" is located on the anterior surface, with the "cork" and secondary tube at the posterior. The rod-shaped nucleus has a concavity in both the anterior and posterior surfaces. The midpiece has seven, parallel mitochondria twisted about the central axis. The centriolar remnant with an indistinct ninefold symmetry in cross section is located posterior to the mitochondria. The prominent central sheath extends from the mitochondria to the end of the flagellum, which has no end piece. The axoneme has glycogen granules located along its length, and an helical marginal fiber. A review of descriptions of spermatozoa of *Branchiobdella* shows some inconsistencies. The observations of two anterior nephridial pores and an eversible penis completes the anatomical characters of *B. kozarovi* used in current branchiobdellidan taxonomy.

Introduction

Sawyer (1986) used, among other criteria, the ultrastructural morphology of spermatozoa from the branchiobdellidan *Branchiobdella pentodonta* Whitman, 1882 to support his argument that the Branchiobdellida (= subclass Branchiobdellidea of Sawyer, 1986: 647) is a member of the class Hirudinea. He assumed that the only species to have had its spermatozoan ultrastructure described was morphologically representative of the branchiobdellidans. A subsequent study by Ferraguti & Gelder (1991) of the spermatozoa from four branchiobdellidans, each from a different genus, revealed that each species had a characteristically different morphology. They reported that although some of the branchiobdellidan spermatozoan characteristics were hirudinean-like, many others were oligochaete-like. This demonstrates that more studies of spermatozoan ultrastructure in branchiobdellidans are needed to ascertain the degree of intergeneric and interspecific variability, and in turn establish the morphological range within the Branchiobdellida. In addition these data must be related to the differences in species anatomy, microhabitat on the host and geographic range. However, the value of such investigations is dependent upon the fact that the spermatozoan descriptions originate from unequivocally identified species.

Following this theme, and given that the spermatozoon of *B. pentodonta* appears to have a highly aberrant morphology when compared to the accepted clitellate arrangement, *Branchiobdella kozarovi* Subchev, 1978 was chosen to confirm the intergeneric consistency and elucidate interspecific variability. The anatomy of the male reproductive apparatus and the relatively restricted range of *B. kozarovi* are different from those in *B. pentodonta*. The taxonomy of the *Branchiobdella* and the existing description of their spermatozoa require a brief review so that anomalies and omissions in the literature can be placed in overall context.

[*Note*: Holt (1965) created the order Branchiobdellida and it contained the single family Branchiobdellidae, so that the accepted common name for both taxa remained branchiobdellid. When he created four additional families in the order (Holt, 1986), branchiobdellid remained the common name for the Branchiobdellidae, but branchiobdellidan was coined as the collective common name for the order (Holt, pers. comm.)].

Brief review of Branchiobdella *taxonomy*

The taxonomy of *Branchiobdella* in the Branchiobdellida has been addressed in a comprehensive synopsis by Holt (1986). His paper is recognized as the current, authoritative work on branchiobdellidan taxonomy, and so attention must be drawn to an error regarding the figures purporting to be of the type species, *Branchiobdella astaci* Odier, 1823. Holt (1986) redrew Figures 3B, E, & C from Dorner (1865), and placed them under the heading of *B. astaci*. In fact, they show the male genitalia (B) and the everted penis (E) of *Branchiobdella parasita* Henle, 1835, and the spermatheca (C) of *Branchiobdella hexodonta* Gruber, 1883. Holt simply repeated the error made by Dorner (1865). A review of the European species of *Branchiobdella* by Pop (1965) concluded that there were four species: *B. astaci, B. parasita, B. hexodonta*, and *B. pentodonta*, the last with three subspecies: *pentodonta, italica* Canegallo, 1923, and *orientalis* Pop, 1965. Karaman (1967, 1970) recognized all seven taxa as separate species and reinstated the name *balcanica* Moszynski, 1937 due to its seniority over *orientalis*. He also described a new subspecies, *Branchiobdella balcanica sketi* Karaman, 1967. With the addition of *B. kozarovi*, the taxonomy proposed by Karaman (1970) is the one we accept. The type description of *B. kozarovi*, and additional branchiobdellidan observations by Boshko (1983), show this to be a new species of branchiobdellidan. However, the number of anterior nephridial pores was not reported, and the penis was described as "protrusible". These two characters raise the possibility that the species may not be a member of the *Branchiobdella* (see Holt, 1986). A reexamination of this species will resolve this point and provide additional anatomical detail for future studies.

Brief review of spermatozoa in the Branchiobdella

The spermatozoa in branchiobdellidans are composed of four portions: acrosome, nucleus, midpiece, and tail; an endpiece may be present or absent (Ferraguti & Gelder, 1991). At the light microscopical level the nucleus and midpiece have frequently been grouped together, or added to the acrosome measurement and then termed "head" (Table 1). The first dimensions of spermatozoa in a branchiobdellidan were given by Dorner (1865) from *B. hexodonta*. Voigt (1883) reviewed the details on spermatozoa then available and added some observations of his own. [*Note*: Voigt published another paper in 1883, in the same journal and with the same title, however the text and page numbers are different for the one cited above.] An

Table 1. Dimensions of the spermatozoa in the *Branchiobdella*.

	Acrosome	Nucleus	/ midpiece	Tail	Total
B. astaci					
Voigt, 1883: 140	25	6		250	281
B. hexodonta					
Dorner, 1865: 486, 492	40	–		330	370
B. kozarovi	16	4.5	1.5	158	180
B. parasita					
Dorner, 1865: 486, 492	110	–		220	330
Voigt, 1883: 193	71	7		320	398
	76	6		300	382
Ostroumoff, 1883: 77	90	–		310	400
B. pentodonta					
Voigt, 1883: 139	25	6		225	256
	30	5		210	245
Bondi, 1963: 44, and Bondi & Farnesi, 1976: 68	35	5	1	300	341
Ferraguti & Gelder, 1991: 1947	33	4.4	1.4	220	258

All measurements are given in micrometers (μm). Where the nucleus/midpiece measurement is absent, a total "head" length is given in the acrosome column.

In Voigt (1885) mature spermatozoa from *B. astaci* (Figures 36–38) and *B. hexodonta* (Figure 39) are illustrated, but no dimensions were given in the text. Dimensions for *B. kozarovi* spermatozoa are included for ease of comparison.

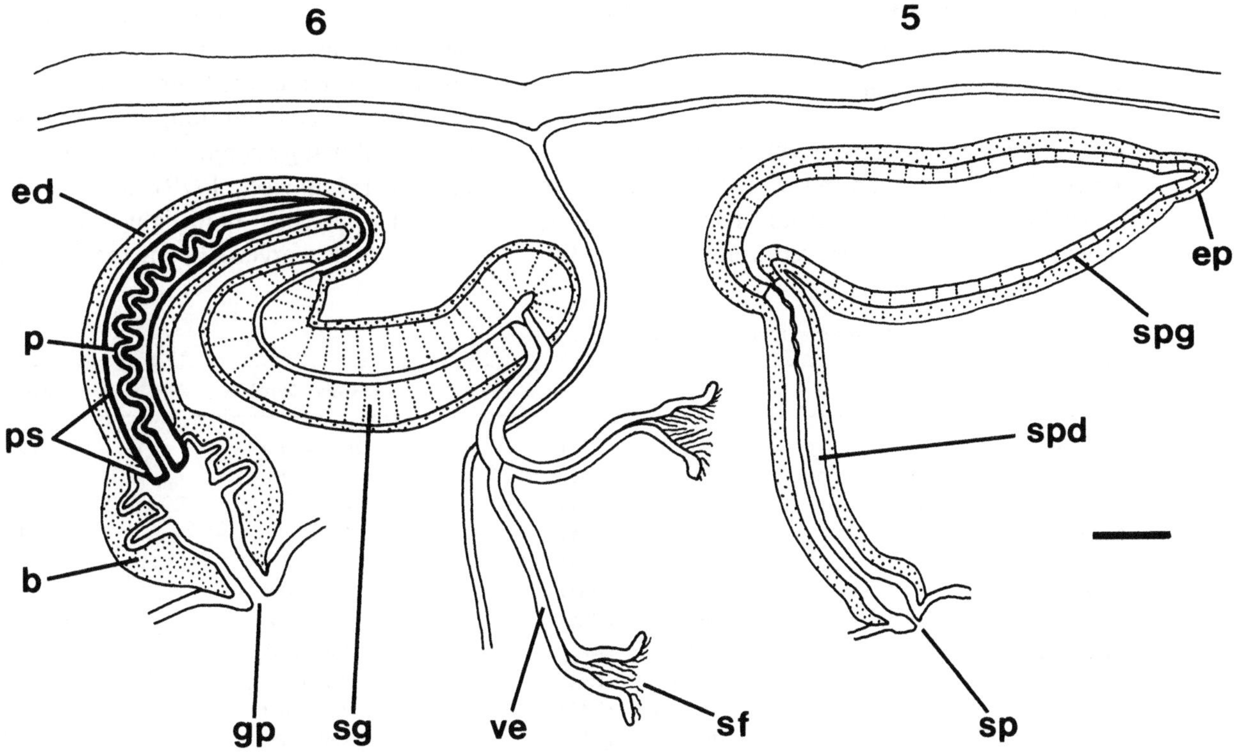

Fig. 1. Diagram of the reproductive organs from a ventral oblique view. Legend: b, bursa; ed, ejaculatory duct; ep, ental process; gp, genital pore; p, penis; ps, penial sheath; sf, sperm funnel; sg, spermiducal gland; sp, spermathecal pore; spd, spermathecal duct; spg, spermathecal gland; ve, vas efferens; 5 and 6, segment numbers; bar = 20 μm.

ultrastructural description of spermatozoa, with some light microscopical observations, from *B. pentodonta* were given by Bondi (1963) and followed later by a re-examination of the spermatozoa ultrastructure (Bondi & Farnesi, 1976). An investigation, including *B. pentodonta*, into the comparative spermatozoan ultrastructure of four species of branchiobdellidans representing four genera, showed that each was significantly different from the others (Ferraguti & Gelder, 1991).

The present study is part of a wider investigation to establish the extent of inter-generic and inter-specific differences of spermatozoan ultrastructure in branchiobdellidans. A practical application of these data would be confirmation of the specific identity of specimens where the anatomical characters of a species appear equivocal to the investigator. When a critical number of species have had their spermatozoa described to produce a meaningful data base, phylogenies will be constructed and compared with those based on comparative anatomy for both the class Branchiobdellae and sister groups in the Clitellata.

Materials and methods

Astacus leptodactylus (Eschscholz, 1823) were collected from the man-made fish ponds near Razdelna, District of Varna, Bulgaria. Living specimens of *B. kozarovi* were removed from the oral and ventral surfaces of the hosts, and kept *in vitro*. The internal anatomy of living worms was examined using a microscope with bright-field illumination. The anatomical terminology used in this work follows that described by Gelder & Brinkhurst (1990). Specimens for light microscopical study were fixed in acetic acid-formalin-alcohol (AFA) (Brinkhurst & Gelder, 1991) or Karnowsky's (1965) fixative with 0.1M cacodylate buffer (pH 7.4) (see below) for the electron microscopical examination of spermatozoa.

Spermatozoan smears were made from living specimens by incising the body wall of segment 5 to release the developing spermatozoa onto a microscope slide. The worm was then transferred into AFA fixative for future reference. The released coelomic contents were diluted with one drop of distilled water to spread the spermatozoa over a larger area of the slide, and then left to air-dry. Spermatozoa were examined with a compound microscope using phase contrast illumination and drawn with the aid of a drawing tube. The lengths of the acrosome, nucleus plus midpiece and tail, and total length were recorded from 20 spermatozoa, five from each of four individuals. The nucleus and midpiece lengths were measured, following staining with the Acridine Orange method (Humason, 1979), under a fluorescence microscope using blue light excitation: nucleus red, midpiece green. From these measurements

Table 2. Dimensions of the spermatozoa in *B. kozarovi.*

			Range (μm)	Mean $\pm$ SD (μm)
Acrosome	length	(LM)	16.8 –20.0	15.69 $\pm$ 2.54
	diameter	(EM)	0.8–0.9	
Nucleus and midpiece	length	(LM)	4.8–7.2	6.04 $\pm$ 0.75
Nucleus	length	(FM)	4.4 –5.5	4.68 $\pm$ 0.37
		(EM)	4.5–4.8	
anterior	diameter	(EM)	0.45 –0.9	
posterior	diameter	(EM)	0.72–0.9	
Midpiece	length	(FM)	1.0 –1.65	1.43 $\pm$ 0.36
		(EM)	1.0–1.28	
	diameter	(EM)	0.54 –0.76	
Tail	length	(LM)	150–168	158.18 $\pm$ 5.91
TOTAL	length	(LM)	177 –186	180.22 $\pm$ 6.53

LM = light microscopy, FM = fluorescence microscopy and EM = electron microscopy, number (n) of spermatozoa measured was 20.

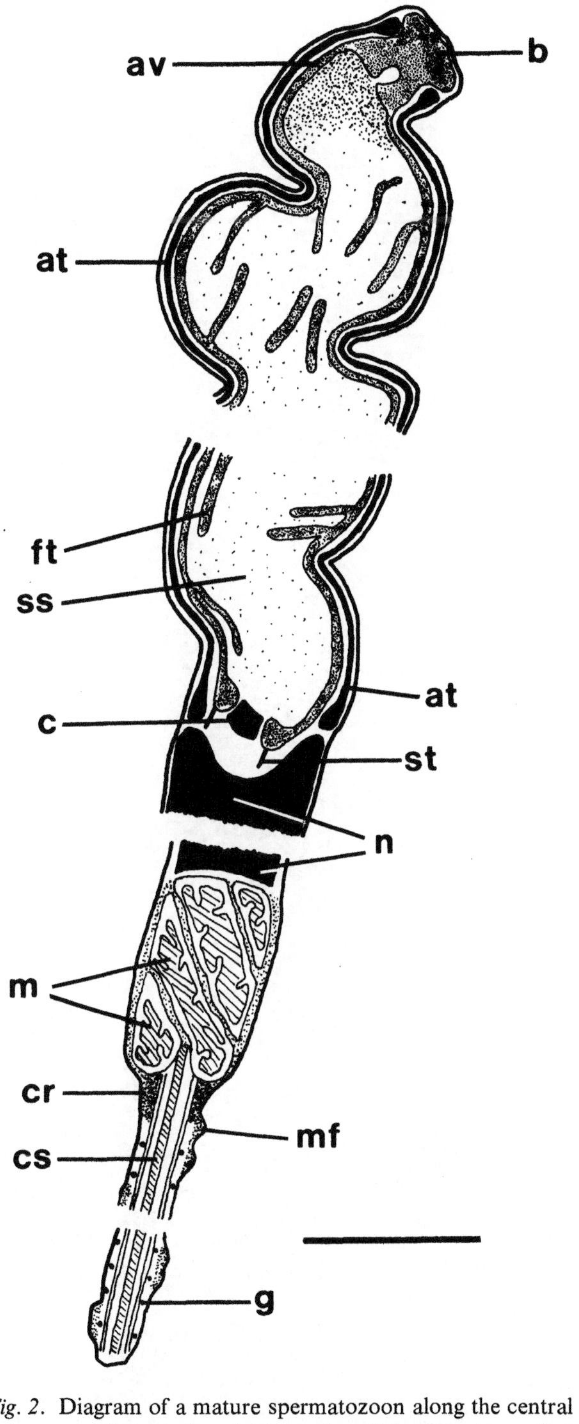

Fig. 2. Diagram of a mature spermatozoon along the central axis, except through the midpiece where it is off-center so that the mitochondrial twisting can be shown more clearly. The breaks in the acrosome, nucleus and tail enable essential detail to be shown without unnecessary partition of the figure. Legend: a, acrosome; at, acrosomal tube; av, acrosomal vesicle; b, "button"; c, cork; cr, centriole remnant; cs, central sheath; f, flagellum; ft, flattened tubules; g, glycogen granules; m, mitochondria; mf, marginal fiber; n, nucleus; ss, subacrosomal space; st, secondary tube; bar = 1 μm.

the mean and standard deviation were calculated, as well as the overall length.

[*Validation of spermatozoan smear technique*: live spermatozoa were released from *Cambarincola* sp. onto a microscope slide, coverglassed and measured using a reticule-calibrated compound microscope with dark-field illumination. Although the beating motion of the tail made accurate measurements very difficult, spermatozoa were observed to be approximately 114 μm long. The coverglass was removed, one drop of distilled water was added, and the coverglass replaced. The spermatozoa became straight and moved with only an occasional beat. Ten randomly selected live specimens were measured to give mean lengths and standard deviations in μm: head 23.5 ± 0.86, tail 88.47 ± 1.2, endpiece 1.76 ± 0.32, total 114.63 ± 2.36. The coverglass was removed and the sperm allowed to air-dry before recovering with the coverglass, and repeating the measurements with dark-field and then phase contrast illumination for each specimen: head 23.5 ± 0.62, tail 88.31 ± 1.29, endpiece 1.76 ± 0.32, total 113.76 ± 1.41.]

Specimens remained in the Karnowsky (1965) fixative for approximately five weeks, then they were washed overnight in 0.1M cacodylate buffer (pH 7.4), post-fixed in 1% osmium tetroxide in the same buffer for 2h, washed in distilled water, *en bloc* stained for a few hours in the dark in 2% aqueous uranyl acetate, dehydrated in a graded ethanol series, and embedded in Spurr's resin (Spurr, 1969). Thin sections were cut and stained by the triple staining technique (Daddow, 1983), carbon coated, and observed with a JOEL 100 XS transmission electron microscope.

Observations

Anatomical characters

The observations from living and preserved specimens described in this section include only

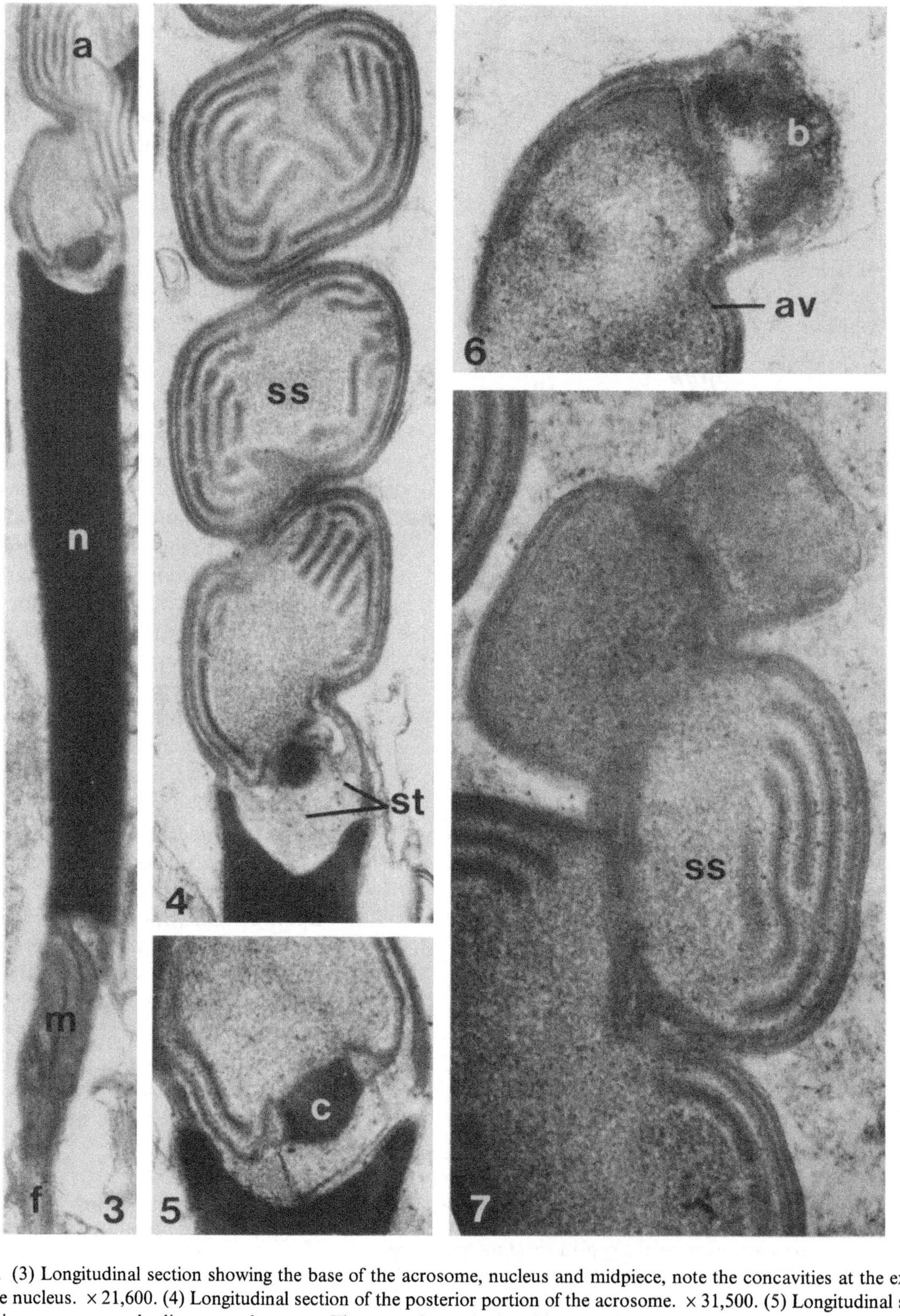

Figs 3–7. (3) Longitudinal section showing the base of the acrosome, nucleus and midpiece, note the concavities at the extremities of the nucleus. × 21,600. (4) Longitudinal section of the posterior portion of the acrosome. × 31,500. (5) Longitudinal section of posterior acrosome and adjacent nuclear area. The secondary tube is evident. × 48,600. (6) Longitudinal section of anterior

previously unreported features and new structural details of known characters in *B. kozarovi.* A coherent set of characters observed from the preserved specimens is then presented in an emended brief description of the species.

The ventral peristomial lip has a median emargination but the margin of the dorsal lip may vary in form depending upon the degree of contraction. The surface can appear smooth, have a median emargination, or four slight protrusions below each of which is a cluster of gland cells. All three conditions were also observed in living specimens, but permanent lobes are absent from this species. There are three pairs of pharyngeal sulci; the largest pair extends across the middle of the pharynx, with two smaller pairs located just posterior to the jaws and just before the posterior end of the pharynx. The major annuli of the body lack dorsal ridges even when a specimen is contracted. A pair of anterior nephridial pores is located on the dorsad of segment 3. The lateral surfaces of segments 8 and 9 each have short, conical lobes. Each lobe on segment 8 contains two pairs of glands, compared to only one pair of glands in the lobes on segment 9. A visual comparison of the width of the head to that of segment 1, and the width of the posterior disc to segment 10 was made on both live and preserved specimens. The lack of consistency in the ratios, with one character being equal to or greater than the other, resulted in omitting these criteria from the emended species description.

The single pair of testes is located in segment 5, and the male reproductive apparatus extends on either side of the gut to approximately the mid-lateral region in segment 6. In segment 5, mature spermatozoa are arranged parallel to each other in two tightly packed blocks with the acrosomes pointing into the sperm funnels (Fig. 1: sf). Each funnel is bell-shaped with cilia on its internal surface, and is connected to a vas efferens (ve). The two vasa efferentia join together to form a common, short vas deferens in segment 6 which then enters the spermiducal gland (sg) subterminal to the ental end (Fig. 1). The gland is short, about 1/4 the diameter of the segment, and bent in the ectal and ental subterminal regions; a prostate gland is absent. The spermiducal gland is composed of an outer layer of circular muscle cells (stippled) and an inner layer of columnar gland cells; a narrow lumen may or may not be observed. The ejaculatory duct (ed) is subequal in diameter and length to the spermiducal gland. The duct wall (stippled) is composed of an outer layer of circular muscle cells overlying a more obvious set of longitudinal muscle cells. The muscle layer surrounds and usually lies in contact with the penial sheath (ps, black). However, a space (blank) can be seen sometimes as shown in Figure 1. The penial sheath protrudes slightly into the atrium of the bursa before inverting to become the "eversible" penis (p). The penial sheath and penis is formed from a tube of squamosal epithelial cells. The penis is tightly packed in folds inside the sheath and then merges with the sheath in the ental portion of the ejaculatory duct (Fig. 1). There is a space (blank) between the penial sheath and the inverted penis. The bursa (b) is subspherical, its diameter is approximately 1/4 of the spermiducal gland length, with an internally folded atrium connecting the tip of the penial sheath to the exterior genital pore (gp). Muscular-epithelial cells comprise the bursal mass. The cuticle of the body invaginates through the genital pore to cover the surface of the atrium, exposed tip of the penial sheath, and penis lining; hooks are absent from the latter region.

The spermatheca is located in segment 5. It consists of a duct, and an elongate, sacculate gland (spg) of equal or lesser length, with a small, nipple-like ental process (ep) (Fig. 1). The shape of the gland varies somewhat depending upon the number of spermatozoa in the lumen, condition at the time of fixation, and probably other factors too. Mature spermatozoa are often found with their acrosomes in the ental process, but some-

acrosome, showing the "button" and the area devoid of branching acrosome vesicles. × 56,000. (7) Similar to Fig. 6, but cut in another longitudinal plane, the characteristic shape of the acrosome vesicle is evident. In this area the acrosome vesicle contacts cell membrane, and has been called "button" in the other clitellates because of its shape. × 56,000 (see Fig. 2 for legend).

times the spermatozoa appear randomly orientated in the lumen of the gland. Cuboid or squamosal cells line the lumen and, with the exception of the ental process, this layer is surrounded by a circular muscle cell layer (stippled). The narrow, ectal end of the gland bends before entering the spermathecal duct. The duct (spd) is lined with a thin layer of cuticle, a tube of cuboidal epithelial cells, surrounded in turn by a circular muscle cell layer. The normal state of contraction in the ectal portion of the duct causes the cuticle-covered epithelial layer to form vertical folds, which result in a small papilla, the pore being at its center (sp).

An emended brief description for *Branchiobdella kozarovi* Subchev, 1978:

Body 1.1 to 3.5 mm long, cylindrical, slight expansion of segments 5 to 7 and slight ventral flattening, dorsal ridges absent; with dorsal and ventral lips, dorsal lobes absent; 16 oral papillae; jaws isomorphic, homognathus, 5/5, occasionally 6/5 or 7/5; vas deferens enters spermiducal gland subterminal to ental end, gland length about 1/4 diameter of segment (short) with subterminal ectal and ental bends; ejaculatory duct length subequal to gland, containing penial sheath, eversible penis, hooks absent; subspherical bursa diameter 1/4 gland length; spermatheca duct length 1/4 segment diameter, elongate sacculate gland length subequal, ental process > 1:10 of duct.

Spermatozoan ultrastructure

The spermatozoa consist of an acrosome (16 μm long), nucleus (4.5 μm long), midpiece (1.5 μm long), and tail (158 μm long), giving an overall length of approximately 180 μm. The dimensions with standard deviation of these component parts of the spermatozoa observed at the light and electron microscopic levels are given in Table 2.

The acrosome is a thick, twisted-columnar structure measuring up to 19 μm long, and about 0.9 μm in diameter. Anteriorly the pitch of the tube is 1.5 μm, narrowing to 1.25 μm posteriorly (Fig. 2), with as many as 15 gyres. The acrosome tube (at) has thickened margins at both ends, and surrounds most of the acrosome vesicle (av) (Figs 2 & 6). The vesicle forms a circular plug that protrudes slightly from the anterior opening of the acrosome tube to touch the plasma membrane (Fig. 6), a ring of electron-dense material is located within the slight protuberance. The posterior surface of the plug has a central, spherical invagination. The vesicle's side extends posteriorly, parallel to the acrosome tube, with the inner membrane folding repeatedly to form small flattened tubules (ft) (Fig. 2). The tubules branch and are tightly packed in the central, subacrosomal space (ss). The material in the subacrosomal space is more electron-dense in the anterior than the posterior portion. The posterior margin of the acrosome vesicle is thickened and its remaining aperture is almost filled by the electron dense cork (c) (Fig. 5). Extending from the thickened vesicle margin is a short, secondary tube (st) (Fig. 5). The nucleus (n) is a rod, about 4.5 μm long, and 0.8 μm in diameter at the anterior end tapering to about 0.55 μm at the posterior. Both anterior and posterior surfaces of the nucleus have a concavity but the former is more distinct (Figs 2 & 3). The short midpiece is about 1.5 μm long and 0.8 μm in diameter, containing seven mitochondria (m) arranged parallel to each other and twisted about the central axis of the spermatozoon (Figs 8 & 9). The long tail, about 158 μm long, consists of a flagellum, but no endpiece. Immediately posterior to the mitochondria is the centriolar region which consists of an electron-dense cylinder and the centriolar remnant (cr) (Figs 2 & 8), that has an indistinct, ninefold symmetry in cross section. At its center, the prominent central sheath (cs) surrounds the central doublet, extending from the mitochondria to the end of the flagellum. The axoneme has glycogen granules (g) located along its length and an helical marginal fiber (mf) of electron-dense material lying just under the plasma membrane of the flagellum (Figs 2 & 11).

Discussion

The taxonomy of *B. kozarovi* has been confirmed by our observations of two anterior nephridial

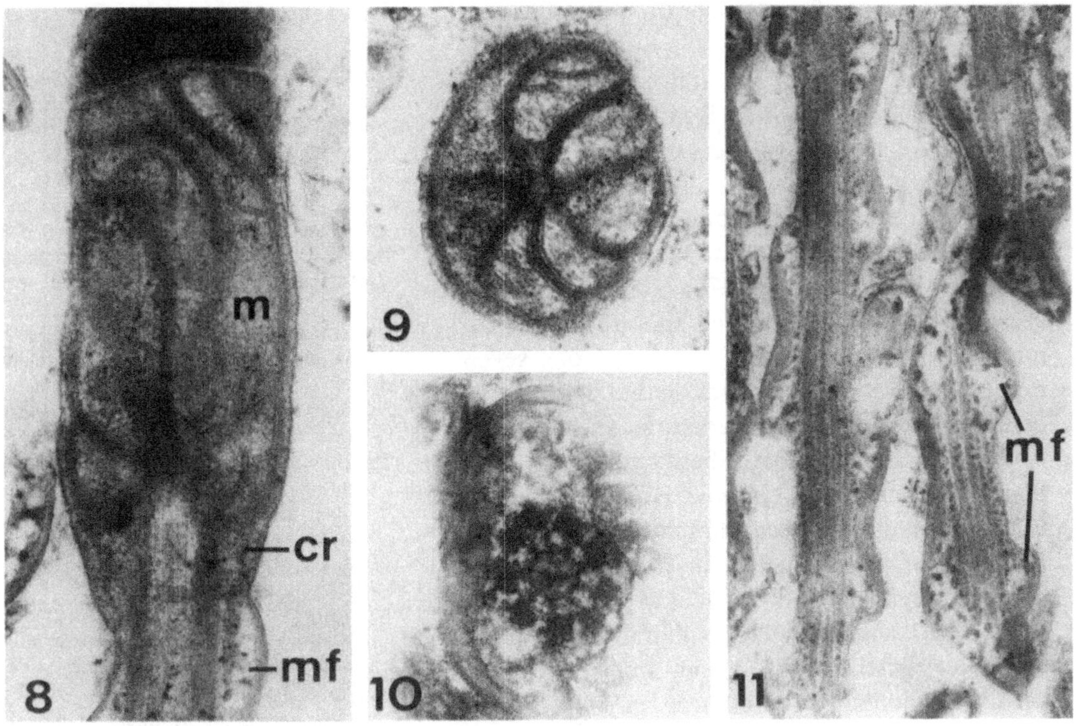

Figs 8–11. (8) Longitudinal section of the midpiece. Note the prominent central sheath passing through the center of the centriole remnant, and the marginal fiber starting at the base of the flagellum. × 53,000. (9) Transverse section of the mid piece, showing seven twisted mitochondria. × 58,500. (10) Transverse section of the flagellum, with the central sheath and marginal fiber. × 72,000. (11) Longitudinal section of the flagellum. × 52,000 (see Fig. 2 for legend).

pores and an eversible penis (Holt, 1986: 686). The previous status of knowledge of the anterior nephridial pores left the possibility of this species being assigned to the wrong family. The report of a protrusible penis in *B. kozarovi* by Subchev (1978) rather than an eversible penis, would have necessitated a new genus being created.

The body shape and length, and predominantly pentodontous jaws in *B. kozarovi* are similar to those found in *B. pentodonta*, *B. italica*, *B. balcanica*, and *Branchiobdella minuta* Pierantoni, 1912. From published descriptions, the size and shape of the spermatheca appear to vary to different degrees within every species. However, an elongate sacculate spermathecal gland with an ental process in *B. kozarovi* does appear to be consistent and is unique in the genus. Dorner (1865) described and illustrated the penial apparatus in *B. parasita* and our description is the only other such detailed report of a species in this genus.

The measurements of spermatozoa in *B. astaci* and *B. hexodonta* (Table 1) are solely the observations made by Voigt (1883) and Dorner (1865), respectively. Where consistency occurs in the reported dimensions in the other two species, those are the dimensions that we accept. The dimensions of the anterior region and tail for *B. parasita* given by Voigt (1883) and Ostroumoff (1883) are very close to each other and we cannot account for the discrepancy of those provided by Dorner (1865). The observation of spermatozoa in *B. pentodonta* given by Voigt (1883) and Ferraguti & Gelder (1991) coincide, but differ from those cited in the studies of Bondi (1963: 44) and Bond & Farnesi (1976: 68). The two descriptions of spermatozoa by Dorner (1865) for *B. parasita* and Bondi (1963) for *B. pentodonta* both differ from the accepted dimensions for the respective species due to the length of tail. However, the tail in *B. parasita* is about 100 μm shorter, and in *B.*

pentodonta about 80 μm longer than the accepted dimensions.

In the description of spermatozoan ultrastructure from *B. pentodonta* by Bondi (1963), his figures 11 & 12 show a spherical midpiece which he named the "basal body", and Fig. 12 clearly shows nine empty membrane-bound structures, which are almost certainly washed-out mitochondria. Bondi & Farnesi (1976: 68) corrected the earlier misnomer and stated that the spherical midpiece is composed of, "six or seven mitochondria". This number of mitochondria is consistent with those observed in spermatozoa from *B. pentodonta* described by Ferraguti & Gelder (1991) and *B. kozarovi*. However, the difference in a subspherical midpiece described by Bondi (1963) compared to the rod-like midpiece observed by Ferraguti & Gelder (1991) and in *B. kozarovi* remains. One explanation for these differences is that two species may have been studied, although the differences may also be due to the effect of a different fixation technique.

The mature spermatozoa from *B. kozarovi* have a significantly shorter acrosome and tail compared to other members of the genus. The presence of terminal, thickened rings at each end of the acrosome tube, the spherical invagination of the subacrosomal space into the anterior acrosome vesicle region, large portion of subacrosomal space devoid of vesicles, and the absence of the unorganized membrane ring in the posterior acrosome vesicle extremity, are clear, if subtle, differences between the spermatozoa of *B. kozarovi* and those from *B. pentodonta*. This study has demonstrated that the two species can be recognized from the morphology of their spermatozoa, and one more branchiobdellidan species has been added to the collection for a future phylogeny based on spermatozoan characters.

Acknowledgements. This work has been supported by Faculty Development and Mini-Grant funds from the University of Maine at Presque Isle, U.S.A. to S.R.G., the Institute of Zoology, Bulgarian Academy of Sciences, Bulgaria to M.A.S. and S.R.G., and Consiglio Nazional della Ricerche (Rome) Italy as a Bilateral Project to M.F. Constructive criticism of the manuscript was provided by Dr Andrea M. Groman and reviewers.

References

Bondi, C., 1963. Osservazioni sulla morfologia e sulla fine struttura del nemasperma di *Branchiobdella pentodonta* (Whitman) esasminato al microscopio elettronico. Riv. Biol. 56: 21–48.

Bondi, C. & R. M. Farnesi, 1976. Electron microscope studies of spermatogenesis in *Branchiobdella pentodonta* Whitman (Annelida, Oligochaeta). J. Morph. 148: 65–88.

Boshko, E. G., 1983. Crayfish oligochaetes in the waterbodies of the Ukraine. Report I. Branchiobdellidae. Vestn. Zool. 7: 13–20.

Brinkhurst, R. O. & S. R. Gelder, 1991. Chapter 12. Annelida: Oligochaeta and Branchiobdellida. In J. H. Thorpe & F. Covitch (eds), Ecology and Classification of North American Freshwater Invertebrates. Academic Press, New York: 401–435.

Daddow, L. Y. M., 1983. A double lead stain method for enhancing contrast of ultrathin sections in electron microscopy: a modified multiple staining technique. J. Micros. 129: 147–153.

Dorner, H., 1865. Ueber die Gattung *Branchiobdella*. Z. wiss. Zool. 15: 464–494.

Ferraguti, M. & S. R. Gelder, 1991. The comparative ultrastructure of spermatozoa from five branchiobdellidans (Annelida: Clitellata). Can. J. Zool. 69: 1945–1956.

Gelder, S. R. & R. O. Brinkhurst, 1990. An assessment of the phylogeny of the Branchiobdellida (Annelida: Clitellata) using PAUP. Can. J. Zool. 68: 1318–1326.

Holt, P. C., 1965. The systematic position of the Branchiobdellidae (Annelida: Clitellata). Syst. Zool. 4: 25–32.

Holt, P. C., 1986. Newly established families of the order Branchiobdellida (Annelida: Clitellata) with a synopsis of the genera. Proc. biol. Soc. Wash. 99: 676–702.

Humason, G. L., 1979. Animal Tissue Techniques. 4th ed. Freeman and Co., San Francisco, 661 pp.

Karaman, S. M., 1967. Branchiobdellidae Jugoslavija. Buletin i Punimeve Shkencore te Fakultetit Filozofik te Prishtines, IV: 39–64.

Karaman, S. M., 1970. Beitrag zur Kenntnis der europäischen Branchiobdelliden (Clitellata, Branchiobdelloidea). Int. Revue ges Hydrobiol. 55: 325–333.

Karnowsky, M. J., 1965. A formaldehyde-glutaraldehyde fixative of high osmolality for use in electron microscopy. J. Cell Biol. 27: 137A.

Ostroumoff, A., 1883. Ueber die Art der Gattung *Branchiobdella* Odier auf der Kiemer der Flusskrebs. (*Astacus leptodactylus* Eschh.). Zool. Anz. 6: 76–78.

Pop, V., 1965. Systematische Revision der europaischen Branchiobdelliden (Oligochaeta). Zool. Jb. Syst. Okol. Geogr. Tiere 92: 219–238.

Sawyer, R. T., 1986. Leech Biology and Behaviour. Clarendon Press, Oxford, 1065 pp.

Spurr, A. R., 1969. A low viscosity resin embedding medium for electron microscopy. J. Ultrastruct. Res. 26: 31–43.

Subchev, M., 1978. A new branchiobdellid – *Branchiobdella kozarovi* sp. n. (Oligochaeta, Branchiobdellidae) from Bulgaria. Acta zool. bulg. 9: 78–80.

Voigt, W., 1883. Die Varietaten der *Branchiobdella astaci* Odier. Zool. Anz. 6: 139–143.

Voigt, W., 1885. Ueber Ei- und Samenbildung bei *Branchiobdella*. Arb. zool.-zoot. Inst. Wurzburg 7: 300–368.

Hydrobiologia **278**: 27–34, 1994.
T.B. Reynoldson & K.A. Coates (eds), Aquatic Oligochaete Biology V.

Oligochaete fauna of Lake Baikal

Ljudmila N. Snimschikova[1] & Tamara W. Akinshina[2]
[1] *Limnological Institute, Irkutsk Scientific Centre, Siberian Division of the USSR Academy of Sciences, Ulhan-Batorskaya Street, 3, 664033 Irkutsk, Russia;* [2] *Scientific Research Institute of Biology of Irkutsk State University, Lenin Street, 3, P.O. Box 24, 664003 Irkutsk, Russia*

Key words: Oligochaeta, distribution, endemism, radiation of species

Abstract

Oligochaetes are one of the most abundant groups of invertebrates in Lake Baikal. They compose up to 70–90% of the biomass and numbers of zoobenthos and are distributed from the water edge to the maximum depth. There are 207 identified species and subspecies belonging to 44 genera and seven families, of which 160 species and 13 genera are endemic. Many are relict species. The main peculiarities of oligochaete fauna of Lake Baikal are: immiscibility of its ecologically different complexes (Palaearctic and Baikalian), its antiquity and heterogeneity, and intense radiation of the species.

Introduction

Baikal is one of the most thoroughly and comprehensively studied lakes in the world but all the scientific achievements on Baikal are only a stage to a more profound knowledge of the Lake which is a remarkable natural phenomenon (Kozhov, 1963). The more thorough is the knowledge, the more riddles and secrets are opened up to the investigators by the amazing nature of Baikal. During the last decade interest in this Siberian 'inland sea' grew not only in Russia but in the whole world.

The oligochaetes of Lake Baikal are one of the most abundant and complicated groups of invertebrates and play an active role in the ecology of the lake. However, up to 1987 the Baikal oligochaete fauna had not been studied as a whole, and no list of common species existed. Previously, either individual families (Lumbriculidae, Naididae) or the fauna of separate regions of the lake (Selenga Region, north end, Maloye More) were investigated to determine their oligochaete composition. Information accumulated during recent decades prompted Timm & Finogenova (1987) to compile a list of aquatic oligochaetes of the USSR which included those of Baikal. A major work has also appeared on the origin and evolution of the Baikalian oligochaetes (Semernoy, 1987).

We undertook the task of compiling a systematic list and table of distribution of the oligochaetes in Baikal which will be published in the new edition of the book 'Lake Baikal and its life' (Kozhov, 1963). The information from this list is the basis for the present review. The limited size of our article here does not enable us to cite it completely (see Appendix 1).

Materials and methods

All the available literature sources on the subject were used. Insufficient data on oligochaete distribution in Baikal are to some extent completed by the authors' unpublished observations, obtained from the benthic collections made by the Limnological Institute and the Institute of the Biology of the Irkutsk University. These collections were made from 1972–1989 in South Baikal in the vicinity of Solzan and Bolshie Koty; in Middle Baikal in the straits of Maloye More and Olkhon Gate and in the Mukhor Gulf or Selenga Shoals: in North Baikal in the vicinity of Sosnovskaya Bank (Fig. 1).

Samples were collected with a Petersen bottom grab (0.10 m^2 or 0.025 m^2), with a dredge and a trawl. The animals were preserved in formaldehyde or ethanol.

Results

Oligochaetes compose up to 70–90% of the biomass and numbers of individuals of bottom communities of the lake and are distributed from the water's edge to the maximum depth (1620 m). Near the shore in open Baikal, not less than 20–30% of the total benthic biomass is oligochaetes. In silty-sands at a depth of 10–20 m about a half of the total biomass is oligochaetes. On silty-sands at a depth over 50 m, oligochaetes form up to 90% of the biomass and occasionally the benthic population consists almost completely of oligochaetes. On Selenga Shallows, the most productive region of the lake (enriched by the organic substances brought down the River Selenga), the maximum benthic biomass is 360 g m^2, the biomass of oligochaetes is 233 g m^2 and their abundance is 14,460 individuals m^2. In North Baikal, in the areas influenced by the rivers Verhknyaya Angara and Kichera, the benthic biomass is 50 g m^2, the biomass of oligochaetes 28 g m^2 and their abundance 20,870 individuals m^2. In South Baikal in the vicinity of Solzan the average benthic biomass is 31 g m^2, with the oligochaete biomass 17 g m^2. The biomass of the benthos decreases to 5 g m^2 with increasing depth, and the proportion of oligochaetes increases to 90% (Bekman, 1983; Snimschikova, 1987; Akinshina & Lezinskaya, 1980).

The species diversity of the oligochaetes of Baikal is astonishing – up to 30 species per sample of $\leq$ 1 m^2. The species diversity is especially high in sandy-stony littoral (depth of 5–20 m), where 75 species and subspecies have been observed, and on silty sediment at depths of 50 m where 63 species and subspecies were identified.

At present 207 species and subspecies of oligochaete have been counted in Baikal from 44 genera and seven families, out of which 160 species and 13 genera are endemic. This does not complete the diversity of the Baikalian oligochaetes. Study of the fauna is continuing and it is certain that new discoveries will be made.

The Lumbriculidae is represented in Baikal by 59 species and subspecies (this is one third of the whole world's lumbriculid fauna) of which 51 are endemic. Lumbriculidae dominate the biomass in the abyssal. Among them are the largest Baikal species: *Rhynchelmis brachycephala* Michaelsen, 1901, up to 185 mm long and *Teleuscolex baicalensis* (Grube, 1873) 75 mm, but dwarf species also occur, including *Lamprodrilus achaetus* Isossimov, 1962 and *Lamprodrilus pygmaeus* Michaelsen, 1902. Several truly abyssal species with specific morphological adaptations have been found: *Lamprodrilus inflatus* Michaelsen, 1905, *Lamprodrilus bythius* Michaelsen, 1905.

Tubificidae are more diverse in Baikal than Lumbriculidae. To date 92 species and subspecies have been identified, 72 of which are endemic. They are most abundant in shallow areas (Snimschikova, 1987). Many small-sized species were discovered among tubificids: *Baikalodrilus exilis* Snimschikova, (2.0–2.3 mm long); *Baikalodrilus kozovi* (Hrabê) (1.2 mm); *Lymphachaeta pinnigera* Snimschikova, (1.8–2.3 mm). Many tubificid species do not exceed 10–12 mm in size. Giant forms are *Baikalodrilus crassus* Snimschikova, (32–35 mm) and *Baikalodrilus discolor* Snimschikova, (28–30 mm). Materials were obtained proving the eurybathic nature of tubificids. Some specimens from abyssal depths were described as *Rhyacodriloides abyssalis*, Ĉekanovskaya, and others (Ĉekanovskaya, 1975; Snimschikova, 1987).

Of Naididae 47 species and subspecies are

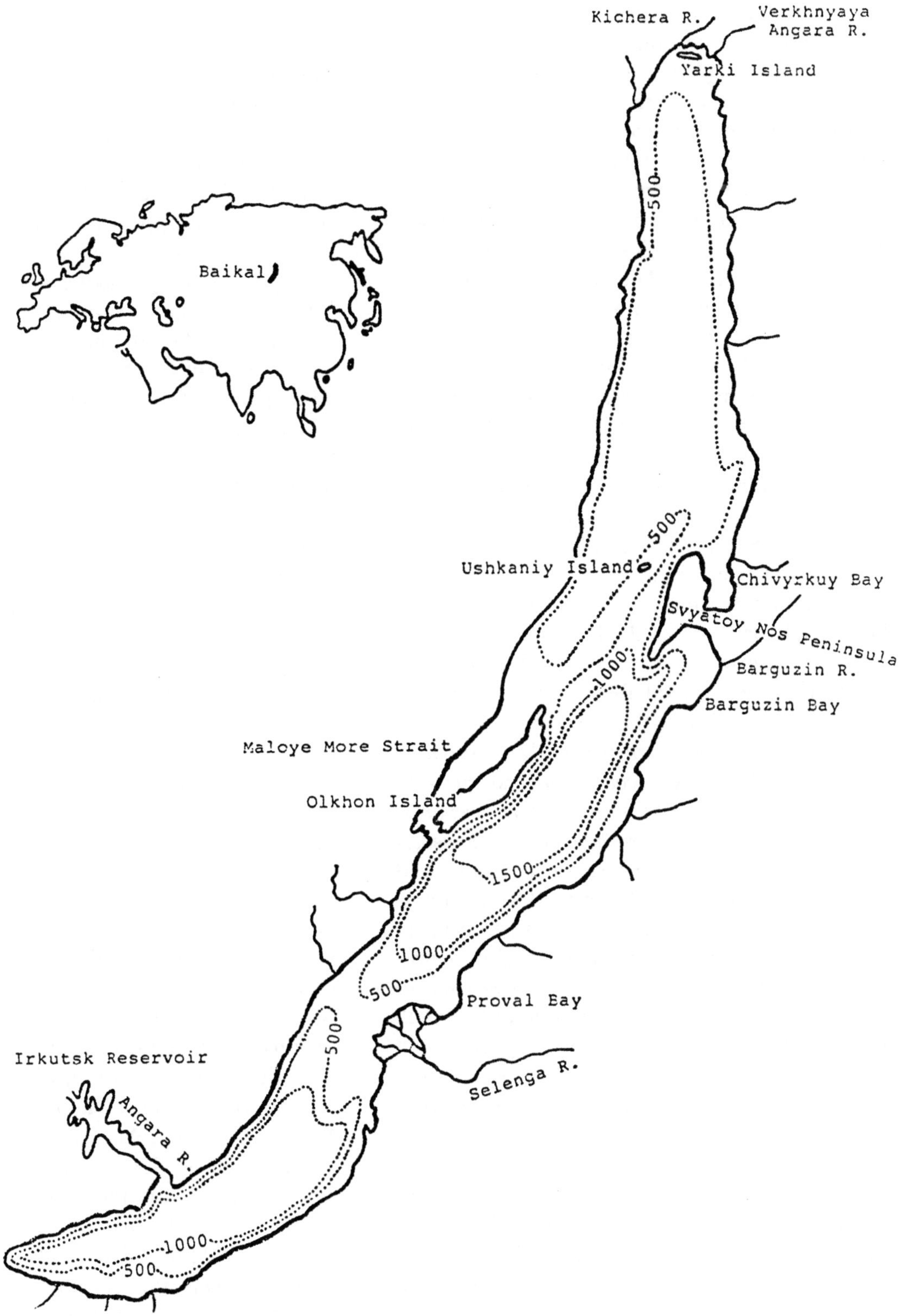

Fig. 1. Schematic map of Lake Baikal.

known in Baikal including 28 endemics. They mainly inhabit the coastal zone and only a few are found at depths (*Nais koshovi* Sokolskaja, was found at a depth of 93 m). Of special interest is the discovery of *Nais abissalis* Semernoy, at a depth of 700 m (Semernoy, 1984).

Enchytraeidae is represented by 3 species (2 endemics), Propappidae by 2 species (1 endemic), Haplotaxidae by 2 species and Aeolosomatidae by 2 endemic species. The latter family has traditionally been included in Oligochaeta, although Brinkhurst & Jamieson (1971) withdrew them from this subclass and included it in Archiannelida, while Timm (1987) placed them into subclass Aphanoneura of the class Clitellata. Description of a representative of Branchiobdellidae in Baikal (Snimschikova, 1987) was erroneous, as it turned out to be a new endemic leech (Finogenova & Snimschikova, 1991).

There are few abundant species of oligochaetes that are found distributed over the whole lake. In the coastal-sor ('Sor' is the local name for half-isolated shallow bays of Lake Baikal) zone and in the open Baikal the palaearctic species *Propappus volki* Michaelsen, 1915, as well as the endemic species *Rhyacodrilus korotneffi* (Michaelsen, 1905), *Rhyacodrilus multispinus multispinus* (Michaelsen, 1905), *Isochaetides arenarius* (Michaelsen, 1926), *Isochaetides baicalensis* (Michaelsen, 1902), *L. pygmaeus pygmaeus* and *Teleuscolex korotneffi* Michaelsen, 1901 are abundant. However, these species do not occur at depths over 260–300 m. *Lycodrilides schizochaetus* Michaelsen, 1905 and *Styloscolex asymmetricus* Isossimov, 1962 are often found in small numbers or individually in the sors, bays and open waters of Baikal. The above endemic species except *L. pygmaeus pygmaeus* have dispersed from Baikal and are found in the Angara River, which discharges from the Lake, while *R. korotneffi* and *L. schizochaetus* are in the Yenisei River. *Lamprodrilus pygmaeus pygmaeus* was introduced into the lakes of Karelia (North Europe) where it has formed populations by asexual reproduction (Popchenko, 1972).

In the open gulfs (Mukhor, Proval) and in the lake proper *Tasserkidrilus acapillatus* (Finogenova, 1972) is numerous. Its distribution is limited to the Caspian Sea, Lake Peipsi and Lake Baikal. As well, the endemic *L. achaetus achaetus*, *Lamprodrilus wagneri* Michaelsen, 1901, and *Stylodrilus opisthoannulatus* (Isossimov, 1962) are abundant in these habitats. *Lycodrilus dybowskii* Grube, 1873, is frequent but not numerous. These species do not appear to enter the sors although *T. acapillatus* and *L. wagneri* are known from Angara and *S. opisthoannulatus* is known from Yenisei. The above-mentioned species are found to depths of 500–860 m.

It is difficult to speak confidently about common species of oligochaetes distributed over the whole Lake. It would be more correct to describe the common species of a certain region of the lake or even of a biocoenosis, for example, of the Selenga region or of the stony littoral. The lake regions differ considerably from each other in their hydrological and hydrochemical conditions, bottom sediment, underwater landscape etc. The biocoenoses of river mouth shoals which have predominantly soft sediments on relatively gentle slopes cannot be compared to the biocoenoses of the stony steep drops. Such areas differ considerably in their species compositions. Typical species of soft sediments in the Selenga region are *I. baicalensis*, different species of *Baikalodrilus*, *L. dybowskii*, and *L. wagneri*. At the North Baikal where the biotopes are more diverse and the change of biocoenoses with increasing depth is faster, *P. volki*, *T. acapillatus*, *Rhyacodrilus isossimovi* Ĉekanovskaya, 1975, *Svetlovia maculata* Ĉekanovskaya, 1975, *L. achaetus* and species of *Stylodrilus* predominate. Still more different is the region of Bolshie Koty (South Baikal) with largely a sandy-stony substratum, where on the periphyton between stones Naididae (*Nais baicalensis* Sokolskaja, 1962, *Nais tygrina* Isossimov, 1962, *Uncinais minor* Sokoloskaja, 1962), small-sized Enchytraeidae of the genus *Mesenchytraeus*, *P. volki*, *L. achaetus*, *L. pygmaeus*, species of the genera *Stylodrilus* and *Isochaetides*, and *L. shizochaetus* are very abundant.

Differences in the ecological conditions decrease as the deep water zone (250 m and more) is approached and the species composition also becomes more uniform. Palaearctic species disappear and subsequently the endemic species that are abundant at middle depths gradually disappear. Typical inhabitants of the abyssal zone are *R. brachycephala, L. wagneri longus, T. baicalensis, R. abyssalis* and *B. discolor* with subspecies.

Discussion

Together with the unusual species richness and high degree of endemism, the Baikalian fauna is defined by relative immiscibility of the two ecologically and historically different complexes of which it consists. The coastal-sor zone is inhabited by a palaearctic species complex widely distributed in Siberia while the open lake is inhabited mainly by the endemics composing the *Baikalian complex* (Wereschagin, 1935; Levanidova, 1948; Kozhov, 1962).

Among the oligochaetes, in contrast to molluscs, amphipods, and other benthic groups, there is not a sharply defined immiscibility barrier between the Palaearctic and Baikalian species. In sors, the Baikalian complex is represented only by its eurybiontic component but composes 63% of the total species number (68 species and subspecies are accounted for in the Baikal sors and bays). On the other hand, the palaearctic species *Rhyacodrilus coccineus* (Vejodovský), *Rhyacodrilus sibiricus* Semernoy, *T. acapillatus, P. volki*, and *Haplotaxis gordioides* (Hartmann) have colonized the open Lake to depths up to 200–270 m. Nevertheless, the fact of immiscibility of the Baikalian and the Palaearctic complexes also occurs in this group. Over 10 palaearctic and cosmopolitan species inhabit only sors and do not venture into the open Lake. This is confirmed by distribution of the papillate tubificids. The palaearctic *Spirosperma ferox* Eisen distributed in the North-Baikalian Sor could be transported by undercurrents of the rivers, Verkhnyaya Angara and Kichera into the North Baikal shallows, whereas they are not found in either the river mouths or beyond the sand spit that separates the sor from Baikal proper. Beyond this spit in the open Lake, on sandy shoals at depths of 2–20 m, and even on silted sand at depths of 20–50 m, the papillate tubificids of the endemic genus *Baikalodrilus* are distributed in masses. No one species from this genus was found in the North-Baikalian Sor.

An exception is the invasion of the cosmopolitan *Tubifex tubifex* (Müller) into open Lake Baikal. Akinshina & Lezinskaya (1978) have also recorded it in South Baikal, in the vicinity of Solzan in sediments polluted by sewage waters at depths of 5–130 m where its frequency increased with an increase in the degree of pollution. The future behaviour of *T. tubifex* in Baikal can only be conjectured. At best, if the public pressure for protection of Baikal achieves closure of the Baikal pulp mill, the discharge of sewage will terminate. The disturbed biocoenoses will be rehabilitated and *T. tubifex* may disappear. However, if the pollution continues for a long period, conditions will be favourable for wider distribution of *T. tubifex*, and a decision whether this species should be left unchanged or subjected to biological irradication will be required.

Immigration of freshwater species into Baikal has taken place throughout the history of the lake, i.e. 28 $\times$ 10^6 years. The faunal elements of the preceding geological epochs were not only maintained but evolved rapidly. Therefore the modern Baikal fauna is historically heterogeneous. Martinson (1958, 1967) has distinguished in it palaeolimnical, mesolimnical and neolimnical elements. The palaeolimnical fauna was wide spread from the early Mesozoic over the whole Asian continent and is represented in modern Baikal by the Palaearctic (Siberian) complex. In the oligochaete group the palaeolimnical element is represented by several species of the genera *Rhyacodrilus, Nais* and *Haplotaxis*, and by *Styloscolex kolmakovi* Burov, *Teneridrilus hubsugulensis* (Semernoy) and others. The mesolimnical element arose during the Cretaceous and Palaeogene in the large lake systems of Central Asia. Such a lake system existed on the area occupied by modern Baikal. These pre-Baikalian water bodies were the source of many modern lake inhabitants: species of *Svetlovia, Isochaetides, Baikalodrilus, T. bazikalovae* group and many Lumbriculidae. The neolimnical element, according to Martinson

(1958), includes later immigrants from northern marine habitats which are not found among the Oligochaeta of Lake Baikal. In the process of intralake speciation, numerous species came into existence. The most prominent examples of biological radiation of Baikal's naidids are 10 species of genus *Chaetogaster*, of tubificids 22 species of genus *Baikalodrilus* and of the lumbriculids many species of the genus *Lamprodrilus*. The antiquity of Baikal, the developmental history of its basin, the history of the mutual influence of Baikal and the surrounding water bodies, diversity of the biotopes, ecological isolation of the abyssal part of the lake, constant influence of many factors of the aquatic environment (temperature, light, ground, biotic relations) have formed a peculiar fauna that is comparable in its species richness to marine systems.

Perspectives of investigations

An inventory of the Baikalian oligochaete fauna must be continued and completion of a species list is a long term objective. The faunal composition of only a few regions has been investigated: the north end, Maloye More, Selenga region, Bolshie Koty, Solzan, and the large sors. Very little is known of the southern part of North Baikal, Barguzinsky and Chivyrkuisky gulfs, underwater hills, the abyssal zone, the rocky slopes that predominate along the shoreline extending down to abyssal depths. Investigation of these areas requires the use of divers and manned submersibles, which although used in Baikal for several years are not yet available to biologists. The traditional methods of comparative morphology must be supplemented by the use of molecular biology, cladistics and phenetics. The Baikalian fauna will be gratifying material for such investigations.

Oligochaetes inhabiting widely different biotopes of the lake play a large role in the circulation of material, determining to a notable degree the rates of sediment accumulation and mineralization of the bottom deposits. Their significance in the self-purification processes may grow in relation to the danger of Baikal's pollution, but this has not been studied in this lake. Work on the developmental biology of Baikalian oligochaetes has hardly begun. Solutions to all these problems will need the organized efforts of a group of well-trained specialists.

Acknowledgements

The authors thank V.V. Tokmakov for the translation of the article into English and T.E. Timm who took the trouble of reading the manuscript and made valuable suggestions and corrections.

List of Oligochaeta found in Lake Baikal

Fam. Aeolosomatidae
Aeolosoma arenicola Semernoy, 1987 nom. nud.; *singulare* Sem., 1982.

Fam. Naididae, subfam. Naidinae
Stylaria fossularis Leidy, 1852; *lacustris* (L., 1767). *Ripistes parasita* (Schmidt, 1847). *Vejdovskyella dilucida* Snimschikova, 1982; *intermedia* (Bret., 1896) + ssp *sublitorale* Sem., 1987 nom. nud. + ssp *schizodentata* Sem., 1982; *Vejdovskyella* sp./ Gavrilov, 1950/. *Slavina appendiculata* (Ud., 1855). *Dero obtusa* Ud., 1855. *Nais abissalis* Sem., 1984; *baicalensis* Sok., 1962; *barbata* Müller, 1773; *bekmanae* Sok., 1962; *communis* Pig., 1906; *elinguis* Müller, 1773; *koshovi* Sok., 1962; *pseudobtusa* Pig., 1906; *similis* Sem., 1984; *simplex* Pig., 1906; *tatijanae* Sem., 1984; *tygrina* Is., 1962; *variabilis* Pig., 1906; *Nais* sp./ Noskova, 1963/; *Nais* sp./Čerepanov *et al.*, 1977/. *Neonais elegans* Sok., 1962. *Specaria josinae* (Vejd., 1883). *Uncinais minor* Sok., 1962; *uncinata* (Oer., 1842); *Uncinais* sp./ Noskova, 1963/.

Subfam. Chaetogastrinae
Amphichaeta magna Sok., 1962; *propinqua* Sem., 1987 nom. nud.; *rostrifera* Akinshina, 1984; (?*sannio* Kall., 1862)/ Gavrilov, 1950/. *Chaetogaster cannibalus* Sem., 1985; *crocodilus* Sem., 1985; *diaphanus* (Gruit., 1828) + ssp *litoralis* Sem., 1985; *dissetosus* spp Sem., 1985; *gavrilovi* Sem., 1985; *grandisetosus* Sem., 1985; *ignotus* Sem., 1985; *intermedius* Ak., 1984; multisetosus Sem., 1985; paucus Sem., 1985; Chaetogaster sp. (?*diastrophus*)/Gavrilov, 1950/.

Fam. Tubificidae, subfam. Rhyacodrilinae
Rhyacodrilus coccineus (Vejd., 1875) + ssp *inaequalis* Mich., 1905, + ssp *selengaensis* Noskova, 1967 nom. nud.; *isossimovi* Ček., 1975; *korotneffi* (Mich., 1905); *korjakovi* Sem., 1987 nom. nud.; *multispinus* (Mich., 1905) + ssp *multiovatus* Burov, 1936; *propus* Sem., 1987 nom. nud.; *sibiricus* Sem., 1971; *sokolska-*

jae Sem., 1971; *stephensoni* Ĉern., 1942; *subtilis* Sem., 1987 nom. nud. *Hrabeus magnus* Sem., 1987 nom. nud.; *tortus* Sem., 1987 nom. nud. *Pararhyacodrilus aspersus* Snim., 1986. *Svetlovia golyschkinae* Snim., 1985; *maculata* Ĉek., 1975; *majusculata* Snim., 1985. *Rhyacodriloides abyssalis* Ĉek., 1975 + ssp *irexoconensis* Snim., 1985.

Subfam. Tubificinae
Tubifex bazikalovae Ĉek., 1975 + ssp *aliquantulus* Snim., 1985, + ssp *crassiseptus* Sem., 1982, + ssp *grandis* Snim., 1987; *excavatus* (Hr., 1982); *eximius* (Sem., 1982); *penicraspedifer* Sem., 1982; *taediosus* Ĉek., 1975; *tubifex* (Müller, 1773); *Tubifex* sp./Snimschikova, 1987/; *Tubifex* sp./Snimschikova, 1989/. *Tasserkidrilus acapillatus* (Fin., 1972); *kessleri baicalensis* (Sem., 1982); *mirandus* (Snim., 1982); (? = *Tubificoides*) *proprius* (Sem., 1987) nom. nud. *Teneridrilus hubsugulensis* (Sem., 1980) n. comb.; *minutus* (Ĉek., 1975) n. comb.; *necopinatus* (Snim., 1985) n. comb. *Spirosperma ferox* Eisen, 1879. *Embolocephalus velutinus* (Grube, 1879). *Baikalodrilus bekmanae* (Snim., 1984); *bifidus* Snim., 1989; *crassus* Snim., 1989; *cristatus* (Snim., 1982); *digitatus* Holm., 1979; *discolor* (Snim., 1984) + ssp *acinacifer* Snim., 1989, + ssp *brevipectinatus* Snim., 1989; *exilis* Snim., 1982; *falcatus* Snim., 1982; *inflatus* (Mich., 1901); *intermedius* Snim., 1991; *kozovi* (Hrabê, 1969); *malevici* (Ĉek., 1975); *medianus* Snim., 1991; *multicrystallifer* Snim., 1989; *paradoxus* (Snim., 1984); *parilis* (Sem., 1987) nom. nud.; *phreodriloides* (Mich., 1905) n. comb.; *scaphoideus* Snim., 1989; *solitarius* (Snim., 1982); *undatus* Snim., 1989; *werestschagini* (Mich., 1933). *Haber vetus* (Sem., 1982) n. comb. *Limnodrilus hoffmeisteri* Clap., 1862; *profundicola* (Verr., 1871). *Isochaetides adenodicystis* Sem., 1982; *arenarius* (Mich., 1926) + ssp *inaequalis* (Mich. et Wer., 1930); *baicalensis* (Mich., 1902) + ssp *selengaensis* (Noskova, 1967); *michaelseni* (Last., 1936); *werestschagini* Ak. et Snim. 1991.; *Isochaetides* sp. n. 1/Snimschikova 1987/; *Isochaetides* sp. n. 2/Snimschikova, 1987/; *Isochaetides* sp.n. 3/Snimschikova, 1987/*Isochaetides* sp. n. 4/Snimschikova, 1987/; *Isochaetides* sp. n. 1/Snimschikova, 1989; *Isochaetides* sp. n. 2/Snimschikova, 1989; *Isochaetides* sp./Ĉerepanov *et al.*, 1977/. *Psammoryctides albicola* (Mich., 1907); *Psammoryctides* sp./Semernoy, 1987/; *Psammoryctides* sp./Snimschikova 1987/. *Potamothrix moldaviensis* Vejd. et Mr., 1902. *Ilyodrilus* sp./Snimschikova 1989/. *Lycodrilus dybowskii* Grube, 1873 (= *Tubipenifer tendens* Sem., 1982); *grubei* Mich., 1905; *nitens* (Sem., 1982) n. comb.; *Lycodrilus* sp./Snimschikova, 1987/. *Lycodrilides schizochaetus* (Mich., 1905).

Subfam. Aulodrilinae
Lymphachaeta pinnigera Snim., 1982.

Fam. Propappidae
Propappus glandulosus Mich., 1905; *volki* Mich., 1915.

Fam. Enchytraeidae
Mesenchytraeus bungei Mich., 1901. *Marionina aliger* (Mich., 1930). *Enchytraeus* sp./Ĉerepanov *et al.*, 1977/.

Fam. Lumbriculidae
Lumbriculus variegatus (Müll., 1774); *Lumbriculus* sp./ Ĉerepanov *et al.*, 1977/. *Lamprodrilus achaetus* Is., 1962 + ssp *hemiachaetus* Snim., 1987; *ammophagus* Mich., 1905; *bythius* Mich., 1905; *bulbosus* Is., 1962; *decathecus* Mich., 1901; *dithecus* Mich., 1901; *dybowskii* Mich., 1905; *glandulosus* Mich., 1905; *inflatus* Mich., 1905; *isoporus* Mich., 1901; *isossimovi* Sem. in litt.; *melanotus* Is., 1962; *nigrescens* Mich., 1905; *novikovae* Hr., 1982; *pallidus* Mich., 1905; *polytoreutus* Mich., 1903; *pygmaeus* Mich., 1902 + ssp *oligosetosus* Is., 1962, + ssp *sulcatus* Is., 1962; *satyriscus* Mich., 1903; *semenkewichi* Mich., 1901; *stigmatias* Mich., 1901; *tetrathecus* Mich., 1905; *wagneri* Mich., 1901 + ssp *longus* Is., 1962. *Teleuscolex baicalensis* (Grube, 1873); *glaber* Hr., 1982; *grubei* Mich., 1901; *korotneffi* Mich., 1901 + ssp *gracilis* Mich., 1901; *Teleuscolex* sp./Noskova, 1963/. *Agriodrilus vermivorus* Mich., 1905. *Stylodrilus asiaticus* (Mich., 1901); *cerepanovi* Sem., 1982; *crassus* Is., 1962 + ssp *crassior* Is., 1962; *gracilis* Sem., 1987 nom. nud.; *mirandus* (Hr., 1982); *mirus* (Ĉek., 1956); *opisthoannulatus* (Is., 1962); *parvus* (Hr. & Ĉek., 1927). *Rhynchelmis brachycephala* Mich., 1901 + ssp *tentaculata* Is., 1962; *dissimilis* Sem., 1987 nom. nud.; *minimaris* Sem., 1987 nom. nud; *spermatochaeta* Sem., 1982; *tetratheca* Mich., 1920. *Pseudorhynchelmis olchonensis* (Burov & Kozov, 1932); *Pseudolycodrilus parvus* (Mich., 1905). *Styloscolex asymmetricus* Is., 1962; *baicalensis* (Mich., 1901); *chorioidalis* Is., 1962; *kolmakovi* Burov, 1931; *solzanicus* Hr., 1982; *swarczewskii* Burov, 1931; *tetrathecus* Burov, 1931.

Fam. Haplotaxidae
Haplotaxis ascaridoides Mich., 1905; *gordioides* (Hartm., 1821).

References

Akinshina, T.W. & I.F. Lezinskaya, 1978. Oligokhety rykhlykh gruntov Baikala (rayon Utulik-Murino) kak pokazatel ego antropogennogo izmeneniya. Gidrobiologicheskie i ikhtiologicheskie issledovaniya v Vostochnoi Sibiri. Irkutsk: 184–189.

Akinshina, T.W. & I.F. Lezinskaya, 1980. K faune oligokhet rykhlykh gruntov Yuzhnogo Baikala v rayone Utulik-Murino. Gidrofauna i gidrobiologiya vodoemov basseina ozera Baikal i Zabaikalya. Ulhan-Ude: 3–6.

Bekman, M.J., 1983. Bentos priustyevykh uchastkov rek. Limnologiya Severnogo Baikala. Nauka, Novosibirsk: 103–108.

Brinkhurst, R.O. & B.G.M. Jamieson, 1971. The Aquatic Oligochaeta of the World. Edinburgh.

Ĉekanovskaya, O.V., 1975. Novye tubificidy (Oligochaeta, Tubificidae) iz abissali ozera Baikal. Novoe o faune Baikala. Nauka, Novosibirsk: 112–130.

Finogenova, N.P. & L.N. Snimschikova, 1991. *Dagarobdella zelenskiji* – a new genus and species of Piscicolidae from Baikal. Zoologicheskii zhurnal 70: 133–136 (in Russian, with English summary).

Kozhov, M.M., 1962. Biologiya ozera Baikal. Nauka, Moskva: 1–313.

Levanidova, I.M., 1948. K voprosu o prichinakh nesmeshivaemosti baikalskoi i palearkticheskoi fauny. Trudy Baikalskoi limnologicheskoi stantsii AN SSSR 12: 57–81.

Martinson, G.G., 1958. Proiskhozhdenie fauny Baikala v svete paleontologicheskikh issledovanii. Doklady AN SSR 120: 1155–1158.

Martinson, G.G., 1967. Problema proiskhozhdeniya fauny Baikala. Zoologicheskii zhurnal 46: 1597–1598.

Popchenko, V.I., 1972. Maloschetinkovye chervi vodoemov Karelii. Vodnye maloschetinkovye chervi. Yaroslavl: 51–52.

Semernoy, V.P., 1984. K morfologii i sistematike roda *Nais* Müller (Oligochaeta, Naididae) v ozere Baikal. Izvestiya Sibirskogo otdeleniya AN SSR, ser. biol. 1: 69–75.

Semernoy, V.P., 1987. Proiskhozhdenie i evolutsiya oligokhet ozera Baikal. Yaroslavl: 1–73. Manuscript deposited with VINITI, No. 8041–B87.

Snimschikova, L.N., 1987. Oligokhety Severnogo Baikala. Nauka, Novosibirsk.

Timm, T., 1987. Maloschetinkovye chervi (Oligochaeta) vodoemov Severo-zapada SSSR. Valgus, Tallin.

Timm, T.E. & N.P. Finogenova, 1987. Spisok vodnykh oligokhet SSSR. Vodnye maloschetinkovye chervi. AN Latviiskoi SSR, Riga: 3–11.

Wereschagin, G.J., 1935. Dva tipa biologicheskikh kompleksov Baikala. Trudy Baikalskoi limnologicheskoi stantsii AN SSSR 6: 199–212.

Hydrobiologia **278**: 35–51, 1994.
T.B. Reynoldson & K.A. Coates (eds), Aquatic Oligochaete Biology V.

A preliminary review of the taxonomic characters used for the systematics of the genus *Trichodrilus* Claparède (Oligochaeta, Lumbriculidae)

Pilar Rodriguez [1] & Narcisse Giani [2]
[1] *Dpto. Biología Animal y Genética, Facultad de Ciencias, Universidad del Pais Vasco, Apdo. 644, 48080 Bilbao, Spain*; [2] *Laboratoire d'Hydrobiologie, Université Paul Sabatier, 118, Route de Narbonne, 31062 Toulouse Cedex, France*

Key words: Oligochaeta, Lumbriculidae, *Trichodrilus*, taxonomy

Abstract

The taxonomic characters of the 34 known species of the genus *Trichodrilus* Claparède, 1862 are discussed as a basis for a systematic revision of the genus. Some additions to the species descriptions are made subsequent to examination of type collections. *Trichodrilus campoyi* Rodriguez, 1988, originally described as a subspecies of *T. strandi* Hrabě, 1936, is emended to specific rank. *Trichodrilus icenorum* Beddard, 1920 is regarded as a synonym of *T. allobrogum* Claparède, 1862. The extension of pharyngeal glands is demonstrated to be a fixed or a variable character depending on the species. Presence or absence of posterior lateral blood vessels or pattern of anterior blood vessels are not well documented for a significant proportion of the species and these characters need to be tested for environmentally induced variation. Taxonomic value of structural traits is strengthened against characters related to size. Thus, a key to the species of the genus is provided, emphasising such structural characters.

Introduction

The lumbriculid genus *Trichodrilus* was first described by Claparède in 1862. Taking as reference the work of Brinkhurst & Jamieson (1971) on the Aquatic Oligochaeta of the World, 20 species were described before 1971. Fourteen species have been described in the following years. In addition, the authors have found five new species in caves and hyporheic biota in Spain and France; descriptions of these new species will be published elsewhere. The taxonomic characters generally used in the systematics of the genus are discussed in order to establish a basis for a subsequent revision of this taxon, previously revised by Cook (1967a, 1971a). A key is provided for identification of the species of the genus.

Material and methods

We studied type specimens from the British Museum (BMNH), the United States National Museum (USNM), Museum National d'Histoire Naturelle de Paris (MNHN), and authors' collections. Some species were studied from photographs sent by D. G. Cook and from personal collections borrowed from T. Timm (Tartu, Estonia) and J. Juget (Lyon, France), A. Camacho (Museo Nacional de Ciencias Natural,

MNCN, Madrid, Spain) and R. Rouch (Lab. Moulis, Toulouse, France).

List of species studied

Trichodrilus allobrogum Claparède, 1862. Type collection not seen. Collection of N. Giani; numerous specimens collected during 1979 from the river Riou Mort, a tributary of the river Lot, Massif Central, France.

Trichodrilus pragensis Vejdovsky, 1875. Type collection not seen. Collection of P. Rodriguez; one specimen from Pozo del Infierno, La Hermida, Cantabria, Spain, 21.I.1989. *Leg.* Ana Camacho (MNCN).

Trichodrilus intermedius (Fauvel, 1903). Type collection not seen. One whole-mounted specimen in Canada balsam found in the late Pr. Vandel collection at Toulouse, France; *Trichodrilus pragensis* Vej., Piguet det., Source de la Folie, Gometz la Ville, Région de Chevreuse, France, 18/04/1920; further dissected and identified by N. Giani. One fluid preserved specimen from MNHN, labelled *Trichodriloides intermedius* Fauvel, well near Paris, France, M. André det., 17/04/1928; identified by N. Giani as *Haplotaxis gordioides*.

Trichodrilus cantabrigiensis (Beddard, 1908). BMNH 1920.10.04. Syntypes (1 whole mounted, 1 complete in alcohol) and Schizosyntype (1 indiv., sections on slide).

Trichodrilus icenorum Beddard, 1920. BMNH 1920. 10.29.2. Syntypes (3 indiv., sections on slide).

Trichodrilus strandi Hrabě, 1936. Photographs of the types loaned by D. G. Cook (Fisheries and Oceans, Canada).

Trichodrilus tatrensis Hrabě, 1937. Photographs of the types loaned by D. G. Cook (Fisheries and Oceans, Canada).

Trichodrilus leruthi Hrabě, 1937. Type collection not seen. Four mature specimens loaned by J. Juget (University Claude Bernard, Lyon I, France); hyporheic floor of Rhône river, near Lyon; 3 localities: Lône des Pêcheurs, Canal de Miribel and Lône du Grand Gravier.

Trichodrilus stammeri Hrabě, 1937. Photographs of the types loaned by D. G. Cook (Fisheries and Oceans, Canada).

Trichodrilus tenuis Hrabě, 1960. Type collection not seen. Three mature specimens from wells between Velanizan and Barca (Soria, Spain), 13.IX.1976. *Leg.* R. Rouch (Lab. Moulis, France).

Trichodrilus hrabei Cook, 1967b. BMNH 1967.1.–10. Syntypes. (?8 indiv., 10 slides, one sectioned specimen, one dissected specimen, one immature, one with rudimentary genitalia and four mature specimens).

Trichodrilus allegheniensis Cook, 1971b. USNM 43491 Paratype (1 whole mounted specimen).

Trichodrilus culveri Cook, 1975. USNM 50868 Paratype (1 whole mounted and 1 dissected specimen).

Trichodrilus seirei Timm, 1979. Holotype (cross sections), Esna Spring 25.Feb.1974; Paratype (cross sections), Esna Spring, 23.Sept. 1973; and other material, as cross sections, from Esna Spring, 23.Sept.1973, Lake Kariste, 22.Jul.1979, and Lake Peipsi, 2.Jun.1990). All borrowed from T. Timm.

Trichodrilus diversisetosus Rodriguez, 1986. Holotype, from R. Butron, Spain, 30.XII.1980; 3 mature indiv., Cave S. Pedro de Busturia, Spain, 24.IX.1985. One mature specimen from the hyporheic flow of the R. Rhône, near Lyon, France, 1988, loaned by J. Juget.

Trichodrilus campoyi Rodriguez, 1988 *nov. status Trichodrilus strandi campoyi* Rodriguez, 1988: 209, 211, fig. 5B. Collection of P. Rodriguez: dissected material from several localities of the Basque Country, Spain. Sampled from 1981 to 1986. Collection of N. Giani: dissected specimens from Ruisseau des Cros, 29/4/82, and Ruisseau Assou, Massif Central, France, 22/03/87.

The following species of the genus have not been studied from type material or other collections, thus they are discussed on the basis of their descriptions. They are *Trichodrilus spelaeus* Moszynski, 1936, *Trichodrilus cernosvitovi* Hrabě, 1937, *Trichodrilus claparedei* Hrabě, 1937, *Trichodrilus moravicus* Hrabě, 1938, *Trichodrilus macro-*

phorophorus Hrabě, 1954, *Trichodrilus medius* Hrabě, 1960, *Trichodrilus ptujensis* Hrabě, 1963, *Trichodrilus sketi* Hrabě, 1963, *Trichodrilus tacensis* Hrabě, 1963, *Trichodrilus pauper* Finogenova, 1973, *Trichodrilus montenegrinus* Karaman, 1973, *Trichodrilus itchaensis* Sokolskaya, 1973, *Trichodrilus aporophorus* Popchenko, 1976, *Trichodrilus gordeevi* Popchenko, 1976, *Trichodrilus drimi* Karaman, 1986, *Trichodrilus macedonicus* Karaman, 1986, *Trichodrilus metohiensis* Karaman, 1986, and *Trichodrilus isabellae* Popchenko, 1988.

In the present work, we also mention five new species of the genus as follows: *Trichodrilus* sp. 1 to sp. 5. These new species will be described in a separate publication but they are keyed out with the currently described species and their most outstanding characteristics are also discussed.

List of characters studied

The size of animals: Length. Number of segments. Body diameter.
The setae: Form. Size. Position.
Prostomium.
Epidermal glands.
Pharynx.
Pharyngeal glands.
Vascular system: Posterior lateral blood vessels. Anterior lateral blood vessels. Longitudinal blood vessels.
Genitalia: Form and size of atrium. Atrial musculature. Penis. Prostate. Vas deferens. Spermathecae (number, form and position). Position of genitalia.

Results

Complements to the description of species

Examination of the material listed above allowed us to complement, in some cases, previous descriptions of the taxa.
1. *T. allobrogum*. Table 1 shows the range of some taxonomic characters in a French population.
2. *T. cantabrigiensis*. (Figs 1 & 2): Body length $\approx$20 mm, body diameter (at clitellum) $\approx$481 μm.
3. *T. icenorum* (Fig. 3): Pharyngeal glands extend to the anterior part of segment VIII, dorsal to the gut. Dimensions of atria are: length: 84–123 μm; maximum diameter: 75–90 μm; atrial musculature: 6–15 μm. The prostatic glands are of two types, long and pedunculated or grouped in bundles and short. The diameter of the vas deferens is 18–51 μm.
4. *T. hrabei* (Fig. 4): Pharynx is developed both dorsally and ventrally. Prostomium is rounded and very elongate (l: 180–210 μm, w: 120–135 μm).
5. *T. leruthi* (Fig. 5): The material from Juget's collection shows an important range of dimensions of the atria. Length of atria (including ampulla, bulb and penis): 406–895 μm; maximum width: 132–150 μm. Atrial ampulla pear-shaped to tubular, with a muscular layer including both circular and longitudinal muscular fibers of 21–36 μm maximum thickness. This orientation of the fibers contrasts with the original description where it is said that muscles run in all directions. Internal epithelium of atria observed only for one specimen (18–21 μm thick). Prostatic glands diffuse and short (36–81 μm high). Penial bulb is well developed (l: 270–281 μm;, w: 150–210 μm). The degree of extension of the penis of the studied specimens varied from 0 up to 330 μm. The vasa deferentia join the atrial ampulla in its distal half, except for one subapical junction, as in Hrabě's description. Efferent duct is relatively thin (diameter: 30 μm). In the study material, it could not be ascertained that the posterior vas deferens penetrates into segment XI, nor that there are posterior lateral vessels. We probably could not make these observations because of the mounting medium used (Apathy's syrup) which macerates tissues. In the original description, Hrabě points out that the posterior vas deferens penetrates into XI then comes *immediately* back to X, therefore it may not be an easy character to see.
6. *T. stammeri*. The photographs loaned by Cook did not allow us to make a more detailed study of this species.
7. *T. culveri* (Fig. 6): In both paratypes the dorsal region of the pharynx is completely everted.

Table 1. Some taxonomic characters in a population of *Trichodrilus allobrogum* Clap. and different individuals of *T. campoyi* Rodriguez.

Specimen	*T. allobrogum* [1]			*T. campoyi* [2]		
Characters [3]	Range	x̄	sd	Range	x̄	sd
Body ∅ (μm)	275–495	396.6	70.1	235–614	388.9	175.2
Prostomium length (μm)	164–235	192.5	26.6	73–207	135.5	33.4
Prostomium diam (μm)	196–235	218.6	16.7	–	–	–
Length of setae (μm)	99–120	106.2	9.5	73–112	85.3	11.2
Pharyngeal g. (up to)	VIII	–		V, VI or VII	–	–
Tegument:						
Longitudinal muscl. (μm)	6.6–19.8	12.9	4.1	6.8–42.5	22.1	10.6
Epidermis (μm)	8.8–15.4	11.6	2.4	(2.5)12–25	17.1	4.4
Genitalia						
Atrium length (μm)	101–134	117	13.9	80–188	141.7	31.6
Atrium diameter (μm)	75–106	87.6	12.3	39–98	72.5	20.7
Atrial muscles (μm)	6.6–15.4	9.9	2.6	≈2–4.2	2.8	0.9
Atrial epithelium (μm)	6.6–11	7.8	2.1	15–41	25.7	9.5
Prostate max. length (μm)	22–45	33.5	11.4	14–93	39.3	27.6
Vasa deferentia to XI	+			+		
Spermatheca I						
Ampulla length (μm)	84–176	130.3	44.6	136–406	216.7	96.1
Ampulla max. diameter (μm)	40–123	71.1	27.2	98–196	140	37.8
Duct length (μm)	56–84	64.9	12.9	119–≈308	179.9	60.3
Duct diameter (μm)	18–56	26.3	14.8	–	–	–
Ampulla position	–	–	–	X, XI or XII	–	–
Spermatheca II						
Ampulla length (μm)	114–168	145.3	22.9	–	–	
Ampulla diameter (μm)	46–79	68	15.4	–	–	
Duct length (μm)	66–75	69.4	4.9			
Duct diameter (μm)	20–28	22.5	4.7	–	–	

[1] Number of studied specimens: 10.
[2] Number of studied specimens: 12.
[3] Not all the characters could be measured on every specimen.

This may be due to fixation or may be the result of some peculiar habit of this species. Very thin body wall musculature (9 μm thick in VII) and transparency characterize this small (body diameter <300 μm) species. Anterior body has some sediment particles attached to the tegument.

Plate I Figs 1–9. (Scale 50 μm).

1. *T. cantabrigiensis.* Longitudinal section of atrium of a type specimen. a, atrium; d, vas deferens; m, muscular ring; pf, porophore; pr, prostate cells.
2. *T. cantabrigiensis.* Posterior lateral blood vessels of a whole mounted type specimen.
3. *T. icenorum.* Longitudinal section of a type specimen showing pharyngeal glands in segments VII and VIII. pg, pharyngeal glands; ss, sperm sac.
4. *T. hrabei.* Atria from type specimen. a, atria; mp, male pores.
5. *T. leruthi.* Penis from French material.
6. *T. culveri.* Prostomium and everted pharynx from paratype.

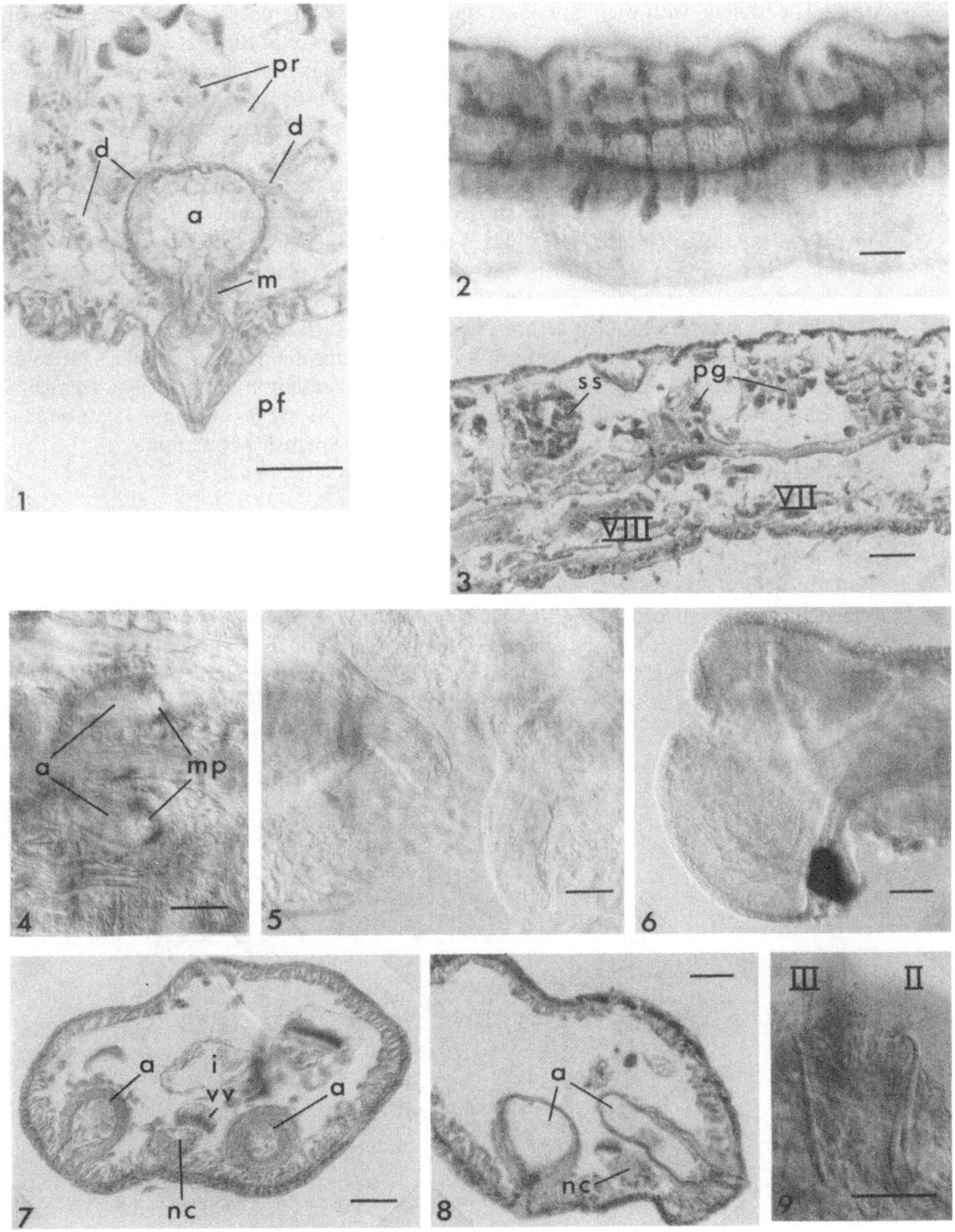

7. *T. seirei*. Transverse section of holotype. a, atrium; i, intestine; nc, nerve cord; vv, ventral blood vessel.
8. *T. seirei*. Transverse section of a paratype. a, atria; nc, nerve cord.
9. *T. diversisetosus*. Ventral seta of segment II.

Atrium is petiolate, with very short duct opening into a simple pore, without a penis, and with a globular ampulla.

8. *T. seirei* (Figs 7 & 8): The spermathecal ducts may have a small dilation in both holotype and paratype which may contain a mass of sperm and assume a bulbous appearance. This kind of spermathecal duct looks like that illustrated for *Trichodrilus moravicus* (Hrabě, 1938: figs 5 & 8).

9. *T. diversisetosus* (Fig. 9): This species was redescribed by Rodriguez and Giani (1987) and we only add that setae of segment II are displaced towards the lateral line. Body length: 8–9 mm.

10. *T. allegheniensis* (Fig. 10): In the paratype the clitellum is not elevated but is marked by the presence of glandular cells in segments X to XIII. The atrial duct forms a very muscular bulb, close to the male pore, located on a porophore. The second pair of spermathecae may penetrate into segment XIII.

11. *T. campoyi*: This species was formerly described as a new subspecies of *T. strandi* Hrabě (1942). It is now elevated to specific rank mainly due to the characteristics of the atrium: absence of an atrial duct and a penis. Number of segments in the only complete breeding individual is 61. Setae of the first segments are somewhat smaller than the following ones (II: 72–77 μm, III and IV: 80–85 μm, following preclitellar setae: 84–96 μm, postclitellar setae: 80–91 μm). Those of II to III or IV are simple-pointed (sometimes, only ventrally) and the following ones are bifid. Prostomium round (l: 150–165 μm). Tegument in segments IV to VIII shows an epidermis (including circular musculature) 12–21 μm thick and a longitudinal musculature 9–15 μm thick. Atrial epithelium is very granulated and high (up to 1/3 of the diameter of ampulla). No lateral blood vessels present in the posterior segments. Table 1 shows the range of some taxonomic characters in a Spanish population.

The taxonomic characters of the genus Trichodrilus

The size of animals

Length and number of segments: A significant number of species of *Trichodrilus* have been described from incomplete specimens (just anterior fragments). Also, body length may vary greatly

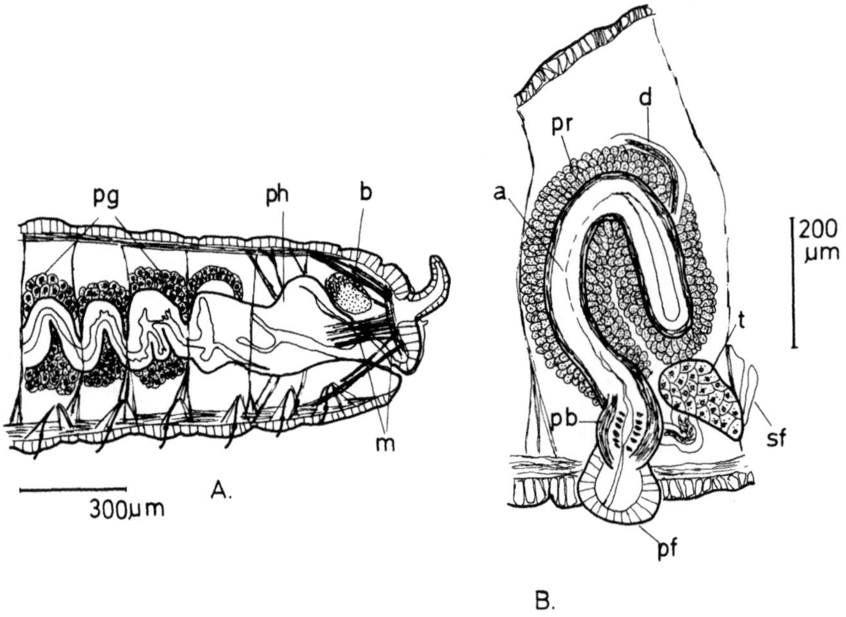

Fig. 10. T. allegheniensis. Paratype. A. Anterior region of the body: b, brain; m, muscular fibers; pg, pharyngeal glands; ph, pharynx. B. Male segment: a, atrium; d, vas deferens; pb, penial bulb; pf, porophore; pr, prostatic cells; sf, sperm funnel; t, testis.

depending on the state of relaxation of the worm when fixative is added.

The number of segments is independent of fixation and therefore it is a better descriptor of the species than length. We should note that even if there is not an absolute correlation between length and number of segments, the two numbers vary closely in most cases. The range of number of segments is very broad, from 19 in *T. seirei* to more than 100 in *T. icenorum* and *T. isabellae*.

Body Diameter: Of all characters describing size, body diameter, measured in the preclitellar or clitellar region, is the most informative. Based on this criterion the species of the genus could be grouped for convenience into 3 size classes: small species ($\varnothing < 0.300$ mm), medium species ($0.3 < \varnothing < 1$ mm) and large species ($\varnothing > 1$ mm) (see Table 2).

The relation between length of atrium and body diameter is of interest, being greater than one to three and usually greater than one to two for the small species. In contrast, this ratio does not exceed one to four for the large species.

Body diameter, however, must be used with caution as it can vary between different populations and even within a population. Hrabě (1938) reported a wide range of body diameters in two isolated populations of *T. moravicus* (0.26 to 0.56 mm). Cook (1967b, 1971a) synonymized *Trichodrilus lengersdorfi* (Michaelsen, 1934) with *T. icenorum* so that it now has a range of diameter from 0.5 to 1.6 mm. Another species with a wide range of body diameter is *T. seirei* (0.3–0.7 mm). The variation within populations of *T. allobrogum* (0.19–0.24 mm) and *T. campoyi* (0.24–0.61 mm) is shown in Table 1.

Table 2. Trichodrilus species grouped by their body size (diameter) and presence/absence of posterior lateral blood vessels [1].

Body size	Posterior laterial blood vessels			
	Present	Absent	Incomplete	Unknown
Small	ptujensis	culveri	cernosvitovi	
(0 < 300 μm)	aporophorus	tatrensis	pauper	
			tenuis	
Median size	leruthi	hrabei		gordeevi
($300 < \varnothing < 1000$ μm)	allobrogum	sketi		
	icenorum	strandi		
	ptujensis	campoyi		
	macroporophorus	seirei		
	pragensis	claparedei		
	cantabrigiensis	allegheniensis		
	isabellae [2]	stammeri		
		tacensis		
		medius		
		moravicus		
		diversisetosus		
		macedonicus		
		metohiensis		
		drimi		
Large	intermedius			itchaensis
(0 > 1000 μm)	icenorum			

[1] No data from *T. spelaeus* Moszynski, 1936 and *T. montenegrinus* Karaman, 1973.

[2] It is not clear in the description if the transversal vessels are the posterior ones.

The setae

Form: The setae of species of *Trichodrilus* are bifid or simple-pointed. Species with bifid setae form a small group composed by *T. strandi, T. diversisetosus, T. campoyi* and *T. isabellae*. In *T. diversisetosus* the setae of the first segments (II to III or IV) are simple-pointed ventrally, with very curved tips (Fig. 9). *Trichodrilus campoyi* also shows simple-pointed setae in the ventral bundles of segment II and sometimes III. The undescribed *Trichodrilus* sp. 4 stands out from the rest of the species, which have simple-pointed setae, by the presence of hair-like setae in the dorsal bundles of postclitellar segments (Fig. 11). These setae are thinner than ventral ones, with a pronouncedly proximal nodulus.

Modified sexual setae are scarce within the

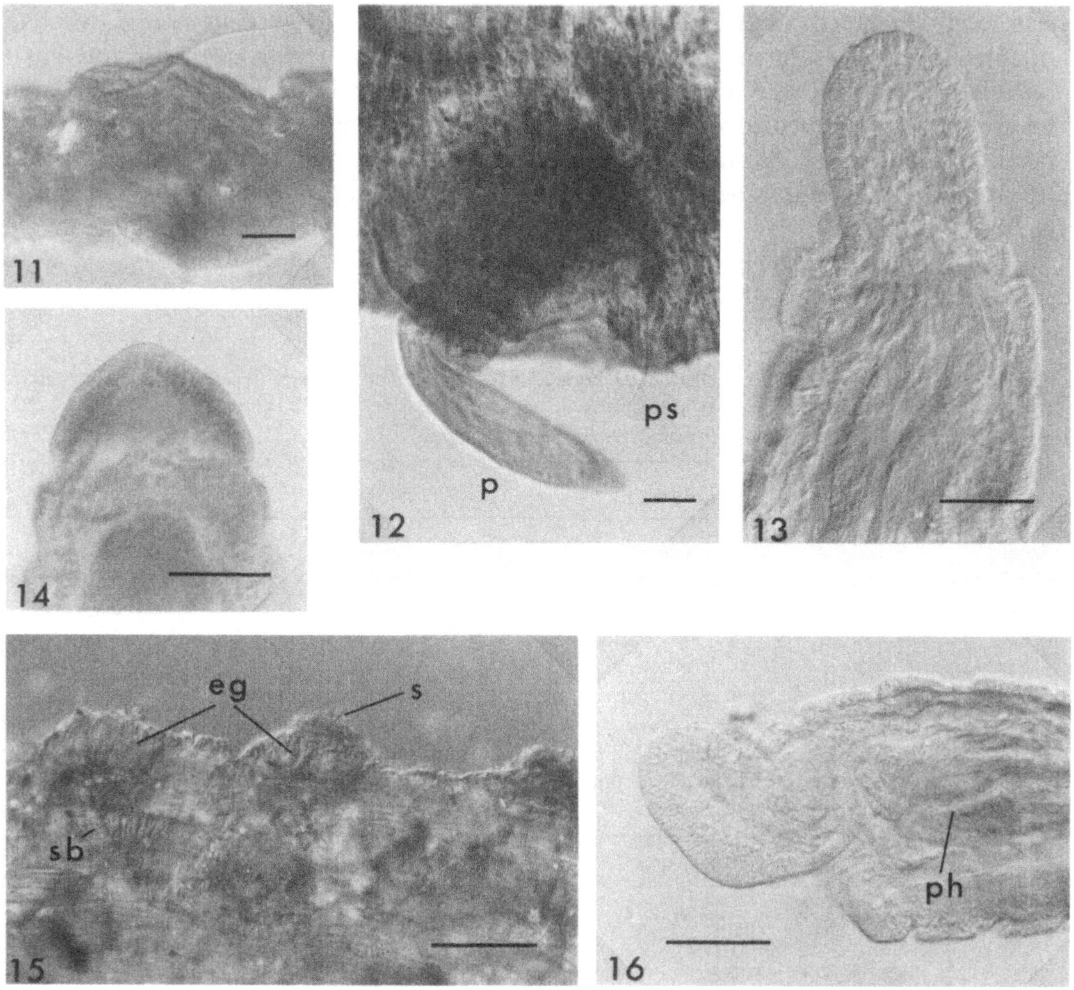

Plate II Figs 11–16. (Scale 50 μm)

11. *Trichodrilus* sp. 4. Hair-like dorsal seta in postclitellar region.
12. *Trichodrilus* sp. 5. p, penis; ps, penial seta.
13. *T. hrabei*. Round prostomium.
14. *Trichodrilus* sp. 4. Conical prostomium.
15. *Trichodrilus* sp. 3. Epidermal glands in dorsal region of posterior segments. eg, epidermal glands; s, secretion; sb, dorsal setal bundles.
16. *T. hrabei*. ph, pharynx.

family and have not yet been reported in the genus. However, *Trichodrilus* sp. 5 has thinner, slightly modified ventral setae close to the male pores (in X) (Fig. 12).

Size: Setae are usually of similar size, although in *T. claparedei, T. leruthi* and *T. moravicus* the dorsal setae are smaller than ventral ones. In *T. diversisetosus* the anterior ventral setae are much larger than the posterior ventral setae but the anterior dorsal setae are much smaller than posterior dorsal setae (Rodriguez & Giani, 1986).

Position: The anterior setae of segment II in *T. diversisetosus* are clearly shifted towards the lateral line.

Prostomium

Trichodrilus species can be divided into those that have round vs. conical prostomia (Figs 13 & 14). However, *T. seirei* is described as having a round or conical prostomium (Timm, 1979) and it remains to be established if the form of the prostomium may be modified by fixation. The original descriptions of 10 species of the genus do not mention this character. Only *T. allegheniensis* is clearly separated from the other species of *Trichodrilus* by the presence of a proboscis-like prostomium associated with a strong musculature (Fig. 10A).

Trichodrilus medius has been distinguished (Cook, 1971a) from other related species (*T. moravicus, T. hrabei* and *T. tenuis*) by the presence of a transverse dorsal furrow in the prostomium. It has not been possible to examine the type material to ascertain the validity of this character but a similar furrow observed in some specimens of other species has been interpreted as a particular state of contraction at the time of death. Thus, the identity of *T. medius* should be established on the basis of other characters.

Epidermal glands

Trichodrilus sp. 3 has characteristic glandular crests in the dorsal postclitellar region of the body (Fig. 15). Similar ventral, glandular formations, interpreted as adhesive organs, have been described in interstitial forms of some taxa (Swedmark, 1964), including such oligochaetes as *Marionina glandulifera* Jansson, 1961 (Kossmagk-Stephan, 1983; Coates, 1983).

Pharynx

The presence of a pharynx with a dorsal, thickened pad is characteristic of Oligochaeta. However, some haplotaxids, lumbriculids and naidids have a muscular pharynx developed both dorsally and ventrally. This kind of pharynx has been related to carnivorous or predatory habits (Brinkhurst, 1991; Brinkhurst & Gelder, 1989). *Trichodrilus* sp. 2 presents this kind of pharynx although no remains of other animals have been found in the gut. *Trichodrilus hrabei* has dorsal and ventral pads of columnar cells in the pharynx (Fig. 16).

Pharyngeal glands

The extension of the pharyngeal glands has been used as a taxonomic character for distinguishing *Trichodrilus* species. Cook (1971a: 224) suggests that the extension of the glands may depend on the size of the animal and also on the state of maturity. Some variability of this character is indicated for *T. macroporophorus* (up to VII or VIII), *T. tacensis* (VII or VIII) and *T. pragensis* (VIII or IX). Cook (1967b: 363) also indicates that this is a variable character in other lumbriculids. However, this has been considered an important character for distinguishing two very similar species: *T. icenorum* (up to VII) and *T. allobrogum* (up to VIII). Our study of the type material of *T. icenorum* has shown that pharyngeal glands may extend up to the anterior part of segment VIII (Fig. 3). Thus, these two taxa, are regarded as synonyms.

This character is very stable for the studied material of *T. allobrogum*; however, it is variable for *T. campoyi*. Therefore, the validity of the character must be established independently in every case since there could be variations between isolated populations of the same species. In any case, it seems highly improbable that populations that only differ in the extension of their pharyngeal glands belong to different species.

Vascular system

Posterior lateral blood vessels (Fig. 2): In *Trichodrilus* there are species with and without posterior lateral blood vessels (Table 2) and this character has acquired some importance in their classification. However, the vessels are, in many cases, difficult to observe due to the presence of sediment particles in the intestine. The lateral blood vessels occur from segment XXX or XL posteriad and, therefore, it is impossible to ascertain their presence in incomplete specimens. The loss of tails occurs with a high frequency in *Trichodrilus* and some species even have been described from anterior fragments of worms. These conditions indicate that this character should not be a key element for the discrimination of different groups of species in the genus, but used as an additional element in the diagnosis of the species. Cook (pers. comm.) suggests that the basic plan of these blood vessels may be a useful, conservative character but not branching or proliferation of the lateral vessels which possibly depends on environmental variables. This criterion needs to be studied further, comparing animals of different ages, states of maturity or in different environmental conditions, in order to establish its stability.

Trichodrilus tenuis is described in Cook (1971a) as lacking the posterior lateral vessels, however, in the original description Hrabě (1960: 271, 273) says that this feature has not been studied because his material lacked its posterior part. In the description of *T. montenegrinus* this character is not mentioned, but Timm (1979: 163) later says that the species lacks these vessels.

Anterior lateral blood vessels: In *T. moravicus*, lateral vessels in the anterior region of the body join the ventral blood vessel in the segment immediately posterior to that in which they originate from the dorsal vessel. Hrabě (1938: 77–78) supposes that this might be a widely distributed character among lumbriculids. Karaman (1973: 357) found that the anterior lateral vessels had their origin and termination in the same segment in *T. montenegrinus*. Hrabě (1938) also described this confinement of the anterior lateral vessels in *Rhynchelmis* and Cook (1967a: figs 1.2.D & 2.3.E) illustrated this pattern for *Lumbriculus variegatus* (Müller, 1774) as well as *Eclipidrilus lacustris* (Verrill, 1871). It may also be a variable character as in *T. culveri* where the lateral vessels cross the posterior septum of segment III but do not cross from segment II.

Longitudinal blood vessels: Popchenko (1976) points out the presence of longitudinal blood vessels in the posterior body region of *T. aporophorus*. This structure is difficult for us to interpret due to the lack of an illustration.

Genitalia

Form and size of atrium: Cook (1971a: 224) suggests that the dimensions of sexual organs can be subject to wide variation depending on the sexual maturity or the activity of the animal. In that sense; *T. seirei* includes wide variations in the size of atria (l: 75–113 μm) and a range of thickness of more than four fold in the atrial musculature (4–18 μm) and atrial epidermis (4–25 μm) which is associated with a change in form of the atria from spherical to pear-shaped (Figs 7 & 8). In the literature, *T. pauper* also presents a range of dimensions for the atrium of more than two-fold. Type material of *T. icenorum* also shows marked variation in atrial size.

Such variations in form and size have been interpreted by Timm (1979) as related to the state of maturity of the worm, those with bigger dimensions and with thinner atrial musculature and epidermis being individuals that have reached the resorption phase. It seems logical that depending on the state of contraction of the atria (derived from the state of maturity or from the condition of the specimen) there will be a relation between smaller atria and thicker atrial musculature, and between larger atria and thinner musculature. Sokolskaya (1973) suggests a relational index between atrial musculature and maximum diameter of the atrium to minimize the variations in size of the worms and their genitalia due to the state of contraction of the animal. This index seems to be useful although sometimes difficult to apply or interpret as in the case of *T. seirei* where the index is, respectively, 0.10 and 0.33 for each of the two forms described by Timm (1979). Is such a difference indicative of two different species?

The form of the atrium ranges from spherical/globular/pear-shaped to tubular-cylindrical. Most of the species of *Trichodrilus* have atria of the first form, with only *T. intermedius, T. leruthi, T. tacensis, T. allegheniensis* and *T. cernosvitovi* having tubular-cylindrical atria. This second group of species, except *T. intermedius*, has atria that are more than one-half as long as the body diameter. In *T. allegheniensis* and *T. cernosvitovi* the atria are longer than the body diameter. *Trichodrilus hrabei* shows a very peculiar laterally and dorsoventrally flattened atrium.

Atrial musculature: Including the data from the literature, there is a group of species that have very thick atrial musculature (> 20 μm): *T. leruthi* (23 μm), *T. claparedei* (23–40 μm), *T. macroporophorus* (20 μm), *T. aporophorus* (27–30 μm), *T. ptujensis* (26 μm), *T. stammeri* (20 μm), *T. itchaensis* (40–50 μm) and *T. diversisetosus* (18–22 μm). *Trichodrilus seirei* could also be included with these species as its atrial musculature reaches up to 25 μm. A range of variation of more than two fold may exist for this muscle layer as is found for *T. campoyi* and *T. allobrogum* (Table 1). We have already discussed the significance of variation of this character and that it must be used with caution.

Penis: Cook (1967b: 356) describes two basic types of penis. The first one, the more common form among *Trichodrilus* species, is formed by a fold of the body wall that may or may not be contained in a penial sac (Fig. 17A, B & C), and the second one is formed by the ectodermal lining cells in the ectal part of the atrium which in extended state, extrude through the male pore (Fig. 17E & G). In the first case, there can be a porophore where the males pores open.

Among the species having the first penial form, *T. cantabrigiensis* is unique due to the presence of an atrial duct which is constricted at its base by a muscular ring separating it from the ampulla (Fig. 1). This constriction gives the duct a spindle or bulbous shape. The second type of penis also has been described in the genera *Lumbriculus* (Hesse, 1902) and *Dorydrilus* (Cook, 1967b). The penes of *T. leruthi* and *Trichodrilus* sp. 5 show this second penial structure and, depending on the

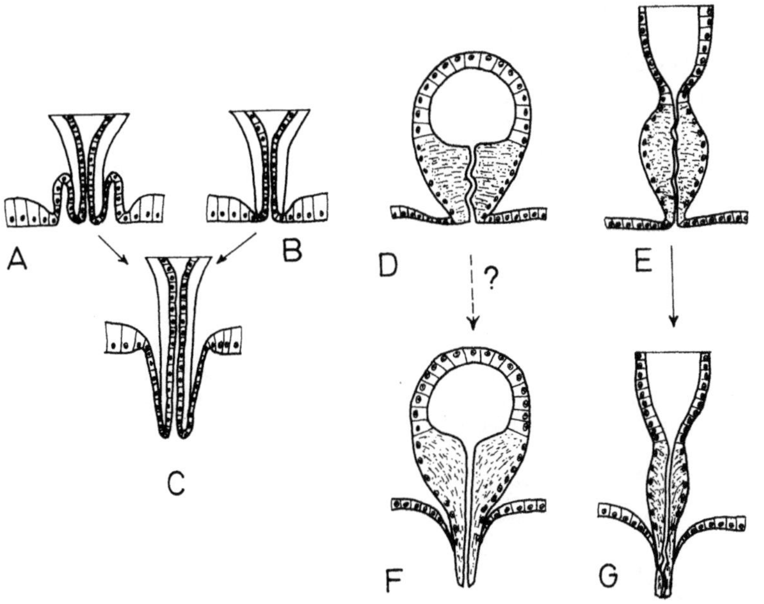

Fig. 17. Types of penes in *Trichodrilus* Type 1: A. Atrial duct opening into a penial sac. B. Atrial duct opening into a simple pore. C. everted position of penes A. and B. Type 2: D. Atrium with distal half filled with cells E. Atrium with protrusible ectodermal lining cells. F. and G. everted penis (or pseudopenis) of penes D. and E., respectively.

state of extension, the penis can reach to a considerable length (Figs 5 & 12). The copulatory apparatus in *T. sketi* could be regarded as a type 2 penis. This species was described as lacking a penis but the distal half of the atrial ampulla is filled by high epithelial cells that could work as a penis (Fig. 17D & F). *Trichodrilus pragensis* and *T. sketi* have been considered by Hrabě (1971) to be very close, probably synonymous species, however, the structure of the penis is of type 1 in the former and of type 2 in the latter. *Trichodrilus leruthi* and *T. allegheniensis* have well-developed penial bulbs in the base of the atria (Fig. 10).

Some *Trichodrilus* species lack penes and the atria open into simple pores. *Trichodrilus campoyi, T. culveri, T. metohiensis* and *T. sketi* have been described as lacking penes.

Prostatic glands: Most species diagnoses lack data on the characteristics of the prostate. However, there are three different types of covering: continuous or diffuse, discrete as a few bundles of cells, and with two kind of cells (pedunculate bundles and small continuous) (Fig. 18). The presence of bundles of prostatic cells is mentioned for *T. sketi, T. hrabei, T. montenegrinus, T. macedonicus, T. metohiensis, T. drimi* and *T. tatrensis.* However, from drawings of species, we can determine that this prostatic gland type also occurs in *T. moravicus* (Hrabě, 1938: fig. 6). The presence of two kinds of prostatic cells is known in *Trichodrilus* sp. 5.

Vas Deferens: The posterior pair of vasa deferentia can penetrate into the postatrial segment, and this has been interpreted by Hrabě (1983, 1984) as an important vestige in Lumbriculidae of the plesiopore ancestral condition. This point of view is shared by Brinkhurst (1989). In *Trichodrilus*, posterior vasa deferentia penetrate into the postatrial segment in most species. *Trichodrilus ptujensis, T. tacensis, T. tatrensis, T. cantabrigiensis* and *T. diversisetosus* are exceptions. In *T. leruthi, T. pragensis* and *T. strandi* just a short section of the posterior vas deferens penetrates into XI. This character is not mentioned for *T. montenegrinus, T. stammeri, T. spelaeus* and *T. cernosvitovi.*

Spermathecal number: The genus *Trichodrilus* includes species with one or two (or four in *Trichodrilus* sp. 2) pairs of spermathecae. This character was considered by Cook (1971a: 225) of low taxonomic importance because *T. pragensis* was originally described by Vejdovsky (1875 and 1876) as having one or two pairs, depending on the state of maturity of individuals. However, Hrabě (1971) redescribed this species as having just one pair of spermathecae, therefore, this character has a real value for the distinction of groups of species in *Trichodrilus.*

Spermathecal form: Many descriptions lack information on the form of the spermathecal duct and ampulla, even though it can be an important diagnostic character for species. We have concluded from the study of a number of individuals of *T. campoyi* that the presence of vacuolate cells

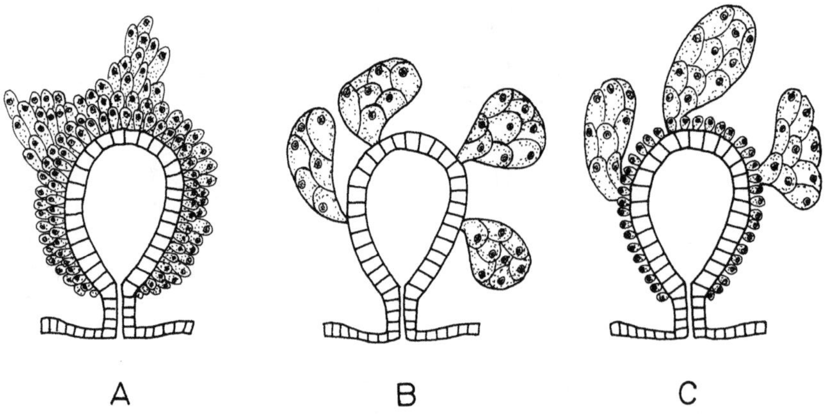

Fig. 18. Types of prostate in *Trichodrilus* A. Diffuse. B. Discrete groups. C. Small cells with discrete groups.

in the wall of the ampulla seems to be related to the state of maturity of the animal. Spermathecal ducts range from short (as in *T. cernosvitovi, T. culveri, T. ptujensis* and *T. tenuis*) to moderate (as in *T. strandi, T. sketi, T. campoyi, T. diversitosus, T. moravicus* and *T. seirei*) or very long (about the body diameter or longer) (as in *T. aporophorus* and *T. allegheniensis*). Some species have a cavity in the spermathecal duct resulting in a bulbous appearance (*T. moravicus, T. seirei, T. diversisetosus* and *T. campoyi*). With respect to the ampulla, some species have large ampullae (as in *T. pragensis, T. culveri, T. tenuis, T. stammeri* and *T. intermedius*) that can penetrate into the following segment, whereas in other species these are rather small (as in *T. allobrogum, T. moravicus* and *T. seirei).*

Spermathecal position: There is little information available on this point, so that we will just indicate the lateral position of spermathecal pores in *T. ptujensis* (near the lateral line), and far posterior, close to the female funnel, in *T. diversisetosus.*

Position of the genitalia: Two groups can be distinguished: *T. strandi, T. diversisetosus, T. campoyi, T. metohiensis* and *T. isabellae* have genitalia shifted one segment anterior so that the male pores are on segment IX. Some individuals of *T. isabellae* may even have the genitalia shifted two segments anteriad. The remaining species have the male pores on X, which is the characteristic lumbriculine condition.

Trichodrilus spelaeus is described as having one male pore on IX and the other on X. Several anomalies in the position of the genitalia have been described within species in the Lumbriculidae (Mrazek, 1906; Holmquist, 1976; Timm, 1979; Rodriguez, 1988). Therefore, we support the status of *species inquirenda* for *T. spelaeus* Moszynsky, 1936, regarding it as a probable abnormal specimen, of some nominal species (Cook, 1971a).

Trichodrilus macedonicus has one pair of spermathecae in segment XII, the postovarian segment. This character state is not found in other *Trichodrilus*, therefore the taxonomic position of this species should be reviewed.

Taxonomy

In the Lumbriculidae, specimens with two pairs of testes, one pair of ovaries, one pair of vasa deferentia per atrium and one or more pairs (two or four) spermathecae from the postatrial segment forwards, are classified in the genus *Trichodrilus.* Some species of the *Lumbriculus* complex also present this configuration of genitalia. Those, however, usually are distinguished by the structure of the atrium and penis and lateral or dorsal position of the spermathecal pores.

Key to the species

A new key of the species of *Trichodrilus* is provided which emphasises anatomical features, structure of genitalia and setal characteristics. Size and form of atria are only considered when they are sharply differentiated. *Trichodrilus macedonicus* is not keyed out because spermathecae are absent in the postatrial segment, a non-trichodriline character.

1. One pair spermathecae 2
   More than one pair spermathecae 11
2. Setae bifid 3
   Setae single-pointed 5
3. Ejaculatory duct and penis absent ....
   .... *T. campoyi*
   Ejaculatory duct present 4
4. Posterior vasa deferentia penetrate into X
   .... *T. strandi*
   .... *T. isabellae*
   Posterior vasa deferentia do not penetrate into X .... *T. diversisetosus*
5. Atria tubular or elongate (length:width >2) 6
   Atria spherical or globular (length:width <2) 7
6. Length of atria 2 to 3 times the body diameter, penetrating into segment XI ....
   .... *T. cernosvitovi*
   Length of atria no more or little more than body diameter, not penetrating into segment XI .... GROUP I
7. Genitalia shifted one segment anteriad ....

.... *T. metohiensis*
Genitalia in the typical lumbriculine position 8
8. Atria not petiolate, distal half of the atria filled with cells that form an atrial duct .... *T. sketi*
.... *Trichodrilus* sp. 1
Atria petiolate 9
9. Atrial musculature very thick (62 μm). Spermathecal duct not distinctly separated from ampulla .... *T. drimi*
Atrial musculature very thin (5–10 μm maximum). Spermathecal duct distinct 10
10. Spermathecal ampulla divided into two regions, thin-squamous proximal epithelium and thick, highly vacuolate distal epithelium. Spermathecal duct conical, very short .... *T. culveri*
Spermathecal ampulla not divided into regions, spermathecal duct tubular .... *T. pragensis*
11. Two pair spermathecae 12
Four pair spermathecae .... *Trichodrilus* sp. 2
12. Atria tubular (l:w = 2.4–3.2) 13
Atria pear-shaped or spherical 16
13. Atria longer than body diameter. Penial bulb developed 14
Atria shorter than body diameter. No penial bulb 15
14. Prostomium forming a proboscis. Male pore on a porophore .... *T. allegheniensis*
Prostomium without proboscis. Penis protrusible very long (up to 300 μm), formed by the internal atrial lining cells .... *T. leruthi*
15. Posterior vasa deferentia penetrate into XI .... *T. intermedius*
Posterior vasa deferentia do not penetrate into XI .... *T. tacensis*
16. Glandular pads on the dorsal tegument, sometimes appearing as bulky masses, on the posterior segments .... *Trichodrilus* sp. 3
Tegument without this structure 17
17 Postclitellar dorsal setae modified into hair-like setae, very fine with proximal nodulus .... *Trichodrilus* sp. 4
All setae more or less alike 18
18. Atria very compressed laterally .... *T. hrabei*
Atria not laterally compressed 19
19. Posterior vasa deferentia do not penetrate into XI. Ejaculatory duct narrowed at its base by an annular muscle resulting in a bulbous duct .... *T. cantabrigiensis*
Posterior vas deferentia penetrate into XI 20
20. Penis protrusible, very long (up to 140 μm), formed by the internal atrial lining cells. Modified penial setae in fully matured specimens .... *Trichodrilus* sp. 5
Penis not of this form. Unmodified penial setae .... GROUP II

Group I

Group I is formed by *T. ptujensis, T. tatrensis, T. stammeri, T. itchaensis* and *T. gordeevi*. Among these species only *T. tatrensis* has atria that may be just longer than the body diameter (Hrabě, 1960). The species of this group can be separated from each other only on the basis of characters that have large ranges of variation, such as the thickness of atrial muscles, the position of the junction of vasa deferentia with ampulla or the size and form of atrium. The absence of posterior lateral vessels has not been definitively confirmed in *T. tatrensis* (observed up to segment XLVII). *T. stammeri* (observed to segment LVI) and in *T. itchaensis* and *T. gordeevi* (character not mentioned in the description). The species showing the extreme values for the atrial musculature are *T. tatrensis* (5 μm) and *T. itchaensis* (up to 50 μm). Two subgroups can be separated by the penetration of posterior vasa deferentia (*T. gordeevi* and *T. itchaensis*) or not (*T. ptujensis* and *T. tatrensis*) into the postatrial segment. There is no mention of this character in the description of *T. stammeri* (Cook, 1971a).

Group II

Group II is formed by *T. moravicus, T. claparedei, T. seirei, T. allobrogum, T. icenorum. T. macroporophorus, T. aporophorus, T. medius, T. montenegrinus, T. tenuis* and *T. pauper*. These species can just barely be separated on the basis of characters that either are variable within the populations, such as thickness of atrial muscles and the size and form of atrium, or have not been definitively diagnosed by their authors. Some of them can be separated into two different groups based on the absence (*T. moravicus, T. claparedei, T. medius* and *T. seirei*) or presence (*T. allobrogum, T. icenorum, T. macroporophorus, T. aporophorus*) of posterior lateral blood vessels. *Trichodrilus medius* is distinguished from the other species of Group II that lack posterior lateral blood vessels primarily by the presence of a transvere furrow on the prostomium, which probably is not a sound character. This species is illustrated with a singular atrial structure, resembling that of *T. sketi*, although it was incompletely described by Hrabě (1960). *Trichodrilus macroporophorus* has a porophore protruded backwards, very close to septum X/XI. *Trichodrilus aporophorus* has ramified, instead of the usual simple, posterior lateral vessels. The classification of *T. montenegrinus* is uncertain as the description does not mention whether or not vasa deferentia penetrate into XI and whether there are posterior lateral vesssels. Information on the presence or absence of posterior lateral vessels is also lacking for *T. tenuis* and *T. pauper* since descriptions were made on incomplete specimens.

Conclusions

The various characters reviewed in this work have proved to be of varied interest when used for taxonomic purposes. Unless it is further demonstrated by other means (biochemically, ethologically, physiologically, etc.), differences between individuals or populations based on characters relative to body size, atrial size, and so on must not be considered sufficient to establish a new species. Although in some cases they may reveal real taxonomic differences, characters related to size of different organs must be considered of low value as they have great ranges of variation between individuals. More emphasis must be made on structural characters, such as presence or absence of an atrial duct and a penis, penis type, posterior vas deferens penetrating the postatrial segment, number and form of the spermathecae, relative sizes of spermathecal duct and ampulla and presence of a vestibulum in the spermathecal duct.

Setae exhibit few forms and distributions in lumbriculids, but in the genus *Trichodrilus* the three types of setae can be very useful, and also a sound character, for systematics. There are four species with bifid setae. Among these, *T. isabellae* does not differ from *T.strandi* except for the size of the atria and the general structure of genitalia is identical in both taxa. The existence of this group of species reinforces the need for a revision of the genus *Lumbriculus* (including *Thinodrilus*) because it is not clearly separated from *Trichodrilus*.

Presence or absence of a proboscis is a valid character, but the distinction of conical from round prostomia sometimes leads to confusing results. Characters such as epidermal glands and pharynx would be better described on the basis of histological work in order to clearly differentiate structural details. The extension of pharyngeal glands can be a fixed or very variable intraspecific character. The structure of the vascular system has been widely applied in the classification of the genus but the degree of variability depending on environmental differences is largely unkown and unstudied.

This preliminary revision of taxonomic characters allow us to advance studies on the systematics of the genus and to pose new designations for some taxa. *Trichodrilus icenorum* (Beddard, 1908) is regarded as a junior synonym of *T. allobrogum* Claparède, 1862, since both may have pharyngeal glands in segment VIII. These species were formerly distinguished only by this character (Cook, 1967b: 363). This entity also includes *T. lengersdorfi* synonymized with *T. icenorum* by Cook (1971). *Trichodrilus campoyi* was first described as a subspecies of *T. strandi* (Rodriguez, 1988). However, differences in the structure of atria (lack of atrial duct, penis and porophore) and spermathecae (presence of a bulb at the base of the duct) have lead us to propose the former as a species. The presence of fissures in the sides of the clitellum in the type material of *T. strandi* is probably an artefact due to fixation, as was stated by Hrabě himself (Hrabě, 1942). *Trichodrilus spelaeus* Moszynski, 1936 is treated here as *species inquirenda* as suggested by Cook (1971). Its particular configuration of genitalia may be due to teratological events, that occurred in the only specimen studied by Moszynski. *Trichodrilus macedonicus* Karaman, 1986 is not considered here as a member of the genus *Trichodrilus* due to the absence of spermathecae in the postatrial segment, a character that is shared by other lumbriculid genera.

Group I species probably include some synonyms, but limited information on the posterior lateral blood vessels does not allow us to go further. Group II is even more poorly known and probably additional synonymies will arise when type collections or new material are studied. On the basis of their descriptions. *T. aporophorus, T. montenegrinus, T. pauper* and *T. seirei* do not show, in our opinion, distinctive structural characters that differentiate them as independent taxa.

Acknowledgements

We are very much endebted to Dr D. G. Cook (Ontario, Canada) for the loan of photographic material; to Dr R. Rouch (Toulouse, France) for the valuable collection from interstitial waters of Spain; to Mr J. Juget (Lyon, France) and Dr T. Timm (Tartu, Estonia) for sending collections of some species; to C. F. Bright (United States National Museum) and A. I. Muir (British Museum of Natural History) for facilitating access to type material. To Dr Kathyrn Coates and an unknown reviewer who made many useful suggestions on the manuscript. P. Rodriguez has received financial support through a project of the Basque Country University 154.310–E156/90.

References

Beddard, F. E., 1908. A note on the occurrence of a species of *Phreatothrix* (Vejd.) in England, and on some points in its structure. Proc. Zool. Soc. Lond. 1908: 365–369.

Beddard, F. E., 1920. On the genus *Trichodrilus* and on a British species of the genus. Ann. Mag. Nat. Hist. 6: 227–239.

Brinkhurst, R. O., 1989. A phylogenetic analysis of the Lumbriculidae. Can. J. Zool. 67: 2731–2739.

Brinkhurst, R. O., 1991. The anatomy of the pharynx of two predatory aquatic oligochaetes. Can. J. Zool. 69: 669–675.

Brinkhurst, R. O. & S. R. Gelder, 1989. Did the lumbriculids provide the ancestors of the branchiobdellidans, acanthobdellidans and leeches?. Hydrobiologia 100: 7–17.

Brinkhurst, R. O. & B. G. M Jamieson, 1971. Aquatic Oligochaeta of the World. Oliver & Boyd, Edinburgh, 860 pp.

Claparède, E., 1862. Recherches sur les oligochètes. Mem. Soc. Phys. Hist. Nat. Genève 16: 217–291.

Coates, K. A., 1983. New records of marine *Marionina* (Oligochaeta, Enchytraeidae) from the Pacific Northeast, with a description of *Marionina klaskisharum* sp. nov. Can. J. Zool. 61: 822–831.

Cook, D. G., 1967a. Studies on the Lumbriculidae (Oligochaeta). Ph.D. Thesis, University of Liverpool, U.K., 269 pp.

Cook, D. G., 1967b. Studies on the Lumbriculidae (Oligochaeta) in Britain. J. Zool. Lond. 153: 353–368.

Cook, D. G., 1971a. Lumbriculidae. In R. O. Brinkhurst & B. G. M. Jamieson (eds), Aquatic Oligochaeta of the World. Oliver & Boyd, Edinburgh: 200–285.

Cook, D. G., 1971b. *Trichodrilus allegheniensis* n.sp. (Oligochaeta, Lumbriculidae) from a cave in southern Tennessee. Trans. Amer. Micros. Soc. 90: 381–383.

Cook, D. G., 1975. Cave-dwelling aquatic Oligochaeta (Annelida) from the Eastern United States. Trans. Amer. Microsc. Soc. 94: 24–37.

Fauvel, P., 1903. Un nouvel oligochète de puits (*Trichodriloides intermedius* n.g. n.sp.). C. R. Ass. Fr. Avanc. Sci. 32: 221.

Finogenova, N., 1973. New species of Oligochaeta from the Caspian Sea. Zool. Zh. 52: 121–124.

Hesse, R., 1902. Zur Kenntnis der Geschlechtsorgane von *Lumbriculus variegatus*. Zool. Anz. 25: 620–622.

Holmquist, C., 1976. Lumbriculids (Oligochaeta) of Northern Alaska and Northwestern Canada. Zool. Jb. Syst. 103: 377–431.

Hrabě, S., 1936. *Trichodrilus strandi* n. sp., ein neuer Vertreter der Höhlen Lumbriculiden. Festschr. f. Prof. Dr. Embrik Strand, Riga 1: 404–407.

Hrabě, S., 1937. Etudes biospéologiques. V. Contribution à l'étude du genre *Trichodrilus* (Oligochaeta:Lumbriculidae) et description de deux espèces nouvelles. Bull. Mus. Hist. Nat. Belge 13: 1–23.

Hrabě, S., 1938. *Trichodrilus moravicus* und *claparedei*, zwei neue Lumbriculiden. Zool. Anz. 121: 73–85.

Hrabě, S., 1942. Zur Kenntnis der Brunnen und Quellen fauna aus den Slowakei. Sbor. Kl. prir. Brno 24: 23–30.

Hrabě, S., 1954. *Trichodrilus macroporophorus* n.sp. eine neue Lumbriculiden-Art aus Osterreich. Zool. Anz. 153: 185–188.

Hrabě, S., 1960. Oligochaeta limicola from the collection of

Dr. Husmann. Publ. Fac. Sci. Univ. Brno 415: 245–277.
Hrabě, S., 1963. Oligochaeta limicola from Slovenija. Biol. Vestnik 11: 67–75.
Hrabě, S., 1971. On *Trichodrilus pragensis* Vejd. (Oligochaeta, Lumbriculidae). Vest. Cs. Spol. Zool. 35: 205–208.
Hrabě, S., 1983. Evolution of the family Lumbriculidae. Hydrobiologia 102: 171–173.
Hrabě, S., 1984. Two atavistic characters of some Lumbriculidae and their importance for the classification of Oligochaeta. Hydrobiologia 115: 15–17.
Karaman, S., 1973. Beitrag zur Kenntnis der Oligochaetenfauna des Skadarsees. Zool. Anz. 190: 351–358.
Karaman, S., 1986. Three new species of lumbriculid worms of genera *Trichodrilus* Claparède (Oligochaeta-Lumbriculidae) from Yugoslavia. Montenegrin Acad. Sci. Arts. (Nat. Sci.), 5: 217–226.
Kossmagk-Stephan, K., 1983. Marine Oligochaeta from a sandy beach of the island of Sylt (North Sea) with description of four new enchytraeid species. Mikrofauna des Meeresboden 89: 1–28.
Michaelsen, W., 1933. Über Höhlen Oligochaeten. Mitt. Höhl., -u. Karstforsch. 1933: 1–19.
Moszynski, A., 1936. Ein neuer Vertreter der Gattung *Trichodrilus* Claparède (*Trichodrilus spelaeus* n. sp.) aus dem stollen in Neu-Klessengrund. Beitr. z. Biol. d. Glatzer Schneeberges 2: 214–216.
Mrazek, A., 1906. Die Geschlechtsverhältnisse und die Geschleschtsorgane von *Lumbriculus variegatus* Gr. Zool. Jahrb. (Anat.) 23: 381–462.
Popchenko, V. I., 1976. A new species of the genus *Trichodrilus* (Oligochaeta, Lumbriculidae) from water bodies of Karelia. Zool. Zh. 55: 139–141.
Popchenko, V. I., 1988. Aquatic Oligochaete (Oligochaeta Limicola) of North Europe. Academy of Sciences of USSR. Leningrad.*
Rodriguez, P., 1988. Sur certaines espèces de Lumbriculidae du Nord de la Péninsule Ibérique. Annls. Limnol. 24: 203–211.
Rodriguez, P. & N. Giani, 1986. Description de trois nouvelles espèces d'oligochètes aquatiques du Pays Basque (Espagne). Hydrobiologia, 139: 269–276.
Rodriguez, P. & N. Giani, 1987. Sur deux espèces de Lumbriculidae (Oligochaeta) cavernicoles: *Cookidrilus speluncaeus* n. gen. n. sp. et *Trichodrilus diversisetosus* Rodriguez, 1986. Bull. Soc. Hist. Nat. Toulouse 123: 45–49.
Sokolskaya, N. L., 1973. A new species of *Trichodrilus* (Oligochaeta, Lumbriculidae) from Kamchatka. Zool. Zh. 52: 1714–1716.
Swedmark, B., 1964. The interstitial fauna of marine sand. Biol. Rev. 39: 1–42.
Timm, T., 1979. The Estonian Lumbriculidae. Eesti NSV Teadoste Akad. Toimetised (Biol.) 28: 158–171.
Timm, T. & V. I. Popchenko, 1978. The aquatic Oligochaeta of the Murmansk Region. Hydrobiol. Invest. Inst. Zool. Bot. Acad. Sci. (Estonian SSR) 7: 71–132.
Vejdovsky, F., 1875. Beiträge zur Oligochaetenfauna Böhmens. Sber. K. Böhm. Ges. Wiss. 1875: 191–201.
Vejdovsky, F., 1876. Ueber *Phreatothrix*, eine neue Gattung der Limicolen. Z. wiss. Zool. 27: 541–554.

Hydrobiologia **278**: 53–66, 1994.
T.B. Reynoldson & K.A. Coates (eds), Aquatic Oligochaete Biology V.

The Enchytraeid fauna of North Africa

Emilia Rota[1,2] & Brenda Healy[1]
[1] *Department of Zoology, University College, Belfield, Stillorgan Road, Dublin 4, Ireland*
[2] *Dipartimento di Biologia Evolutiva, Via Mattioli 4, 53100 Siena, Italy*

Key words: Enchytraeidae, taxonomy, distribution, North Africa

Abstract

As part of a survey of the Mediterranean enchytraeids, large numbers of samples were collected from a wide range of habitats at 38 localities of Tunisia and Algeria in 1989. Twenty species were identified belonging to six genera. Five species are new records for Tunisia and 17 are new for Algeria. Most of the recorded species are widely distributed throughout Europe, or even cosmopolitan, but some have a more restricted range. *Fridericia caprensis* has a circummediterranean distribution, *Fridericia sardorum* seems to be endemic to the western Mediterranean and *Fridericia berninii* is known only from the Balearic Islands and Algeria. *Enchytraeus minutus bisetosus* n. ssp. is described. A critical revision of some species is also provided.

Introduction

In spite of the fact that enchytraeids are readily collected in a wide range of habitats, faunistic surveys and zoogeographic studies of these worms present many difficulties. This is due mainly to severe handicaps imposed by the need, in most cases, to make observations on living specimens for quick and reliable identifications. This is particularly true for the common terrestrial genera *Fridericia* and *Enchytraeus*. Real problems arise when a field investigation takes place far from facilities for extraction and microscopic examination because the transport of large quantities of soil is often not feasible.

In autumn 1989, we were invited to participate in a collecting expedition to Algeria and Tunisia, organised by the Italian programme "Biogeographic and Faunistic Survey of Mediterranean Countries". The only enchytraeid taxa known for the area were three species recorded from the neighbourhood of Algiers by Černosvitov (1933) and seven genera (four species identified) recorded during more recent surveys of freshwater oligochaetes in western Algeria (Gagneur *et al.*, 1986) and Tunisia (Boumaiza *et al.*, 1986; Martinez-Ansemil & Giani, 1987). We succeeded in carrying out an intensive survey by extracting worms in the field and maintaining cultures until we could complete careful observations of live material.

Methods and materials

Samples were collected between October 17th and November 5th from 38 localities. Habitats included seashore, margins of freshwater bodies and salt lakes (chotts) as well as grassland, woodland, scrub, agricultural land and pre-Saharian oases. Most of the country traversed was very arid and it was necessary to seek out moist soils, *e.g.* on north facing slopes, under rocks and logs, in depressions and among plant roots.

Soil samples of approximately 0.5 kg were taken from one or more sites at each locality and kept in plastic bags. A portion of each sample was used to provide immediate information about the presence of enchytraeids, using a portable extraction apparatus described by Healy & Rota (1992). This allowed a more efficient choice of sampling locations and samples to retain, thus reducing the amount of soil to be transported. At some sites, enchytraeids were also obtained by dry sieving of leaf litter. About half the worms extracted in the field were kept alive on soil-agar culture plates (Springett, 1964), the remainder were preserved in 70% alcohol. Measurements of pH were made in the field using a 1:1 soil-water mixture (Duchaufour, 1965). Soil not extracted in the field was returned to Italy and stored at 4 °C for subsequent extractions. Extracted worms were kept on culture plates at 15 °C to allow maturation of young specimens. Except for *Buchholzia appendiculata* (Buchholz, 1862), only mature worms were used for identification. Representatives of all recorded species were fixed in 70% alcohol or in Bouin's fluid and kept in the authors' collection.

Most species described by earlier authors were identified according to Nielsen & Christensen's (1959, 1961, 1963) revised descriptions; in some cases, however, reference to the original or subsequent, consistent descriptions was made to integrate information (*e.g.* on the structure of coelomocytes of *Fridericia* species). Redescriptions and illustrations of *Fridericia caprensis* Bell, 1947 and *Fridericia sardorum* Cognetti, 1901 provided recently by Dózsa-Farkas (1988, 1989), were used to identify with certainty these two 'southern' species. Reference to the original description was primarily made for *Fridericia paroniana* Issel, 1904, since discrepancies were found in Nielsen & Christensen's (1959) and subsequent descriptions.

A complete description of the material examined is given herein only in the case of doubtful identifications. In all other instances, the observations are restricted to morphometric data and to additions or deviations from the reference description.

All measurements of body length and width given herein were taken on living specimens placed under a cover-slip. Indications of the setal arrangement are as in Kasprzak (1986): dorsal preclitellar – dorsal postclitellar: ventral preclitellar – ventral postclitellar ('÷' = from .. to ..). Types of peptonephridia for *Fridericia* species are according to Nielsen & Christensen (1959). Types of coelomocytes for *Fridericia* species are according to Möller (1971).

Results

Enchytraeids were present at 26 (68%) of the localities sampled (Fig. 1; Table 1). They were absent from very dry soils and from moist or wet mud at the edge of chotts. Twenty species were identified from six genera. Five species are new records for Tunisia and 17 are new for Algeria. *Fridericia* and *Enchytraeus* dominated at most sites. Only *Enchytraeus buchholzi* Vejdovský, 1879 lived, at very low density, in the poor, dry soil at the edge of cultivated fields.

Description of the fauna

Genus ***Achaeta*** *Vejdovský, 1877*

Achaeta bohemica (Vejdovský, 1879) *sensu* Nielsen & Christensen, 1959

Material examined: 2 specimens.

Site: Algeria, 7a.

Observations from material: Length 8 mm. Width at clitellum 250 μm. Segments 36, 38. Dorsal setal follicles up to 87 μm long. Clitellar gland cells arranged in 30 regular transverse rows; no dorso-lateral hyaline fields were observed. Seminal vesicles absent. Sperm funnel 187–218 by 68.5 μm; collar as wide as the funnel. Spermathecae confined to V in both the specimens examined. One egg mature.

Remarks: As conceived by Nielsen & Christensen (1959), the species is distinguished in having only dorsal setal follicles, dorsal vessel origi-

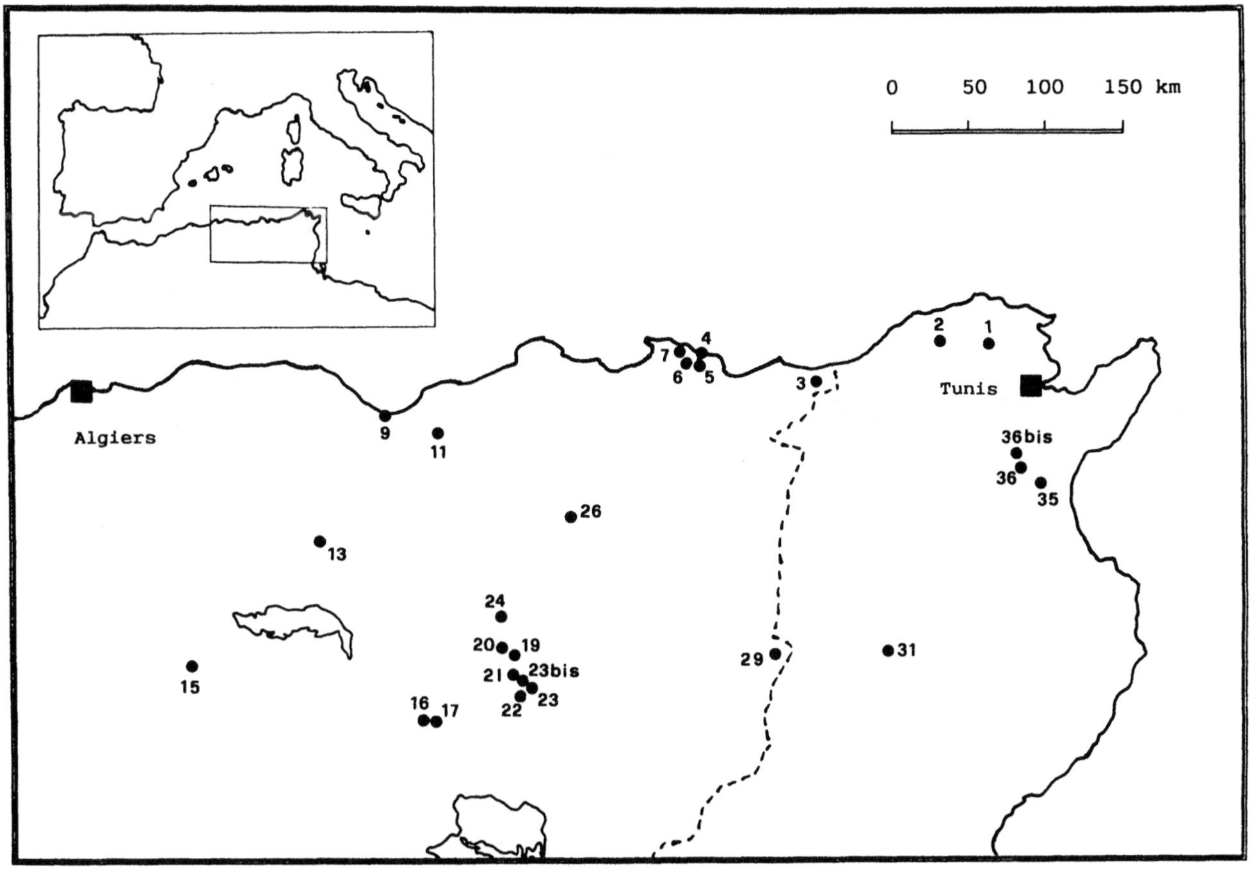

Fig. 1. Map of sampling localities

nating in VII, no seminal vesicles, ectal spermathecal duct not perpendicular to the body axis. Algerian specimens conform to Nielsen & Christensen's (1959) description except for a shorter length of the spermathecal mid-duct.

Distribution: European [Great Britain, Ireland, Denmark, The Netherlands, Poland, Czechoslovakia, Germany, Spain, France, Switzerland, Italy; records from Belgium (Černosvitov, 1939) and Sweden (Backlund, 1947) have been omitted since they refer to a different species described under the same name from the Russian Carpathians by Černosvitov (1928; see also Kasprzak, 1972)]. New for Algeria.

Genus ***Henlea*** *Michaelsen, 1889*

Henlea perpusilla Friend, 1911 augm. Černosvitov, 1937

Material examined: 10 specimens.

Sites: Algeria, 6a, 11b, 13, 15a.

Observations from material: Colour whitish to pale yellow. Length 4–6 mm. Width 280–295 μm, at clitellum 310 μm. Segments 30–33. Setae $2 \div 7 – 2 \div 5 : 2 \div 7 – 2 \div 6$. Coelomocytes light, finely granulated, about 30 μm long; a few vacuoles may be present around the nucleus. Two pairs (dorsal and ventral) of sponge-like oesophageal appendages in VI. Clitellum elevated; granular gland cells small, irregularly scattered with large clear interspaces. Seminal vesicles absent. Sperm funnel small, 63–125 by 38–50 μm. Spermathecae typical of cytotype 2x, A1 (acc. to Nielsen & Christensen, 1959); no glands at the external orifice. One or two mature eggs.

Remarks: The species is known to be morphologically and cytologically heterogeneous (Nielsen & Christensen, 1959). The present specimens were quite homogeneous as regards body size,

Table 1. Sampling sites.

No.	Habitat (altitude)	Soil	pH
TUNISIA			
1	Edge of a wet ditch.	Sticky clay. Wet.	–
2	Steppe on limestone (200 m)	(a) Sticky clay beside a water trickle. Wet.	7.8
31	Edge of an oued (550 m)	Yellow-brown sandy mud. Damp.	7.8
35	Dry oued with oleanders and few trees (210 m)	(a) Clay with debris of reeds. Moist-wet.	7.7
		(b) Clay with palm debris. Moist.	7.6
36	Pine wood on hillside (450 m)	(a) Damp sandy soil with wood litter in torrent bed.	7.3
		(b) Pine litter on sandy loam. Just moist.	7.4
		(c) Sandy loam with roots, moss, liverworts. Slightly moist.	7.3
		(d) Fine, loose red-brown soil. Rather dry.	7.2
36bis	Pine plantation and mixed wood on hillside (500 m)	(a) Dark coarse humus and litter. Moist.	7.8
		(b) Loose, fine humus with woody debris. Dry.	7.5
ALGERIA			
3	*Quercus suber* wood with eucalyptus and acacia	(a) Dark sandy loam under grass and rotten logs. Moist.	5.9
		(b) Dark sandy humus with litter and fibrous roots. Moist.	5.7
4	Lake near beach (10 m)	Dark sandy soil with roots. Wet	7.2
5	*Q. suber* forest (350 m)	Dark, fine sandy humus with litter and liverworts. Moist.	7.1
6	*Q. afares* forest on steep hillside (500 m)	(a) Coarse leaf litter, deep in channels. Moist.	–
		(b) Dark, fine humus with moss. Rather dry.	6.4
		(c) Dark, deep, loose sandy humus at roadside.	6.5/7.1
7	Mixed oak forest with a stream (800 m)	(a) Dark sandy humus with liverworts and roots. Wet.	5.8
		(b) Dark fine humus and litter. Rather dry.	6.2
8	Pine plantation with eucalyptus near the beach	(a) Dark sandy loam with grass roots. Very moist.	7.6
		(b) Dark, coarse humus with sticks and loam. Moist.	7.4
11	Mixed *Cedrus atlantica* wood on steep slope crossed by a stream (1250 m)	(a) Cedar and juniper litter on dark humus. Rather dry.	7.2
		(b) Yellow-brown gravelly clay with roots. Wet.	7.7
		(c) Humus and *Q. ilex* litter. Fairly dry.	7.4
12	Floodbank of a stream in a gorge with oleanders (1150 m)	Reddish fine soil with woody debris. Dry.	7.6
15	Oasis in steppe with cultivated patches fed by a river (800 m)	(a) Red-brown sandy loam with grass under olive trees.	7.5
		(b) Mixed leaf litter in wet patches.	7.4
16	Date palm grove beside a nearly dry oued (120 m)	Brown clayey loam with creeping plants. Wet.	7.7
17	Palm grove with olive trees and water pools (110 m)	(a) Wet clay with grass. Sticky loam.	7.9
		(b) Yellowish fine crumb with olive litter and dark loam.	8.3
19	Grass at the edge of a stream beside an orchard (110 m)	Brown gravelly clay. Wet.	7.8
20	River bed in steppe; isolated juniper and pine trees (1200 m)	*Juncus* roots and debris in sand and silt. Moist-wet.	7.7
21	Steppe with a few juniper and oaks (1525 m)	Dark soil under rocks.	7.7
22	Steppe with few juniper (1000 m)	Stony, silty sand with woody litter. Rather dry.	7.9
23	Nearly barren steppe with few isolated juniper (1450 m)	(a) Dark soil in shade of rocks and trees. Moist-wet.	7.9
		(b) Sticky soil with limestone fragments in river bed.	7.6
23bis	Steppe with low scrub on mountain pass (1800 m)	Moist soil under stones with grass.	7.9
24	Shaded hillside with a few juniper and *Q. coccifera* (1650 m)	Dark reddish-brown soil with litter. Rather dry.	7.9
26	Spring and stream issuing from under a limestone knoll (1000 m)	(a) Grass at downstream edge. Fairly dry.	7.9
		(b) Water edge near the source. Wet.	7.9
29	Stream below pine forest on steep limestone slope (1175 m)	Pine litter with fungal growth. Dry.	7.5

segment number and shape of the spermathecae; the major variation concerned the size of the sperm funnel.

Distribution: Holarctic [Great Britain, Ireland, Iceland, Spitzbergen, Norway, Sweden, Finland, Denmark, The Netherlands, Germany, Spain, France, Austria, Italy, Poland, Czechoslovakia, Hungary, Russia, Lebanon, Japan, Greenland, Canada]. New for Algeria.

Henlea ventriculosa (Udekem, 1854)

Material examined: 2 specimens.

Site: Algeria, 3a.

Observations from material: Length 10–13 mm. 43 segments in both specimens. Setae straight, narrow, 4÷6–3÷6 : 6÷8–4÷6. Three rows of equal sized cutaneous gland cells in each anteclitellar segment. Coelomocytes larger than the chloragogen cells, opaque, filled with brown granules. Oesophageal appendages present in VI. Seminal vesicles absent. Sperm funnel small, collar broad covered with dark spermatozoa.

Remarks: The material conforms with Nielsen and Christensen's (1959) revised description. The species is easily recognized by having 4 intestinal pouches closely applied to the gut in VIII, dorsal vessel arising in IX, spermathecae not merging entally.

Distribution: Cosmopolitan [Great Britain, Ireland, Iceland, Spitzbergen, Norway, Sweden, Finland, Denmark, The Netherlands, Belgium, Germany, Spain, France, Austria, Switzerland, Italy, Poland, Czechoslovakia, Hungary, Romania, Russia, Tibet, Japan, New Zealand, Greenland, Canada, USA, Patagonia]. New for Algeria.

Henlea nasuta Eisen, 1878

Material examined: 12 specimens.

Sites: Algeria, 11b, 15a.

Observations from material: Colour white-yellow. Length 15–22 mm. Width at clitellum 810 μm. Segments 54–60. Setae straight, 2÷7–4÷7: 2÷8–3÷8. Brain 145–160 long, 100 μm wide. Dorsal vessel with 3 pulsating hearts in VI–VIII. Coelomocytes finely granulated, 38–55 μm long. Two unpaired (dorsal and ventral) ear-like oesophageal appendages in IV. Oesophageal appendages in VI present. Clitellum $\frac{1}{2}$XI–$\frac{1}{2}$XIII; gland cells small, granular, irregularly scattered with clear interspaces. Seminal vesicles absent. Sperm funnel partly covered with opaque glandular tissue, 2–3 times longer than wide (in one spec. from loc. 15a: 368 by 118 μm). The spermathecae may have small glands at the external orifice.

Remarks: Up to the present *H. nasuta* is the only species of the genus known to possess oesophageal appendages in IV. Nielsen & Christensen (1959) describe these organs as "plurilobed appendages with a rather wide and much coiled canal". In our specimens they appear laterally flattened, with an oval, rather regular outline and a coiled, clearly visible canal; no lobes or branches were observed. The same form has been observed recently from Italian and Irish material (Rota, in prep.).

Distribution: Holarctic [Great Britain, Ireland, Iceland, Spitzbergen, Norway, Sweden, Finland, Denmark, The Netherlands, Belgium, Germany, Portugal, Spain, France, Austria, Italy, Yugoslavia, Crete Is., Poland, Czechoslovakia, Hungary, Russia, Greenland, Canada]. Already known from Algeria (Černosvitov, 1933).

Genus ***Buchholzia*** *Michaelsen, 1887*

Buchholzia appendiculata (Buchholz, 1862)

Material examined: 23 specimens (11 mature + 12 immature).

Sites: Tunisia, 2a, 36bis a/b; Algeria, 6a/c, 7a, 15a/b, 19, 21, 23b, 24, 26b, 29.

Observations from material: Thin, yellow worm. Length 9–13 mm. Width 280 μm. Segments 43–56. Setae 2÷4–3,4 : 2÷6–3÷5. One to five rows of pale, quadrangular cutaneous gland cells in each anteclitellar segment. Body wall transparent. Brain truncate posteriorly, 112–130 by 100–110 μm. Supernumerary pair of septal glands at 7/8. Chloragogen cells from VI, brown, as large as the larger coelomocytes. Clitellum VIII–$\frac{1}{2}$IX or IX–$\frac{1}{2}$X. Seminal vesicles in VII–VIII. Sperm funnel in VII or VIII, 125–200 μm long, twice the width. Sperm duct irregularly coiled, 7.7 μm wide. Male pore in VIII or IX. Spermathecae typical with long ental ducts; two large glands at the external orifice.

Remarks: Mature specimens conform to Nielsen & Christensen's (1959) description. Immature specimens were identified by the supplementary pair of septal glands and the shape of the intestinal diverticula.

Distribution: Holarctic [Great Britain, Ireland, Iceland, Norway, Sweden, Finland, Denmark, The Netherlands, Belgium, Germany, France, Austria, Italy, Yugoslavia, Poland, Czechoslovakia, Hungary, Russia, Greenland, Canada]. New for Tunisia and Algeria.

Buchholzia fallax Michaelsen, 1887 (Fig. 2a)

Material examined: 3 specimens.

Site: Algeria, 11b.

Observations from material: Colour yellowish. Length 9–11 mm. Segments 37–40. Setae 3÷5–3,4: 5÷7–5,6. Three rows of irregular cutaneous gland cells in each anteclitellar segment. Brain indented posteriorly. Chloragogen cells less dense from VIII to X. Seminal vesicles absent. Sperm duct wound in even coils (Fig. 2a). Male pore in XII. Spermatheca with two small glands at the external orifice.

Remarks: Distinguishing characters of this species are the unpaired intestinal diverticulum and the arrangement of spermatozoa within the ampulla (Nielsen & Christensen, 1959). The regular winding of the sperm duct was noticed also by Michaelsen (1887:375): "Die Samenkanäle sind lang, regelmässig zusammengelegt, ungefähr so, wie Schiffstaue zusammengelegt werden".

Distribution: European [Great Britain, Ireland, Sweden, Finland, Denmark, The Netherlands, Belgium, Germany, Spain, France, Switzerland, Austria, Italy, Poland, Czechoslovakia]. Known

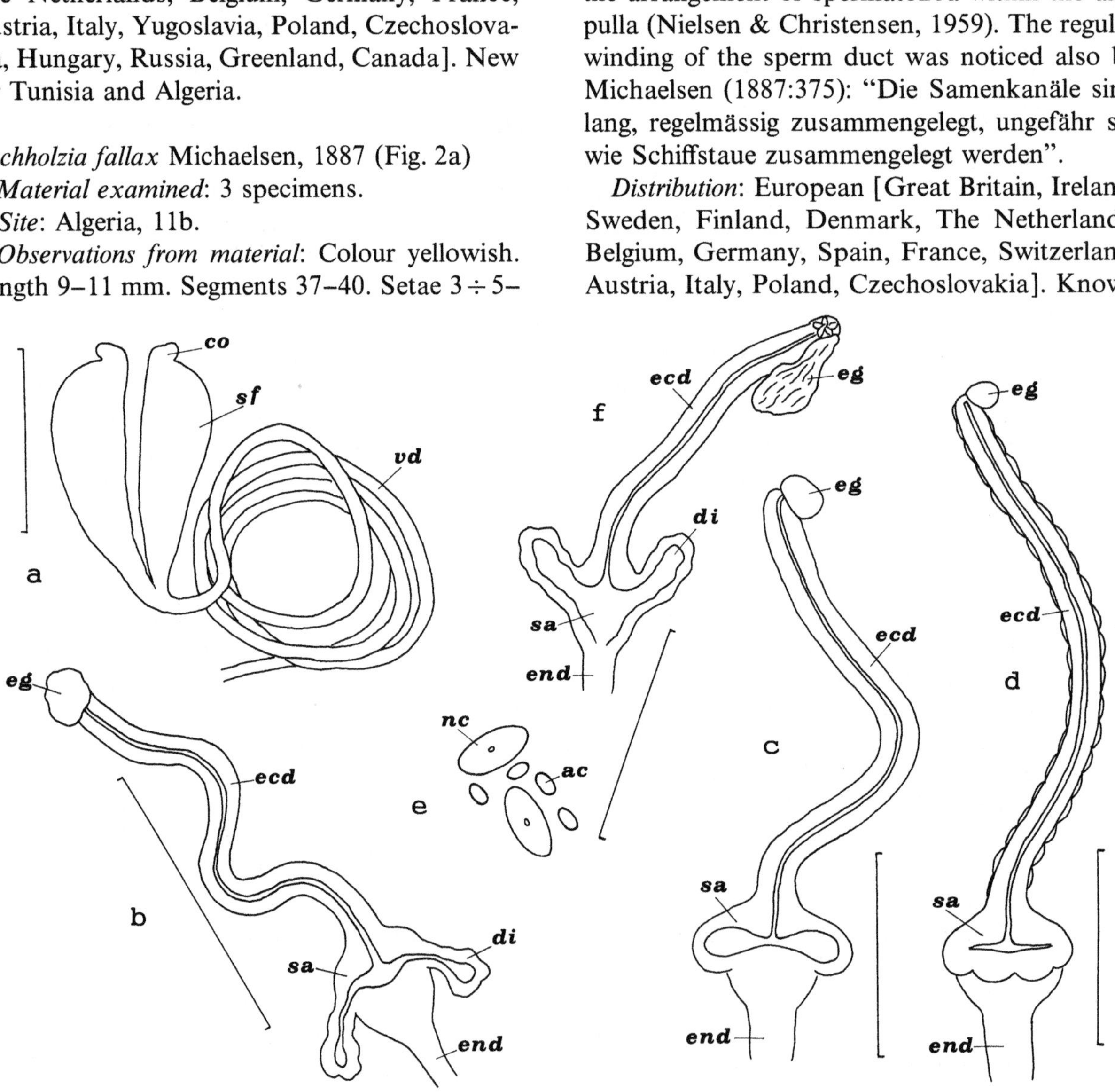

Fig. 2. (a) *Buchholzia fallax*, sperm funnel. (b) *Fridericia berninii*, spermatheca. (c) *F. bulbosa* form *b*, spermatheca. (d) *F. bulbosa* form *c*, spermatheca. (e) *F. paroniana*, coelomocytes. (f) *F. paroniana*, spermatheca. Scales = 100 μm; (e) and (f) drawn to same scale. *ac*, anucleate coelomocyte; *co*, collar of sperm funnel; *di*, diverticulum of ampulla; *ecd*, spermathecal ectal duct; *eg*, ectal gland; *end*, spermathecal ental duct; *nc*, nucleated coelomocyte; *sa*, spermathecal ampulla; *sf*, sperm funnel; *vd*, sperm duct.

from Tunisia (Boumaiza *et al.*, 1986). New for Algeria.

*Genus **Fridericia** Michaelsen, 1889*

Fridericia ? aurita Issel, 1905

Material examined: 1 specimen.

Site: Algeria, 17b.

Observations from material: Colour brownish. Length 15 mm. Segments 51. Setae 4–2÷4 : 4÷5–2÷5, maximum length 115 μm. 3–4 rows of cutaneous gland cells per segment anteclitellarly. Thick body wall. Peptonephridia long, coiled and with few branches. Dorsal vessel from XIX. Chloragogen cells from V, small (22 μm long), filled with yellow-green granules. Coelomocytes: nucleated type *a*, up to 37.5 μm long; anucleate 7.5–10 μm long. Clitellum over XII–$\frac{1}{2}$XIII. Seminal vesicles absent. Sperm funnel tapering entally to ectally, 175 by 75 μm. Male duct long, coiled, 10 μm wide. Penial bulb large (200 by 135 μm). Spermatheca with cylindrical ampulla and two sessile ear-like diverticula; ectal duct 150 μm long, 22.5 μm thick; one small gland at the external orifice.

Remarks: The features of this specimen fall within the range of variation of *F. aurita* in the original (Issel, 1905) and following descriptions (Southern, 1907; Nielsen & Christensen, 1959); however the ratio in size of the two kinds of coelomocytes and the type of habitat are quite unusual for the species which is elsewhere (Italy: Issel, 1905; Italy and Ireland: Rota, pers.obs.) characterized by minute anucleate coelomocytes and typically found in wood litter.

Distribution: European [Ireland, The Netherlands, Belgium, Germany, France, Austria, Italy, Yugoslavia, Bulgaria, Poland, Czechoslovakia]. Already recorded (with some doubts) from Algeria (Cernosvitov, 1933).

Fridericia ? berninii Dózsa-Farkas, 1988 (Fig. 2b)

Material examined: 2 specimens.

Site: Algeria, 23a.

Observations from material: Segments 36, 39. Setae 2–0÷2 : 2÷4–1,2. Three or four rows of cutaneous gland cells in each anteclitellar segment. Brain truncate posteriorly, 125 by 75 μm. Peptonephridia type *b*. Dorsal vessel from XIII. Chloragogen cells from V, large (30–32 μm long). Coelomocytes: nucleate type *b*, 25–30 μm long; anucleate small. Clitellum over XII–$\frac{1}{2}$XIII, gland cells arranged in regular transverse rows. Spermathecae with subspherical ampulla (diameter: 25–30 μm) and two stalked, irregularly shaped diverticula bent backwards (Fig. 2b); ampullae communicating separately with oesophagus. Ectal duct 200–225 μm long, 12 μm wide; one medium gland at the external orifice.

Remarks: The two specimens differ from the original description (Dózsa-Farkas, 1988) in the type of nucleate coelomocytes (*a* in Dózsa-Farkas' material) and in that the ampullae do not merge entally. As these differences concern characters which are relevant in the taxonomy of *Fridericia*, the identification of the Algerian sample remains uncertain until more material is collected.

Distribution: Balearic Islands. New for Algeria.

Fridericia bulboides Nielsen & Christensen, 1959

Material examined: 8 specimens.

Site: Algeria, 3a.

Observations from material: Length 5–8 mm. Width at clitellum 300 μm, at mid body 235 μm. 34–39 segments. Setae straight, 2÷4–2÷4: 3,4–2÷4, maximum length 65.5 μm (caudally). 1–2 rows of cutaneous glands in each anteclitellar segment. Brain 125 μm long. Dorsal vessel from XVI–XVII. Chloragogen cells from V, rather large, filled with grey-brown or olive-green granules. Coelomocytes: nucleated with few coarse granules arranged peripherally (type *b*); anucleate medium sized. Clitellum with clear interspaces between scattered gland cells. Seminal vesicles absent. Sperm funnel small, tapering entally to ectally, 75 μm long, 31 μm wide. Spermathecal ectal duct as long as the inner diameter of the worm: 187 μm. One mature egg.

Remarks: The species is easily recognizable by the structure of peptonephridia (type *b*) and type of nucleated coelomocytes (see also Möller, 1971; Chalupský, 1986; Kasprzak, 1986).

Distribution: Holarctic [Great Britain, Ireland, Iceland, Spitzbergen, Norway, Sweden, Finland, Denmark, The Netherlands, Germany, Spain,

France, Austria, Italy, Poland, Czechoslovakia, Hungary, Romania, Russia, Greenland, Canada]. Recorded also from SW Australia (Springett, 1971). New for Algeria.

Fridericia bulbosa (Rosa, 1887) *sensu lato* (Fig. 2c, d)

Three forms were recognizable (Table 2) and probably each deserves specific rank. *Material examined*: Form *a*: 23 specimens. Form *b*: 1 specimen. Form *c*: 8 specimens.

Sites: Form *a*: Tunisia, 36a/c, 36bis a/b; Algeria, 8b, 15b, 23bis, 24. Form *b*: Algeria, 6c. Form *c*: Algeria, 5, 6c, 8b.

Remarks: The principal distinguishing characters between the three forms are body size, segment number, setal complement, type of nucleated coelomocytes and shape of the spermathecal ampullae.

Form *a* agrees with the original description by Rosa (1887) and with the redescription by Nielsen & Christensen (1959). Form *b* differs from both forms *a* and *c* in having only two setae in the postclitellar bundles, compressed spermathecal ampullae (Fig. 2c) and postero-ventral origin of the efferent ducts of the anteclitellar nephridia. Form *c* is characterized by larger body size, a different type of nucleated coelomocyte and knobbed spermathecal ampullae (Fig. 2d). Forms *b* and *c* were collected at the same site and have recently been found, with the identical distinguishing characters, in Italian localities (Rota, in prep.).

Distribution: Cosmopolitan [Great Britain, Ireland, Iceland, Spitzbergen, Norway, Sweden, Finland, Denmark, The Netherlands, Belgium, Germany, Spain, Madeira Is., France, Austria, Italy, Poland, Czechoslovakia, Hungary, Romania, Russia, India, Indonesia, Japan; eastern Africa; SW Australia; Greenland, Canada, USA, Mexico, Brazil]. Known from Morocco (Dózsa-Farkas, 1989), new for Tunisia and Algeria.

Fridericia caprensis Bell, 1947

Material examined: 53 specimens.

Sites: Tunisia, 1, 2b, 31, 35a/b, 36b/c/d, 36bis b; Algeria, 7b, 8c, 11b, 15b, 16, 17a/b, 19, 21, 22, 23a/b, 23bis, 26a.

Observations from material: Colour white-yellow. Length 6.5–16 mm. Width at clitellum 450–660 μm. Segments (33,34),37–46,(51). Setae 3÷6–3÷7: 4÷8–2÷7, maximum length 70–87 μm. Three to 5 rows of pale, rectangular cutaneous glands in each anteclitellar segment. Dorsal pores clearly visible. Brain posteriorly truncate or slightly indented, 137–175 μm long, 70–110 μm wide. Peptonephridia long, with lateral and terminal branches (type *c*). Dorsal vessel from XIII,XIV,(XV). Coelomocytes: nucle-

Table 2. Distinguishing characters of the three forms of *Fridericia bulbosa*.

Fridericia bulbosa	Form *a*	Form *b*	Form *c*
Length (mm)	6–9	9	12–13
Segments	31–41	38	43–46
Setae	2–2 : 2–2 or 2, 3, 4–3, 2 : 2, 3, 4, 5–4, 3, 2	2, 3, 4–2 : 2, 3, 4–2	3, 4–3, 4 : 3, 4, 5–5, 4, 3, 2
Cutaneous glands	3–4 rows, elongate rectangular, light	2 rows, large, rectangular, light	2 rows, irregular, light
Coelomocytes	nucleated: type *a* anucleate small	nucleated: type *a* anucleate medium	nucleated: type *b* anucleate small
Dorsal vessel from	XV–XVII	XV	XVII
Peptonephridia	type *a*, short, unbranched	type *a*, unbranched	type *a*, unbranched or with few short branches
Nephridial duct	mid-ventral	mid-posteroventral	mid-ventral
Clitellar glands	few, sparse, with clear interspaces	regularly scattered	small, irregularly arranged
Eggs	1 mature	1 mature	1–2 mature
Spermathecae:			
ampullae	onion-shaped, often close to each other	compressed, with two lateral lobes	onion-shaped, knobbed
ectal duct	long, up to 7 times the ampulla	long, 3–4 times the ampulla	long, covered with cells
ectal gland	1 small or absent	1 medium	1 small or absent

ated 25–40 μm long, with fine dark grains denser peripherally and around the nucleus; anucleate small, 5–8 μm long. Efferent ducts of anteclitellar nephridia arising antero-ventrally. Clitellum over XII–$\frac{1}{2}$XIII only; small gland cells arranged in 27–35 more or less regular rows. Seminal vesicles absent or small in XI. Sperm funnel cylindrical, 87–200 by 45–112 μm; collar high, open, as wide as or a little narrower than the funnel. Sperm duct 13.7–17.5 μm thick, coiled in XII. Penial bulb compact, 125–150 by 100 μm. Inner wall of spermathecal ampullae "pimpled" when contracted; ectal duct up to 340 μm long (3–5 times the length of the ampulla), 22.5–25 μm thick. One small gland at the ectal orifice. More than 4 mature eggs may be present at a time.

Remarks: The species is easily distinguished by the shape of the spermathecae (ampullae merging entally, no diverticula, ental duct very short or absent) and is one of the commonest enchytraeids in the study area. The variation observed from site to site with regard to body size, setal complement and size of internal structures summarizes, altogether, what has been reported in previous descriptions (Bell, 1947; Nielsen & Christensen, 1963; Dózsa-Farkas, 1988). However, the peptonephridia are here more branched and only one small gland occurs at the spermathecal orifice (the latter was found also in Balearic material by Dózsa-Farkas, 1988).

Distribution: Circummediterranean [Balearic Is., Spain, France, Italy, Turkey (Rota, in prep.)]. Reported dubiously from Canada (Nurminen, 1973). New for Tunisia and Algeria.

Fridericia paroniana Issel, 1904 (Fig. 2e, f)

Material examined: 23 specimens.

Sites: Algeria, 6a/b/c, 7a, 8b, 11a/b/c.

Observations from material: Colour white. Length 7–11 mm. Width at mid body 250–300 μm (300–320 μm at clitellum). Segments 35–47. Setae straight with distinct ental hook, 2 in all bundles; parcels of detached setae in the coelomic fluid of segments VI–IX. Three rows of large, rectangular cutaneous gland cells in each anteclitellar segment, only the central row complete. Body wall transparent. Brain egg-shaped, 131–150 μm long, 69 μm wide. Peptonephridia short, unbranched (type *a*). Dorsal vessel from (XVII), XVIII–XIX,(XX,XXI). Coelomocytes (Fig. 2e): nucleated oval, hyaline (type *a*), 31–40 μm long; anucleate large, 11 μm long. Efferent ducts of anteclitellar nephridia arising mid-ventrally. Clitellar gland cells irregularly scattered with clear interspaces. Seminal vesicles absent or small in XI. Sperm funnel barrel-shaped, 100–150 by 53–75 μm. Sperm duct 7.7 μm thick. Penial bulb compact, 150 μm long. Spermathecae with ovoid ampulla and two, long, irregularly outlined diverticula (Fig. 2f); ectal duct short; a huge (31–50 μm long), stalked gland is always present at the external orifice. Up to 4 mature eggs.

Remarks: Since the redescription by Nielsen & Christensen (1959), the taxonomy of this species has become increasingly confused. We are convinced that all the specimens identified as *F. paroniana* (e.g. Nielsen & Christensen, 1959 (partim?); Möller, 1971; Dózsa-Farkas, 1989) but differing in critical characters (setal complement, coelomocytes, shape of the spermatheca and size of the ectal gland) from the original description, belong to different taxa. The specimens collected in Algeria fit the original description (Issel, 1904) and agree exactly with material collected recently in Turkey, Tuscany and Ireland (Rota, in prep.; for the Irish record see also Southern, 1909). Further, they are fairly similar to the specimens from Morocco described as *Fridericia maculata macroglandulosa* by Dózsa-Farkas (1989).

Distribution: Euro-Mediterranean [present with certainty in Ireland, Italy, Turkey (Rota, in prep.)]. New for Algeria.

Fridericia perrieri (Vejdovský, 1877)

Material examined: 30 specimens.

Sites: Algeria, 3a/b, 5, 6a/c.

Observations from material: Colour yellow-brownish. Length 9–20 mm. Width at clitellum 455–617 μm. Segments 41–65. Setae straight without distinct ental hook, $3 \div 6 – 2 \div 5 : 4 \div 7 – 2 \div 6$. Up to five rows of pale cutaneous gland cells in each anteclitellar segment, the median rows larger. Body wall opaque. Brain posteriorly truncate, 156 μm long, 75 μm wide. Dorsal vessel arising from XVIII–XXII. Chloragogen cells brown, small, dense from VII. Coelomocytes: nucleated oval, 31 μm long, with smooth surface (type *a*) and small, evident nucleus; anucleate small (5–6 μm). Efferent ducts of anteclitellar nephridia arising mid-ventrally. Sperm funnel tapering or cylindrical, 175–200 by 60–65 μm. Penial bulb compact, large. Spermathecal ectal duct muscular. More than 4 mature eggs may be present.

Remarks: The species is well characterized by the structure of the spermatheca as described by Nielsen & Christensen (1959). A third spermathecal diverticulum and small seminal vesicles were found occasionally also in the present material.

Distribution: Euro-Ethiopian [Great Britain, Ireland, Iceland, Norway, Sweden, Finland, Denmark, The Netherlands, Germany, Spain, France, Switzerland, Austria, Italy, Yugoslavia, Poland, Czechoslovakia, Hungary, Bulgaria, Romania, Palestine, eastern and southern Africa]. New for Algeria.

Fridericia ratzeli (Eisen, 1872) complex

Material examined: 5 specimens.

Site: Algeria, 3b.

Observations from material: Colour white-greyish or yellowish. Length 20–25 mm. Width at clitellum 588 μm. Segments 63–67. Setae straight, $4 \div 6–4 \div 6 : 4 \div 7–4 \div 6$. Many rows of irregularly shaped cutaneous gland cells in each anteclitellar segment. Body wall thick. Chloragogen cells brown-yellow, dense from VII. Coelomocytes: nucleated up to 50 μm long, with evident nucleus; anucleate up to 11 μm long. Efferent ducts of anteclitellar nephridia arising anteroventrally. Seminal vesicles red-brown in (IX),X–XI. Sperm funnel cylindrical, 500–514 by 147–150 μm; collar as wide as the funnel. Sperm duct 14 μm wide. Penial bulb compact, large. Spermathecal ampulla with 6–7 pouch-like, sessile diverticula; one or two small glands at the ectal orifice.

Remarks: This species is known to be heterogeneus both on cytological and morphological bases (Nielsen & Christensen, 1959).

Distribution: Holarctic [Great Britain, Ireland, Iceland, Norway, Sweden, Finland, Denmark, The Netherlands, Germany, Belgium, France, Switzerland, Austria, Italy, Yugoslavia, Poland, Czechoslovakia, Hungary, Romania, Russia, Greenland, Canada, USA]. New for Algeria.

Fridericia sardorum Cognetti, 1901

Material examined: 4 specimens.

Site: Algeria, 8b.

Observations from material: Colour yellow. Length 20–25 mm. Width at clitellum 750 μm. Segments 60–65. Setae straight $6 \div 8–5 \div 7 : 7 \div 9–5 \div 8$, maximum length 150 μm. Three or four rows of cutaneous gland cells in each anteclitellar segment. Head pore 100 μm long. Body wall thick (62.5 μm). Brain slightly indented posteriorly, 225 μm long, 155 μm wide. Dorsal vessel from XV–XVII. Chloragogen cells brown, dense from V. Coelomocytes: nucleated up to 50 μm long, with smooth surface (type *a*) and small, evident nucleus; anucleate small (6–9 μm). Efferent ducts of anteclitellar nephridia arising antero-ventrally. Seminal vesicles absent. Sperm funnel small, 150–175 μm long (less than half the inner body diameter); collar as wide as the funnel, distinct. Spermatheca heart-shaped with conical ampulla and two sessile diverticula; ectal duct short (300 μm) and thick; two large (68.5 μm long) ectal glands.

Remarks: Although the original description by Cognetti (1901) was rather detailed, the species was considered doubtful by Nielsen & Christensen (1959). It has been revalidated recently by Dózsa-Farkas (1989) on the basis of Moroccan material. The Algerian specimens are in good agreement with both descriptions, the major discrepancies being in the higher segment number and smaller sperm funnel size found in our material. It should be noted that Cognetti (1901) gives a range of body size of 15–25 mm, not 15.2 mm only as quoted by Dózsa-Farkas (1989).

The species may be more widespread and common in the study area than our results appear to indicate. A number of large-sized, intensely yellow, immature specimens of *Fridericia* were extracted from samples of several other localities but unfortunately did not survive till maturity so as to allow identification. The same poor tolerance of manipulation and culture conditions was observed in mature specimens of *F. sardorum*.

Distribution: West-Mediterranean [Sardinia, Sicily (Rota, in prep.), Morocco (Dózsa-Farkas, 1989)]. New for Algeria.

Genus ***Enchytraeus*** *Henle, 1837*

Enchytraeus buchholzi Vejdovský, 1879 *sensu lato*

Material examined: 66 specimens.

Sites: Tunisia, 2b, 31, 35a, 36a/d; Algeria, 3b, 6a, 7b, 8a/b/c, 11b, 16a/b, 17b, 19, 23bis, 24, 26b, 29.

Observations from material: Colour whitish. Length 3–13 mm. Segments 26–41. Setae 2,3–2,3 : 3–3. Brain 125–200 μm long, 75–100 μm wide. Peptonephridia anteriorly cylindrical, then flat and with much coiled canal; posteriorly narrow, curled up and with irregular outline. Chloragogen tissue filled with more or less dense oil droplets from IV–V. Coelomocytes smooth or granular. Clitellum over XII–½XIII or XII–XIII; the gland cells may be irregularly scattered. Sperm funnel 2–3 times longer than broad (length 62.5–125 μm, width 25–62.5 μm). Penial bulbs small, solid, with smooth surface. Spermathecal ectal duct of variable length, covered with glands along half or all of its length. Three to 5 eggs mature at a time.

Remarks: Both Tunisian and Algerian specimens varied conspicuously with regard to body size, structure of coelomocytes, length of spermathecal ectal duct, sperm funnel size, and in some cases this seemed to go beyond the variation allowable for a single species. However, in spite of much effort, we found it impossible to distinguish homogeneous groups on the basis of conventional morphological characters. Such a puzzling situation is not new for *E. buchholzi* (Healy, pers. obs.; Chalupský, 1992) and may affect other species of the genus *Enchytraeus* (Westheide & Graefe, 1992). Recently, a similar problem of species discrimination has been approached successfully by combining light microscopy with ultrastructural and biochemical methods (Brockmeyer, 1991; Gabrich *et al.*, 1991; Jacob *et al.*, 1991; Schlegel *et al.*, 1991; Westheide *et al.*, 1991; Westheide & Graefe, 1992).

Distribution: Cosmopolitan [Great Britain, Ireland, Iceland, Norway, Sweden, Finland, Denmark, The Netherlands, Belgium, Germany, Portugal, Spain, Balearic Is., France, Switzerland, Austria, Italy, Greece, Poland, Czechoslovakia, Hungary, Romania, Turkey, Palestine, Russia, Japan, Greenland, Canada, Brazil, Lake Titicaca, Argentina]. Already known from Tunisia (Boumaiza *et al.*, 1986), new for Algeria.

Enchytraeus bulbosus Nielsen & Christensen, 1963

Material examined: 18 specimens.

Sites: Tunisia, 2a; Algeria, 6a, 8a, 11b, 19, 22.

Observations from material: Colour white. Length 3–6 mm. Segments 27–35. Setae 2,(3)–(2),3 : 3–3. Brain 175 μm long, 87 μm wide. Peptonephridia anteriorly wide, straight and with a much coiled canal; posteriorly narrow, loose and with irregular outline. Chloragogen tissue filled with large oil droplets from IV–V, not so dense as to mask the spermathecae. Coelomocytes opaque, covered with dark refractile granules [nearly as in *Marionina argentea* (Michaelsen, 1889)]. Clitellum over XII–½XIII or XII–XIII, interrupted ventrally; gland cells may be irregularly scattered with clear interspaces. Sperm funnels 2–3 times longer than broad (length 50–196 μm, width 30–98 μm); collar narrower than the funnel. Sperm duct 7.5 μm wide. Penial bulbs of variable size (length 70–125 μm, width 50–100 μm: up to one-third the body diameter), with smooth surface; they are compact but not steady, their bulging mass being pushed to and fro by the coelomic fluid. Spermatheca with large, spherical ampulla; ectal duct muscular, with glands near the ectal orifice. One to 2 eggs mature at a time.

Remarks: The species appears more variable than indicated by Nielsen & Christensen (1963). Sperm funnels and penial bulbs, particularly, vary in size from population to population. Specimens from sites 2a (Tunisia) and 22 (Algeria) had the largest funnels and penial bulbs.

Distribution: European [Finland, Czechoslovakia, Hungary, Spain]. New for Tunisia and Algeria.

Enchytraeus minutus bisetosus n. ssp. (Fig. 3)

Material examined: 39 specimens.

Sites: Tunisia, 31; Algeria, 4, 8a/b, 16, 17b, 19, 20, 23bis.

Type material: Holotype, whole mounted specimen from Plage d'Aokas, Gouvernatorat de Bejaia, Algeria (loc. 8a, Fig. 1, Table 1), 21.x.89, deposited at the Museo Civico of Rome (Italy); paratypes, two whole mounted specimens from type locality and a whole mounted specimen from the neighbourhood of Arris, Gouvernatorat de Batna, Algeria (loc. 20, Fig. 1, Table 1), 28.x.89, deposited at the Museo Civico of Rome (Italy).

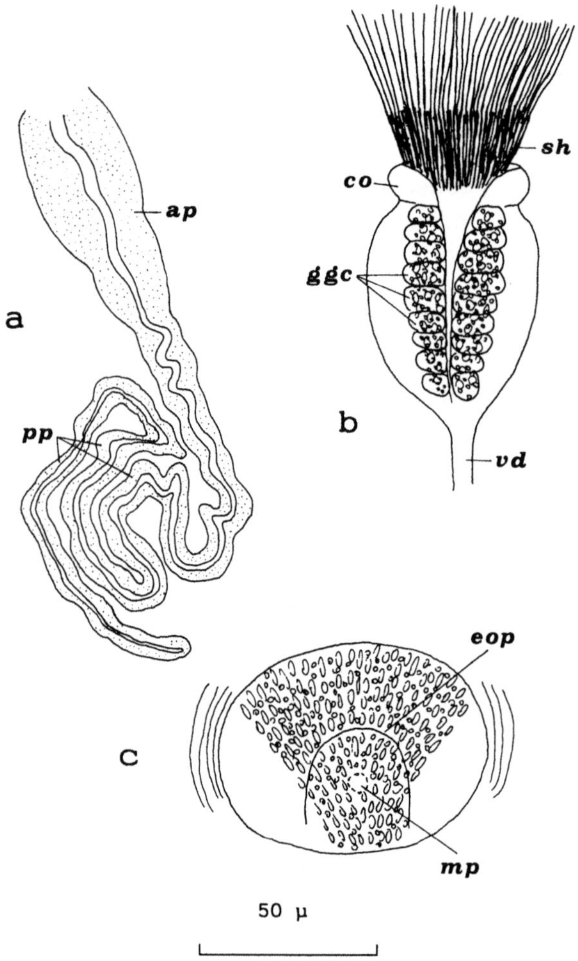

Fig. 3. Enchytraeus minutus bisetosus n. ssp. (a) peptonephridium. (b) sperm funnel. (c) penial bulb. All drawn to same scale. *ap*, anterior portion of peptonephridium, attached to the back of pharynx; *co*, collar of sperm funnel; *eop*, external opening of penial bulb; *ggc*, granular gland cells; *mp*, male pore; *pp*, posterior coils of peptonephridium; *sh*, heads of spermatozoa; *vd*, sperm duct.

Description: Colour whitish. Length 4–8 mm. Width 300–370 μm. Segments 26–34. Setae straight, 2 in all bundles, 25–48 μm long. Brain about 2 times longer than broad (125–175: 75–100 μm); posterior end truncate or convex. Peptonephridia long and narrow, posteriorly much coiled and bulging into V (Fig. 3a). Three pairs of septal glands, none of them merging dorsally; each pair possessing small ventral lobes. Dorsal vessel from XIII; blood colourless. Chloragogen tissue present from V, filled with grey-brown oil droplets but less opaque than in *E. buchholzi*. Coelomocytes oval or pear-shaped, 25–55 μm long, covered with light granules. Nephridia with anteseptal consisting of funnel only; oval postseptal with efferent duct arising posteroventrally. Clitellum over XII–$\frac{1}{2}$XIII; granular gland cells at full maturity arranged in regular transverse rows with clear interspaces. Seminal vesicles small in XI. Sperm funnels small (length 42–125 μm, width 25–100 μm), partly covered with granular cells. Collar somewhat narrower than the funnel; the heads of the spermatozoa form a thick dark brush 2–2.5 times higher than the collar (Fig. 3b). Penial bulbs small (75–125 × 50–100 μm), solid, with granular gland cells on surface (Fig. 3c). Up to 3 mature eggs present. Spermatheca with long ectal duct, covered for half its length with small fused glands. Ampulla ovoid or spherical, half as long as the ectal duct. A number of otherwise sexually mature specimens lacking spermathecae were observed.

Remarks: *Enchytraeus minutus bisetosus* differs from the nominal subspecies in the following characters: setal complement; each pair of septal glands possessing ventral lobes; glands covering half the length of the spermathecal ectal duct.

Distribution: *Enchytraeus minutus* Nielsen & Christensen, 1961 is a Holarctic species [Great Britain, Ireland, Iceland, Norway, Sweden, Finland, Denmark, The Netherlands, France, Austria, Italy, Poland, Hungary, Lebanon, Russia, Greenland, Canada]. *Enchytraeus m. bisetosus* has only been found in northern Africa.

Genus ***Marionina*** *Michaelsen, 1889*

Marionina argentea (Michaelsen, 1889)

Material examined: 2 specimens.

Site: Algeria, 29.

Observations from material: Colour white. Length 2 mm. Width 103–117 μm. Segments 21. Coelomocytes dark in transmitted light, 15.5–19.3 μm long. One egg mature.

Remarks: Representatives of this species are usually shorter than indicated by Nielsen & Christensen (1959) [see also Johnson & Ladle (1989)]. In all other respects, the Algerian specimens fit Nielsen & Christensen's (1959) description.

Distribution: Holarctic [Great Britain, Ireland, Iceland, Spitzbergen, Norway, Sweden, Finland, Denmark, Germany, Spain, France, Switzerland, Italy, Poland, Czechoslovakia, Hungary, Romania, Lebanon, Greenland, Canada]. New for Algeria.

Marionina riparia Bretscher, 1899 augm. Černosvitov, 1928

Material examined: 4 specimens.

Site: Tunisia, 2a.

Observations from material: Length 5–5.5 mm. Segments 29–30. Two or three rows of orange cutaneous gland cells in each anteclitellar segment. Setae 3,4–2,3: 3,4–2,3.

Remarks: The species is easy to recognize by the shape of the setae, the coloured cutaneous glands and the characteristic arrangement of the spermatozoa within the ampulla as described by Nielsen & Christensen (1959).

Distribution: Palearctic [Great Britain, Ireland, Sweden, Finland, Denmark, Germany, Spain, France, Switzerland, Austria, Italy, Poland, Czechoslovakia, Romania, Lebanon]. Already known from northern Africa (Morocco, Algeria, Tunisia: Martinez-Ansemil & Giani, 1987).

Discussion

In spite of an extensive sampling programme covering most of the major habitats likely to be inhabited by enchytraeids, the number of species recorded is small compared with similar sized regions in Europe. This confirms the impression gained from previous investigations on enchytraeids and other oligochaete families (Martinez-Ansemil & Giani, 1987; Omodeo & Martinucci, 1987) that the fauna of Algeria and Tunisia is poor in species and genera. The aridity of the soils and the degradation of the landscape due to intense grazing and browsing by cattle, sheep and goats may be partly responsible. Another reason may be a relative uniformity of soil types and especially an absence of the acid soils (Table 1) preferred by some enchytraeid species. Most of the recorded species are widely distributed throughout Europe, or even cosmopolitan, but some have a more restricted range. *Fridericia caprensis* has a circummediterranean distribution (Rota, in prep.), *F. sardorum* seems to be endemic to the western Mediterranean and *F. berninii* is known only from the Balearic Islands and Algeria. *Enchytraeus minutus bisetosus* may be an endemic subspecies.

Acknowledgements

Support for this research was received from the Italian Programme "Biogeographic and Faunistic Survey of Mediterranean Countries" (Ministero Università e Ricerca Scientifica e Tecnologica). We are much indebted to Prof. P. Omodeo who kindly invited us to join the expedition to Algeria and Tunisia. We are also grateful to all our other travelling companions and particularly to Prof. D. Caruso, head of the Programme, for assistance in the field.

References

Backlund, H. O., 1947. Swedish Enchytraeida. II. Acta Univ. Lund, Adv. 2, 43(8): 1–31.

Bell, A. W., 1947. Some new enchytraeids (Oligochaeta) from the old world. Trans. am. micr. Soc. 66: 190–202.

Boumaiza, M., E. Martinez-Ansemil & N. Giani, 1986. Les Oligochètes et Aphanoneura des eaux courantes de Tunisie. I – Données faunistiques. Ann. Limnol. 22: 231–237.

Brockmeyer, V., 1991. Isozymes and general protein patterns for use in discrimination and identification of *Enchytraeus* species (Annelida, Oligochaeta). Z. zool. Syst. Evolut.-forsch. 29: 343–361.

Černosvitov, L., 1928. Die Oligochaetenfauna der Karpathen. Zool. Jb., Abt. Syst. 55: 1–28.

Černosvitov, L., 1933. Note sur les Oligochètes d'Algérie. Bull. Soc. Hist. nat. Afrique du Nord 24: 255–258.

Černosvitov, L., 1939. Etudes biospéologiques. X. Catalogue des Oligochètes hypogés. Mus. Roy. Hist. nat. de Belgique 15(22): 1–92.

Chalupský, J., 1986. Czechoslovak enchytraeids (Oligochaeta, Enchytraeidae). I. Enchytraeids from an apple orchard by Bavorov in South Bohemia. Vest. cs. Spolec. zool. 50: 13–21.

Chalupský, J., 1992. Terrestrial Enchytraeidae (Oligochaeta) and Parergodrilidae (Polychaeta) from Sweden, with description of a new enchytraeid species. Zool. Scr. 21: 133–150.

Cognetti, L., 1901. Gli Oligocheti della Sardegna. Boll. Mus. Zool. Anat. comp. Torino 16: 1–26.

Dózsa-Farkas, K., 1988. *Fridericia berninii* sp. n. und weitere Angaben über die Enchytraeiden-Fauna (Oligochaeta) der Balearen. Acta zool. hungarica 34: 339–344.

Dózsa-Farkas, K., 1989. *Fridericia* Arten aus Marokko (Oligochaeta: Enchytraeidae). Acta zool. hungarica 35: 29–39.

Duchaufour, P., 1965. Précis de pédologie. Masson & Cie, Paris, 481 pp.

Gabrich, A., P. P. Jaros & V. Brockmeyer, 1991. Application of immunological methods for the taxonomic study of two selected animal taxa: *Tisbe* (Crustacea, Copepoda) and *Enchytraeus* (Annelida, Oligochaeta). Z. zool. Syst. Evolut.-forsch. 29: 381–392.

Gagneur, J., N. Giani & E. Martinez-Ansemil, 1986. Les Oligochètes aquatiques d'Algérie. Bull. Soc. Hist. Nat., Toulouse. 122: 119–124.

Healy, B. & E. Rota, 1992. Methods for collecting Enchytraeidae during expeditions. Proceedings of ISEE4, Avignon, June 1990. Soil Biol. Biochem. 24: 1279–1281.

Issel, R. 1904. Due nuove *Fridericia*. Atti Soc. ligust. Sc. nat. geogr., Genova 15: 31–39.

Issel, R. 1905. Oligocheti inferiori della fauna italiana. I. Enchitreidi di Val Pellice. Zool. Jb., Abt. Syst. 22: 451–476.

Jacob, J., H.-P. Hanssen, B. Ziemsen, V. Brockmeyer & S. Willig, 1991. Chemical composition of lipids in populations and species of *Enchytraeus* (Annelida, Oligochaeta) and *Tisbe* (Arthropoda, Crustacea). Z. zool. Syst. Evolut.-forsch. 29: 373–380.

Johnson, P. & M. Ladle, 1989. The Enchytraeidae (Oligochaeta) of streams of Southern England. Ann. Limnol. 25: 121–129.

Kasprzak, K., 1972. *Achaeta seminalis* sp. n., a new species of Enchytraeidae (Oligochaeta) from Poland and notes on other species of the genus *Achaeta* Vejdovský, 1877. Bull. Acad. Polon. Sci., Cl. II, 20: 187–191.

Kasprzak, K., 1986. Skąposzczety wodne i glebowe, II. Rodzina: Wazonkowce (Enchytraeidae). Państwowe Wydawnictwo Naukowe, Warszawa, 366 pp.

Martinez-Ansemil, E. & N. Giani, 1987. The distribution of aquatic oligochaetes in the south and eastern Mediterranean area. Hydrobiologia 155: 293–303.

Michaelsen, W., 1887. Enchytraeiden-Studien. Arch. f. mikr. Anat. 30: 366–378.

Möller, F., 1971. Systematische Untersuchungen an terricolen Enchytraeiden einiger Grünlandstandorte im Bezirk Potsdam. Mitt. zool. Mus. Berlin 47: 131–167.

Nielsen, C.O. & B. Christensen, 1959. The Enchytraeidae, critical revision and taxonomy of European species. Natura jutl. 8–9: 1–160.

Nielsen, C.O. & B. Christensen, 1961. The Enchytraeidae, critical revision and taxonomy of European species. Supplement 1. Natura jutl. 10: 1–23.

Nielsen, C.O. & B. Christensen, 1963. The Enchytraeidae, critical revision and taxonomy of European species. Supplement 2. Natura jutl. 10: 1–19.

Nurminen, M. 1973. Enchytraeidae (Oligochaeta) from the vicinity of Montreal, Canada. Ann. zool. fenn. 10: 399–402.

Omodeo, P. & G. Martinucci, 1987. Earthworms of Maghreb. In: A. M. Bonvicini Pagliai & P. Omodeo (eds), On earthworms. Selected Symposia and Monographs U.Z.I., 2, Mucchi, Modena: 235–250.

Rosa, D., 1887. Il *Neoenchytraeus bulbosus* n.sp. Boll. Mus. Zool. Anat. comp., Torino 2(29): 1–3.

Schlegel, M., G. Steinbrück, M. Kramer & V. Brockmeyer, 1991. Restriction fragment patterns as molecular markers for species identification and phylogenetic analysis in the genus *Enchytraeus* (Oligochaeta). Z. zool. Syst. Evolut.-forsch. 29: 362–372.

Southern, R., 1907. Contributions to the natural history of Lambay, County Dublin. Oligochaeta. Irish Nat. 16: 68–82.

Southern, R., 1909. Contributions towards a monograph of the British and Irish Oligochaeta. Proc. R. Irish Acad. 27, Sec. B: 119–182.

Springett, J.A., 1964. A method for culturing Enchytraeidae. Oikos 15: 175–177.

Springett, J.A., 1971. The Enchytraeidae (Oligochaeta) of South Western Australia: the genus *Fridericia* Michaelsen 1889. J. Roy. Soc. West. Australia 54: 17–20.

Westheide, W. & U. Graefe, 1992. Two new terrestrial *Enchytraeus* species (Oligochaeta, Annelida). J. nat. Hist., Lond. 26: 479–488.

Westheide, W., G. Purschke & K. Middendorf, 1991. Spermatozoal ultrastructure of the taxon *Enchytraeus* (Annelida, Oligochaeta) and its significance for species discrimination and identification. Z. zool. Syst. Evolut.-forsch. 29: 323–342.

Hydrobiologia **278**: 67–78, 1994.
T.B. Reynoldson & K.A. Coates (eds), Aquatic Oligochaete Biology V.

Propappidae and aquatic Enchytraeidae (Oligochaeta) from the farthest southeast of Russia

Tarmo Timm
Institute of Zoology and Botany of the Estonian Academy of Sciences, Vanemuise 21, EE2400 Tartu, Estonia

Key words: Propappidae, Enchytraeidae, new species, streams, East Asia

Abstract

Two mountain streams north and northeast of Vladivostok were studied in 1983–1984. In the Komarovka Stream, a great variety of *Mesenchytraeus* occurred in the spring zone, among them *M. crenobius* sp. n. with giant setae in the ventral bundles of VI–VIII and *M. vshivkovae* sp. n. with numerous spermathecal diverticula. In the stream zone of the Frolovka Stream *Cernosvitoviella pensau* sp. n., characterized by a complicated penial bulb and a very long spermathecal ampulla, was the dominant oligochaete. *Propappus arhynchotus* Sokolskaya was the second most abundant oligochaete in this stream. It proved to be a true member of this genus and not synonymous with any other species. These are the first records of aquatic enchytraeids for the Primorski Region of Russia.

Introduction

Enchytraeidae is a mostly terrestrial family of oligochaetes although several genera and individual species are amphibious or aquatic. Among them, many representatives of the genera *Mesenchytraeus* and *Cernosvitoviella* inhabit the cold, oxygen-rich freshwaters in northern or mountainous areas (Eisen, 1904; Giani, 1976; Piguet & Bretscher, 1913; Welch, 1919; Timm & Popchenko, 1978; etc.). The genus *Propappus*, isolated from the Enchytraeidae as the family Propappidae by Coates (1986), is fully aquatic. I recently studied the diverse oligochaete material collected at the Far-East Scientific Centre (Vladivostok) in two torrential streams of the Primorski Region (Primorye), Russia (Timm, 1990, 1991). In this paper the species of *Propappus, Mesenchytraeus*, and *Cernosvitoviella* found in these streams will be discussed. Several other enchytraeids, belonging to terrestrial genera, are ignored. No earlier data on the aquatic enchytraeids in the Primorski Region existed.

Materials and methods

The Komarovka (formerly Suputinka) is a first-order tributary of the River Razdolnaya (Suifun) flowing into the Japan Sea west of Vladivostok (Fig. 1). The stream drains the southern ranges of the Sikhote-Alin Mountains and is 70 km long. The first 18 km flow in the submountainous region through the V.L. Komarov Ussuriisk State Nature Reserve and are clean, while the lower reaches are slightly eutrophied. The stream begins with a shallow spring pool (3.5–10 °C). The uppermost 36 km were studied in 1983–1984 by the staff of the Far East Scientific Centre (Vladivostok), monthly benthos samples were taken from

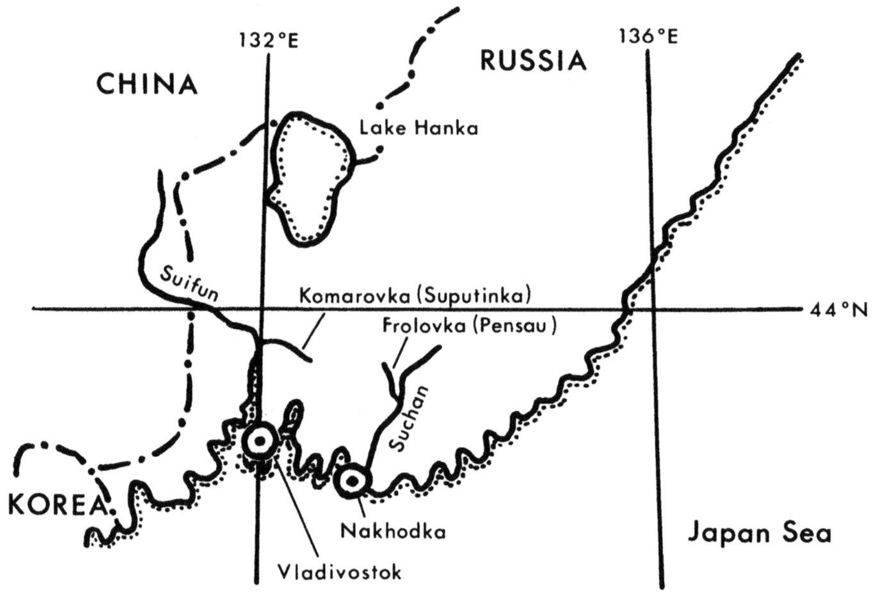

Fig. 1. Location of the sampling area.

May to November at 12 stations using a Levanidov benthometer (analogous to the Surber sampler, capturing area 30 × 40 cm), a simplified, small benthometer (25 × 25 cm), or by the kick-sampling method. In all, 51 samples with oligochaetes were obtained and preserved either in 4% formalin or 70% ethanol.

The Frolovka (Pensau) is a clean first-order tributary of the River Partizanskaya (Suchan) flowing into the Japan Sea east of Nakhodka (Fig. 1). The Frolovka is situated higher and has a more rapid current than the Komarovka; it is only 22 km long, with the source lying at a height of 1060 m, and the mouth at 100 m. The Frolovka was studied at 11 stations from July to October, 1984 with methods analogous to those applied in the Komarovka. Oligochaetes were found in 36 samples.

Thirty-two specimens of aquatic enchytraeids of the genera *Mesenchytraeus* and *Cernosvitoviella* (only 3% of the total oligochaetes) were collected in the Komarovka. In the Frolovka, the specimens of these genera together with Propappidae consisted of 1,385 individuals accounting for 63% of the oligochaete fauna. The worms were studied as whole mounts in glycerine or Canada balsam and some as serial sections. The material including the type specimens is deposited at the Võrtsjärv Limnological Station (VLS), Tartumaa District, Estonia.

Abbreviations used in Figures: a – atrium; ag – atrial glands; cl – clitellum; dv – dorsal vessel; e – egg; ei – epidermal invagination; eig – glands on the epidermal invagination; es – egg sac; i – intestine; m – retracting muscles of the epidermal invagination; mp – male pore; nc – nerve cord; o – ovary; p – penis; pb – penial bulb; phg – pharyngeal glands; sa – spermathecal ampulla; sd – spermathecal duct; sdi – spermathecal diverticula; sf – sperm funnel; sp – spermatozoa; ss – sperm sac; vd – vas deferens.

Taxonomy

Propappus arhynchotus *Sokolskaya, 1972 emend. 1983. Figs 2–8*

Material examined. Frolovka Stream, 403 specimens (about a half of them mature) from all 11 stations, from July to October, 1984. 10 specimens studied in sections, 80 as whole mounts. Lake Ushkovskoye, several specimens, identified by N. Sokolskaya.

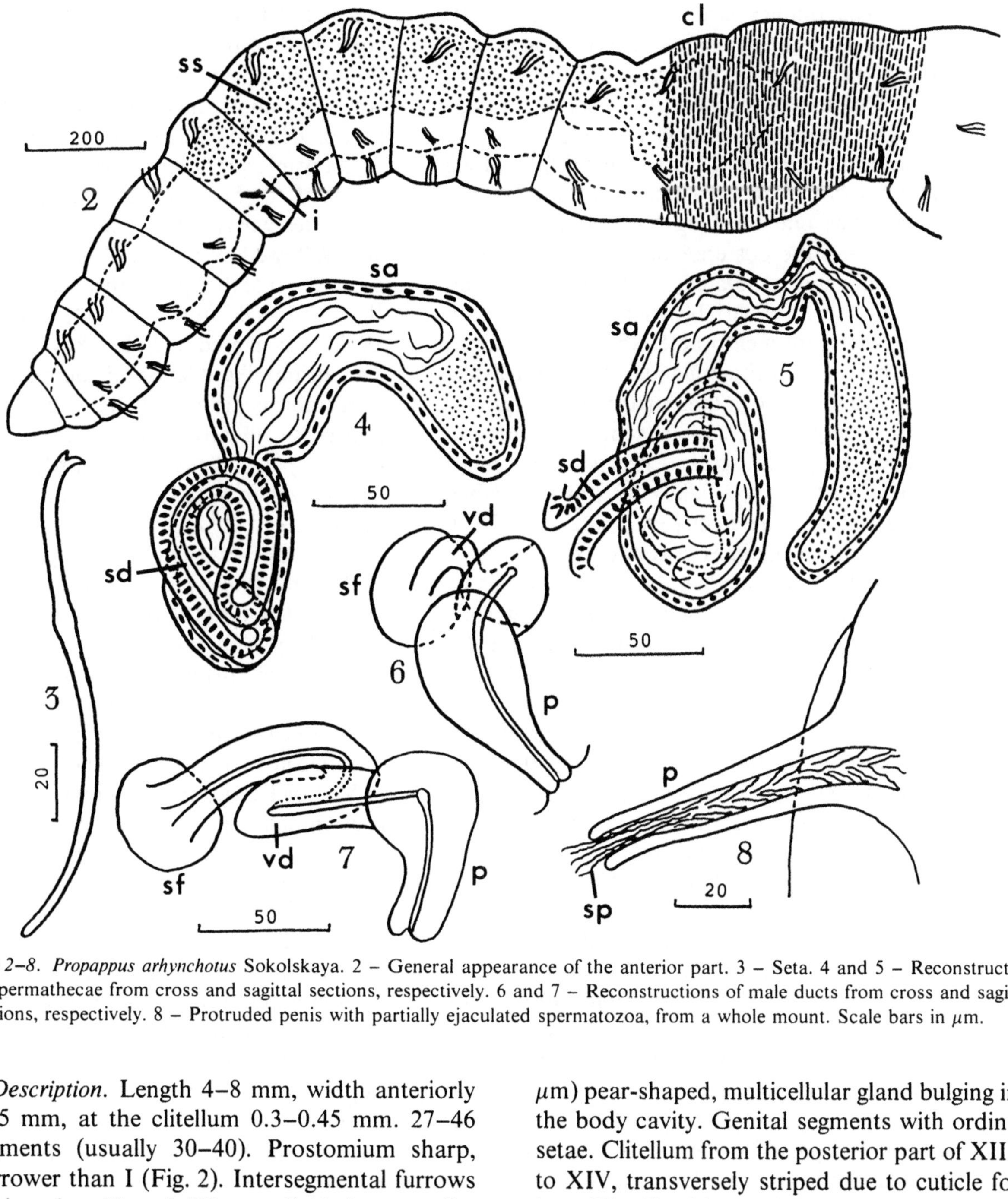

Figs 2–8. Propappus arhynchotus Sokolskaya. 2 – General appearance of the anterior part. 3 – Seta. 4 and 5 – Reconstruction of spermathecae from cross and sagittal sections, respectively. 6 and 7 – Reconstructions of male ducts from cross and sagittal sections, respectively. 8 – Protruded penis with partially ejaculated spermatozoa, from a whole mount. Scale bars in μm.

Description. Length 4–8 mm, width anteriorly 0.25 mm, at the clitellum 0.3–0.45 mm. 27–46 segments (usually 30–40). Prostomium sharp, narrower than I (Fig. 2). Intersegmental furrows distinct but II and III not divided externally. Cuticle with dense transverse folds (the result of fixation?). No pigmentation. Setae mostly 3 per bundle, rarely 1, 2, 4, or (when being replaced) even 6, about 100 μm long (70–80 μm in anteriormost segments), with a short upper tooth (Fig. 3). Posterior to every bundle a large (45 × 30 μm) pear-shaped, multicellular gland bulging into the body cavity. Genital segments with ordinary setae. Clitellum from the posterior part of XII up to XIV, transversely striped due to cuticle foldings (Fig. 2) with glandular epithelium about 15 μm thick. Male pores anterior to ventral setae of XII, on the anterior edge of clitellum, inconspicuous. Spermathecal pores lateral, at 4/5.

Brain deeply incised posteriorly. Pharynx with thick-walled dorsal pad in II–III and with a thickened wall on all sides in posterior of III.

Esophagus (thinner part of intestine) in IV–XIII. Pharyngeal and esophageal glands sparse and diffuse, the beginning of chloragogen tissue indistinct. The first nephridium in V, nephridia consisting of apparently entangled tubules.

Testes in XI, cavity of this segment full of long filiform spermatozoa. Two unpaired sperm sacs, the anterior one reaching V–VIII, usually VII (Fig. 2), and the posterior one XII, full of spermatogonia and spermatocytes. Sperm funnels on 11/12 large and of variable shape, at least 50 μm wide, length not measurable. Short, thick-walled vasa deferentia without glands, most bent Z-like in XII, 16 μm wide in their proximal part, gradually widening up to 25–30 (sometimes 40) μm distally. Large muscular, pear-shaped penes about 40–50 μm wide and 100 μm long when contracted; vasa deferentia enter laterally (Figs 6 & 7). Penes sometimes partially extruded, at a right angle or directed anteriad, the projecting parts being 70–100 μm long and 20–25 μm wide, their cavity containing spermatozoa (Fig. 8). Ovaries in XIII, numerous eggs in XIII–XIV, in egg sac.

Spermathecal ampulla free, sausage-shaped, slightly sinuous, 40–60 μm wide, mostly in V. Duct shorter than ampulla, distinct, 30–40 μm wide proximally and 20 μm at the orifice, its lumen being 5–15 μm wide. Ampulla filled with a mass of filiform spermatozoa, its proximal end can also contain coarse granules (Figs 4 & 5).

Remarks. These worms are smaller than those described by Sokolskaya (1972, 1983), namely 4–8 mm long rather than 6–12 mm, without any papillae or secondary annulation (as a result of different preservations?). Setae are longer in my material, usually 3 per bundle (according to Sokolskaya 3–6 anteriorly), and the clitellum begins in the posterior part of XII (according to Sokolskaya in the anterior part). I checked several specimens from Lake Ushkovskoye (Kamchatka), determined by N. Sokolskaya, and was convinced of their setal number and position of the clitellum being identical with my material. Her statement about the ovaries lying in the posterior part of XII (Sokolskaya, 1972) is probably based on a misinterpretation of her material that corresponded to the situation with many other oligochaetes where a testicular segment is immediately followed by an ovarial segment. The original description of the spermathecal ampullae as being short and oval, situated entirely in V (Sokolskaya, 1972), was altered by Sokolskaya herself in 1983. The atrium-like dilation of the male duct with a retractable penis described by Sokolskaya in both papers is evidently the same organ interpreted by me as a penis. Thus, the Frolovka population of *Propappus* can be regarded as conspecific with the Kamchatkan *P. arhynchotus.*

This species is not synonymous with the Baikalian *Propappus glandulosus* Michaelsen, 1906, as Coates (1986) discussed. The vas deferens in *P. glandulosus* is wider in its ental part, and only a small atrium-like dilation terminates the male duct, instead of the pronounced penis found in *P. arhynchotus.* The beginning of the widened intestine in VIII, the situation of the first nephridia in VIII or IX and the proximal part of spermathecal ampullae in VI are other characters distinguishing *P. glandulosus* from *P. arhynchotus.* These characters are consistent in *P. arhynchotus* (pers. obsv.) but their variation in *P. glandulosus* has not been studied.

Distribution and ecology. This species is common in various types of freshwater bodies of the Kamchatka Peninsula (Sokolskaya, 1983). One more finding originates from a reservoir on the upper Kolyma River (Morev, 1990). It also is common and abundant in the Frolovka Stream. Mature specimens occurred in all months observed (July-October).

Figs 9–17. Mesenchytraeus crenobius sp. n. 9 – General appearance of the anterior part (holotype). 10 and 11 – Ventral setae from V and VI, respectively (holotype). 12–17 – Cross sections with reproductive organs (paratype): 12 – Spermathecal ducts and ampullae (V), 13 – Transition between spermathecal ducts and ampullae, 14 – Spermathecal ampullae (VI), 15 – Sperm funnels (XI/XII), 16 – Vasa deferentia, atria, and epidermal invaginations (XII), 17 – Vasa deferentia and atria in the egg sac (XIII). Scale bars in μm.

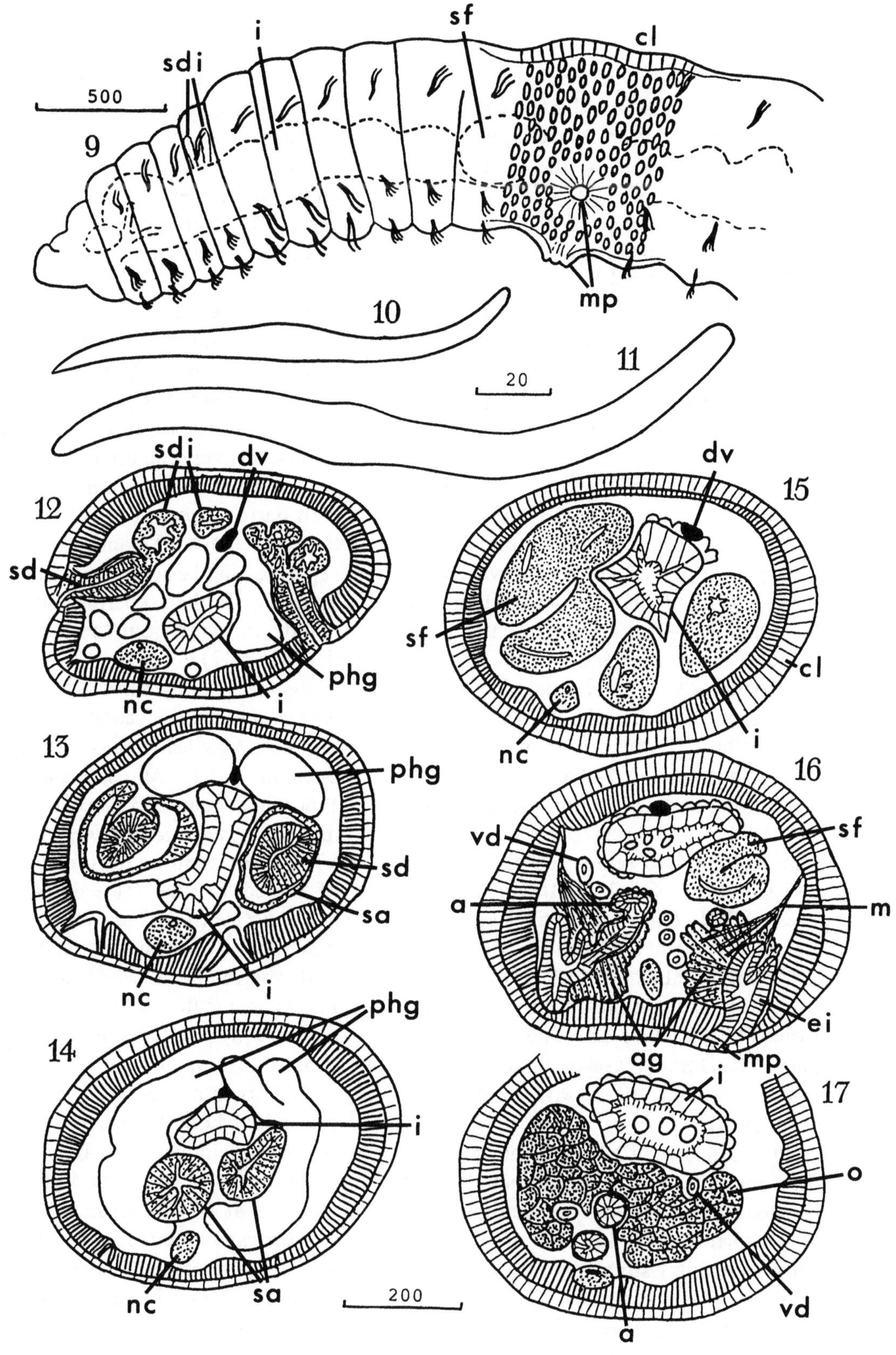

500
9
sdi
i
sf
cl
mp
10
11
20
12
sdi
dv
sd
nc
i
phg
13
phg
sd
sa
nc
i
14
phg
i
nc
sa
200
15
dv
sf
cl
nc
i
16
vd
sf
a
m
ei
ag
mp
17
i
o
vd
a

Mesenchytraeus crenobius *sp. n. Figs 9–17*

Material examined. Holotype (VLS 14–1, whole mount in Canada balsam) and paratypes (VLS 14–2 and 14–3, series of cross sections) are clitellate worms collected from the Komarovka Stream, station 0, sample 145, April 26, 1984. Other material (14 specimens, 5 of them mature or maturing, among them two studied as sections) was collected in 1983–1984 from the type locality, one specimen at the next station downstream and another in the source of the neighbouring Imanny Stream.

Etymology. Crenobius, "living in springs" from the Greek krenos (from a spring) + bios (life or living).

Description. Length of adults 8–15.5 mm, width anteriorly about 0.8 mm, up to 1.1 mm at clitellum; 47–75 segments. Prostomium conical, with dorsal head pore. Intersegmental furrows distinct (Fig. 9). Setae sigmoid, blunt, with distal nodulus (Fig. 10). Either 2, 2–4 or 3 setae in anteclitellar dorsal bundles of separate individuals, and 2–5 (mostly 3) posteriorly. Ventral setae of most segments 4–6, rarely 3 per bundle, 90–160 μm long and 7–10 μm thick, as the dorsal ones. On VI–VIII enlarged ventral setae, 1–2 or 1–3 per bundle, 140–220 μm long and 11–20 μm thick, the largest on VI (Fig. 11). They are weakly developed in juvenile individuals. No setae on XII in mature worms. Clitellum on XII–1/2XIII, the wrinkled male pores on XII (Fig. 9). Indistinct spermathecal pores in 4/5 above the ventral setal line.

Primary pharyngeal glands in 3 pairs (IV–VI), without any dorsal junctions; secondary glands reach from VI to VIII or IX. The first pair of nephridia in VII. Dorsal vessel originating in XIII. Coelomocytes discoid, 10–18 μm in diameter, granular, nucleate.

Testes in XI, ovaries in XII. Egg sac extending to XXX, containing posteriorly directed sperm sac. Long (length not measurable), winding sperm funnels with narrow collar 160–170 μm wide, filling XI and sometimes stretching into sperm sac (Figs 15 & 16). Vasa deferentia 22–30 μm wide with lumen only 1–2 μm, very long and coiled, passing through several segments posteriad inside egg sac (Figs 16 & 17). Vas deferens passes gradually into a long tubular atrium, 60–80 μm wide, with a complicated wall (external epithelium about 6 μm thick, muscular layer 3 μm, internal epithelium 15 μm), surrounded by a mass of elongate multicellular glands (not forming any compact penial bulb) in its distal part (Figs 16 & 17). Atria open into the voluminous folded epidermal invaginations of the body wall (Fig. 16).

Spermathecae in V–VI. Ampullae bulky but empty, pear-shaped, up to 200 μm wide distally, where they partially embrace the ectal ducts (Fig. 13). Narrower proximal ends (Fig. 14) join esophagus close together on the ventral side in VI. Ectal ducts vertical, up to 180 μm long, about 100 μm wide in their proximal part and 60 μm distally, with narrow lumen (5 μm), thick epithelium and thin muscular and glandular layers (Fig. 12). No separate glands at the spermathecal pores. Two internally folded, sperm-containing diverticula, 60–70 μm wide, join the proximal end of each duct (Fig. 12).

Remarks. Only three known species of *Mesenchytraeus* have enlarged setae in ventral bundles: *Mesenchytraeus monochaetus* Bretscher, 1900 in V–VII, *Mesenchytraeus kontrimavichusi* Piper, MacLean and Christensen, 1982 in V–IX, and *Mesenchytraeus tetrapodus* Timm, 1978 in VI–VII. In addition to the location and numbers of paired bundles of enlarged setae, both *M. kontrimavichusi* and *M. tetrapodus* differ from the new species by the construction of their male duct and spermatheca while *M. monochaetus* (not known in a mature state) has a terminal head pore and sharply pointed coelomocytes, different from those in *M. crenobius*.

The organ named in *M. crenobius* as an atrium may not be homologous to the atria of Tubificidae or Lumbriculidae. Atria are simply sections of the male duct with associated glands, however, presumably all of ectodermal origin.

Distribution and ecology. Found only in the sources of two neighbouring mountain streams, Komarovka and Imanny. Apparently breeding in winter, since clitellate specimens were found only in April, while juvenile, maturing and post-reproductive individuals also occurred in May, July, September, and October.

Mesenchytraeus vshivkovae *sp. n. Figs 18–25*

Material. Four clitellate individuals, Komarovka Stream, station 0, sample 145, April 26, 1984. Holotype: VLS 15–1, whole mount, larger specimen. Paratypes: smaller specimen on the same slide, and two series of cross sections, VLS 15–2 and 15–3.

Etymology. Named in honour of Dr T.S. Vshivkova, collector of the material.

Description. Length 5–12.5 mm, width 0.5–0.7 mm, 59–61 segments. Prostomium small, with dorsal head pore. Intersegmental furrows distinct only in anteclitellar segments (Fig. 18). Setae 2–3 in anteriormost bundles and 3–5 elsewhere, bluntly pointed, sigmoid with distal nodulus, 80–130 μm long (shortest in the first bundles) and about 8 μm thick (Fig. 19). No ventral setae on XII. Clitellum on 1/2XI–XIII, with small gland cells; male pores on XII, knob-like. Spermathecal pores on 4/5, lateral (Fig. 18).

Brain anteriorly bifid, posteriorly blunt. Small primary pharyngeal glands in V and VI, secondary glands in V–VII. Chloragogen tissue as a thick layer on the intestine from VIII. The first pair of nephridia in VII. Dorsal vessel traceable from XIII anteriad. No coelomocytes visible.

Testes in XI, ovaries in XII. Anterior sperm sac, sometimes containing an anterior egg sac, can reach X (Fig. 23), paired or bifurcate posterior egg sacs extending to XX–XXIII, paired posterior sperm sacs to XX. Sperm funnels (overall dimensions not measurable in sections) sometimes bulging into anterior sperm sac, short, irregularly cylindrical, hollow; no collar found (Fig. 23). Vasa deferentia 25–30 μm wide, very long, coiling, partially in egg sac in XIII; widening gradually to long, tubular, ciliated atria, 50–60 μm wide, with narrow lumen, thick internal epithelium and thinner muscle layer, without any external glands (Figs 24 & 25). Atria enter into the median side of the slit-like epidermal invaginations which adhere to the body wall and receive numerous, long, multicellular glands, without compact penial bulbs (Figs 24 & 25).

Spermathecae in V–VI. Ampullae pear-shaped, up to 140 μm wide in their distal part (Fig. 21); narrower ental ends join esophagus dorsally in the middle of VI (Fig. 22). Ampullae thin-walled (6–12 μm), contain spermatozoa. Ectal ducts also pear-shaped, 220 μm long and up to 120 μm wide at bulging distal end, with negligible lumen, very thick epithelium and thin muscular cover, without any glands (Figs 18 & 20). A crown of at least six sacculate, sperm-containing diverticula surrounds spermatheca between ampulla and ectal duct (Figs 20 & 24).

Remarks. The new taxon differs from the other, approximately, 70 known species of *Mesenchytraeus* in having the most diverticula on a spermatheca. Most of the other species have 1–2 diverticula or none, only *Mesenchytraeus solifugus* (Emery, 1898) has 3, and *Mesenchytraeus mirabilis* Eisen, 1878 has 4–5 (Čejka, 1914; Welch, 1919).

Distribution and ecology. Known only from the type locality. Probably winter-breeding (found with clitellum in April). Immatures are indeterminable.

Mesenchytraeus *spp*

Representatives of several more species (9 individuals from the Komarovka and 67 from the Frolovka) were collected but not determined due to insufficient material.

A variable set of 8 individuals, sympatric with *M. crenobius* and *M. vshivkovae* (among them two whole-mounted and two sectioned mature ones) revealed a tendency to enlargement of the ventral setae of V–VII, as in *M. kontrimavichusi*, expressed in mature specimens only. The reproductive organs of the sectioned worms differ from the three species named above (small bent sperm funnels, long free spermathecal ampulla without any diverticula). As the pattern of "normal" setae also varied (even in specimens with enlarged setae), the conspecificity of this group remains problematical.

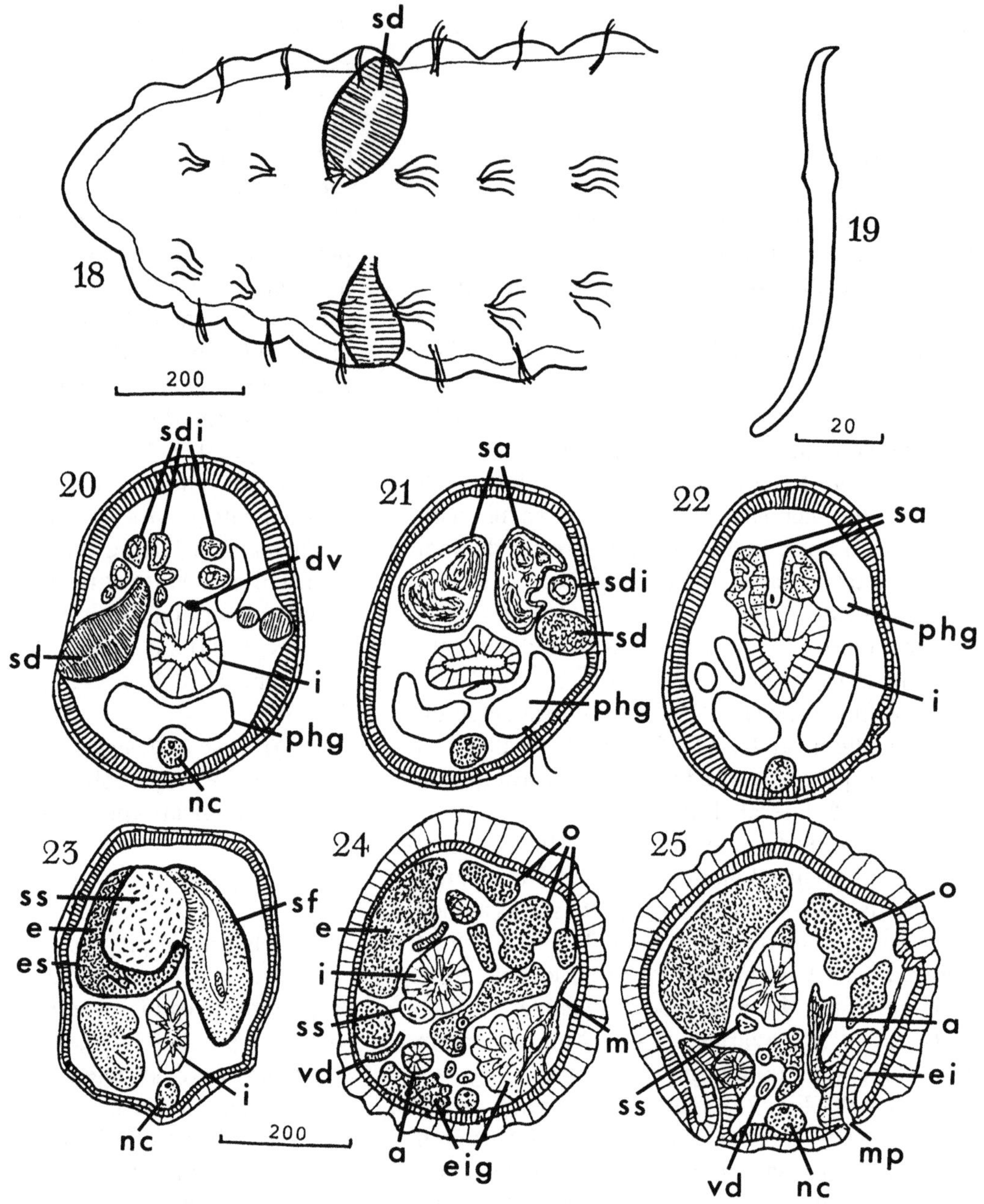

Figs 18–25. Mesenchytraeus vshivkovae sp. n. 18 – General appearance of the anterior end (holotype). 19 – Seta (holotype). 20–25 – Cross sections with reproductive organs (paratype): 20 – Spermathecal ducts and diverticula (V), 21 – Spermathecal ampullae (V), 22 – Connection of spermathecal ampullae with the esophagus (VI), 23 – Sperm funnels, the anterior sperm sac and egg sac (X), 24 and 25 – Vasa deferentia, atria, epidermal invaginations with their glands, ovaries, ripe eggs, and the posterior sperm sac (XII). Scale bars in μm.

A single small juvenile found farther downstream in the Komarovka has only 1–2 setae in its dorsal bundles and 2–3 in the ventral ones.

Only immature individuals were met with in the Frolovka Stream (July-August). Judging by their setal pattern, they can belong to 5–6 different species. One of them (52 individuals in one sample) has 1–2 thickened (7 μm) setae in dorsal bundles of V–VII, like *Mesenchytraeus armatus* (Levinsen, 1884) and *Mesenchytraeus konyamensis* Michaelsen, 1916.

Cernosvitoviella pensau *sp. n. Figs 26–29*

Material. Holotype (VLS 16–1) and 5 paratypes on the same slide are whole-mounted in Canada balsam, another paratype (VLS 16–2) is serially cross-sectioned. Type locality: Frolovka Stream, station 3, sample 14, July 24, 1984. In addition, 930 specimens (about 150 of them mature or maturing) from 19 samples gathered at 8 different stations of Frolovka Stream from July till October, 1984. Four of them were studied as sections, the rest as whole mounts.

Etymology. After the vernacular name of the stream.

Description. Mature worms 5–5.5 mm long, anteriorly 0.2–0.3 and at clitellum up to 0.37 mm wide, tail considerably thinner. 32–37 segments (median 35); intersegmental furrows distinct. Prostomium small, variable, usually oval. I very short, II–IV inflated, the thinnest part of forebody at 4/5 (Fig. 26). Setae 5–8 (usually 6) per bundle, sigmoid, sharp, with distal nodulus (Fig. 27). Clitellum covers XII, a small part of XI and most of XIII, with oval gland cells in regular transverse rows (Fig. 26). No setae on XII in clitellate worms. Male pores on XII, as simple longitudinal slits (Figs 26 & 28). Spermathecal pores in 4/5 in line of dorsal setae, inconspicuous.

Pharynx in III, small. Primary pharyngeal glands on 4/5 and 5/6, small, not joining dorsally; secondary glands from V to the end of VI, sometimes reaching VII. Gradual dilation of intestine from VIII. Origin of dorsal vessel in XIII. Epidermis of body wall about 2 μm thick, up to 10 μm in clitellum, thinner between male pores. Coelomocytes not seen.

Testes in XI, ovaries in XII. Sperm funnels in XI oval, flattened, about 150 μm long, with narrow collar. Vasa deferentia 8–13 μm wide, long and coiled, entangled into one common ball in XII. Paired penial bulbs in XII about 150 μm long, flat; consisting of an infolded (probably eversible) invagination of the body wall partially surrounded by a mass of glands and muscles; on one side (medial or lateral) of the invagination lies an almost spherical body, about 70 μm long, with a dense muscular wall, the inner space sparsely filled with muscle fibres and granular secretions (Fig. 28). Eggs numerous, extending from XII into XIII.

Spermathecal ducts about 30 μm wide, long and smooth, without glands. Behind pharyngeal glands, in VII or VIII, the duct widens abruptly into a long, cylindrical, thin-walled (about 2 μm), spacious, free ampulla, usually reaching X and filled with spermatozoa (Fig. 29). In one specimen an asymmetrical reproductive system was found, with a single male duct (without sperm funnel) and unequal spermathecae, reaching X and VIII, respectively.

Remarks. Fourteen or 15 valid species have been described in the genus *Cernosvitoviella*, either in Europe (Healy, 1979; Giani, 1979; Gadzińska, 1974; Klungland & Abrahamsen, 1981) or North America (Dash, 1970; Nurminen, 1973a, 1973b). None were known either in East Asia or on the Pacific Coast of North America. The new species differs from all the known congeners by its very long and coiled vas deferens, by the complicated penial bulb and the extraordinarily long spermathecal ampulla. Only in *Cernosvitoviella ampullax* Klungland & Abrahamsen, 1981 does the ampulla sometimes reach X, but it is pear-shaped. Elongate spermathecal ampullae are also recorded for *Cernosvitoviella briganta* Springett, 1969 and *Cernosvitoviella palustris* Healy, 1979 but they have not been observed to reach as far posterior as X, nor are the ampullae swollen, as in the new species.

Distribution and ecology. Cernosvitoviella pensau sp. n. is the most abundant oligochaete in Frolovka Stream (about 43% of all specimens) but was not observed anywhere else. It inhabits the stream zone, is lacking in the spring zone and at the lowermost station located in the village. Juveniles were always dominant whereas clitellate worms occurred only in the upper reaches and only in July and August. In October, in lower reaches of the stream only, several post-reproductive or maturing individuals were found to-

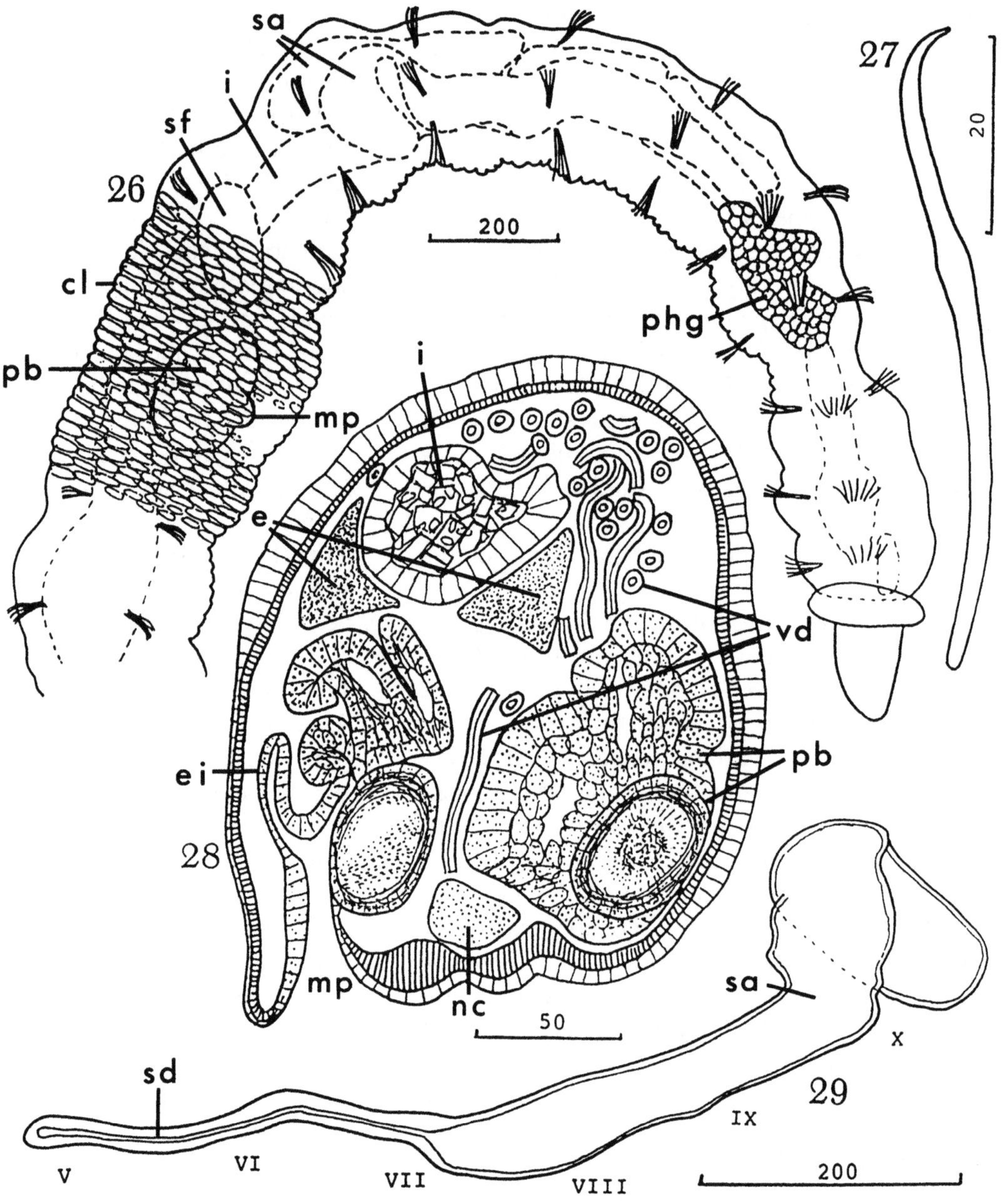

Figs 26–29. Cernosvitoviella pensau sp. n. 26 – General appearance of the anterior part (holotype). 27 – Seta (holotype). 28 – Cross section of XII with vasa deferentia and penial bulbs (paratype). 29 – Reconstruction of a spermatheca after sagittal sections. Scale bars in μm.

gether with abundant juveniles. Apparently, the species breeds mainly in springtime.

Cernosvitoviella *spp*

Two more, smaller species were found as incomplete, partially clitellate individuals but not identified further since there are no satisfactory sections.

Four worms from Frolovka Stream (October 11, 1984) have 6–8 setae per bundle, about 30 μm long. Pharyngeal glands are similar to those of *C. pensau* but the sperm funnels are small and round.

Two worms from the Komarovka Stream (September 21, 1984) were the smallest, only up to 0.18 mm wide at the clitellum and have 3 setae per bundle on anteclitellar segments, and 4–5 on mid-body.

Discussion

One species of *Propappus (P. arhynchotus)*, possibly 5–6 species of *Mesenchytraeus* (among them *M. crenobius* and *M. vshivkovae*), and 3 species of *Cernosvitoviella* (among them *C. pensau*) were found in the two streams studied. All three genera were recorded for the first time in Primorye. In the case of the genus *Cernosvitoviella*, these are also the first findings in East Asia. The two other species of *Propappus* are known from Lake Baikal, the Kamchatka Peninsula and several large rivers; *Mesenchytraeus* species are reported from a number of terrestrial biotopes in northeast Asia (Michaelsen, 1905; Čejka, 1914; Sokolskaya, 1972, 1983; Piper *et al.*, 1982). The propappid and enchytraeid fauna of the Sikhote-Alin mountain streams lives far southward of the previously known east Asian habitats and may be an assembly of cold-water relicts.

A strong tendency to endemicity, together with some faunistic connection with the northeast, were revealed as already noted in the same streams for Naididae and Tubificidae (Timm, 1990). Among the four species determined, *P. arhynchotus* and *C. pensau* were found only in the Frolovka while *M. crenobius* and *M. vshivkovae* were found only in the Komarovka. As several species and juvenile individuals remain undetermined, the existence of some species of *Mesenchytraeus* common to both streams is also possible. One of the four species, *P. arhynchotus*, is shared with the Kamchatka Peninsula and the upper Kolyma River.

Propappidae and Enchytraeidae inhabit colder biotopes than most other oligochaetes found in this study area. They are dominant only in the spring pool in the Komarovka, however, they prevail at all stations in the higher-lying Frolovka. In both streams, all species seem to breed in winter or spring but the reproductive period ends earlier at the lower stations of the Frolovka Stream.

Acknowledgements

I am greatly indebted to Drs. Tatyana S. Vshivkova and Iya M. Levanidova (Far East Scientific Centre, Vladivostok) for the delivery of the oligochaete material collected in the Komarovka and Frolovka Streams, respectively, as well as to Mrs. Ester Jaigma for her linguistic help.

References

Čejka, B., 1914. Die Oligochaeten der Russischen in den Jahren 1900–1903 unternommenen Nordpolarexpedition III. Zapiski Akademii Nauk 29: 1–24.

Coates, K. A., 1986. Redescription of the oligochaete genus *Propappus*, and diagnosis of the new family Propappidae (Annelida: Oligochaeta). Proc. biol. Soc. Wash. 99: 417–428.

Dash, M. C., 1970. A taxonomic study of Enchytraeidae (Oligochaeta) from Rocky Mountain forest soils of the Kananaskis region of Alberta, Canada. Can. J. Zool. 48: 1429–1435.

Eisen, G., 1904. (1910). Enchytraeidae of the west coast of North America. Harriman Alaska expedition series. Vol. XII. Annelids. Smithsonian Institution, New York: 1–166.

Gadzińska, E., 1974. *Cernosvitoviella parviseta* sp. n. – a new species of Enchytraeidae (Oligochaeta) from the Polish Tatra Mountains. Bull. Acad. polon. sci., biol., Cl. II, 22: 403–404.

Giani, N., 1976. Les oligochètes aquatiques du Sud-Ouest de la France. Annal. Limnol. 12: 107–125.

Giani, N., 1979. Description de deux nouvelles espèces d'Enchytraeidae des Pyrenées. Annal. Limnol. 15: 107–112.

Healy, B., 1979. Three new species of Enchytraeidae (Oligochaeta) from Ireland. Zool. J. Linn. Soc. 67: 87–95.

Klungland, T. & G. Abrahamsen, 1981. *Cernosvitoviella ampullax* sp. n. (Oligochaeta, Enchytraeidae) from Hardangervidda, S. Norway. Zool. Scripta 10: 251–253.

Michaelsen, W., 1905. Die Oligochaeten des Baikal-Sees. Wissenschaftliche Ergebnisse einer Zoologischen Expedition nach dem Baikal-See 1900–1902. R. Friedländer und Sohn, Berlin. 69 pp.

Morev, A.P., 1990. Macrozoobenthos of the Kolyma Reservoir. Biologiya vnutrennikh vod. Informatsionny byulletin (Leningrad) 88: 56–61 [in Russian]

Nurminen, M., 1973a. Enchytraeidae (Oligochaeta) from the vicinity of Montreal, Canada. Ann. zool. fenn. 10: 399–402.

Nurminen, M., 1973b. Enchytraeidae (Oligochaeta) from the Arctic archipelago of Canada. Ann. zool. fenn. 10: 403–411.

Piguet, E. & K. Bretscher, 1913. Oligochètes. Catalogue des invertébrés de la Suisse 7. Genève, 1913. 214 pp.

Piper, S. R., S. F. MacLean & B. Christensen, 1982. Enchytraeidae (Oligochaeta) from taiga and tundra habitats of northeastern U.S.S.R. Can. J. Zool. 60: 2594–2609.

Sokolskaya, N. L., 1972. New data on the fauna of aquatic Oligochaeta from Kamchatka. Sbornik trudov Zoologicheskogo muzeya Moskovskogo universiteta 12: 74–90. [in Russian]

Sokolskaya, N. L., 1983. Freshwater Oligochaeta from Kamchatka and the Koryak Mountains. Sbornik trudov Zoologicheskogo muzeya Moskovskogo universiteta 20: 22–119. [in Russian]

Timm, T., 1990. Aquatic Oligochaeta from the farthest south-east of the USSR. 1. Naididae and Tubificidae. Proc. Estonian Acad. Sci., Biol. 39: 55–67.

Timm, T., 1991. Branchiobdellida (Oligochaeta) from the farthest south-east of the USSR. Zool. Scripta 20: 321–331.

Timm, T. & V. Popchenko, 1978. The aquatic Oligochaeta of the Murmansk Region. Hydrobiological Researches (Tartu) 7: 71–132. [in Russian]

Welch, P. S., 1919. Further studies on North American mesenchytraeids (Oligochaeta). Trans. Amer. Micros. Soc. 38: 175–188.

Hydrobiologia **278**: 79–84, 1994.
T.B. Reynoldson & K.A. Coates (eds), Aquatic Oligochaete Biology V.

Oligochaetes (Naididae, Tubificidae, Enchytraeidae and Alluroididae) of Guyana, Peru and Ecuador

Kathryn A. Coates & Donald F. Stacey
Invertebrate Zoology, Royal Ontario Museum, 100 Queen's Park, Toronto, Ontario, Canada

Key words: freshwater oligochaetes, Enchytraeidae, Guyana, Peru, Ecuador

Abstract

More than 41 species in 23 genera of the microdrile oligochaete families Tubificidae, Naididae, Opistocystidae, and Enchytraeidae and the freshwater megadrile family Alluroididae have been identified in recent collections made in Peru, Guyana and Ecuador. Just less than 70% of our species records are new for one or more of these countries and one is a new, albeit tentative, generic record for the South American continent. About 16 species new to science remain to be described. One of these is only the second reported species of *Brinkhurstia* (Alluroididae) and possesses unusual, single, very elongate penial setae. All of our species records are pertinent to tests of different hypotheses about historical and phylogenetic relationships among organisms of northern and southern South America and North America. The species, including new ones, with limited distributions are of particular significance to such hypotheses.

Introduction

The microdrile and other aquatic oligochaetes of South and Central America are poorly known. As such, little can be said with confidence about the distributions or even presence of particular species. Nonetheless, we feel reasonably certain, based on recent but limited studies, that the species diversity of oligochaetes in these areas is high, including many endemics. Thus, these species can contribute valuable information to hypotheses of relationships among taxa of various zoogeographical regions and to hypotheses regarding areas of origin and areas of endemism that are significant to conservation issues.

The recent publication of an identification guide to the freshwater oligochaetes of South and Central America (Brinkhurst & Marchese, 1989) has greatly facilitated the study of worm collections from Central and South America. Enchytraeid species have been omitted from the guide. However, they are even more poorly studied than the other oligochaete families with freshwater species and keys beyond genera would certainly be premature. In addition, most of the enchytraeids reported from South America are only semi-aquatic or are fully terrestrial.

With the aid of the guide (Brinkhurst & Marchese, *op. cit.*) and Nielsen's and Christensen's (1959) revision of the Enchytraeidae, supplemented by the primary literature, we have completed the first level identifications of terrestrial and freshwater microdriles and one freshwater megadrile from collections made in Peru, Guyana and Ecuador over the last seven years. Overall, our study is still "a work in progress" and species-level identifications and descriptions of enchytraeids, of a species of *Unci-*

nais (Naididae), and of a freshwater megadrile of the genus *Brinkhurstia* (Alluroididae) are not complete. Specimens that were identified to species were not observed to differ in any significant way from published descriptions (Brinkhurst & Jamieson, 1971; Brinkhurst & Marchese, 1987).

The new species descriptions as well as any necessary comments on the known species will be published subsequently, along with details of the habitats and distributions.

Materials and methods

All mature specimens and representative specimens of naidids were sorted from the collections, stained with borax carmine, dehydrated in an alcohol through turpineol graduated series and mounted permanently in Canada balsam. For the enchytraeids, with a relatively thick cuticle, and for larger, thick-bodied specimens, we found this dehydration and clearing series much less satisfactory than an ethanol to xylene series. Some specimens had to be reprocessed through the latter series in order to make observations of internal organs.

Identifications were made on a compound photomicroscope with interference contrast optics. Videomicroscopy was used so that both investigators could simultaneously observe a specimen. The high resolution of both the video camera (Dage MTI 65) and of the monitor (Hitachi VM 1201) allowed critical observations to be made even from the video image. Thirty-five mm black and white photography was done with a Wild MPS 46/52 Photoautomat on Kodak Techpan at ASA 100.

Collections

The specimens we identified came from seven collections housed at the Royal Ontario Museum. The Royal Ontario Museum, Invertebrate Zoology, accession numbers, dates of collection, collectors and area descriptors are as follows:
1987–022 and 1988–127, February 1985 by J. Römbke, Peru, at Panguana (area descriptor, PP);
1988–167, July and August 1988 by K. A. Coates, Peru, at Panguana and around Pucallpa (area descriptor, PP);
1986–080, June 1986 by R. O. Brinkhurst and D. Kathman, Peru, Lake Titicaca and Lake Umayo in the southeast and Yanayako River near Iquitos (area descriptors, PT, PU, PY, respectively);
1986–077, June and July 1986 by K. A. Coates, Peru, Maranon River and northern provinces (area descriptor PN);
1990–115, September and October 1990 by D. F. Stacey, Guyana, Potaro and Essequibo Rivers (area descriptor, G); and
1991–055, 1985 or 1986, Ecuador, near Portoviejo (area descriptor, E).
The Römbke collections are all from primarily terrestrial habitats, (see Römbke & Verhaagh, 1992; Righi & Römbke, 1987) Coates and Stacey collections are from both terrestrial and freshwater habitats, and Brinkhurst and Kathman collections and the collection from Ecuador are strictly freshwater, including both lakes and rivers.

Results

From our small, geographically discrete, collections (Fig. 1) we have recognized more than 41 species in 24 genera of the families Alluroididae, Tubificidae, Naididae, Opistocystidae, and Enchytraeidae (Table 1). The collections also include numerous specimens of earthworms that are available for outside study. The species found in the Peruvian collection of Brinkhurst and Kathman (Table 1, PT, PU and PY) were all reported in a previous publication (Brinkhurst & Marchese, 1987), usually from the same, respective, locations if not the same samples. Even these records are important, however, because they include species for which very few specimens are known, such as *Epirodrilus antipodum* Cernosvitov, 1939 and *Paranidrilus descolei* Gavrilov, 1955.

The collection from Guyana (Table 1, G) unquestionably represents all new species records for that country because there are no previously published studies of the oligochaete families we

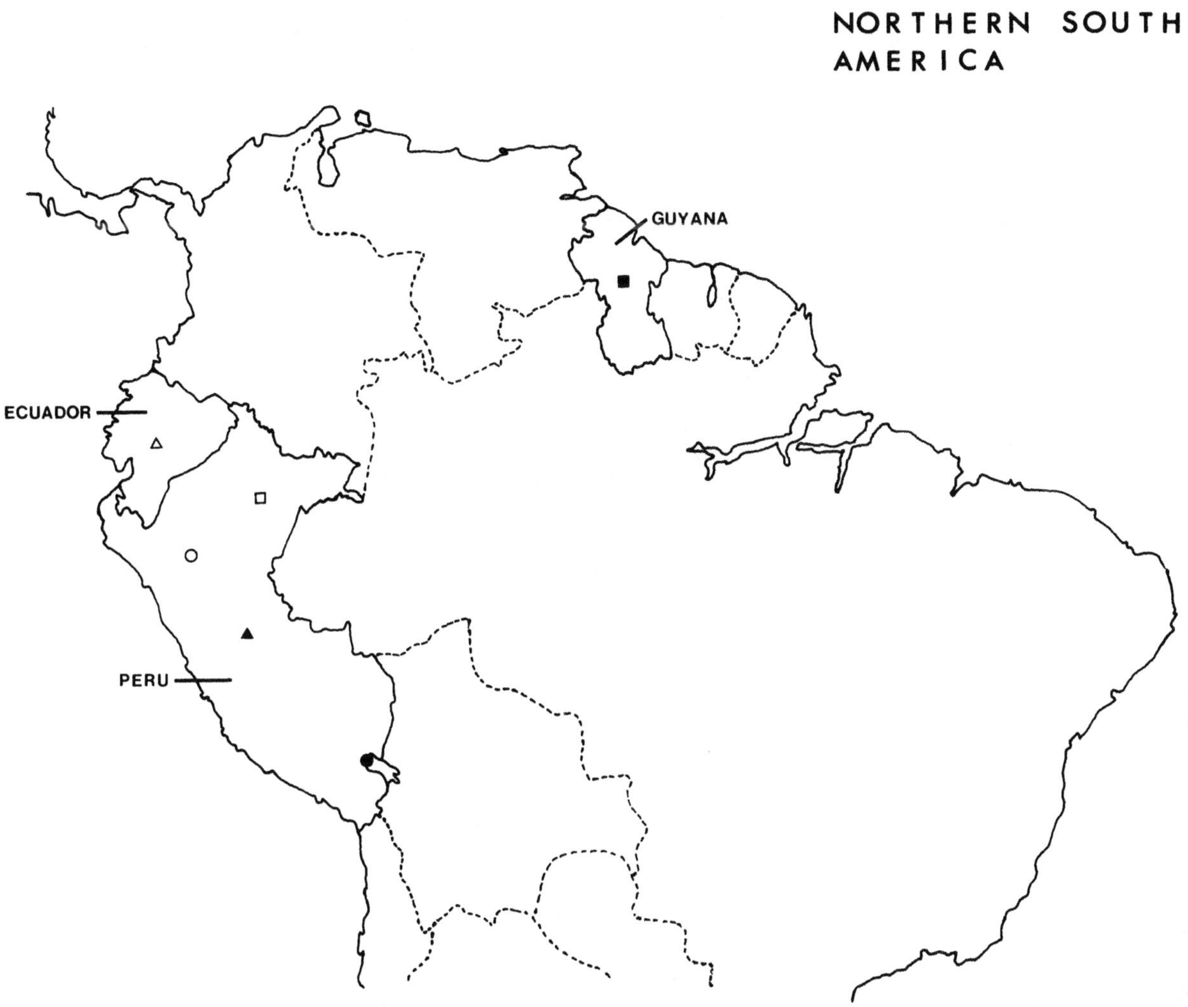

Fig. 1. Locations of major collections sites in Peru, Ecuador and Guyana. ●, Lakes Titicaca and Umayo, Accession # 1986-080; ▲, Panguana and Pucallapa, Accession #'s 1987-022, 1988-127 and 1988-167; ○, Maranon river and northern provinces of Peru, Accession # 1986-077; □, Yanayako River near Iquitos, Accession # 1986-080; ▵, Ecuador near Portoviejo, Accession # 1991-055; and, ■, Potaro and Essequibo Rivers, Guyana, Accession # 1990-115.

recorded. Many of the enchytraeids, from all of the collections, will prove to be new species because enchytraeids have been studied only sporadically throughout South America. At the level of genus, all the enchytraeid records are new for Peru and Guyana. Similarly, all our species records, only naidids were found, are new for Ecuador.

A very distinctive species of *Brinkhurstia* (Alluroididae) was found at sites on the Potaro and Essequibo Rivers in Guyana. The only other recorded species of *Brinkhurstia* is from Argentina and the province of Amazonas in Brazil. The new species is clearly distinguishable from the other by the presence of very elongate penial setae (Fig. 2a, b, c) that extend from just posterior to the dorsal pore of the spermatheca (Fig. 2b) to emerge at or near the penial pore (Fig. 2a, c).

Discussion

The species of the enchytraeid genera *Hemienchytraeus*, *Guaranidrilus*, and *Tupidrilus*, once determined, will be used in continuing studies (Coates, 1987, 1989) of the phylogenetic relation-

Table 1. Taxon list for all samples, with country occurrences for this study. E, Ecuador, G, Guyana, PN, northern Peru and Maranon drainage, PP, Peru in the vicinity of Panguana, PT, Lake Titicaca, PU, Lake Umayo, PY, Yanayako River in northeastern Peru; ^{nr}, new record for country, *, new record for South America.

Family	Genus/Species	Location
Alluroididae	*Brinkhurstia* sp.	G^{nr}
Enchytraeidae	*Enchytraeus*	PP^{nr}, PN
	Fridericia	PP^{nr}, PN
	Guaranidrilus	PP^{nr}
	Hemienchytraeus > 3 spp	PP^{nr}, PN, G^{nr}
	Henlea	PN^{nr}
	Marionina	PU^{nr}, PP
	Mesenchytraeus	G^{nr}
	?Tupidrilus	PP^{nr}, $?G^{nr}$
Naididae	*Allonais inaequalis* (Stephenson, 1911)	E^{nr}
	Dero borellii Michaelsen, 1911	E^{nr}
	Dero digitata (Müller, 1773)	E^{nr}
	Dero furcatus (Müller, 1773)	G^{nr}
	Dero pectinata (Aiyer, 1929)	E^{nr}
	Dero ?raviensis (Stephenson, 1914)	E^{nr}
	Dero sawayai Marcus, 1943	E^{nr}
	Nais communis Piguet, 1906	PN, E^{nr}
	Pristina aequiseta Bourne, 1891	E^{nr}
	Pristina biserrata Chen, 1940	E^{nr}
	Pristina synclites Stephenson, 1925	E^{nr}
	Pristinella acuminata (Liang, 1958)	PP
	Pristinella jenkinae (Stephenson, 1931)	PP
	Pristinella longidentata (Harman, 1965)	PP^{nr}
	Pristinella osborni (Walton, 1906)	PN
	?Uncinais	PP^{nr*}
Opistocystidae	*Opistocysta funiculus* Cordero, 1948	G^{nr}
Tubificidae	*Aulodrilus pigueti* Kowalewski, 1914	PY
	Bothrioneurum sp.	PT, PN
	Epirodrilus antipodum Cernosvitov, 1939	PT
	Limnodrilus hoffmeisteri Claparede, 1862	PN
	Limnodrilus neotropicus (Cernosvitov, 1939)	PN^{nr}
	Paranadrilus descolei Gavrilov, 1955	PY
	Potamothrix bavaricus (Oschmann, 1913)	PT, PN
	Tubifex siolii (Marcus, 1947)	G^{nr}
	Tubifex tubifex (Müller, 1774)	PN

ships and speciation processes among the neotropical enchytraeids. In fact, all of the taxa found can be used to investigate new ideas about the historical and phylogenetic relationships among northern and southern South America and North America (Crisci *et al.*, 1991).

The genera *Hemienchytraeus* and *Guaranidrilus* are both reported in North America from Florida (Healy, 1989) and the latter also from the Columbia River in Washington state (Coates & Diaz, 1988). Both genera are also reported from other sites on the southern continents plus India and northern Africa well as from Europe. As far as we know, *Tupidrilus* is restricted to South America.

Preliminary analyses of the phylogenetic relationships among these taxa (Coates, 1989) indicate that the lineages of *Hemienchytraeus* and *Guaranidrilus* are evolutionarily independent, symplesiomorphically South American, and with taxonomically independent occurrences in Europe and North America. The presence of both these taxa in Europe and southeastern North

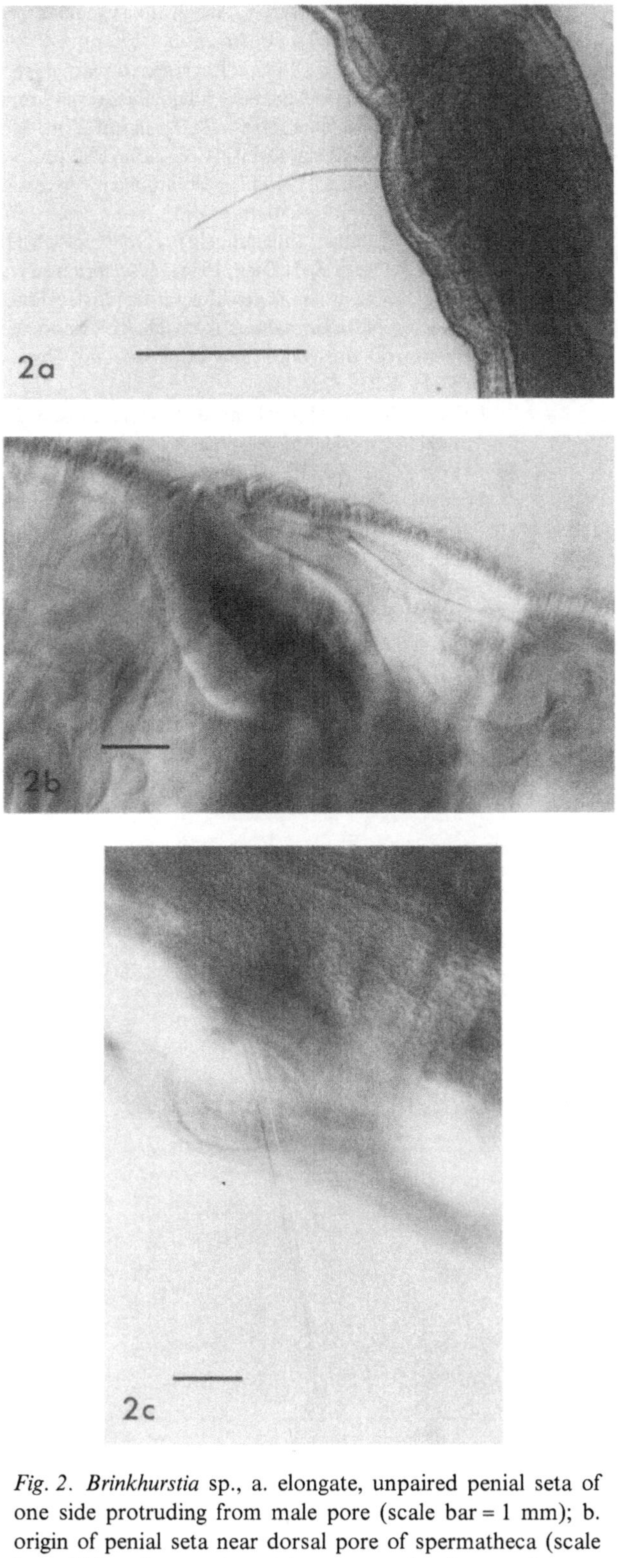

Fig. 2. Brinkhurstia sp., a. elongate, unpaired penial seta of one side protruding from male pore (scale bar = 1 mm); b. origin of penial seta near dorsal pore of spermatheca (scale bar = 100 μm); c. and penial seta emerging from male pore (scale bar = 100 μm).

America may be the outcome of shared vicariant events. As far as we can determine the area relationships predicted by the species distributions are congruent.

One of the species of *Hemienchytraeus* reported from Florida (Healy, 1989) was identified as *Hemienchytraeus bifurcatus* Nielsen and Christensen, 1959, the single species known from Europe, and the other species from Florida is the single widespread species *Hemienchytraeus stephensoni* (Cognetti, 1927). The sister relationships among the species of *Guaranidrilus* are not fully resolved and species are not shared between the two regions. The terrestrial enchytraeids of Central America, the historically important link between North and South America, are virtually unknown.

The other genera of enchytraeids are widely distributed, although some species in each are not. Refined understanding of the phylogenetic relationships of individual species can allow us to discern meaningful biogeographic information even from species that have become widespread due to human intervention.

The taxa in other families recorded here (Table 1) include both widespread and species known only from a few sites in South America as well as two naidids, *Pristina biserrata* Chen, 1940 and *Pristinella acuminata* Liang, 1958, that are recorded only from North and South America and China. As the phylogenetic relationships among the species of different genera of Naididae and Tubificidae and in the other families found are not known, very little can be said about their evolutionary or biogeographic histories.

Acknowledgements

This research was assisted by an individual NSERC research grant ØGP0046464 to K. A. Coates and by grants from the Royal Ontario Museum Science Field Studies Committee. Field work in Peru was also supported by the Javier Prado Museo de Historia Natural of Universidad Mayor de San Marcos in Lima, by their field station IVITA (Instituto Veterinario de Investigaciones Tropicales y de Altura Salamanca) near

Pucallpa, by the Universidad National de Cajamarca, by CUSO-Universidad National de Cajamarca Centro de Desarrollo y Promocion Piscicola, and by Prof. H.-W. Koepcke, owner of Panguana, and the Modena family, the custodians of Panguana. Field work in Guyana was assisted by Youth Challenge International. Numerous individuals from each of these institutions and organizations participated actively and were indispensible to the success of the collecting programs. We thank them all.

References

Brinkhurst, R. O. & B. G. M. Jamieson, 1971. Aquatic Oligochaeta of the World, Oliver and Boyd, Edinburgh, 860 pp.

Brinkhurst, R. O. & M. Marchese, 1987. A contribution to the taxonomy of the aquatic Oligochaeta (Haplotaxidae, Phreodrilidae, Tubificidae) of South America. Can. J. Zool. 65: 3154–3165.

Brinkhurst, R. O. & M. Marchese, 1989. Guide to the freshwater aquatic Oligochaeta of South and Central America. Coleccion CLIMAX 6. Asociacion Ciencias Naturales del Litoral. Santo Tome, Argentina, 179 pp.

Coates, K. A., 1987. Phylogenetic analysis of some Enchytraeidae (Annelida: Oligochaeta):parsimony analysis of structural characters. PhD. dissertation. University of Victoria, Victoria, British Columbia, 392 pp.

Coates, K. A., 1989. Preliminary investigations of hybridization/reticulate evolution in *Guaranidrilus* (Enchytraeidae: Oligochaeta). Hydrobiologia 180: 35–45.

Coates, K. A. & R. J. Diaz, 1988. Description of *Guaranidrilus oregonensis* n.sp. (Oligochaeta: Enchytraeidae), the first species of *Guaranidrilus* from North America, with comments on the genus and other species. Proc. biol. Soc. Wash. 101: 773–783.

Crisci, J. V., M. M. Cigliano, J. J. Morrone & S. Roig-Junent, 1991. Historical biogeography of southern South America. Syst. Zool. 40: 152–171.

Healy, B., 1989. Preliminary report on the Enchytraeidae (Oligochaeta) of West Florida. Hydrobiologia 180: 47–56.

Nielsen, C. O. & B. Christensen, 1959. The Enchytraeidae: critical revision and taxonomy of European species. Studies on Enchytraeidae VII. Nat. Jutl. 8–9: 1–160.

Righi, G. & J. Römbke, 1987. Alguns Oligochaeta da Bolívia e do Peru. Rev. Brasil. Biol. 47: 523–533.

Römbke, J. & M. Verhaagh, 1992. About earthworm communities in a rainforest and an adjacent pasture in Peru. Amazoniana 12: 29–49.

Hydrobiologia **278**: 85–86, 1994.
T.B. Reynoldson & K.A. Coates (eds), Aquatic Oligochaete Biology V.

Oligochaetes of the River Rhine. Preliminary records

Rüdiger M. Schmelz
Zoologisches Institut der Universität Freiburg im Breisgau, Albertstrasse 21 a, D-79104 Freiburg im Breisgau, Germany

Key words: Rhine, oligochaetes

Abstract

Oligochaetes from bottom surveys of the Rhine River from Lake Constance to the Dutch border were identified to species in order to specify the response of communities to river pollution and impoundment. In these preliminary surveys, 37 species of Naididae and Tubificidae were determined, far more than the few "easily identifiable" species normally identified from such surveys. These studies will allow comparison to the oligochaete populations of other large European river systems.

Introduction

The species composition and distribution of oligochaetes in the River Rhine is still largely unknown, although there have been numerous general surveys of the Rhine macrozoobenthos (see the compilation in Kinzelbach, 1990). Except for easily identifiable species such as *Eiseniella tetraedra* (Savigny, 1867), *Stylaria lacustris* (Linné, 1767) or *Stylodrilus heringianus* (Claparède, 1862) the collected material has rarely ever been identified to the species level, even at sites where oligochaetes outnumber all other macrobenthic animals.

Results

My preliminary investigations of the oligochaete fauna of the Rhine River bottom from Lake Constance to the Dutch border (Fig. 1) consisted of identifying specimens originating from three different surveys on the Rhine invertebrate fauna, performed in 1989/1990 (Schmelz & Schöll, 1993). I have found 37 species, the Tubificidae (17 species) and Naididae (12 species) being predominant. *Propappus volki* (Michaelsen, 1905) and *Limnodrilus hoffmeisteri* (Claparède, 1862) are most abundant. The former species is confined to the sand bed of the main channel where the current is strong; this is typical of large rivers in general (Bird, 1982). The latter species dominates in silty areas close to the banks and near dams where the current is low. *Stylodrilus heringianus* is dominant and abundant in the upper sections of the river, from Lake Constance to Basel. Throughout the course of the river *E. tetraedra* is the most frequent species. The distribution of all species except *P. volki* is patchy, but longitudinal and horizontal zonations mainly determined by the current regime and the substrate composition can be found.

Discussion

Future studies will aim at completing the species list with respect to interstitial habitats and back-

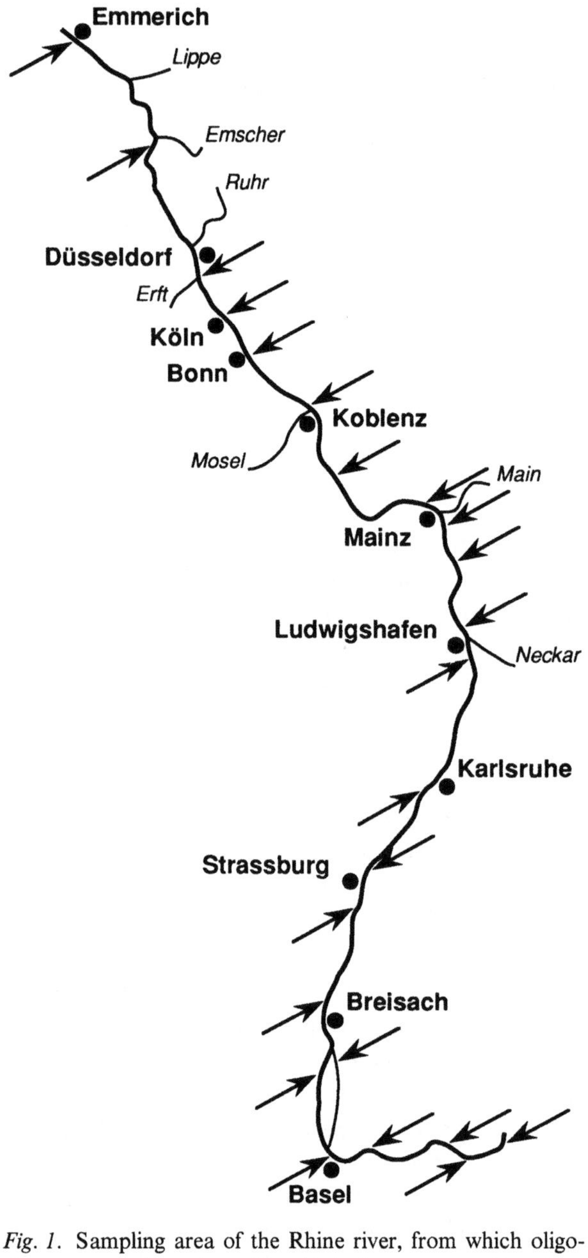

Fig. 1. Sampling area of the Rhine river, from which oligochaete material has been identified. Arrows: sampling sites; tributaries in italics.

waters for comparison with other large rivers such as the Rhône and the Danube (Lafont & Juget, 1976). This is of special zoogeographical interest and will, as the Rhine is the most heavily industrialized river in Europe, help to specify the response of oligochaete communities to pollution and river impoundments.

Acknowledgements

Part of the oligochaete material was kindly provided by the Bundesanstalt für Gewässerkunde (BfG), Koblenz. I also thank the Institut für angewandte Hydrobiologie (IfaH), Litzelstetten, for providing me with the specimens collected in a 1990 survey on the "Hochrhein" (Lake Constance to Basel).

References

Bird, G. J., 1982. Distribution, life cycle and population dynamics of the aquatic enchytraeid *Propappus volki* (Oligochaeta) in an English chalkstream. Holarct. Ecol. 5: 67–76.

Kinzelbach, R. & G. Friedrich (eds), 1990. Biologie des Rheins. Limnologie aktuell 1. Gustav Fischer, Verlag, Stuttgart, 496 pp.

Lafont, M. & J. Juget, 1976. Les Oligochètes du Rhône. I. Relèves faunistiques generaux. Ann. Limnol. 12: 253–268.

Schmelz, R. M. & F. Schöll, 1993. Über Oligochaeten auf der Stromsohle des Rheins. Lauterbornia 12: 1–10.

Hydrobiologia **278**: 87–92, 1994.
T.B. Reynoldson & K.A. Coates (eds), Aquatic Oligochaete Biology V.

Composition and distribution of bottom oligochaete fauna of a north Italian eutrophic lake (Lake Ledro)

S. Casellato & F. Caneva
Dipartimento di Biologia, Universita di Padova, Via Trieste 75, Padova, Italy

Key words: Oligochaeta, eutrophication, lakes, Anoxia

Abstract

The profundal macroinvertebrates, particularly the oligochaetes, of Lake Ledro (Trento, Italy), that has recently undergone eutrophication, were studied.

A statistical approach of random sampling was used to study the distribution and abundance of the oligochaete species. The optimum sample number was calculated from a preliminary sample series. The oligochaete community was made up of five tubificid species, one naidid and one lumbriculid species that on average represents more than 80% of the macrobenthic community. Population density was correlated with depth and decreasing oxygen concentration. The role of *Tubifex tubifex* as a eutrophic, tolerant species was confirmed; and in fact it was the only species found (although at low density) in the deepest and anoxic zone. No comparable data are available for the lake prior to eutrophication, but these data will be valuable for future comparison once a remediation program for the lake has been implemented.

Introduction

Before 1928 Lake Ledro (Table 1) was a natural basin, it has subsequently undergone hydrological modification with the construction of a tunnel connecting it with Lake Garda (Fig. 1).

In recent years Lake Ledro has shown very rapid cultural eutrophication and in a recent study on the trophic state of Lake Ledro, Casellato (1990) showed the total phosphorus and total nitrogen concentrations to be much higher than its former oligotrophic condition. Other variables,

Table 1. Morphometric characteristics of Lake Ledro.

S_L (km^2) (lake area)	2.177
S (km^2) (drainage basin area)	97
S/S_L	44.5
V (volume) (m^3)	75775×10^3
Z (m) (maximum depth)	47
Z (m) (mean depth)	35
T_w (years) (time of whole mixing)	0.8

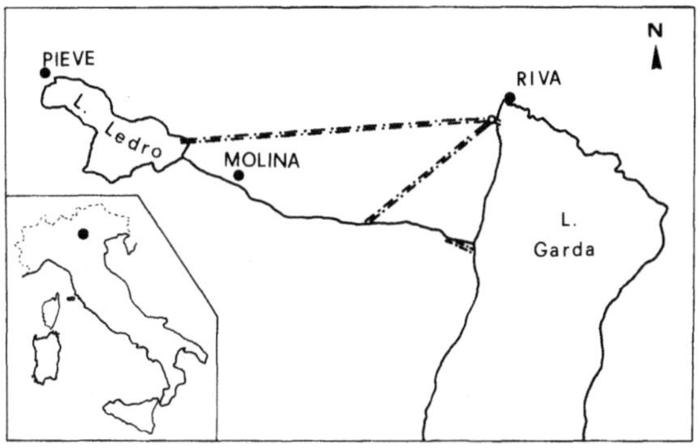

Fig. 1. Lake Ledro with the tunnel connection with Lake Garda.

such as zooplankton and phytoplankton biomass, confirmed the high degree of eutrophication of this lake. Blooms of *Oscillatoria rubescens* occurred frequently during the year. Bottom water in the central area of the lake, at depths greater than 40 m, has become anoxic.

Unfortunately, no data on profundal benthic fauna are available, for the oligotrophic period. Only studies on plankton and physico-chemical water parameters have been reported in the literature (Tonolli, 1956; Barbato, 1977).

Methods

In the first phase of our research we collected seasonal benthic samples at four fixed stations (Fig. 2-A,B,C,D) taking three replicates per station with an Ekman grab (250 cm^2 sampling area). Selected water quality variables (Table 2) were measured once a month. The low benthic community diversity in the central area of the lake required us to collect numerous samples, in a random sampling program, to define the distribution of oligochaetes in this lake.

The optimum number of stations was calculated from the four seasonal station results using the relationship among mean, variance and standard error according to Elliott (1977):

$$n = \frac{S^2}{D^2 x^2},$$

where

n = sample number
$\bar{x}$ = mean
S^2 = variance
D = standard error.

From this relationship we determined that 65 stations be sampled. The locations were selected using a 16.5 × 22.5 cm grid based on random coordinates. Field locations were retraced topographically as accurately as possible.

Some of the 65 points were in the littoral zone and were excluded (Fig. 2) as the littoral zone benthic fauna differs from the profundal fauna. Moreover, it was impossible to sample with the Ekman grab because of the different grain size of sediment and the presence of vegetation and stones.

Oxygen concentration and other chemical parameters in the deepest water layers, in the central part of the lake, were checked along radial transects. Water samples were collected at the bottom every 100 m, starting from the deepest point.

Results

After exclusion of the stations in the littoral zone we collected samples from 49 points (Fig. 3). The 49 stations sampled appear to form three faunal zones. Thirty-four stations form a zone of varying depth from 46 to 35 m, where there are only a few *Tubifex tubifex*. In this area there was complete oxygen depletion, in addition $N-NH_4$ compounds were present at concentrations often higher than 1 mg/1, and sulphide is present with a mean concentration of 0.4 mg/1.

In the second zone, at depths of between 35 and 25 m, other species of tubificids occurred such as *Limnodrilus hoffmeisteri* and *Limnodrilus profundicola*, but *T. tubifex* is still the most abundant and reached its highest population density. As well as oligochaetes we found leeches (*Batrocobdella paludosa*) and a small number of chironomids (*Polypedilum* gr. *nubeculosum* and *Criptochironomus* spp).

In the stations comprising the third zone, between 10 and 25 m, we found the greatest diversity but the macrobenthic community is still poor in species and *T. tubifex* is replaced by *L. hoffmeisteri* as the most abundant species. Other tubificids, *Aulodrilus pluriseta, Psammoryctides barbatus* and *L. profundicola* are present. Rare specimens of unidentified lumbriculids and the naidid *Slavina appendiculata* have been found, at one station near the littoral zone at a depth of 19 m (Table 3).

A gradual reduction not only in the number of species but also of the population density of the whole community was observed with depth.

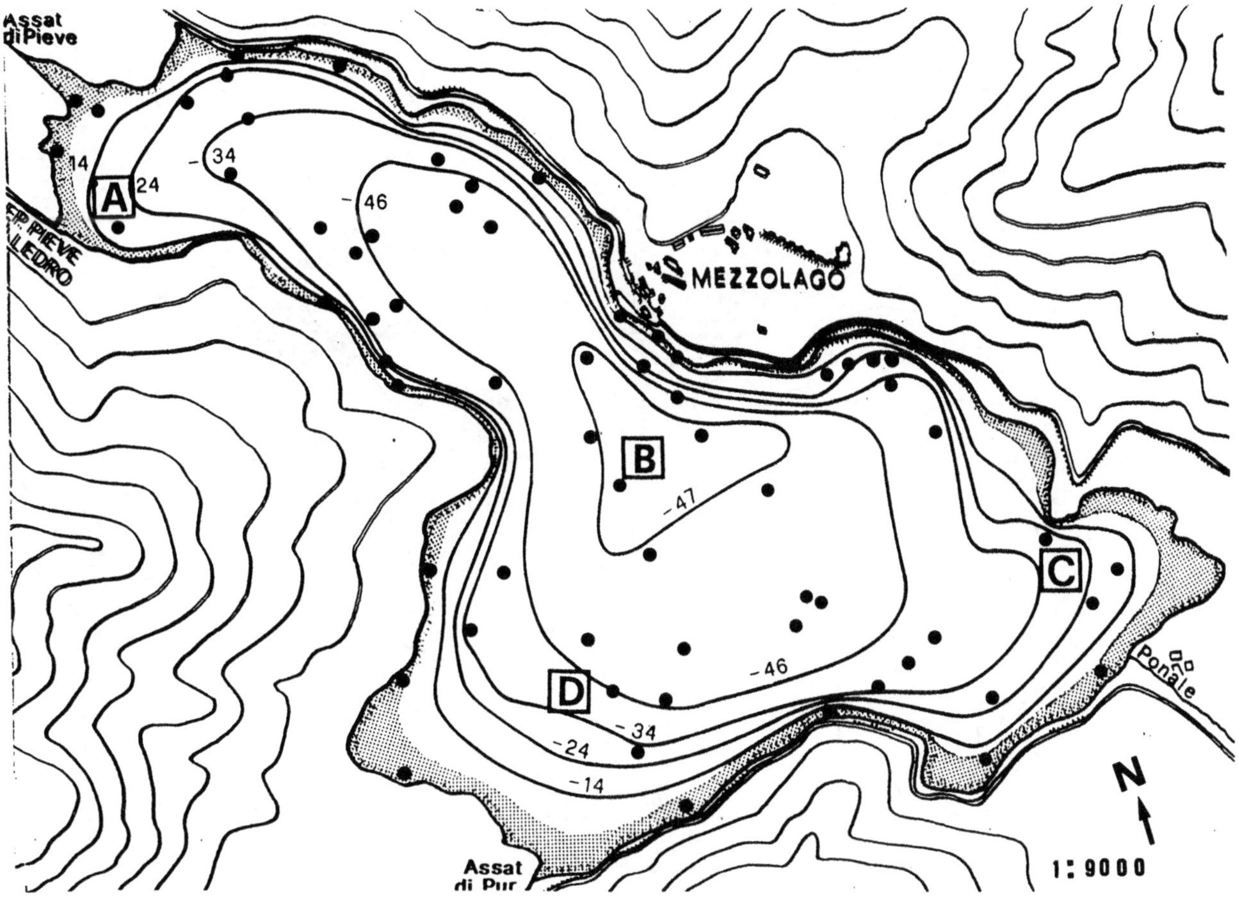

Fig. 2. Location of fixed sampling stations (A, B, C, D) in the first phase of research and random sampling points (65) statistically calculated: 16 of them fall into the littoral zone (shadow drawn) and were excluded. Depths in meter.

Variance analysis and χ^2 for spatial distribution study

To confirm whether the areas in the lake were distinct and if spatial dispersion of oligochaete populations was random or contagious, the Index of Dispersion was tested for departure from unity (Elliott, 1977). This was done on the entire data set (49 stations) and separately for the three station groups. A χ^2 test for agreement with a Poisson series was also performed using the ratio: $\chi^2 = s^2(n-1)/\bar{x}$. These results (Table 4) show ex-

Table 2. Minimum, maximum, mean and confidence limits of important water physic-chemical parameters from monthly samplings, in four different zones of Lake Ledro.

Area	% sat. O_2		BOD5 (mg/l)		$N\text{-}NH_4$ (mg/l)		Tot. N (mg/l)		Tot. P (μg/l)	
	min–max	(mean ±)	min–max	(mean ±)	min–max	(mean ±)	min–max	(mean ±)	min–max	(mean ±)
A)	44.9–100.8	(80.9 ± 32.47)	0.4–1.5	(1.08 ± 0.082)	0.01–1.04	(0.24 ± 0.196)	0.82–5.26	(2.56 ± 1.19)	13–93	(45.5 ± 37.5)
B)	0.0–5.4	(3.13 ± 1.71)	0.5–5.4	(3.0 ± 1.18)	0.41–1.7	(0.908 ± 0.3)	1.1–4.02	(2.68 ± 0.61)	55–375	(164 ± 68.62)
C)	3.9–113.5	(40.35 ± 25.1)	0.3–2.2	(0.85 ± 0.7)	0.0–0.88	(0.212 ± 0.18)	1.04–4.21	(2.38 ± 1.86)	12–507	(164 ± 112.7)
D)	1.6–61.2	(12.25 ± 12.7)	0.4–3.8	(1.39 ± 0.8)	0.14–1.02	(0.53 ± 0.21)	1.03–4.32	(2.46 ± 0.74)	36–331	(116.27 ± 71.8)

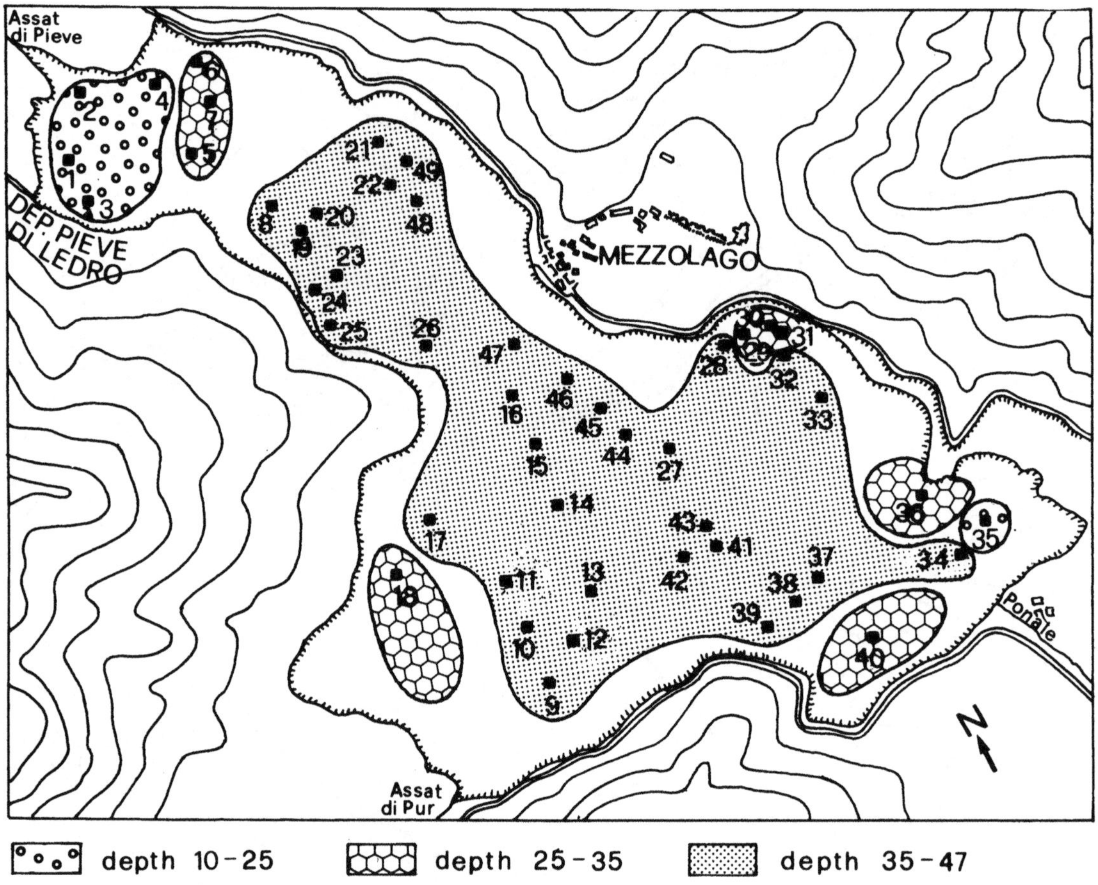

Fig. 3. The sampled stations divided in three groups, according to different depth of bottom.

tremely high values of S^2 and χ^2, particularly when we considered all the stations together. This indicates a strongly contagious distribution. The lowest values were calculated for the zone representing the deepest area, however, many stations in this group have no individuals. If only the stations with animals were considered calculation showed the χ^2 to be much higher than variance. The highest values of χ^2 were calculated for stations in the 10 to 25 m zone, this was expected as these stations occur in a different part of the lake (NE and NW), which differs in the principal chemical parameters, particulary with regard to oxygen concentration (Table 2).

Discussion and conclusions

The community structure of the bottom fauna of Lake Ledro appears to be very simple: the faunistic groups are few and within each group the species number is very low. Below 40 m only one species, *T. tubifex*, the dominant species in the lake, was present and it was most abundant between 25 and 35 m. Unfortunately, we do not have pre-eutrophication data to compare, but the almost complete absence of animals in most of the profundal zone, is a consequence of the recent eutrophication of the lake. This absence of organisms is most likely due to anoxic conditions that occur for most of the year. Low dissolved oxygen and anoxia are an important limiting factor for benthic organisms.

It is known that there are some aquatic oligochaetes able to tolerate prolonged periods of anoxia, such as *T. tubifex*, and other oligochaete species which tolerate high organic loadings. Many researchers have dealt with the use of oligochaetes as indicators (Milbrink, 1978, 1980,

Table 3. Population density (ind/m^2) of oligochetes from Lake Ledro (dashes indicate absence of animals, the stations with complete absence of animals are not reported in the table).

	ZONE 3					ZONE 2									ZONE 1		
Sampling stations	2	1	3	35	4	6	29	30	36	5	40	31	7	18	34	12	11
Depth (m)	12	18	19	19	25	30	31	31	31	33	33	34	35	35	39	42	43
TUBIFICIDAE																	
Tubifex tubifex	787	4134	315	4252	13898	10276	5906	1024	8425	7323	866	551	236	118	79	158	158
Limnodrilus hoffmeisteri	945	3425	1890	5984	–	–	–	118	59	–	–	–	–	79	–	–	–
Limnodrilus profundicola	158	–	–	354	39	–	–	118	–	–	–	–	–	–	–	–	–
Limnodrilus juvenes	158	118	–	1220	512	354	158	–	391	158	–	391	–	–	–	–	–
Aulodrilus pluriseta	276	79	118	79	–	–	–	–	–	–	–	–	–	–	–	–	–
Psammoryctides barbatus	279	–	–	906	591	–	–	–	–	–	–	–	–	–	–	–	–
LUMBRICULIDAE n.d.	–	–	79	–	–	–	–	–	–	–	–	–	–	–	–	–	–
NAIDIDAE																	
Slavina appendiculata	–	–	79	–	–	–	–	–	–	–	–	–	–	–			
Tot. ind.	2603	7756	2481	12795	15040	10630	6064	1260	8875	7481	866	942	236	197	79	158	158

1983, 1987; Lang & Lang-Dobler 1980; Sarkka, 1987). These authors stress the tolerance of *T. tubifex*, which is the dominant species of the profundal benthic community of other European lakes (Ruggiu *et al.*, 1972; Probst, 1987). *Limnodrilus hoffmeisteri*, which is also known as a tolerant and eutrophic species, is not able to colonize the deepest zones of this lake but it is dominant in samples collected between 10 and 25 m. All the other oligochaete species found are typical of enriched habitats, except *P. barbatus*, but they are less abundant.

From a quantitative perspective, the gradual decrease in population density with depth is a phenomenon which has been observed in many other lakes which are not eutrophic, *e.g.*, Lake Garda (Bonomi, 1974). In Lake Ledro this phenomenon is pronounced, below 25 m *P. barbatus* and *A. pluriseta* were not found. *L. profundicola*, although reported as a species occurring in deeper water (Probst reported it at depth in Lake Constance, 1987), is absent below 30 m, which is related to oxygen depletion which occurs below that depth.

The spatial dispersion of benthic macroinvertebrates in this lake, shows a high degree of contagion as indicated by the high variance and χ^2. A contagious distribution has not always been observed in other lakes. In Lunzer Untersee the oligochaete populations showed a tendency to

Table 4. Statistical analysis was performed using numerical data, for each dredge sample, not transformed in ind/m^2.

	n	$\bar{x}$	S^2	$I = (S^2/\bar{x})$	X^2
Station group 10–25 m	5	206	22.05	107.08	428.33
Station group 25–35 m	9	101.2	11.187	110.62	884.13
Station group 35–46 m	35	0.31	1.22	3.89	132.29
Station group 35–46 m (considering only stations with animals)	3	3.66	2.33	0.636	1.273
All stations	49	39.84	8425.08	211.48	10,151.22

randomness (Ravera, 1966) and the variance and χ^2 values calculated by Ruggiu and Saraceni (1972) for Lake Mergozzo were not as high as in Lake Ledro. Evidently, this lake, besides presenting a strong depth gradient, also has a horizontal one.

Environmental factors rarely follow a random distribution but the spatial distribution in Lake Ledro is very contagious and this phenomenon could be related to local water quality conditions.

Acknowledgements

This study was supported by a financial grant of Provincia Autonoma di Trento.

References

Barbato, G., 1977. Note idrobiologiche sul Lago di Ledro. Ann. Mus. Civ. St. Nat. Brescia 14: 92–109.

Bonomi, G., 1974. Indagini sul Lago di Garda. Benton profondo. IRSA CNR, 18: 211–223.

Casellato, S., 1990. Il Lago di Ledro: Valutazione del suo stato trofico. Provincia Autonoma di Trento, 111 pp.

Elliot, J. M., 1977. Some methods for the statistical analysis of samples of benthic invertebrates. Freshwat. Biol. FBA 25. 160 pp.

Lang, C. & B. Lang-Dobler, 1980. Structure of tubificid and lumbriculid communities, and three indices of trophy based upon these communities, as descriptors of eutrophication level of Lake Geneva (Switzerland). In R. O. Brinkhurst & D. G. Cook (eds), Aquatic Oligochaete Biology. Plenum Publishing Corporation, New York: 457–470.

Milbrink, G., 1978. Indicator communities of oligochaetes in Scandinavian lakes. Verh. int. Ver. Limnol. 20: 2406–2411.

Milbrink, G., 1980. Oligochaete communities in pollution biology. The European situation with special reference to lakes in Scandinavia. In R. O. Brinkhurst & D. G. Cook (eds), Aquatic Oligochaete Biology. Plenum Publishing Corporation, New York: 433–455.

Milbrink, G., 1983. An improved environmental index based on the relative abundance of oligochaete species. Hydrobiologia 102: 89–97.

Milbrink, G., 1987. Biological characterization of sediments by standardized tubificid bioassays. Hydrobiologia 155: 267–275.

Probst, L., 1987. Sublittoral and produndal Oligochaeta fauna from the lake Constance (Bodensee-Obersee). Hydrobiologia 155: 277–282.

Ravera, O., 1966. Stability and pattern of distribution of the benthos in different habitats of an alpine oligotrophic lake: Lunzer Untersee. Verh. int. Ver. Limnol. 16: 233–244.

Ruggiu, D. & C. Saraceni, 1972. A statistical study of the distribution and abundance of the bottom fauna of a lake undergoing accelerated eutrophication (Lake Mergozzo, N. Italy). Mem. Ist. ital. Idrobiol. 29: 169–187.

Sarkka J., 1987. The occurrence of oligochaetes in lake chains receiving pulp mill waste and their relation to eutrophication on the trophic scale. Hydrobiologia 155: 259–265.

Tonolli, V., 1956. Il Lago di Ledro nelle sue vicende di lago naturale, lago serbatoio e lago rifornito. Mem. Ist. ital. Idrobiol. 9: 25–48.

Hydrobiologia **278**: 93–105, 1994.
T.B. Reynoldson & K.A. Coates (eds), Aquatic Oligochaete Biology V.

Habitat preferences and species associations of shallow-water marine Tubificidae (Oligochaeta) from the barrier reef ecosystems off Belize, Central America

Robert J. Diaz [1] & Christer Erséus [2]
[1] *Virginia Institute of Marine Science, School of Marine Science, College of William and Mary, Gloucester Point, Virginia 23062 U.S.A.;* [2] *Department of Invertebrate Zoology, Swedish Museum of Natural History, Box 50007, S–104 05, Stockholm, Sweden*

Key words: aquatic Oligochaeta, Tubificidae, Caribbean fauna, animal–sediment relations, coral reefs, gutless fauna

Abstract

The marine tubificid oligochaete fauna of Belize is the most diverse that has been described for the Caribbean Sea, with records for over 50% of all known Caribbean species. Tubificids were sampled at 77 stations around the outer barrier of the coral reefs off the Belize mainland. A total of 1,529 individuals representing 52 species were collected. Species distribution patterns were controlled mainly by sediment type and the presence or absence of mangroves. Six species (*Phallodrilus molestus, Tubificoides bermudae, Inanidrilus leukodermatus, Heterodrilus flexuosus, Bathydrilus formosus, Smithsonidrilus hummelincki*) with diverse life history characteristics were common and had broad habitat preferences being found in various types of heterogeneous sandy sediments. *Heterodrilus flexuosus* was the only common species in these samples known only in Belize. The other five common species range from Belize to Florida or Bermuda. Most other species have narrower habitat preferences and are limited to specific habitats. For example, one group of gutless species was restricted to organically enriched fine to medium sands, a group of Limnodriloidinae occurred only in mud around mangrove cays, and a group of meiofaunal species were restricted to saline groundwater.

Introduction

The total species diversity of marine Tubificidae (Oligochaeta) has surpassed that of the freshwater Tubificidae. Compared to the freshwater species, however, little is known of the ecology and habitat preferences of the marine species. Some work has been done on the autecology of individual tubificid species from temperate areas (Brinkhurst, 1964; Hunter & Aurthur, 1978; Birtwell & Arthur, 1980; Diaz, 1980; Bamber & Spencer, 1984; Erséus & Diaz, 1989) and on temperate marine tubificid assemblages (Cook, 1971; Baker, 1984; Davies, 1985; and Diaz *et al.*, 1987) but little is known of subtropical and tropical assemblages. Taxonomic studies of shallow subtropical and tropical marine tubificid oligochaetes indicate that they are very diverse and abundant (Erséus, 1990 and references therein).

The marine tubificid fauna of the Caribbean region (Bermuda, Florida, Barbados, and Belize) has received the most taxonomic attention with over 100 species described. Belize appears to be the most species rich area of the Caribbean with about 50% of these species (Erséus, 1990). In this paper we have taken the data on the 'Belizean' fauna collected by Erséus (1990) and analyzed them for ecological patterns and habitat preferences.

The belize barrier reef ecosystems

The barrier reef ecosystem off the Belize mainland is the largest continuous reef in the Caribbean Sea. It is 10 to 32 km wide and 250 km long (Rützler & Macintyre, 1982a). Carrie Bow Cay, the center of our collection area, sits on top of the barrier reef proper. The area to the west of Carrie Bow Cay forms a shallow *Thalassia* dominated lagoon, less than 5 m deep (Rützler & Macintyre, 1982a). To the east of Carrie Bow Cay the reef system rapidly grades from reef crest to inner fore reef, 2 to 12 m, to outer fore reef, 12 to 40 m, to outer barrier reef platform over a distance of less than a kilometer (Burke, 1982). Further eastward, about 22 km, is a large platform reef (Glovers Reef) that protects the Carrie Bow Cay area from open ocean waves. This protection is a principle factor that has allowed the development of highly diverse communities in the Carrie Bow Cay area (Burke, 1982).

The tide around Carrie Bow Cay has a mean range of 15 cm and is mixed semidiurnal (Kjerfve *et al.*, 1982). Channels to the north and south of Carrie Bow Cay are the main exchange routes between the reef system and ocean. While tidal currents are weak the almost constant wind-wave action supports the movement of materials on and off the reef. The complex physical structures of the reef system interacting with the water currents and protection from ocean waves have led to an extremely high diversity of habitats. This physical diversity of habitats is further enhanced by the organisms occupying and building the reefs. Details of the physical and biological aspects of the Belize barrier reef ecosystems can be found in (Rützler & Macintyre, 1982b).

Materials and methods

Erséus (1990), with the assistance of others, collected marine tubificid oligochaetes at 85 stations around the outer barrier of the coral reefs off the Belize mainland. While the purpose of these collections was primarily taxonomic, sufficient material was collected to evaluate ecological and habitat relationships. At each station approximately equal amounts of sediment were examined for Tubificidae. Because of taxonomic problems with immature individuals only sexually mature individuals were preserved and identified to species. Qualitative notes were also made on the physical characteristics of the site.

Numerical classification was used to classify species and stations into groups based on species abundance and distribution patterns. The flexible sorting algorithm was used ($\beta = -0.25$) with the quantitative form of the Bray-Curtis similarity coefficient (Boesch, 1977). Both species and station analyses were done on a reduced set of the original data. Species with only one or two collection occurrences were dropped from the analysis. To evaluate the strength of the species and station groups nodal analysis was conducted (Lambert & Williams, 1962). Nodal analysis is a post classification analysis that compares how well species station groups are formed.

Constancy and fidelity statistics (Lambert & Williams, 1962) were used to estimate the strength of the species-station groups. Constancy is a measure of species occurrence and can be expressed as

$$C_{ij} = a_{ij} \cdot 100\% / (n_i n_j) ,$$

where a_{ij} is the total number of occurrences of species in species group i and site group j, and n_i and n_j are total number of species and stations in the respective groups being compared. Expressed as a percentage, 100% constancy means every species group occurred at least once at every station in the site group being considered. Fifty percent constancy is when any combination of species occurrences is half the total possible number of species × station occurrences.

Fidelity measures the average frequency of occurrence of species in a group relative to the rest of the collection. It estimates the preference or avoidance of species for a particular site group and can be expressed as

$$F_{ij} = (a_{ij} \Sigma_j n_j)/(n_j \Sigma_j a_{ij})$$

using the same terms as constancy. The summation terms go across all site groups ($j = 1,...,J$) for a particular species group (i). When fidelity is equal to 1 the species group (i) has no preference for the site group (j), less than 1 indicates avoidance, and greater than 1 preference. Programs COMPAH and NODAL of the Virginia Institute of Marine Science, College of William and Mary were used for these analyses.

Results

Of the 90 stations collected by Erséus (1990) we used 72. Eight of the stations dropped were collected at different times or with different methods. The other ten stations were dropped because of their low information content, as explained below. The physical descriptions of the retained stations are given in Erséus (1990). The location of each sampling area is in Fig 1.

Faunal composition

A total of 1,529 individuals representing 52 tubificid species in 15 genera were collected (Table 1). About half of the genera (7 of 15) were represented by one or two species, accounting for about 20% of the total species. The four most speciose genera, accounting for about half (27 of 52) of the species, were *Heterodrilus*, *Phallodrilus*, *Olavius*, and *Limnodriloides*. The remaining four genera (*Inanidrilus*, *Bathydrilus*, *Thalassodrilides*, and *Smithsonidrilus*) represented about 30% (16 of 52) of the species.

Of the 52 species in the collection 21 (40%) occurred at only one or two stations and were considered rare. A third of the species (17) occurred at three to six different stations and were considered uncommon. Of the remaining species, 14 occurred at eight or more stations and were common. Three species (*Heterodrilus flexuosus* Erséus, 1990, *Phallodrilus molestus* Erséus, 1988, *Inanidrilus leukodermatus* (Giere, 1979)) were very common, occurring at 30 or more stations. These three species also accounted for 42% of the total individuals collected (Table 1). Their average

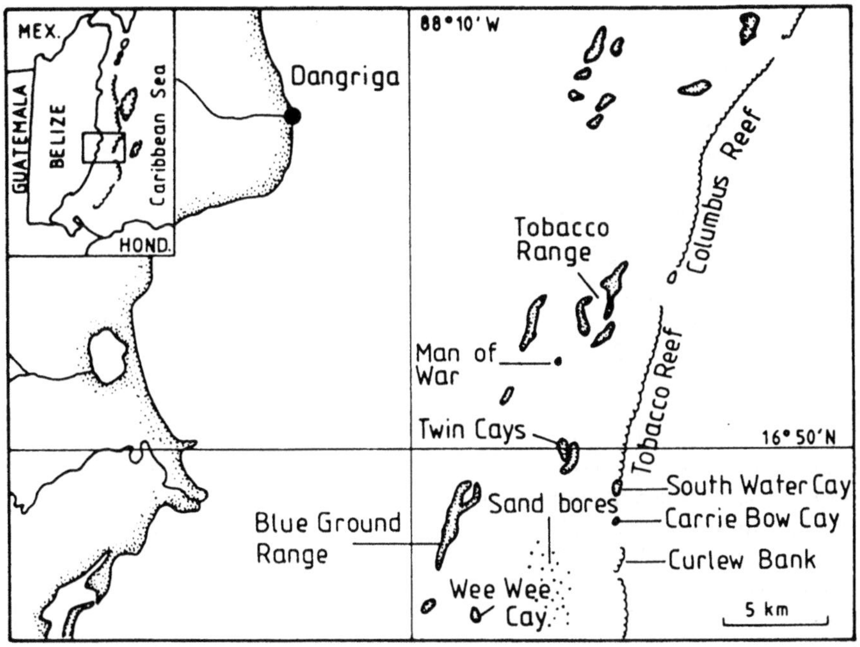

Fig. 1. Location of sampling areas on the barrier reef ecosystems off the coast of Belize.

Table 1. Marine Tubificidae collected at 77 stations located throughout the outer barrier of the coral reef ecosystems off the coast of Belize.

	Station occurrences	Number of individuals
Subfamily Rhyacodrilina		
Heronidrilus gravidus Erséus, 1990	5	8
Heterodrilus flexuosus Erséus, 1990	37	158
Heterodrilus rarus Erséus, 1990	2	5
Heterodrilus paucifascis Milligan, 1987	9	97
Heterodrilus modestus Erséus, 1990	10	34
Heterodrilus pentcheffi Erséus, 1981	17	48
Heterodrilus quadrithecatus (Erséus, 1981)	1	1
Heterodrilus ersei (Giere, 1979)	3	6
Subfamily Phallodrilinae		
Coralliodrilus rugosus Erséus, 1990	1	1
Phallodrilus compactus Erséus, 1990	5	9
Phallodrilus singularis Erséus, 1990	1	1
Phallodrilus deminutius Erséus, 1979	6	8
Phallodrilus molestus Erséus, 1988	33	199
Phallodrilus vicinus Erséus, 1990	1	1
Phallodrilus nastus Erséus, 1990	1	1
Phallodrilus bipartitus Erséus, 1990	1	4
Aktedrilus longitubularis Finogenova & Shurova, 1980	1	1
Aktedrilus parvithecatus (Erséus, 1978)	2	10
Jamiesoniella athecata Erséus, 1981	1	1
Inanidrilus leukodermatus (Giere, 1979)	30	284
Inanidrilus scalprum Erséus, 198	12	18
Inanidrilus belizensis Erséus, 1984	4	4
Inanidrilus aduncosetis Erséus, 1984	3	6
Inanidrilus reginae Erséus, 1990	1	2
Olavius imperfectus Erséus, 1984	6	40
Olavius tenuissimus (Erséus, 1979)	3	4
Olavius finitimus Erséus, 1990	2	5
Olavius vacuus Erséus, 1990	6	12
Olavius (Olavius) tantulus Erséus, 1984	6	18
Olavius (Olavius) pravus Erséus, 1990	1	3
Olavius (Olavius) longissimus (Giere, 1979)	2	8
Duridrilus tardus Erséus, 1983	13	30
Bathydrilus vetustus Erséus, 1990	1	3
Bathydrilus egenus Erséus, 1990	1	9
Bathydrilus adriaticus (Hrabê, 1971)	3	7
Bathydrilus formosus Erséus, 1986	19	96
Subfamily Limnodriloidinae		
Thalassodrilides gurwitschi (Hrabê, 1971)	5	20
Thalassodrilides bruneti Erséus, 1990	4	60
Thalassodrilides ineri (Righi & Kanner, 1979)	2	6
Tectidrilus bori (Righi & Kanner, 1979)	3	3
Limnodriloides monothecus Cook, 1974	1	2
Limnodriloides anxius Erséus, 1990	2	8
Limnodriloides uniampullatus Erséus, 1982	2	2
Limnodriloides sacculus Erséus, 1990	4	5

Table 1. (Continued)

	Station occurrences	Number of individuals
Limnodriloides barnardi Cook, 1974	8	18
Limnodriloides rubicundus Erséus, 1982	6	35
Smithsonidrilus luteolus (Erséus, 1983)	18	40
Smithsonidrilus appositus Erséus, 1990	15	23
Smithsondrilus hummelincki (Right & Kanner, 1979)	20	65
Smithsondrilus involutus Erséus, 1990	3	7
Subfamily Tubuficinae		
Tubificoides bermudae Råsmark & Erséus, 1986	16	92
Tubificoides parviductus Helgason & Erséus, 1987	1	1

abundance was 4.3, 6.0, and 9.5 individuals/ station, respectively. Other species that did not occur as frequently but when they occurred had high average station abundance were *Thalassodrilides bruneti* Erséus, 1990 (15.0 individuals/ station), *Heterodrilus paucifascis* Milligan, 1987 (10.8), and *Bathydrilus egenus* Erséus, 1990 (9). Species with the highest individual station abundances were the gutless *I. leukodermatus* with 64 individuals at station 13, *T. bruneti* with 52 individuals at station 61, and *H. paucifascis* with 42 individuals at station 6. It must be noted, however, that the sampling was semi-quantitative.

For 47 species there were sufficient individuals to get an estimate of size (Table 2). The majority of the species were small ranging from 5 to 10 mm in length. Eight species were of a meiofaunal size and belong to genera that are among the smallest for marine Tubificidae (*Aktedrilus*, *Phallodrilus*, *Coralliodrilus*). The exception was the genus *Heterodrilus* which also contains very large species (Table 2). Two of the largest species were gutless, with *Olavius (Olavius) longissimus* (Giere, 1979) reaching a maximum length of 46 mm. The maximum length for *Olavius vacuus* Erséus, 1990 was 20 mm. The other three large species belong to genera that contain a broad size range of species (*Heterodrilus*, *Bathydrilus*, *Thalassodrilides*).

Species distribution patterns

Cluster analysis was done using a reduced data matrix which included only species that occurred at three or more stations. This reduced the data to 31 species and 74 stations. Eight stations were dropped because they only had 'rare' species (<3 occurrences). If retained, the prevalence of 'rare' species in the collection would tend to weken the cluster analysis and make pattern recognition difficult. After the deletion of rare species two additional stations (20, 34) were dropped from the analysis because they ended up having only a single species. The species analysis at the nine group level (Fig. 2) and the station analysis at the 12 group level (Fig. 3) were interpretable relative to species characteristics and sediment type.

Species groups A and B were the most common species in the collection. The three species (*P. molestus*, *Tubificoides bermudae* Råsmark & Erséus, 1986, *I. leukodermatus*) that composed group A had highest constancy and fidelity with site groups 1 and 2 which were characterized by clean fine-medium sands a meter or less in depth (Figs 4 & 5). A few stations in these site groups were deeper (2, 44) or had finer sediments (2, 26). Species group B (*H. flexuosus*, *Bathydrilus formosus* Erséus, 1986, *Smithsonidrilus hummelincki* Righi & Kanner, 1979) contained the most broadly distributed species. Group B was strongly

Table 2. Size categories and other life history characters of marine tubificids collected on the barrier reef ecosystems off the coast of Belize. Measurements for each species are maximum preserved length in mm and maximum reported collection depth from Erséus (1990). A depth of 0 indicates intertidal or saline groundwater.

	Length	Depth	Gutless	Papillae	Brackish	Circumtropical
UP TO ABOUT 4 mm						
Heterodrilus quadrithecatus	3	15				
Coralliodrilus rugosus	3	0				
Phallodrilus vicinus	4	< 1				
Phallodrilus nasutus	2	0				
Phallodrilus compactus	3	2				
Phallodrilus bipartitus	4	0				YES
Aktedrilus longitubularis	2	0				
Aktedrilus parvithecatus	3	0				YES
ABOUT 5–10 mm						
Heterodrilus flexuosus	10	24				
Heterodrilus rarus	8	1				
Heterodrilus modestus	6	9				
Heterodrilus pentcheffi	11	39				
Phallodrilus deminutius	6	14				
Phallodrilus molestus	7	70				YES?
Inanidrilus leukodermatus	11	16	YES			
Inanidrilus scalporum	11	27	YES			
Inanidrilus belizensis	9	24	YES			
Inanidrilus aduncosetis	13	8	YES			
Olavius imperfectus	9	8	YES			
Olavius tenuissimus	10	141	YES			
Olavius finitimus	7	2	YES			
Olavius (Olavius) pravus	7	1	YES			
Olavius (Olavius) tantulus	6	27	YES			
Bathydrilus adriaticus	10	32				YES
Thalassodrilides gurwitschi	8				YES	YES
Thalassodrilides bruneti	5	0				
Limnodriloides monothecus	9	583			YES	YES
Limnodriloides anxius	9	20			YES	
Limnodriloides sacculus	5	1				
Limnodriloides barnardi	6	150				
Smithsonidrilus luteolus	9	24				
Smithsonidrilus appositus	6	3				
Smithsonidrilus involutus	10	1				
ABOUT 10–15 mm						
Heronidrilus gravidus	11					
Heterodrilus paucifascis	12	39				
Duridrilus tardus	16	39		YES		YES?
Bathydrilus vetustus	12	6				
Tectidrilus bori	11	70				YES?
Limnodriloides rubicundus	16	74				YES
Smithsonidrilus hummelincki	14	21				
Tubificoides bermudae	14	15		YES		
MORE THAN 15 mm						
Heterodrilus ersei	25	9				

Table 2. (Continued)

	Length	Depth	Gutless	Papillae	Brackish	Circumtropical
Olavius vacuus	20	58	YES			
Olavius (Olavius) longissimus	46	2	YES			
Bathydrilus formosus	18	20				YES
Bathydrilus egenus	20	2				
Thalassodrilides ineri	25	2				

associated with site group 5 which was primarily coarse sand and rubble, with the exception of two stations that were muddy (32, 35). Both species groups A and B were poorly represented at site groups 9, 10, and 11. Site group 9 was all coarse sand and rubble around Carrie Bow Cay on the outer barrier side. Site group 10 was all fine sand muddy stations around mangrove cays. Site group 11 was also fine grained muddy sediments and included the deepest stations.

Species group C (*Phallodrilus deminutius* Erséus, 1979, *Smithsonidrilus luteolus* (Erséus, 1983a), *Duridrilus tardus* Erséus, 1983b) was primarily associated with site group 8, which was mainly outer barrier coarse sands with secondary heterogeneity consisting of *Halimeda* flakes, rubble, and *Thalassia*. The species in group C did not occur or appeared to avoid, based on fidelity, site groups with muddy fine grained stations, including groups 4, 10, and 12. Species group D was composed of three *Heterodrilus* species (*paucifascis*, *modestus* Erséus, 1990, *ersei* Giere, 1979) that were strongly associated only with site group 9 which was outer barrier coarse sand rubble stations that likely were exposed to higher wave energies. Group D species also did not occur at site groups with muddy stations, including groups 2, 3, 4, 10, and 12. Species group E was five species (*Smithsonidrilus appositus* Erséus, 1990, *Smithsonidrilus involutus* Erséus, 1990, *Heronidrilus gravidus* Erséus, 1990, *Limnodriloides barnardi* Cook, 1974, *Limnodriloides sacculus* Erséus, 1990) that were associated with only site group 6. This site group was unique being composed of outer barrier coarse sand and *Halimeda* flake sediments. There was only one other coarse sand *Halimeda* flake station in the collection not in site group 6, it was station 30 in group 7. There were four other *Halimeda* flake stations with finer grain sediments. Species group E did not occur at site groups 3, 7, or 9 and was only weakly associated with one other site group, 4 (Fig. 5).

With the exception of *Heterodrilus pentcheffi* Erséus, 1981a species group F was exclusively composed of gutless species. The species in this group were highly associated with the shallow cleaner fine to medium sand stations of site group 1. Group F was, to a lesser degree, also associated with site group 11 which was composed of shallow and deep fine to medium sand stations that were intermediate in organic content. Group F did not occur at the muddy mangrove cays site group 10 or at site group 12. If *H. pentcheffi* is excluded from group F, the remaining species, all gutless, also did not occur at site groups 5 and 9, both of which were coarse-rubble station groups.

Species group G was a weak association of *Phallodrilus compactus* Erséus, 1990 and *Bathydrilus adriaticus* (Hrabě, 1971) that had very high fidelity to the more organic finer sand stations of site group 11. This species group did not occur at half the site groups (2, 6, 7, 8, 9, 10). Species group H was also not well-defined but included two gutless species (*Olavius tenuissimus* Erséus, 1979, *Inanidrilus aduncosetis* Erséus, 1984) and *Tectidrilus bori* (Righi & Kanner, 1979) that were moderately associated with the coarser sediment site group 9 and the fine sand-muddy site group 11. Species group H did not occur at site groups 2, 3, 4, 7, and 8.

Species group I was the most unique of all the species groups. It was composed of three Lim-

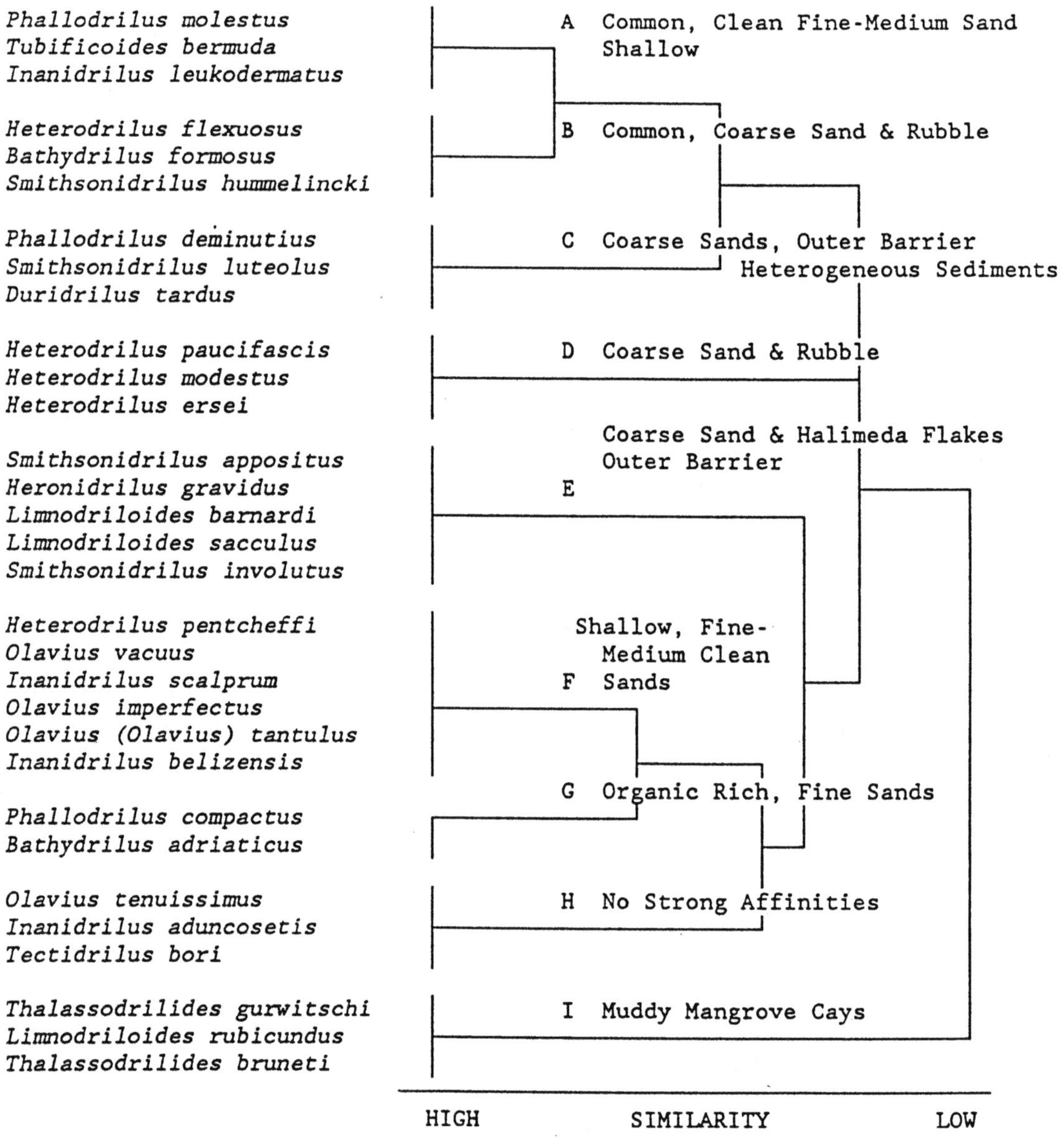

Fig. 2. Species groups defined from cluster analysis of Belize marine tubificid oligochaetes. See text for analysis detail and group characteristics.

nodriloidinae species (*Thalassodrilides gurwitschi* (Hrabê, 1971), *T. bruneti*, *Limnodriloides rubicundus* Erséus, 1982) that had very high constancy and fidelity to the muddy mangrove cays site group 10. Outside site group 10, the only other occurrences of *T. bruneti* (sandy station 43) and *L. rubicundus* (muddy station 52) were at site group 12, which was a mix of outer barrier and mangrove stations.

Rare species

Among the rare species (1 or 2 station occurrences) there were four gutless species, all of which occurred at stations with heterogeneous sediments. *Inanidrilus reginae* Erséus, 1990 occurred in a muddy *Thalassia* bed (station 49) giving it an affinity with the gutless species of species group H. *Olavius (Olavius) pravus* Erséus, 1990, *O.*

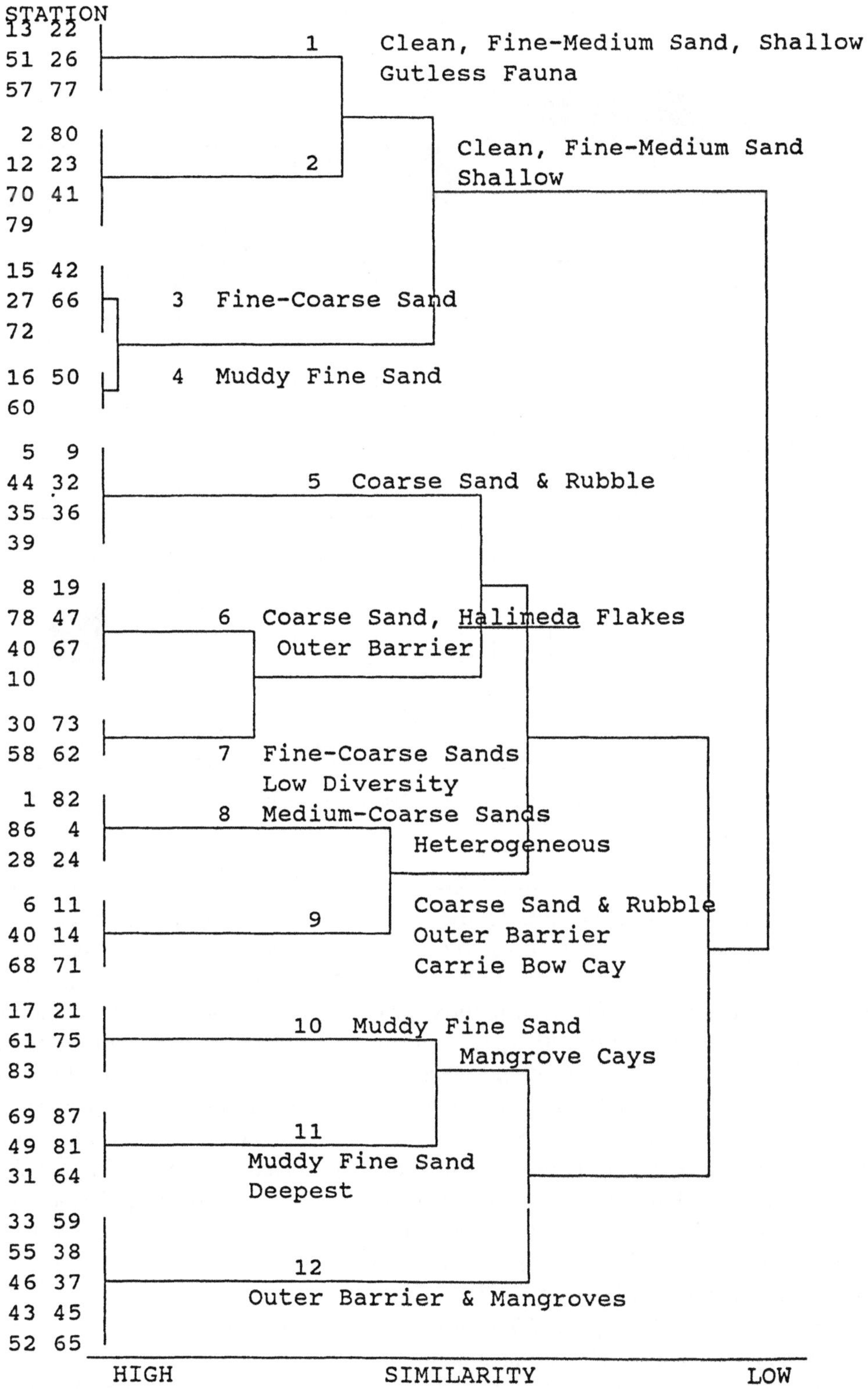

Fig. 3. Site groups defined from cluster analysis of Belize marine tubificid oligochaetes. See text for detail of analysis and group characteristics.

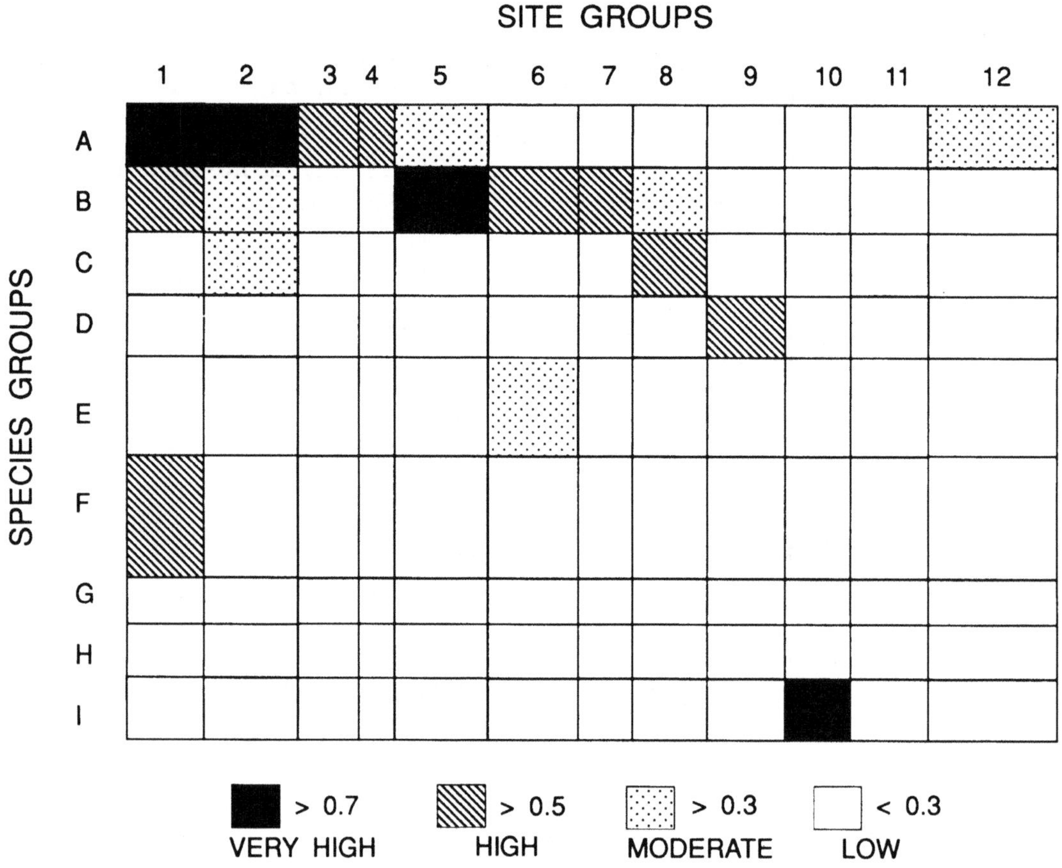

Fig. 4. Nodal consistency in a two-way table of species and site groups as described in Figs 2 & 3.

(O.) longissimus and *Olavius finitimus* Erséus, 1990 were all found in coarse sediments (stations 22, 31, 57, 68, 71) giving them affinity to species group F.

Seven rare species were found in muddy sediments giving them an affinity with species group I and site group 10. *Thalassodrilides ineri* (Righi & Kanner, 1979) occurred in muddy *Thalassia* and mangrove areas (stations 17 and 83), as did *Limnodriloides monothecus* Cook, 1974 (49). The other five muddy sediment species were *Limnodriloides anxius* Erséus, 1990 (52, 61), *Limnodriloides uniampullatus* Erséus, 1982 (61, 75), *Heterodrilus quadrithecatus* (Erséus, 1982a) (50), *Heterodrilus rarus* Erséus, 1990 (22, 25), and *Phallodrilus singularis* Erséus, 1990 (2). The small meiofaunal *Aktedrilus parvithecatus* (Erséus, 1978) (7, 20), *Aktedrilus longitubularis* Finogenova & Shurova, 1980 (54), *Jamiesoniella athecata* Erséus, 1981b (54), *Phallodrilus nasutus* Erséus, 1990 (18), *Phallodrilus bipartitus* Erséus, 1990 (7), and *Coralliodrilus rugosus* Erséus, 1990 (18) occurred in a range of sediments from organic rich mud to medium-coarse sands, but only at or near the high tide line in saline groundwater. None of these five species were found subtidally. The remaining four single-station occurring species were all found in medium to coarse sands and included *Phallodrilus vicinus* Erséus, 1990 (11), *Bathydrilus vetustus* Erséus, 1990 (36), *B. egenus* (9), and *Tubificoides parviductus* Helgason & Erséus, 1978 (82).

Discussion

For the Caribbean, the marine Tubificidae fauna of Belize is the most diverse known with records for over 50% of all Caribbean species. This in-

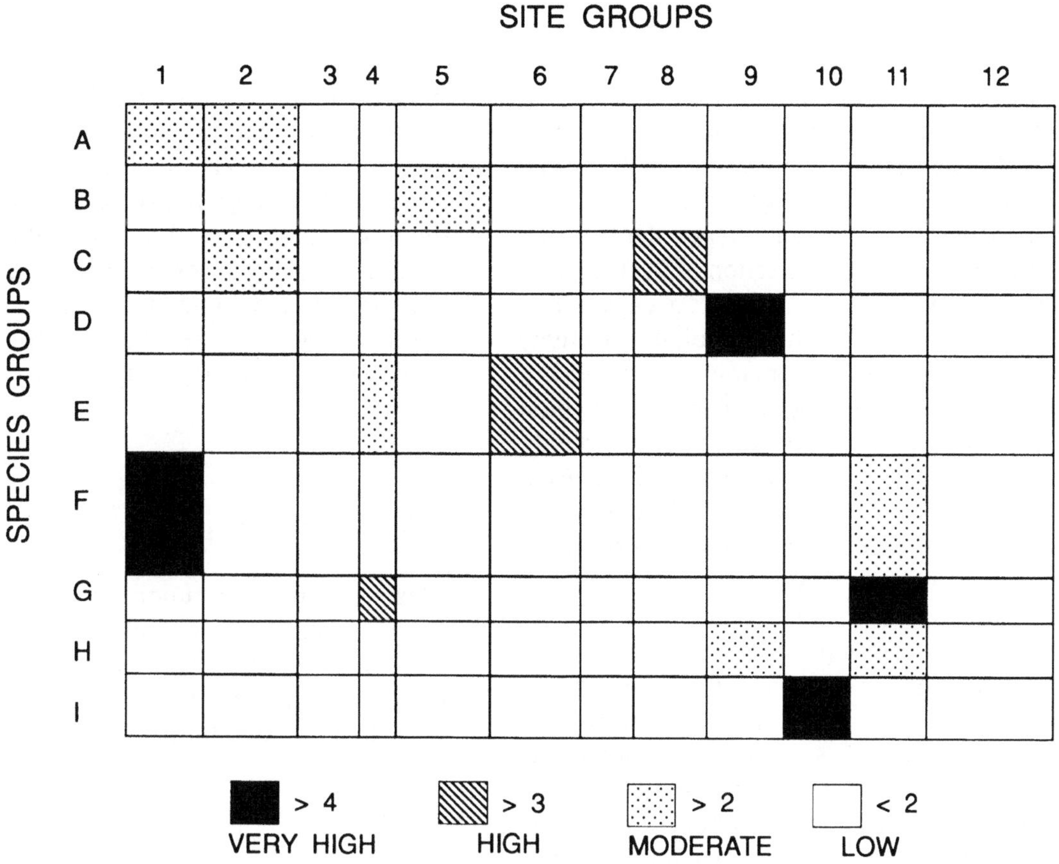

Fig. 5. Nodal fidelity in a two-way table of species and site groups as described in Figs 2 & 3.

cludes the 52 species reported on here plus seven others found by Erséus (1990), that occurred at other stations, compared to a total Caribbean fauna of about 100 species. For other areas of the Caribbean that have been sampled with comparable intensity, Florida is second to Belize with approximately 40% of the species, Bermuda third with 35% and Barbados fourth with about 20%.

The spatial distribution of Belizean marine tubificid species appears to be typical of other benthic fauna with species occurrence and abundance following a negative binomial or Poisson distribution. About 40% of the species were considered rare being represented by a few individuals and occurring at only one or two stations (Table 1). The three most common species (*H. flexuosus*, *P. molestus*, *I. leukodermatus*) occurred at about 60% of the stations and were about 40% of all individuals collected. The habitat preferences of these three abundant species were varied. *Phallodrilus molestus* and *I. leukodermatus* preferred shallow fine-medium clean sands and *H. flexuosus* coarse sand and rubble habitats. *Inanidrilus leukodermatus* is a gutless species that inhabits the aerobic-anaerobic sediment interface (Giere *et al.*, 1982).

The high species diversity and lack of dominance by a single species in the Belize tubificid fauna may be due to a combination of habitat stability through time and habitat heterogeneity. The occurrence of several species per genera also points to a degree of microheterogeneity or niche specialization within or between habitat types (Cohen, 1978). Five genera had five to seven species occurrences (*Heterodrilus*, *Phallodrilus*, *Inanidrilus*, *Olavius*, *Limnodriloides*). *Phallodrilus*, however, is currently being split into a number of smaller genera (Erséus, 1992).

While there were common species that had broad spatial distributions (*P. molestus*, *T. bermudae*, *I. leukodermatus*, *H. flexuosus*, *B. formosus*, *S. hummelincki*, species groups A and B) none of them occurred in all habitat types. Their broad distribution was more a reflection of the predominance of sandy substrates around the cays. These species, however, did show habitat selectivity being completely absent from mud and mangrove associated sites. The highest degree of habitat specificity was shown by the three Limnodriloidinae species (*T. gurwitschi*, *T. bruneti*, *L. rubicundus*, group I) that occurred exclusively in mud and mangrove sites, except for a single occurrence of *T. bruneti* at a sandy station.

The gutless species that were not rare were divided into two groups by their habitat preferences. The majority of them (*O. vacuus*, *Olavius imperfectus* Erséus, 1984, *O. (O.) longissimus*, *Inanidrilus scalprum* Erséus, 1984, *Inanidrilus belizensis* Erséus, 1984) occurred in shallow fine to medium clean sands and did not occur in coarse-rubble sediments. The second group (*O. tenuissimus* and *I. aduncosetis*, species group H) were associated primarily with coarse *Halimeda* sands and rubble sediments and did not occur in other types of sediment. The habitat preferences of the most common gutless species, *I. leukodermatus*, included all the habitats in which the other gutless species occurred with highest affinity for fine to medium clean sands.

The meiofaunal species (*A. parvithecatus*, *A. longitubularis*, *J. athecata*, *P. nasutus*, *P. bipartitus*, *C. rugosus*) consistently occurred at or near the high tide line in saline groundwater. These habitats are particularly favored by meiofaunal oligochaetes from all geographic areas.

The high degree of habitat preference exhibited by the Belizean marine tubificid fauna is not unusual for the Tubificidae. Diaz *et al.* (1987) found a similar high degree of microhabitat partitioning among the Tubificidae of the middle Atlantic Bight. The tubificids in Cape Cod Bay were also very sediment-type specific (Cook, 1971). Slight changes in the sediments in both these areas resulted in changes to the tubificid fauna. The one species that occurred in Cape Cod, the Middle Atlantic Bight, and Belize, *L. monothecus*, was consistently found in fine sands. *Olavius tenuissimus* and *H. pentcheffi* found both in the Middle Atlantic Bight and Belize occupied similar medium to coarse grained habitats. While there are no other Caribbean or tropical studies for comparison the taxonomic literature (summarized by Erséus, 1990) indicates that a high degree of microhabitat partitioning is typical for marine Tubificidae.

Acknowledgements

This paper is contribution no. 401 of the Smithsonian Institution's Caribbean Coral Reef Ecosystems Program, supported in part by the Exxon Corporation and contribution no. 1770 of the Virginia Institute of Marine Science.

References

Baker, H. R., 1984. Diversity and zoogeography of marine Tubificidae (Annelida, Oligochaeta) with notes on variation in widespread species. Hydrobiologia 115: 191–196.

Bamber, R. N. & J. F. Spencer, 1984. The benthos of a coastal power station thermal discharge canal. J. mar. biol. Ass. U.K. 64: 603–623.

Boesch, D. F., 1977. Application of numerical classification in ecological investigations of water pollution. Special scientific report 77, Virginia Institute of Marine Science, Gloucester Pt., Virginia, 114 pp.

Brinkhurst, R. O., 1964. Observations on the biology of the marine oligochaete *Tubifex costatus*. J. mar. biol. Ass. U.K. 44: 11–16.

Birtwell, I. K. & D. R. Arthur, 1980. The ecology of tubificids in the Thames estuary with particular reference to *Tubifex costatus* (Claparede). In R. O. Brinkhurst & D. G. Cook (eds), Aquatic Oligochaete Biology. Plenum Press, New York: 331–381.

Burke, R. B., 1982. Reconnaissance study of the geomorphology and benthic communities of the outer barrier reef platform, Belize. In K. Rützler & I. G. Macintyre (eds), The Atlantic Barrier Reef Ecosystem at Carrie Bow Cay, Belize, I. Structure and Communities. Smithson. Contr. Mar. Sci. 12: 509–527.

Cohen, J. E., 1978. Food webs and niche space. Princeton University Press, Princeton, New Jersey, 189 pp.

Cook, D. G., 1971. The Tubificidae (Annelida, Oligochaeta) of Cape Cod Bay. II. Ecology and systematics with the description of *Phallodrilus parviatriatus* nov. sp. Biol. Bull. 141: 203–221.

Cook, D. G., 1974. The systematics and distribution of marine Tubificidae (Annelida: Oligochaeta) in the Bahia de San Quintin, Baja California, with description of five new species. Bull. S. Cal. Acad. Sci. 73: 126–140.

Davis, D., 1985. The Oligochaeta of Georges Bank (NW Atlantic), with descriptions of four new species. Proc. biol. Soc. Wash. 98: 158–168.

Diaz, R. J., 1980. Ecology of tidal freshwater and estuarine Tubificidae (Oligochaeta). In R. O. Brinkhurst & D. G. Cook (eds), Aquatic Oligochaete Biology. Plenum Press, New York: 319–330.

Diaz, R. J., C. Erséus & D. F. Boesch, 1987. Distribution and ecology of Middle Atlantic Bight Oligochaeta. Hydrobiologia 155: 215–225.

Erséus, C., 1978. Two new species of the little-known genus *Bacescuella* Hrabê (Oligochaeta, Tubificidae) from the North Atlantic. Zool. Scr. 7: 263–267.

Erséus, C., 1979. Taxonomic revision of the marine genus *Phallodrilus* Pierantoni (Oligochaeta, Tubificidae), with descriptions of thirteen new species. Zool. Scr. 8: 187–208.

Erséus, C., 1981a. Taxonomic revision of the marine genus *Heterodrilus* Pierantoni (Oligochaeta, Tubificidae). Zool. Scr. 9: 97–111.

Erséus, C., 1981b. Marine biological investigations in the Bahamas. 20. A new species of *Jamiesoniella* (Oligochaeta, Tubificidae). Sarsia 66: 161–162.

Erséus, C., 1982. Taxonomic revision of the marine genus *Limnodriloides* (Oligochaeta, Tubificidae). Verh. naturw. Ver. Hamb. (N.F.) 25: 207–277.

Erséus, C., 1983a. Taxonomic studies of the marine genus *Marcusaedrilus* Righi & Kanner (Oligochaeta, Tubificidae), with descriptions of seven new species from the caribbean area and Australia. Zool. Scr. 12: 25–36.

Erséus, C., 1983b. *Duridrilus tardus* gen. et. sp.n., a marine tubificid (Oligochaeta) from Bermuda and Barbados. Sarsia 68: 29–32.

Erséus, C., 1984. Taxonomy and phylogeny of the gutless Phallodrilinae (Oligochaeta, Tubificidae), with description of one new genus and twenty-two new species. Zool. Scr. 13: 239–272.

Erséus, C., 1986. Marine Tubificidae (Oligochaeta) at Hutchinson Island, Florida. Proc. biol. Sci. Wash. 99: 286–315.

Erséus, C., 1988. Taxonomic revision of the *Phallodrilus rectisetosus* complex (Oligochaeta: Tubificidae). Proc. biol. Soc. Wash. 101: 67–71.

Erséus, C., 1990. The marine Tubificidae (Oligochaeta) of the barrier reef ecosystems at Carrie Bow Cay, Belize, and other parts of the Caribbean Sea, with descriptions of twenty-seven new species and revisions of *Heterodrilus*, *Thalassodrilides*, and *Smithsonidrilus*. Zool. Scr. 19: 243–303.

Erséus, C., 1992. A generic revision of the Phallodrilinae (Oligochaeta, Tubificidae). Zool. Scr. 21: 5–48.

Erséus, C., & R. J. Diaz, 1989. Population dynamics of *Tubificoides amplivasatus* (Oligochaeta, Tubificidae) in the Oresund, Denmark. Hydrobiologia 180: 167–176.

Finogenova, N. P. & N. M. Shurova, 1980. A new species of the genus *Aktedrilus* (Oligochaeta, Tubificidae) of the lit toral zone of the Sea of Japan. In Pribrezhnyi plankton i bentos severnoi chasti yaponskogo morya. Akademia Nauk SSSR, Vladivostok: 65–69. (In Russian.)

Giere, O., 1979. Studies on marine Oligochaeta from Bermuda, with emphasis on new *Phallodrilus* species (Tubificidae). Cah. Biol. mar. 20: 301–314.

Giere, O., G. Liebezeit & R. Dawson, 1982. Habitat conditions and distributions pattern of the gutless oligochaete *Phallodrilus leukodermatus*. Mar. Ecol. Prog. Ser. 8: 291–299.

Helgason, G. V. & C. Erséus, 1987. Three new species of *Tubificoides* (Oligochaeta, Tubificidae) from the North-East Atlantic and notes on the geographic variation in the circumpolar *T. kozloffi*. Sarsia 72: 159–169.

Hunter, J. B. & D. R. Arthur, 1978. Some aspects of the ecology of *Peloscolex benedeni* Udekem (Oligochaeta: Tubificidae) in the Thames estuary. Estuar. coast. mar. Sci. 6: 197–208.

Hrabê, S., 1971. On new marine Tubificidae of the Adriatic Sea. Scr. Fac. Sci. Nat. Univ. Brno. (Biol.3) 1: 215–226.

Kjerfve, B., K. Rützler, & I. G. Macintyre, 1982. Tides at Carrie Bow Cay. In K. Rützler & I. G. Macintyre (eds), The Atlantic Barrier Reef Ecosystems at Carrie Bow Cay, Belize, I. Structure and Communities. Smithson. Contr. Mar. Sci. 12: 47–53.

Lambert, J. M. & W. T. Williams, 1962. Multivariate methods in plant ecology. IV. Nodal analysis. J. Ecol. 50: 775–802.

Milligan, M. R., 1987. Marine Tubificidae (Oligochaeta) from Puerto Rico with descriptions of two new species, *Tubificoides aguadillensis* and *Heterodrilus paucifascis*. Proc. biol. Soc. Wash. 100: 480–489.

Righi, G. & E. Kanner, 1979. Marine Oligochaeta (Tubificidae and Enchytraeidae) from the Caribbean Sea. Stud. Fauna Curaçao 58: 44–68.

Råsmark, B. & C. Erséus, 1986. A new species of *Tubificoides* (Oligochaeta: Tubificidae) from Bermuda and Bahamas. Proc. biol. Soc. Wash. 99: 612–615.

Rützler, K. & I. G. Macintyre, 1982a. The habitat distribution and community structure of the Barrier Reef complex at Carrie Bow Cay, Belize. In K. Rützler & I. G. Macintyre (eds), The Atlantic Barrier Reef Ecosystems at Carrie Bow Cay, Belize, I. Structure and Communities. Smithson. Contr. Mar. Sci. 12: 9–47.

Rützler, K. & I. G. Macintyre, (eds), 1982b. The Atlantic Barrier Reef Ecosystem at Carrie Bow Cay, Belize, I. Structure and Communities. Smithson. Contr. Mar. Sci. 12: 1-539.

Hydrobiologia **278**: 107–110, 1994.
T.B. Reynoldson & K.A. Coates (eds), Aquatic Oligochaete Biology V.

Communities of oligochaetes in mountain streams of Poland

Elzbieta Dumnicka
Institute of Freshwater Biology, Polish Academy of Sciences, Slawkowska 17, 31-016 Krakow, Poland

Key words: Oligochaete communities, mountain streams, Poland, Tubificidae, Naididae, Lumbriculidae

Abstract

This study compares the structure of oligochaete communities in mountain streams of Poland. These streams can be divided into three groups. First, those dominated by the Enchytraeidae (*Mesenchytraeus armatus*), second where the Naididae, *Nais bretscheri*, *Nais alpina*, *Nais communis* or *Nais variabilis* prevail and third, where the Lumbriculidae, represented by *Stylodrilus heringianus*, and in one case by *Stylodrilus parvus*, are dominant. The nature of the stream bottom is the main factor determining the structure of the community, but geographic location is also important.

Introduction

The bottom fauna of mountain streams is usually dominated by Ephemeroptera, Plecoptera, Trichoptera, and Chironomidae with oligochaetes normally accounting for only a small percentage of the total fauna. However, understanding their distribution is important particularly in cases of advanced eutrophication or when pollution from urban growth and increased agriculture modify mountain stream communities (Szczesny, 1974).

To date there has been no comprehensive study of oligochaete communities in mountain streams of Poland. This study presents results on the composition and structure of oligochaete communities in streams from various mountain ranges of Poland and discusses some of the factors important in the distribution of the fauna.

Fig. 1. The distribution of mountain ranges in Poland.

Results

The results are a synthesis of earlier studies (Dumnicka, 1976, 1982; Dumnicka & Kukula, 1990) and data from other authors (Kahl, 1983, 1986; Kasprzak, 1976, 1977, 1979; Kasprzak & Szczesny, 1976; Szczesny, 1974, 1990) on oligochaetes from mountain streams.

It is difficult to compare these results directly as various samplers were used. Thus, to allow comparison results are expressed as percentage per sample. The communities are characterized by the group of species present in the greatest numbers and a species is considered dominant if it comprises more than 10% of the oligochaete fauna.

The general characteristics of the study streams for which information on the benthic communities are available are shown in Table 1 and based on their oligochaete fauna the streams have been divided into three groups.

The first group is formed from one high mountain stream flowing from Pod Chlopkiem Pass almost 2000 m above sea level (Table 1). The oligochaete fauna is very poor in this stream with *Mesenchytraeus cf. armatus* as the dominant species representing up to 96% of the community (Table 2) and includes species of Enchytraeidae (*Henlea perpusilla, Fridericia spp.*).

The second group of eight streams are lower altitude with their sources between 1400 and 300 m above sea level (Table 1) in which the Naididae dominate. In six of the stream, *Nais variabilis* is the most common species and *N. communis* is also common. These streams originate in the Pieniny and Male Pieniny Mountains, near the source of the Raba River (Olszowy Stream in Gorce Mountains) and Czarna Woda Stream in the Swietokrzyskie Mountains. Also in this group are some of the streams flowing from mountain lakes e.g. Rybi Potok Stream in the Tatry Mountains. These streams have a stoney substrate with a strong current and sandy or muddy sediments occur only on the banks.

The third group of five streams are characterized by an oligochaete fauna dominated by the Lumbriculidae, either *Lumbriculus variegatus* or *Stylodrilus heringianus*. These streams are also lower altitude (1100 to 350 m). The streams of the Karkonosze and Izerskie Mountains are characterized by low pH and gravel-sand type of bottom. The dominant species is *Lumbriculus variegatus* but *S. heringianus* and *Stylodrilus brachystylus* also occur. In the Swiętokrzyskie Mountains, *L. variegatus* together with *S. heringianus* comprise 93% of all the oligochaete fauna in the upper part of Lubrzanka River, which has a substrate similar to that of streams of the Karkonosze Mountains. However, *L. variegatus* was not found in the small streams of this range, with various levels of acidity (minimum pH 3.3 to 6.0) and stony substrates. The few oligochaetes

Table 1. Characteristic of the examinated streams in southern Poland.

Stream	Mountain range	Altitude	Substrate	Stream order	pH
Stream from the pass pod Chłopkiem	Tatry	1900	rocky rubble	I	?
Rybi Potok	Tatry	1180–1140	boulders, stones, pebbles, sandy flooded areas	I	6,2–6,9
Czarny Potok	Tatry	1620–1460	boulders, large stones, sand	I	6,6
Sucha Woda, Roztoka	Tatry	1500–1000	boulders, stones, pebbles and sand	I	7,2–7,4
Stream from the Czarny Staw lake below Rysy	Tatry	1580–1400	stones and rock partially overgrown with moss	I	6,6–6,7
Raba	Gorce	1180–990	stones and gravel	I–II	7,4–8,7
Kamionka	Małe Pieniny	840–580	stones, gravel, clay earth	I–II	7,6–8,0
Biała Woda	Małe Pieniny	690–650	boulders, large stones	I	7,3–8,0
Czarna Woda	Swiętokrzyskie	280	small stones, gravel and sand	I	5,3–6,3
Kryniczanka	Beskid Sądecki	650–600	stones and sand	I–II	7,1–8,0
Wołosatka	Bieszczady	1120–720	stones of different sizes	I–II	?
Lubrzanka	Swiętokrzyskie	350	sand and mud	I	3,2–4,8
Kamienna	Karkonosze	650–600	gravel and sand	I	?
Some streams near the Swieradöw Zdröj town	Izerskie	675–500	gravel and sand	I–III	3,8–5,5

Table 2. Oligochaete in mountain streams (percentage shares).
1 – streams from the pass pod Chłopkiem
2 – stream from the Czarny Staw lake below Rysy
3 – streams nearŚwieradów Zdrój–Izerskie Mts.

Species	I	II								III				
	1	Rybi Pot.	Czar. Pot.	S.W. Rozt.	2	Raba	Kamionka	Biała Woda	Czar Woda	Krynicz	Woło sat.	Lubrzan.	Kamienna	3
Mesenchytraeus armatus (Lev.)	96,0	14,6	4,6	2,0	1,2	4,6		+			+			
Enchytraeidae*	4,0	7,2	6,6	4,6	1,2	17,0	2,1	5,5		5,0	1,1			+
Nais communis Pig.		5,3	3,7	4,5	8,9	67,2	9,9	+		4,3	+	+	+	
– bretscheri Mich.				4,1				43,5		+	+			
– alpina Sperb.		+					28,4	3,9		2,0	+			
– variabilis Pig.		71,3	66,7	70,5	88,2		14,8				+			
– elinguis Mull.								+		35,0			22,4	78,0
– spp. O.F. Muller									85,0					
Stylodrilus heringianus Clap.	+		9,5			+				31,0		43,4	1,3	3,0
– parvus Hr. et Cern.											66,3			
– brachystylus Hrabe													4,0	13,8
Lumbriculus variegatus (O.F. Mull.)			1,5	+	+	+						49,6	36,8	1,2
Propappus volki Mich.			+	11,1	+	+				18,9				
Haplotaxis gordioides (Hartm.)		+		+	+	+				+	+	+	26,9	

* Including Cernosvitoviella spp, Cognettia spp, Henlea spp and Fridericia spp.
\+ Species present.

found in such streams were Enchytraeidae and *S. heringianus*.

Rarely does the structure of the oligochaete community remain constant along the entire length of a stream. The streams in the Bieszczady Mountains are atypical in that *Stylodrilus parvus* is dominant along the entire length. In contrast, in a short reach of the Biala Woda, a stream in the Male Pieniny Mountains, where the distance between stations was 1.5 km, the structure of the community changed considerably (Table 3). Similarly, in the Olszowy stream in the Gorce Mountains there was a change in community structure.

Table 3. Changes of percentages share of oligochaete taxa in mountain streams.

Taxa	Stream station	Wołosatka		Biała Woda		Raba	
		1	6	1	2	0	1
	After	Dumnicka & Kukuła 1990		Dumnicka 1982		Kasprzak & Szczęsny 1974	
Stylodrilus parvus Hr. & Cern.		66.3	91.8				
Cognettia sphagnetorum (Vejd.)		10.7					
Enchytraeidae juv.		7.0		3.9	0.01		
Nais bretscheri Mich.				29.4	7.4		
– alpina Sperb.			1.3	2.0	33.8		
Nais communis Pig.				0.2	7.1	67.2	10.4
Propappus volki Mich.			2.9			0.9	5.2
Cernosvitoviella spp. N. et Ch.						1.8	6.2
Mesenchytraeus spp. Eis.		6.0				4.6	26.0
Henlea spp. Mich.		5.0				15.2	3.1
Fridericia spp. Mich.		5.0					3.1

Discussion

Most previous studies describing the distribution of oligochaete species stresses their connection with habitat. Some of the results obtained in these mountain streams confirm these findings, e.g., in streams with a sandy-gravel substrate and low pH *L. variegatus* occurred, while in stony substrate *S. heringianus* was dominant. The Tubificidae were rare, as species of this family prefer muddy habitats that were not included in this study. *Haplotaxis gordioides*, a species typical of cold-water mountain streams, does not dominate in any of the streams covered by this study although it occurs in almost all the streams in low numbers.

According to the river continuum concept, the streams studied are either 1st and 2nd order, in which detritivorous species should predominate. Some streams had such an oligochaete fauna in particular the high altitude mountain streams and those medium altitude streams dominated by Lumbriculidae. However, in many streams of the same order the algae-eating Naididae dominated. The geographical location of a given stream may also affect the structure of the oligochaete community, e.g., in south-western Poland the common dominant species is *S. brachystylus* while in south-eastern Poland *S. parvus* is dominant.

In the high mountain streams, the genera of Enchytraeidae found were the same as those in the soils of the surrounding regions e.g. *Mesenchytraeus spp., Cognettia spp.* primarily terrestrial, and *Cernosvitoviella spp.* that are primarily aquatic.

References

Dumnicka, E., 1976. Oligochaetes (Oligochaeta) of some streams of the High Tatra Mts and of the River Bialka Tatrzanska. Acta Hydrobiol. 18: 305–315.

Dumnicka, E., 1982. Stream ecosystems in mountain grassland (West Carpathians) 9. Oligochaeta. Acta Hydrobiol. 24: 391–398.

Dumnicka, E. & K. Kukula, 1990. The communities of oligochaetes of the Wolosatka and Terebowiec streams (the Bieszczady National Park, south-eastern Poland). Acta Hydrobiol. 32: 423–435.

Kahl, K., 1983. Materials to the fauna of Oligochaeta of the Holy Cross Mountains (Gory Swiętokrzyskie) and of the Nida Lowland (Niecka Nidzianska). Fragm. Faun. 28: 23–37.

Kahl, K., 1986. The Oligochaeta of the Lubrzanka River in the Swiętokrzyskie Mts (Central Poland). Fragm. Faun. 30: 35–43.

Kasprzak, K., 1976. Materials to the knowledge of the Oligochaeta fauna of Karkonosze Mts. and Izerskie Mts. Przegl. Zool. 20: 323–325.

Kasprzak, K., 1977. New data on the Oligochaeta of the Gorce Mts. and Beskid Sądecki Mts. Przegl. Zool. 21: 27–31.

Kasprzak, K., 1979. Oligochaetes (Oligochaeta) of the Pieniny Mountains. II. Naididae, Tubificidae, Haplotaxidae, Lumbriculidae, Branchiobdellidae. Fragm. Faun. 24: 57–80.

Kasprzak, K. & B. Szczesny, 1976. Oligochaetes (Oligochaeta) of the River Raba. Acta Hydrobiol. 18: 75–87.

Szczesny, B., 1974. The effect of sewage from the town of Krynica on the benthic invertebrates communities of the Kryniczanka stream. Acta Hydrobiol. 16: 1–29.

Szczesny, B., 1990. Benthic macroinvertebrates in acidified streams of the Swietokrzyski National Park (Central Poland). Acta Hydrobiol. 32: 156–169.

Hydrobiologia **278**: 111–123, 1994.
T.B. Reynoldson & K.A. Coates (eds), Aquatic Oligochaete Biology V.

Oligochaeta in *Spartina* stems: the microdistribution of Enchytraeidae and Tubificidae in a salt marsh, Sapelo Island, USA

Brenda Healy [1] & Keith Walters [2]
[1] *Department of Zoology, University College Dublin, Belfield, Dublin 4, Ireland*
[2] *Marine Institute, University of Georgia, Sapelo Island, Georgia 31327, USA*

Key words: enchytraeids, tubificids, *Spartina alterniflora*, Aerenchyma tissue, wet funnel extraction, stem dissection, log-linear models

Abstract

The distribution and abundance of Enchytraeidae and Tubificidae in and around *Spartina alterniflora* plants in a tidal salt marsh on Sapelo Island, Georgia, USA were studied using two different sampling techniques: wet funnel extraction and stem dissection. At least 80% of all worms inhabited leaf sheaths at the bases of *S. alterniflora* plants, and densities were low in sediment, root and surface debris samples. Oligochaete densities were dependent on the position within the marsh, the height on stems and the stage of sheath decay. Six predominant species were identified and included *Marionina appendiculata*, *Marionina spartinae*, *Marionina waltersi*, *Marionina paludis*, and *Monopylephorus parvus*. Individual species were distributed differently on stems and enchytraeids were more common than tubificids on standing-dead and further up *S. alterniflora* stems. Estimates of oligochaete densities in salt marsh habitats are increased dramatically when the numbers of worms on stems are considered. Possible advantages of the stem microhabitat are discussed in relation to the biology and ecology of oligochaetes.

Introduction

Research on marine oligochaetes historically has focused on sediment-associated species, although worms also have been identified from intertidal algal wrack (Backlund, 1945). Benthic oligochaetes are distributed worldwide in a variety of habitats, and maximum densities of individual species typically range between 1×10^4 and 6×10^6 individuals m^{-2} (Giere & Pfannkuche, 1982). In salt marsh sediments, oligochaetes can represent a major fraction of the benthic invertebrate fauna (Cammen, 1979; Kneib, 1984; Jackson *et al.*, 1985).

The emphasis on benthic oligochaetes in previous studies has neglected a potentially significant habitat for worms in many nearshore, coastal environments: above-ground vegetation. Intertidal areas in North and South America and western Europe are covered by extensive salt marsh vegetation (Chapman, 1977). Along the southeastern US Atlantic and Gulf coasts alone there are more than 2.5×10^6 acres of salt marshes (Knox, 1986). The surface area associated with the first 10 cm of salt marsh plant stems can exceed the total sediment surface area, and invertebrate densities on plants often exceed densities in the sediments (Walters, pers. comm.). The aggregation of benthic fauna around marsh plant culms has been documented (Van Dolah, 1978; Radar, 1984), but only recently have the marine invertebrates living directly on salt marsh grasses been studied (Burke, 1976; Fiers & Rutledge, 1990; Rutledge & Fleeger, in press).

In this paper, we investigate the oligochaete assemblage associated with the stems of the salt marsh cordgrass *Spartina alterniflora* Loisel 1907, a habitat previously ignored in studies of oligochaete biology and ecology. *S. alterniflora* is the predominant, intertidal salt marsh grass in the southeastern US. The distribution and abundance of Enchytraeidae and Tubificidae from microhabitats within and around *S. alterniflora* plants are described using two different sampling strategies.

Materials and methods

Sample site

All samples were collected in May 1991 from Kenan Field, an intertidal salt marsh site located along the Duplin River on the western side of Sapelo Island, Georgia, USA (see Kneib, 1987). The substrate consisted of fine silt which remained moist at all stages of the tide during the study period. The salinity of water flooding the marsh varies between an autumn minimum of 15% and a summer maximum of 28% (Pomeroy *et al.*, 1972). In the Duplin River watershed, *S. alterniflora* comprises 95% of the marsh flora (Reimold *et al.*, 1973). *Spartina alterniflora* extends from low to high intertidal elevations within the Kenan Field site, and three distinct vegetation zones may be distinguished: creekside marsh, low marsh, and high marsh (see Pomeroy & Wiegert, 1981). Vegetation zones correspond with the height of *S. alterniflora* plants and the frequency of flooding. The creekside marsh contains the tallest (> 2 m) plants flooded on the lowest of high tides, and the high marsh contains the shortest (0.25 m) plants flooded only on the highest of high tides. Differences in plant growth forms among zones are phenotypic, resulting from a gradient in soil water movement that produces differences in nitrogen availability and salinization (Smart, 1982; Wiegert *et al.*, 1983). Samples of *S. alterniflora* plants were collected from within each vegetation zone.

Spartina alterniflora plants

Spartina alterniflora has the characteristic growth form of grasses, with a hollow stem continuing underground as a rhizome bearing adventitious roots and leaves arising from nodes on the stem. Each leaf consists of two parts, a sheath which is wrapped around the stem and a flat blade carried at an angle to the stem (Fig. 1A). Lower leaf blades become senescent and die while still attached to the growing plant, but sheaths persist for some time. Young plants have several layers of leaf sheaths around the base of the stem. Larger plants produce more leaves, and therefore more sheaths, than smaller plants. In the southeastern US, aerial parts of plants live for up to 18 months (Morris & Haskin, 1990), but stems may remain standing for a year or more after dying. Dead stems typically do not have associated leaf sheaths, except in the early stages of senescence, and are approximately half the diameter of live stems. At the time of sampling, *Spartina* plants at all levels were at three different stages: growing plants without flowering spikes, young shoots with unformed leaf blades, and standing, dead plants. Growing plants on the high marsh were less that 70 cm in height, those on the low marsh were up to 1.1 m, while creekside plants reached 1.5 m. Densities of live plants, young shoots and dead stems were determined during our study within each marsh vegetation zone.

A unique feature of *S. alterniflora* is its well-developed aerenchyma system (Fig. 1B-F). The system consists of a series of parallel tubes (gas passages) running the length of the leaf blade and continuous with wider ducts in the sheath, stem, rhizomes, and roots (Anderson, 1974; Armstrong, 1979). The tubes are cavities in the mesophyll that lack a structured lining and vary in size and shape according to the position in the sheath and plant growth conditions. In creekside plants, tube diameters range from 0.35 to 0.60 mm tangentially and 0.60 to 1.00 mm radially. Tube passages become narrower towards the lateral borders of the sheath and are smaller in smaller leaves. Transverse diaphragms composed of a spongiform layer of irregular cells occur at 4 to 8 mm intervals

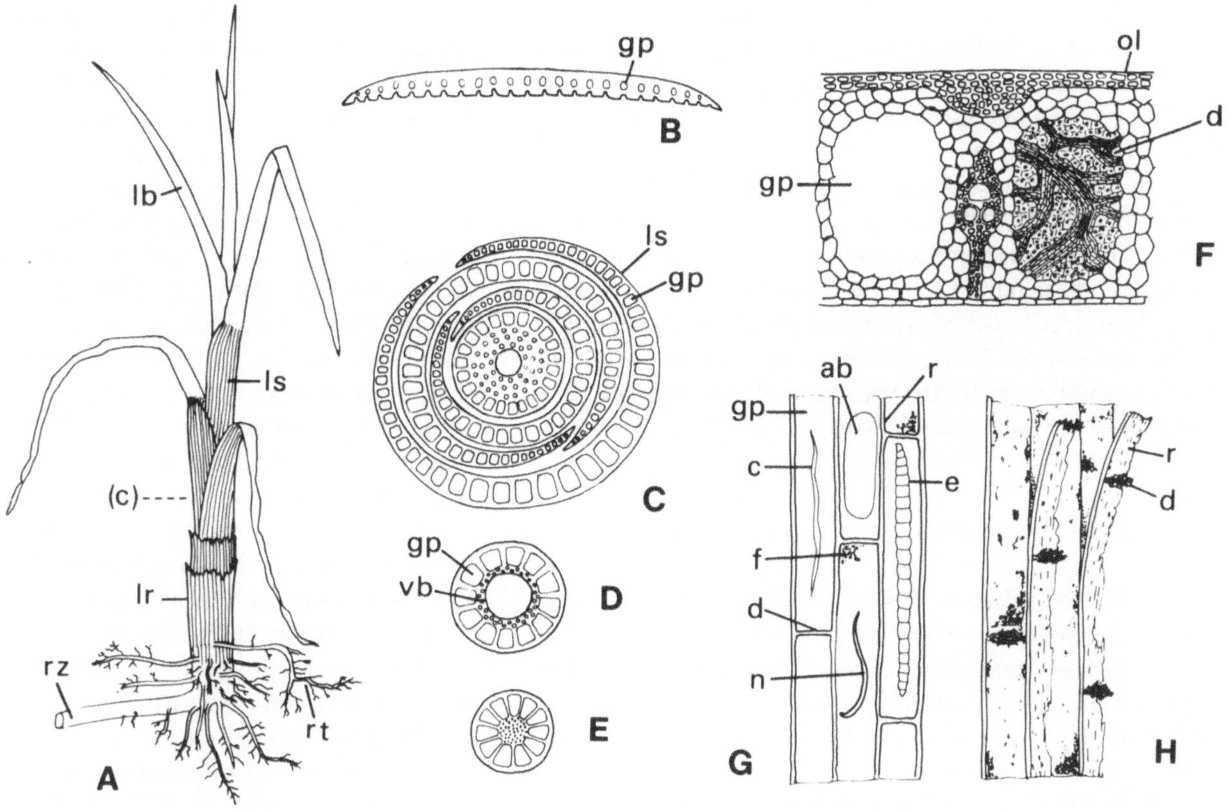

Fig. 1. Anatomy of *Spartina alterniflora*. A. a growing plant (May); B. transverse section of a leaf blade; C. transverse section of a plant at the level indicated on A. showing three leaf sheaths surrounding a hollow stem; D. transverse section of a rhizome; E. transverse section of root; F. part of leaf sheath seen in transverse section showing two gas passages with vascular bundle between; G. portion of a Stage 2 leaf sheath viewed from the inside; H. portion of a Stage 4 leaf sheath with inner walls broken down and ribs between the gas passages becoming detached. Legend: ab – air bubble, c – crack in innner wall allowing entry of water, d – diaphragm, e – enchytraeid, f – frass, gp – gas passage, lb – leaf blade, ls – leaf sheath, lr – leaf sheath remains, n – nematode, ol – outer layer of thick-walled cells, r – rib, rt – root, rz – rhizome, vb – vascular bundle.

within the aerenchyma tubes. Diaphragms are pierced by numerous small pores and strengthened by a dendritic network of thicker-walled, elongated cells. Gas passages in the leaf blades are typically half as wide as those in the sheaths. The aerenchyma system supplies oxygen to plant parts below the sediment surface, and oligochaetes typically were found within tubes in the stem sheaths (see Results).

In order to determine the distribution of oligochaetes on plants, four decompositional stages of sheath tissue were distinguished qualitatively:

Stage 1. Live sheath (before onset of senescence): Color pale pink or almost white; gas passages intact without bubbles (indicating no entry of water); no obvious damage to the internal surface of the sheath.

Stage 2. Early senescence: Color as in stage 1 or faintly yellow; occasional slits or holes on the inner face of gas passages; some bubbles visible (Fig. 1G).

Stage 3. Mid senescence: Color pale yellow with some dark spots and patches; openings on the inner wall of gas passages more frequent and extensive; some accumulation of frass.

Stage 4. Late senescence: Color darker with many spots and patches; inner sheath surface disintegrating; ribs between the gas passages becoming separated from the outer wall; diaphragms blackened, only consisting of a supporting framework; abundant frass and detritus in air passages (Fig. 1H).

Typically, sheaths remained attached to the stem until final disintegration. Stem decomposition is continuous, often making the separation into stages arbitrary. Sheath removal from stems during the extraction of oligochaetes sometimes caused mechanical damage to the inner wall resembling natural damage and confounded the distinction between Stages 1 and 2.

Collection and sample processing

The distribution and abundance of oligochaetes were determined using two different techniques to process samples: wet funnel extraction and stem dissection. Extraction samples were collected by excising live and dead *S. alterniflora* stems at ground level in each marsh zone. Stems were divided into 10 cm sections up to 20 cm for live stems and 30 cm for dead stems from high and low marsh sites, and up to 30 cm for live stems and 50 cm for dead stems from the creekside site. Stem sections were shredded lengthwise by tearing into strips 2–3 mm wide, to make it easier for worms to escape from the air passages, and *ca.* 20 g wet weight of shredded material was used for extraction. Leaf sheaths were collected from different stems in a similar manner to the whole stem sections, classified according to Stages 1–4, and *ca.* 20 g wet weight of sheath used for extraction. At the creekside site, leaf sheaths also were collected from below-ground stem sections. Dead leaves from the marsh surface and root samples consisting of the fibrous roots and attached mud also were collected from each marsh site.

Oligochaetes were extracted from samples using a version of the wet funnel method described by O'Connor (1955). The method was adapted for use in warm climates, where worms are accustomed to high temperatures, by using a greater wattage bulb (100W), and hence a faster extraction, than that recommended for extractions from temperate climates such as northern Europe (Healy & Rota, 1992). The efficiency of the extraction procedure was estimated for live sheaths by first counting worms in the sheaths using transmitted light microscopy and then shredding and extracting. Extraction efficiency for Stage 2 sheaths was about 90%, but is likely to be somewhat higher for older sheaths where inner walls of the gas passages were partly broken down allowing worms to escape more easily. Densities of extracted oligochaetes were determined for either 20 g of stem material or by multiplying numbers per stem by stem densities, m^{-2} of marsh surface.

Only live stems were collected and processed by stem dissection. Three stems within each marsh zone were excised at ground level and the lower 15 cm cut into approximately 5 cm sections. One 5 cm section of stem also was collected from below ground level in each zone. All stem sections were preserved in 10% formalin with Rose Bengal. Numbers of oligochaetes were determined by dissecting each stem section under a microscope at 20 × magnification. Leaf sheaths were removed from each section starting from the very outer sheaths, and worms were removed from aerenchyma tubes, counted and identified. The surface area of each stem section was determined by measuring the length and top and bottom diameters to the nearest 0.5 mm and calculating the area of the subscribed frustum. Dissected oligochaete densities were determined for both stem and marsh surface areas.

Comparison of sampling techniques

The original data from stem extractions and dissections are not directly comparable because of differences between the sampling techniques. Extraction data depended on stem weights, and far fewer creek bank (2–4) compared to high marsh (4–15) stems were needed to produce the same 20 g sample. Dissection data, because of the haphazard way in which stems were collected, was influenced by the decompositional variability between stems; stem sections from the same marsh site could contain different numbers and decompositional stages of sheaths. To compare sampling techniques, the mean number of stems within each marsh site was multiplied by the number of worms per stem, determined from each sampling technique, to estimate the total numbers of worms m^{-2} of marsh surface. Stem densities were low at the creek bank and greater in the high marsh (Table 1). Within each marsh site, estimates of oligochaete densities from stem dissections were 2.1 to 14.2 times greater than from light extractions (Table 2).

The numbers of oligochaetes residing in the sediments surrounding *S. alterniflora* plants were determined quantitatively only from the low marsh site. Preliminary sampling had shown that worms did not penetrate below the top 1 cm of the sediment. Five sediment samples were collected using a 2.1 cm diameter corer, i.e. around the same diameter as that of the plant stems, to a depth of 1 cm. Samples were preserved in 10% formalin with Rose Bengal and the number of worms later determined.

Data analyses

Numbers of oligochaetes from the various marsh microhabitats were analyzed using appropriate hierarchical log-linear models. Procedures for applying log-linear models to categorical data can be found in Fienberg (1980) and Sokal & Rohlf (1981). All data from stem extractions were analyzed using the hiloglinear procedure of SPSS/PC 4.1 (Norusis, 1990). Stem dissection data were analyzed using the stratified jackknife log-linear procedure of CPLX (Fay, 1985) because data were collected from a complex sample design. The significance of terms in each model was

Table 1. Densities (mean m^{-2}) of live and dead *S. alterniflora* plants in May 1991 at Kenan Field, Sapelo Island, GA.

	Live stems	New shoots	Dead stems
High marsh	104	35	155
Low marsh	109	27	195
Creekside marsh	80	8	96

Table 2. Comparison of the number of oligochaetes m^{-2} of marsh surface area calculated from extraction and dissection data for 0–10 cm sections of live and dead stems.

		High marsh	Low marsh	Creekside marsh
Extraction –	live stems	624	4,796	6,080
	dead stems	1,364	644	3,610
Dissection –	live stems	8,840	10,246	14,160

evaluated based on the likelihood ratio chi-square (L^2) or the change in L^2 between models at the 0.05 level of significance. Data from extraction samples were analyzed to determine the dependence of oligochaete numbers on marsh site (creek, low, high), stem type (live, dead), stem position (only 0–10, 10–20 cm sections), sheath stage (1–4) and family (Enchytraeidae, Tubificidae). Dissection data were analyzed to determine the dependence of oligochaetes on marsh site, stem position (–5–0, 0–5, 5–10, 10–15 cm), and sheath stage.

Results

Oligochaete distribution and abundance

Oligochaetes were found on all stems from each marsh site (Figs 2 & 3). Total oligochaete stem densities, calculated from both live and dead stem extraction data, ranged from 22,703 worms m^{-2} of marsh at the creekside site, where numbers on live and dead stems were similar (Fig. 2), to 6,889 worms m^{-2} of marsh at the low marsh site, and 3,584 worms m^{-2} of marsh at the high marsh site. The vertical distribution of worms on stems ranged from at least 50 cm above the sediment surface on creekside dead stems (Fig. 2) to 5 cm below the sediment surface on live stems (Fig. 3). Worms were common within the aerenchyma tubes of the leaf sheath and were recovered from inside tubes of >70% of dissected stems. In live stem sheaths, worms typically lay stretched lengthwise within a tube compartment, but tubificids sometimes extended through the diaphragm into the next compartment. Worms were often so

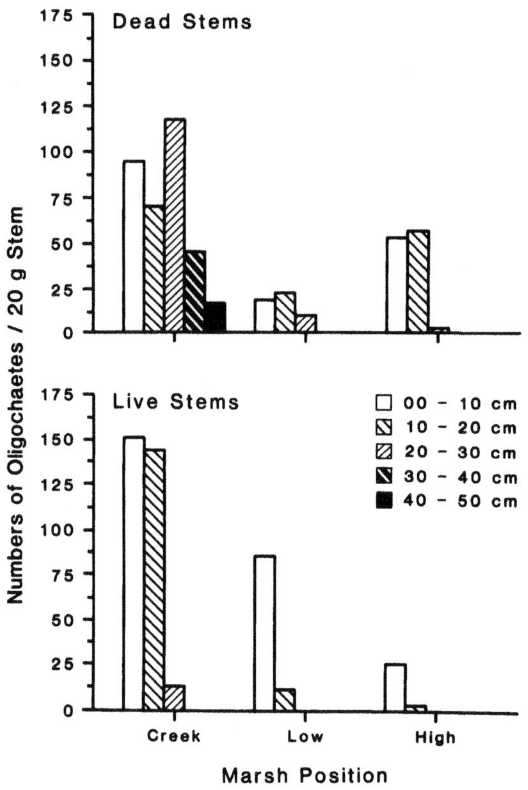

Fig. 2. Numbers of oligochaetes extracted from 10 cm sections of live and dead *S. alterniflora* stems collected in creekside, low and high marsh sites at Kenan Field, Sapelo Island, GA.

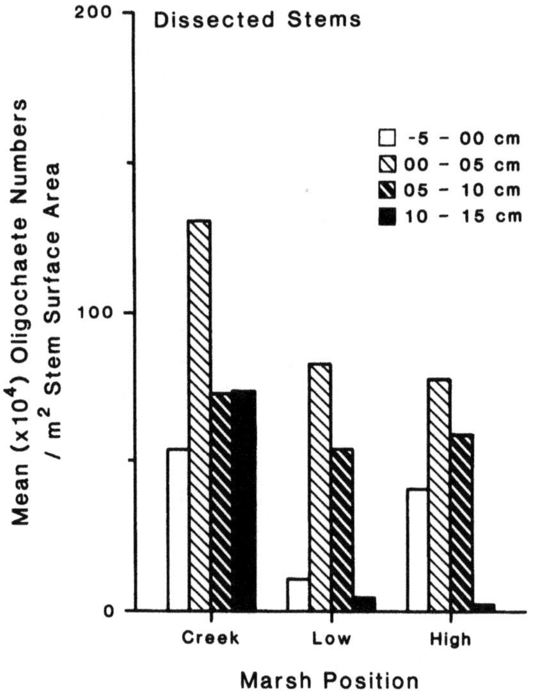

Fig. 3. Mean numbers of oligochaetes dissected from 5 cm sections, beginning 5 cm below the sediment surface, of live *S. alterniflora* stems collected in creekside, low and high marsh sites at Kenan Field, Sapelo Island, GA.

numerous that two or more were present in the same compartment.

Extracted oligochaete densities were dependent on the interaction between marsh site, stem type and stem position ($L^2 = 43.8$, df = 2, $P < 0.001$); how the interaction between stem type and position affected worm numbers was dependent on the different marsh sites. The greatest number of worms were found on live stems at the creekside site within the first 10 cm of stem (Fig. 2). The density of worms in the lower 20 cm of stem was generally less in dead than in live stems but the vertical range was greater in dead stems, especially in creekside plants. There was no interaction between marsh site and stem position for worm densities from dissected stems, but densities were dependent on marsh or stem position individually ($L^2 = 1.67$, df = 6, $P < 0.05$). Oligochaete densities were greatest along the creek bank and within the first 5 cm of stems (Fig. 3).

Oligochaete densities were dependent on the stage of leaf sheath decomposition for extraction stems only ($L^2 = 99.5$, df = 6, $P < 0.001$). Stage 1, the least decomposed, and Stage 4, the most decomposed, sheaths had the fewest numbers of worms (Fig. 4). The stage of sheath decomposition did not influence worm densities in dissected stems, and interactions between sheath stage and stem position ($L^2 = 1.12$, df = 27, $P > 0.05$) or sheath stage and marsh site ($L^2 = 1.20$, df = 36, $P > 0.05$) were not significant. Analysis of the association between oligochaete numbers and sheath stage for dissected stems likely was influenced by the fact that not all haphazardly selected stems contained every stage of sheath. Few high and low marsh dissected stems had either Stage 3 or 4 sheaths.

Few oligochaetes were associated with either dead leaves or roots at any of the marsh sites.

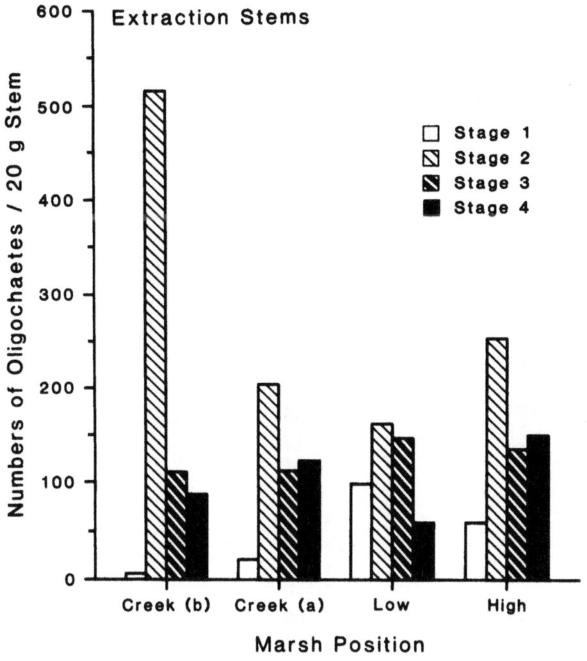

Fig. 4. Numbers of oligochaetes extracted from ca. 20 g of leaf sheath collected from live *S. alterniflora* stems in creekside, low and high marsh sites at Kenan Field, Sapelo Island, GA. Sheath material was classified according to decompositional stage, Stage 1 being the least decomposed and Stage 4 the most decomposed (see Materials and methods). Creek b sheaths were collected from stems below the sediment surface. All other sheaths were collected from stems above the sediment surface.

Total numbers varied between 6 and 15 worms 20 g^{-1} of material. Worm densities also were low in low marsh site sediments. A mean of only 1,156 worms m^{-2}, mainly tubificids, was collected from sediments surrounding stems. Combined with extraction estimates of oligochaete densities on live and dead stems, a total of 8,045 worms m^{-2} of marsh were found in the low marsh site.

Oligochaete species distribution and abundance

Five species of Enchytraeidae and one or two species of Tubificidae were identified from the Kenan Field site. The enchytraeids included four species of *Marionina*: *M. appendiculata* Nielsen & Christensen, 1959, a common species in salt marshes in North America and Europe, and *M. spartinae*, *M. waltersi* and *M. paludis*, all newly described species from *S. alterniflora* marshes of Sapelo Island (Healy, in press). A large species of *Enchytraeus*, possibly *E. albidus* Henle, 1837 was present only as immatures. *Monopylephorus parvus* Ditlevsen, 1904 and another less frequent, unidentified tubificid also were present. *Marionina appendiculata* and *M. paludis* had similar setal arrangements and were difficult to distinguish as immatures. The other species had distinct setal shapes and numbers and all stages could be recognized, even in heavily stained material.

The distribution of oligochaetes on stems was influenced by family affiliation. In extraction data, interactions between marsh site and stem type (L^2 = 13.6, df = 2, $P<0.005$) or marsh site and stem position ($L^2 = 21.1$, df = 2, $P<0.005$) depended on whether worms were enchytraeids or tubificids. Equivalent numbers of enchytraids were collected on both live and dead stems, but tubificids were less than half as abundant on dead stems (Table 3). Tubificids also were not distributed as far up stems, remaining within the first 10 cm (Table 3). In dissected stems, tubificids were greater than 80% of the below-ground oligochaete assemblage, but were only 50–60% of the above ground worms at any marsh site (Table 4).

There were distinct distributional differences among individual oligochaete species also. The predominant enchytraeid on both live and dead stems was *M. spartinae* (Tables 3 & 4). *Marionina spartinae* typically was the only enchytraeid in the inner leaf sheaths, and was more abundant above 10 cm in high and low marsh and above 20 cm in creekside dead stems (Table 3). Compared to dead stems, *M. spartinae* was not distributed as far up live stems in any marsh site (Tables 3 & 4). *Marionina waltersi* was rare in high and low marsh samples, and was associated mainly with the later stages of sheath decomposition, *i.e.* Stages 3 and 4. On creekside dead stems, *M. waltersi* was found 50 cm above the marsh surface. The distribution of both *M. spartinae* and *M. waltersi* on dead stems above 10 cm suggested a limited dependence on the wetting of stems during normal semidiurnal tides. *Marionina paludis* was present on

Table 3. Total numbers for oligochaete species found on 20 g of live or dead stems at 3 marsh sites for 3 stem heights (cm) above the marsh sediments within Kenan Field, Sapelo Island, GA.

Marsh	Creekside marsh		Low marsh			High		
Species	< 10	< 20	> 20	< 10	< 20	> 20	< 10	< 20
M. spartinae	24	32	123	2	14	9	14	40
M. waltersi	9	6	13	1	0	0	0	0
M. paludis	8	6	14	5	1	0	10	0
M. appendiculata	2	2	4	1	0	0	1	0
Marionina immatures	13	12	8	3	0	0	6	2
Enchytraeus sp.	3	1	0	0	0	0	0	0
Tubificidae	35	10	16	6	7	0	22	0
% *Marionina*	63	86	91	67	68	100	58	100
% Tubificids	37	14	9	33	32	0	42	0
Live stems								
M. spartinae	35	47	6	9	5	–	5	1
M. waltersi	1	7	1	0	0	–	0	0
M. paludis	16	24	2	0	0	–	0	0
M. appendiculata	4	4	0	0	0	–	0	0
Marionina immatures	20	27	3	1	0	–	0	0
Enchytraeus sp.	0	0	0	0	0	–	0	0
Tubificidae	76	37	2	77	7	–	21	1
% Marionina	50	75	83	11	42	–	19	50
% Tubificids	50	25	17	89	58	–	81	50

dead stems in all marsh sites, but was only found on creekside live stems. *Marionina appendiculata* were found almost exclusively on creekside stems in relatively low numbers (Tables 3 & 4). *Enchytraeus* sp. was present only in the lowest sections of creekside dead stems and no mature specimens were found. The immature *Enchytraeus* specimens found on stems may have originated from a population in the creek sediments. The tubificid *M. parvus* was the most abundant oligochaete on *S. alterniflora* stems (Tables 3 & 4) and the greatest densities were on live stems, especially in Stage 2 sheaths. *Monopylephorus parvus* was found in the sediment, on dead leaves

Table 4. Total numbers for oligochaete species dissected from live *S. alterniflora* stems in 3 marsh sites for 3 stem heights (cm) above and below the sediments at Kenan Field, Sapelo Island, GA.

Marsh	Creekside marsh			Low marsh			High		
Species	< 0	< 10	> 10	< 0	< 10	> 10	< 0	< 10	> 10
M. spartinae	5	119	47	1	49	0	4	104	2
M. waltersi	9	49	19	0	10	0	2	6	0
M. paludis	4	6	7	0	2	0	0	3	0
M. appendiculata	2	21	1	0	7	0	0	2	0
Marionina immatures	0	10	0	2	12	0	4	8	0
Enchytraeus sp.	0	0	0	0	0	0	0	0	0
Tubificidae	106	414	103	17	157	0	82	124	0
% *Marionina*	16	33	42		34		9	50	100
% Tubificids	84	67	58	99	66		91	50	0

and among roots. Juvenile *M. parvus* were common on stems, suggesting active reproduction within sheaths.

Discussion

The gas passages of *Spartina* leaf sheaths were the most densely populated oligochaete habitat in the Sapelo marsh. *Marionina appendiculata*, which is a widespread species in estuarine muds, and *M. parvus*, which is also known from estuarine habitats (Baker & Brinkhurst, 1981), may be preadapted to conditions on the marsh, but other species may be more specialized. All four species of *Marionina* are of similar size, corresponding to the dimensions of the sheath compartments, and all have sigmoid setae, an unusual character in the genus which may confer some special advantage. Three of the four species are known only from *Spartina* marshes in the southeast US, and were absent from other coastal habitats sampled in Georgia and Florida, including *Salicornia* marshes, sand dune marshes, *Thallassia* beds, and seagrass tidal debris (Healy, in press). None of the Sapelo species were recorded from an intensively studied marsh in Louisiana where a tubificid identified as *Peloscolex* sp. was the only oligochaete recorded (Burke, 1976). In Louisiana, the air passages (which Burke wrongly called water transport tubes) were larger, measuring 1–1.5 mm across, and may be a less suitable habitat for the smaller *Marionina* species compared to the larger tubificid sp. The distributional data and observations on oligochaete ecology suggest that while some oligochaetes in *Spartina* marshes may be opportunists, invading the stems as an extension of the benthic zone, others have evolved in close association with *Spartina* and are stem specialists.

Spartina alterniflora stems clearly are an important habitat for oligochaetes in Sapelo Island marshes. In the low marsh site, worm densities were an order of magnitude greater on stems processed by dissection (10,246 m^{-2}) than in sediments (1,156 m^{-2}). Previous estimates of oligochaete densities in Sapelo Island marshes based only on sediment samples ranged between 1,000 to 17,000 worms m^{-2} for creekside, 6,600 to 8,000 worms m^{-2} for low marsh, and 0 to 4,000 worms m^{-2} for high marsh sites, with maximum densities occurring in winter (Teal, 1962). Oligochaete densities in stems were similar to those recorded in a previous study of sediment-dwelling individuals (Teal, 1962) indicating that total marsh worm numbers had been underestimated from sediment samples alone. In North Carolina marshes, oligochaete densities from sediment samples ranged between 1000 and 8000 worms m^{-2} (McMahan, 1972; Cammen, 1979) and likely underrepresented the total density of worms in the marsh. Oligochaetes were the most abundant macroinfaunal taxa in both North Carolina (Cammen, 1979) and Louisiana (Burke, 1976) salt marshes, and were the second most abundant annelid taxa after capitellids in Sapelo Island marshes (Teal, 1962). When the numbers of worms in standing stems are taken into account, however, the density of oligochaetes in *Spartina* marsh communities is far greater than previously realized.

Oligochaetes probably are present on all *Spartina* marshes, although they often are not mentioned in accounts of salt marsh fauna. Kraeuter and Wolf (1974) suggested that the apparent absence of oligochaetes in salt marshes was the result of a lack of study effort, a view supported by our experiences. Observations in other *Spartina* marshes, two in northern Florida, USA, three in New Brunswick, Canada, and two in Ireland (Healy, unpublished), and an unpublished study from Louisiana, USA (Burke, 1976) indicate that oligochaetes also are living on stems in all those *Spartina* marshes. The proportion of worms found on stems appears to depend on the availability of appropriate stem microhabitats. In Canadian and Irish marshes, *Spartina* plants are smaller and more slender and often have none or only one or two habitable leaf sheaths. Dead stems usually are destroyed during winter storms, and marshes accumulate more leaf debris on the sediment surface. A large proportion of oligochaetes in Canadian and Irish marshes is found in the leaf debris, and far fewer worms reside on stems (Healy, unpublished). In southeastern USA marshes, dead leaves are decomposed quickly and live stems typically have three or four sheaths which can be occupied by worms. The stem habitat appears to be more important to oligochaetes under the conditions found in southeastern marshes where *Spartina* platns grow taller and thicker in the warmer climate.

The presence of oligochaetes in the above ground parts of *Spartina* plants has two obvious explanations: plant stems represent a more favorable physical environment than the marsh sediments and/or a refuge from predation. Conditions within marsh sediments are almost always reducing, with an absence of free oxygen and high concentrations of potentially toxic compounds such as H_2S, Mn^{2+} and Fe^{2+} (Ponnamperuna, 1972; Armstrong, 1975, 1979). Oligochaetes are sensitive to anoxic conditions (Giere & Pfannkuche, 1982), and meiofaunal organisms in marsh sediments typically are distributed within the first few millimeters of the sediment surface (Coull & Bell, 1979). Densities of worms in Kenan Field low marsh sediments were relatively low, and worms were found only within the surface sediments, less than 1 cm deep. Oxygen concentrations can be greater around *Spartina* roots where diffusion creates an oxygen 'halo' in the rhizosphere (Teal & Kanwisher, 1961), but not enough oxygen may be released for respiration in saturated sediments, especially during tidal flooding (Gleason & Zieman, 1981; Mendelssohn & McKee, 1987). Few oligochaetes or other infauna were found associated with the roots and rhizomes in Kenan Field samples, indicating that animals are not able to avail of the increased oxygen concentration in the rhizosphere.

Stem habitats probably do not suffer from the same limited availability of oxygen that is common in sedimentary habitats. Within the aerenchyma system of live *Spartina* plants, O_2 and CO_2 gradients exist created by photosynthesis and the flux of gases between leaves and roots (Teal & Kanwisher, 1966). The concentration of both gases varies according to diurnal and tidal cycles (Gleason & Zieman, 1981). Oxygen concentrations in the upper section of *Spartina* rhizomes were 11% at mid and 15% at high tide (Teal & Kanwisher, 1966). Concentrations in living sheaths certainly would be greater than in the rhizomes, and probably would be sufficient for faunal activity at all times, even during tidal immersion. The oxygenated, intermittently wetted conditions of *Spartina* stems favor the more 'terrestrial' enchytraeids that seem to need comparatively greater concentrations of oxygen (Giere & Pfannkuche, 1982; Healy, 1987). Tubificids generally are adapted physiologically and behaviorally to habitats with low oxygen concentrations. The differential distribution of enchytraeids and tubificids on stems, with enchytraeids extending further up stems compared to tubificids, supports the view that microhabitat conditions, particularly oxygen concentrations, are critical to oligochaete distributions on stems.

In dead stems, oxygen is provided by diffusion and moisture is maintained by capillarity, but conditions will vary both within and between marsh zones and seasons. In May, many of the stems that appeared to be dead (*i.e.*, were dry with broken flower stems, dead and mostly fallen leaf blades, and brown, mud-encrusted leaf bases), were found to have inner leaf sheaths with the appearance and fauna typical of live plant sheaths in decompositional Stage 2 or 3. Leaf sheaths are among the last tissues of the aerial stem to decompose, which may explain the similarity between the fauna on live and dead stems.

Worms inhabiting the first millimetres of marsh sediment surface might avoid anoxic conditions only to be exposed to predation from the many fish and crustacea which move over the marsh and forage during high tide (Kneib, 1988; Rozas & LaSalle, 1990). Fauna found in live or dead stems presumably are less vulnerable to predation from nekton, but we know of no studies that compare predation pressures on stem and sediment fauna. Stem oligochaetes also would be vulnerable to predation from other stem dwelling fauna. Among co-habitants in the sheaths, the most likely predators on oligochaetes would be mites and ceratopogonid larvae, although neither were present in great numbers on either live or dead stems in May.

Although physical stress and predation pressure may be reduced in stems, questions remain about what the worms are feeding on and whether worms are entirely dependent on stems during all life history stages. About 80% of enchytraeids in Stage 1–2 sheaths had empty guts, and the remainder contained pale, amorphous material with no recognizable food items. In enchytraeids from older sheaths and also from dead stems and leaves, the amorphous material was browner with recognizable fungal hyphae and spores, diatoms, and some fragments of plant tissue. The diet of tubificids was difficult to determine because the chloragogen tissue partly masks the gut contents. In a minority of tubificids with solid gut contents, the material appeared similar to that in enchytraeids. The qualitative difference in gut contents between enchytraeids from young and old sheaths suggests a limited movement of stem oligochaetes to feed on detritus-rich zones, and a possible reliance on *in situ* bacterial films, fungal hyphae, plant fragments and cells, or even dissolved organic matter. Enchytraeids are frequently associated with living plants but their ability to consume living plant tissues is disputed. However, the possession of an eversible pharynx capable of grasping solid food makes it theoretically possible for them to ingest the thin-walled cells which line the gas passages, expecially when these have been weakened by microbial activity. The air passages of *Spartina* are lysogenous cavities which form by the breakdown of mesophyll tissue. Broken cell walls can be seen lining the cavity (Fig. 1F), and it is possible that the worms are able to grasp and ingest cell debris. During senescence, the cavities become invaded by fungal hyphae and spores (May, 1974), and recognizable bacterial films, diatoms, protozoa and other organisms can be found in sheaths. Microbes, together with senescent cells and the excretory products of other stem fauna, may constitute a plentiful food source for oligochaetes in older sheaths. The uptake of dissolved organic matter has been recorded in Oligochaeta (e.g. Siebers & Ehlers, 1978) and may be important in the present context.

Oligochaetes appear to be permanent inhabitants and not temporary residents on stems. All sizes of enchytraeids and tubificids were present on stems, and cocoons were frequent in older sheaths. Mature individuals of all four *Marionina* species were common in May 1991. No mature *Enchytraeus* sp. were found however, and juveniles may be recruits from open sediment populations and not typical stem residents. Movement of oligochaetes between stem and sediment environments may be common. In a Louisiana marsh, greater densities of a tubificid were recorded in sediments during the winter and on stems in the summer (Burke, 1976). Migration onto stems in the summer and passive transfer to the sediment following sheath disintegration in the winter could explain observed differences in the tubificid densities. Other possible seasonal changes in oligochaete distributions might be expected given the natural successional changes that occur in *Spartina* plants.

Considerable numbers of enchytraeids and tubificids where found in the early decompositional stages of sheaths, Stage 1 and 2, and it is difficult to understand how worms could gain entry into apparently intact gas passages and pass between compartments. Oligochaetes do not have the mechanical ability to break through several layers of cells which separate air cavities from the outside, therefore, colonizing individuals must find openings made by other organisms. Visible damage to sheaths usually occurred just above the mud surface and stems occasionally were perforated with small holes, possibly made by insects. Another possible point of entry is at the ligule where the dying leaf blade eventually breaks away. On the high marsh, the pulmonate snail *Melampus bidentatus* Say, 1822 often was seen resting during the day in accumulated sediment near the ligule or inside the sheath. The feeding activities of *M. bidentatus*, which is known to consume epidermal tissues (Hausman, 1932), could cause or accelerate the breakdown of the sheath wall. The movement of worms between aerenchyma tube compartments is not understood. Both enchytraeids and tubificids have been seen passing with ease through apparently intact diaphragms, and yet diaphragms examined microscopically appeared, in most cases, to form complete barriers across ducts.

We have examined a number of other plant species growing in both brackish and freshwater habitats to see whether the faunal invasion of aerenchyma is unique to *Spartina* or a general phenomenon in large aquatics. Apart from herbivorous Diptera, few invertebrates were found in stems or leaf bases of *Typha, Iris, Phragmites* or *Scirpus* growing in fresh or brackish water. At one *Spartina* marsh in Ireland, however, enchytraeids were common in sheath remains on *Phragmites* growing at the upper edge of the marsh, but only in plants growing in the wetter areas. The species, a *Lumbricillus*, was the same as that living on the local *Spartina* stems.

Acknowledgements

Our thanks to Christer Erséus and Michael Milligan for identification of tubificid material, and Janet Carton and Eileen Hedick for technical help. Support for our research was provided in part by the Sapelo Island Research Foundation under the University of Georgia Marine Institute's Visiting Scientist Program (B. Healy) and a National Science Foundation grant (OCE-9017807) to K. Walters. This is Contribution No. 719 of the University of Georgia Marine Institute.

References

Anderson, C. E., 1974. A review of the structure in several North Carolina salt marsh plants. In R. J. Reimold & W. H. Queen (eds), Ecology of Halophytes. Academic Press, New York: 307–344.

Armstrong, W., 1975. Waterlogged soils. In J. R. Etherington (ed.), Environment and plant ecology. John Wiley, New York: 181–218.

Armstrong, W., 1979. Aeration in higher plants. Adv. bot. Res. 7: 225–332.

Backlund, H. O., 1945. Wrack fauna of Sweden and Finland. Ecology and chorology. Opusc. Ent. Suppl. 1945: 1–237.

Baker, H. R. & R. O. Brinkhurst, 1981. A revision of the genus *Monopylephorus* and redefinition of the subfamilies Rhyacodrilinae and Branchiurinae (Tubificidae: Oligochaeta). Can. J. Zool. 59: 939–965.

Burke, W. W., 1976. Vertical and horizontal distribution of macroinvertebrates on the cord grass, *Spartina alterniflora* in a Louisiana salt marsh. Unpublished thesis, Louisiana State University, Baton Rouge, 116 pp.

Cammen, L. M., 1979. The macroinfauna of a North Carolina salt marsh. Amer. Midl. Nat. 102: 244–253.

Chapman, V. J., 1977. Introduction. In V. J. Chapman (ed.), Ecosystems of the world. 1. Wet coastal ecosystems. Elsevier, New York: 1–29.

Coull, B. C. & S. S. Bell, 1979. Perspectives of marine meiofaunal ecology. In R. J. Livingston (ed.), Ecological processes in coastal and marine systems. Plenum, New York: 189–216.

Fay, R. R., 1985. A jackknifed chi-squared test for complex samples. J. Am. Statist. Assoc. 80: 148–157.

Fienberg, S. E., 1980. The analysis of cross-classified categorical data. MIT Press, Cambridge, Massachusetts, 198 pp.

Fiers, F. & P. Rutledge, 1990. Harpacticoid copepods associated with *Spartina alterniflora* culms from the marshes of Cocodrie, Louisiana (Crustacea, Copepoda). Bull. l'Inst. Sci. nat. Belg. 60: 105–125.

Giere, O. & O. Pfannkuche, 1982. Biology and ecology of marine Oligochaeta, a review. Oceanogr. mar. Biol. ann. Rev. 20: 173–308.

Gleason, M. L. & J. C. Zieman, 1981. Influence of tidal inundation on internal oxygen supply of *Spartina alterniflora* and *Spartina patens*. Est. cstl. Shelf Sci. 13: 47–57.

Hausman, S. A., 1932. A contribution to the ecology of the salt marsh snail, *Melampus bidentatus* Say. Amer. Nat. 66: 541–545

Healy, B., 1987. The depth distribution of Oligochaeta in an Irish quaking marsh. Hydrobiologia 155: 235–247.

Healy, B., in press. New species of *Marionina* (Oligochaeta, Enchytraeidae) from *Spartina* marshes on Sapelo Island, Georgia, USA. Proc. biol. Soc. Wash.

Healy, B. & E. Rota, 1992. Methods for collecting Enchytraeidae during expeditions. Soil. Biol. Bioch. 24: 1279–1281.

Jackson, D., C. F. Mason & S. P. Long, 1985. Macroinvertebrate populations and production on a salt-marsh in east England dominated by *Spartina anglica*. Oecologia 65: 406–411.

Kneib, R. T., 1984. Patterns of invertebrate distribution and abundance in the intertidal salt marsh: causes and questions. Estuaries 7: 392–412.

Kneib, R. T., 1987. Seasonal abundance, distribution and growth of postlarval and juvenile grass shrimp (*Palaemonetes pugio*) in a Georgia, USA, salt marsh. Mar. Biol. 96: 215–223.

Kneib, R. T., 1988. Testing for indirect effects of predation in an intertidal soft-bottom community. Ecology 69: 1795–1805.

Knox, G. A., 1986. Estuarine ecosystems: a systems approach. CRC Press, Inc., Florida, 289 pp.

Kraeuter, J. N. & P. L. Wolf, 1974. The relationship of

marine macroinvertebrates to salt marsh plants. In R. J.Reimold & W. H. Queen (eds), Ecology of Halophytes. Academic Press, New York: 449–462.

McMahan, E. A., 1972. Relative abundance of three marsh floor organisms in a sewage-affected marsh and a sewage-free marsh. J. Elisha Mitchell Sci. Soc. 88: 61–65.

May, M. S., 1974. Probable agents for the formation of detritus from the halophyte *Spartina alterniflora*. In R. J. Reimold & W. H. Queen (eds), Ecology of Halophytes. Academic Press, New York: 429–440.

Mendelssohn, I. A. & K. L. McKee, 1987. Root metabolic response of *Spartina alterniflora* to hypoxia. In R. M. M. Crawford (ed.), Plant life in aquatic and amphibious habitats. Blackwell, London: 239–253.

Morris, J. T. & B. Haskin, 1990. A 5-yr record of aerial primary production and stand characteristics of *Spartina alterniflora*. Ecology 71: 2209–2217.

Norusis, M. J., 1990. SPSS/PC + advanced statistics 4.0 for the IBM PC/XT/AT and PS/2. SPSS Inc., Chicago: 139–158.

O'Connor, F. B., 1955. Extraction of enchytraeid worms from a coniferous forest soil. Nature 175: 815–816.

Pomeroy, L. R. & R. G. Wiegert, 1981. The ecology of a salt marsh. Springer-Verlag, New York, 271 pp.

Pomeroy, L. R., L. R. Shenton, R. D. H. Jones & R. J. Reimold, 1972. Nutrient flux in estuaries. In G. E. Likens (ed.), Nutrients and Eutrophication. Amer. Soc. Limnol. Oceanogr. Spec. Symp. 1: 274–291.

Ponnamperuna, F. N., 1972. The chemistry of submerged soils. Adv. Agron. 24: 29–96.

Radar, D. N., 1984. Salt-marsh invertebrates: small-scale patterns of distribution and abundance. Estuaries 7: 413–420.

Reimold, R. J., J. L. Gallagher & D. E. Thompson, 1973. Remote sensing of tidal marsh. Photog. Engng. 39: 477–488.

Rozas, L. P. & M. W. LaSalle, 1990. A comparison of the diets of the Gulf killifish, *Fundulus grandis*, entering and leaving a Mississippi brackish marsh. Estuaries 13: 332–336.

Rutledge, P. A. & J. W. Fleeger, in press. Abundance and seasonality of meiofauna, including harpacticoid copepod species, associated with *Spartina alterniflora* stems. Estuaries

Siebers, D. & V. Ehlers, 1978. Transintegumentary absorption of acidic amino acids in the oligochaete annelid *Enchytraeus albidus*. Comp. Biochem. Physiol. A. Comp. Physiol. 61: 55–60.

Smart, R. M., 1982. Distribution and environmental control of productivity and growth form of *Spartina alterniflora* Loisel.. In D. N. Sen & K. S. Rajpurohit (eds), Contributions to the ecology of halophytes. Junk, Amsterdam: 127–142.

Sokal, R. R. & F. J. Rohlf, 1981. Biometry. W. H. Freeman & Co., San Francisco, 857 pp.

Teal, J. M., 1962. Energy flow in the salt marsh ecosystem of Georgia. Ecology 43: 614–624.

Teal, J. M. & J. Kanwisher, 1961. Gas exchange in a Georgia salt marsh. Limnol. Oceanogr. 6: 388–399.

Teal, J. M. & J. Kanwisher, 1966. Gas transport in the marsh grass *Spartina alterniflora*. J. exp. Bot. 17: 355–361.

Van Dolah, R. F., 1978. Factors regulating the distribution and population dynamics of the amphipod *Gammarus palustris* in an intertidal salt marsh community. Ecol. Monogr. 48: 191–217.

Wiegert, R. G., A. G. Chalmers & P. F. Randerson, 1983. Productivity gradients in salt marshes: the response of *Spartina alterniflora* to experimentally manipulated soil-water movement. Oikos 41: 1–6.

Hydrobiologia **278**: 125–127, 1994.
T.B. Reynoldson & K.A. Coates (eds), Aquatic Oligochaete Biology V.

Distribution of Oligochaeta in some lakes and pools of Bolivia

J. Juget[1] & M. Lafont[2]
[1] *Universite Claude Bernard, URA – CNRS 1451 Ecologie des Eaux Douces et des Grands Fleuves 43, Boulevard du 11 novembre 1918, F – 69622 Villeurbanne Cedex, France;* [2] *CEMAGREF, Division Biologie des Ecosystèmes aquatiques 3, bis, quai Chauveau, C.P. 220, F – 69336 Lyon Cedex 09, France*

Key words: oligochaetes, Bolivian lakes, South America

Abstract

This work is a contribution to the taxonomy and ecology of the benthic Oligochaeta of Lake Titicaca, Lake Poopo and smaller lake basins from western and eastern sides of the Bolivian Royal Cordillera. Benthic fauna of the small lakes include several typical taxa known exclusively from South America. In the two great lakes, Titicaca and Poopo, the chemical properties of the water play an important part in the composition of the benthic fauna.

Introduction

A recent study was carried out by ORSTOM (Institut Français de Recherche Scientifique pour le Développement en Coopération: previously, Office de Recherche Scientifique et Technique Outre Mer) on the benthic fauna in the Bolivian part of Lake Titicaca (littoral and profundal zones), Lake Poopo and some smaller lakes, pools, springs and rivers from Altiplano and Amazonian sides of the Bolivian Royal Cordillera at several altitudes between 3750 and 4850 m.

Among the material sampled by C. Dejoux and J. G. Wasson in 1987–1988, more than 30 taxa (excluding the Enchytraeidae which were not identified to species) were recorded (19 species of Naididae, 10 species of Tubificidae, one species of Phreodrilidae, one species of Acanthodrilidae) (Table 1).

These data confirm the presence in the small lakes and pools of typical taxa of South America, for example the tubificids *Isochaetides lacustris* (Cernosvitov, 1939) (= *Isochaeta baicalensis* sensu Gavrilov, 1981), *Epirodrilus antipodum* Cernosvitov, 1939, *Bothrioneurum americanum* Beddard, 1894, also previously recorded from Lake Titicaca (Cernosvitov, 1939; Gavrilov, 1981; Brinkhurst & Marchese, 1987; Lafont & Juget, 1992) and the naidids *Paranais macrochaeta* Cernosvitov, 1939, *Nais andina pectinata* Martinez-Ansemil & Giani, 1986 and *Pristina peruviana* Cernosvitov, 1939 (Cernosvitov, 1939; Gavrilov, 1981; Brinkhurst & Coates, 1985; Dumnicka, 1986; Martinez-Ansemil & Giani, 1986; Harman *et al.*, 1988).

Three eurytopic species of Tubificidae, *Potamothrix bavaricus* (Öschmann, 1913), *Potamothrix hammoniensis* (Michaelsen, 1901) and *Potamothrix heuscheri* (Bretscher, 1900), predominating in the deep areas (> 150 m) of Lago Grande, the largest lake basin of Lake Titicaca which is 284 m deep, were not found elsewhere in Bolivia. This distribution could be related to prolonged deficiencies in oxygen of the hypolimnion in Lago Grande because the vertical circulation from July to September is insufficient to reoxygenate the

Table 1. Distribution of Oligochaeta in some lakes and pools of Bolivia

	1a	1b	2	3	4	5	6	7	8	9	10	11	12	13	14
Naididæ															
Chætogaster diastrophus (Gruithuisen 1828)								*							
Paranais macrochæta Cernosvitov, 1939							*			*		*			
Paranais salina (Cernosvitov, 1939)			*												
Paranais litoralis (Müller, 1784)													*		
Nais communis Piguet, 1906				*	*	*	*	*	*	*					
Nais variabilis Piguet, 1906											*				
Nais elinguis Müller, 1773													*		
Nais pardalis Piguet, 1906		*		*	*			*	*	*	*				
Nais bretscheri Michælsen 1899					*										
Nais simplex Piguet, 1906				*	*				*		*		*		
Nais andina pectinata Martinez-Ansemil & Giani, 1986					*	*				*			*	*	*
Dero cooperi Stephenson, 1932 *Dero sawayai Marcus, 1943*	*	*		*	*			*	*	*					
Dero digitata (Müller 1773)					*(?)										
Pristina longiseta Ehrenberg, 1828															*
Pristina leidyi Smith, 1896												*			
Pristina æquiseta Bourne, 1891												*			
Pristina peruviana Cernosvitov, 1939								*	*						
Pristinella osborni (Walton, 1906)					*	*			*	*	*				
Tubificidæ															
Tubifex ignotus (Stolc, 1886)	*	*													
Potamothrix bavaricus (Oschmann, 1913)	*	*													
Potamothrix hammoniensis (Michælsen, 1901)		*													
Potamothrix heuscheri (Bretscher, 1900)		*													
Limnodrilus hoffmeisteri Claparède, 1862	*	*	*	*	*	*	*	*	*	*		*			
Limnodrilus udekemianus Claparède, 1862	*														
Isochætides lacustris (Cernosvitov, 1939)	*		*	*	*	*(?)	*								
Epirodrilus antipodum Cernosvitov, 1939				*	*	*	*		*	*	*			*	
Bothrioneurum americanum Beddard, 1894		*				*(?)	*	*	*	*				*	
Rhyacodrilus sp.	*				*								*		
Phreodrilidæ															
Phreodrilidæ sp.					*								*	*	
Enchytraeidæ															
g. spp.			*	*	*	*	*		*				*	*	*
Acanthodrilidæ															
Microscolex sp.									*						

1a	Lake Titicaca (littoral zone)	7	Lake Taypi Chaka
1b	Lake Titicaca (profundal zone)	8	Lake Sora Kkota
2	Lake Poopo	9	Lake Allta Kkota
3	Lake Khara Kkota	10 - 11	Small rivers tributaries of the Lake Titicaca
4	Lake Khotia	12	Pools and hot springs (South Lipez)
5	Lake Jankho Khotia	13	Laguna negra (Amazonian side)
6	Lake Wila Lloreta	14	Lake Viscachani (Amazonian side)

deep water layers (Iltis *et al.*, 1992) in a lake system with the special features of low latitude and high altitude making it more vulnerable to eutrophication (Northcote, 1992).

As opposed to small lakes Khara Kkota, Khotia, Jankho Khotia, Wila Lloreta, Taypi Chaka, Sora Kkota, Allta Kkota and Viscachani whose mean conductivities do not rise beyond 50 μS cm^{-1}, Lake Titicaca exhibits a conductivity $\geq$ 1400 μS cm^{-1}, due to high concentrations of Na^{+} and Cl^{-} ions, and a total dissolved salt content close to one gram per litre (Iltis *et al.*, 1992). These data could also explain the dominance, in samples from littoral and profundal zones of Lake Titicaca, of *P. bavaricus* which tolerates high salinities in estuaries (Timm, 1970).

Lake Titicaca (3809 m altitude) overflows via the rio Desaguadero into Lake Poopo (3686 m), the type locality of *Paranais salina* Cernosvitov, 1939. This large but shallow lake (2.5 m deep) is characterized by low rainfall and the absence of permanent superficial drainage channels which leads to increase the affluent water salinity (up to 50 g l^{-1}) (Carmouze *et al.*, 1977).

A single mature specimen of *Microscolex* from Lake Sora Kkota (Overhuyo Valley) is closely related to the cosmopolitan bankdwelling *Microscolex phosphoreus* (Dugès, 1837), originating from South America, (according to Michaelsen 1903, in Bouché 1972).

References

Bouché, M. B., 1972. Lombriciens de France, Ecologie et Systématique. Ann. Zool. Ecol. Anim. numéro hors série: 671 pp.

Brinkhurst, R. O. & K. A. Coates, 1985. The genus *Paranais*. (Oligochaeta: Naididae) in North America. Proc. biol. Soc. Wash. 98: 303–313.

Brinkhurst, R. O. & M. Marchese, 1987. A contribution to the taxonomy of the aquatic Oligochaeta (Haplotaxidae, Phreodrilidae, Tubificidae) of South America. Can. J. Zool. 65: 3154–3165.

Carmouze, J. P., C. Arce & J. Quintanilla, 1977. La régulation hydrique des lacs Titicaca et Poopo. Cah. O.R.S-.T.O.M., ser Hydrobiol. 11: 269–283.

Cernosvitov, L., 1939. VI. Oligochaeta. In H. C. Gilson (ed.), The Percy Sladen Trust Expedition to Lake Titicaca in 1937. Trans. Linn. Soc. London, 1: 81–116.

Dumnicka, E., 1986. Naididae (Oligochaeta) from subterranean waters in West Indian Islands: distribution, taxonomic remarks and description of a new species. Bijdr. Dierkd. 56: 267–281.

Gavrilov, K., 1981. Oligochaeta. In S. H. Hurlbert, G. Rodriguez & N. D. Santos (eds), Aquatic biota of tropical South America, Part 2: Anarthropoda.

Harman, W. J., R. O. Brinkhurst & M. Marchese, 1988. A contribtuion to the taxonomy of the aquatic Oligochaeta (Naididae) of South America. Can. J. Zool. 66: 2233–2242.

Iltis, A., J. P. Carmouze & J. Lemoalle, 1992. Physicochemical properties of the water. In C. Dejoux & A. Iltis (eds), Lake Titicaca: A Synthesis of Limnological Knowledge, Kluwer Academic Publishers, Dordrecht: 89–97.

Lafont, M. & J. Juget, 1992. The benthic Oligochaetes. In C. Dejoux & A. Iltis (eds), Lake Titicaca: Synthesis of Limnological Knowledge, Kluwer Academic Publishers, Dordrecht: 302–306.

Martinez-Ansemil, E. & N. Giani. 1986. Algunos oligoquetos acuáticos de Bolivia. Oecologia aquatica 8: 107–115.

Michaelsen W., 1903. Die Geographische Verbreitung der Oligochaeten. Berlin. 187 pp.

Northcote, T. G., 1992. Eutrophication and pollution problems. In C. Dejoux & A. Iltis (eds), Lake Titicaca: A Synthesis of Limnological Knowledge, Kluwer Academic Publishers, Dordrecht: 551–561.

Timm, T., 1970. On the fauna of the Estonian Oligochaeta. Pedobiologia 10: 52–78.

Hydrobiologia **278**: 129–131, 1994.
T.B. Reynoldson & K.A. Coates (eds), Aquatic Oligochaete Biology V.

Influence of small river conditions on the abundance of Tubificidae

Eustolia I. Lestochova
All-Union Scientific-Industrial Society for Agricultural Use of Sewage 'Progress', Sovetskaya la, 142450 Staraya Kupavna, Moscow Region, Russia

Key words: small rivers, water contamination, Tubificidae, bacteria, nitrogen, phosphorus

Abstract

Two small rivers, the Yerik and Nezhegolyok, were studied seasonally in 1979–1980. Increase in concentration of nitrogen and phosphorus compounds in the river water as well as proliferation of enterococci and enterobacilli favoured the abundance of Tubificidae while, at the same time, their numbers were reduced by chironomids and leeches.

Introduction

This paper describes part of a complex research programme conducted in 1979–1980 on two small rivers, the Yerik and Nezhegolyok, in the Belgorod Region of Central Russia. The arable lands of the Yerik River basin are generally fertilized with liquid manure. The Nezhegolyok River was chosen as a control as its basin is not in an agricultural area.

Material and methods

Samples were taken at the same time in both rivers over the period April to October in 1979 and 1980. Two stations were located on Yerik River, 'A' situated 1 km upstream from the arable lands fertilized with liquid manure, and 'B' 1 km downstream from them.

Samples for water chemistry (144) were analyzed in the laboratory of Belgorod Sanitary and Epidemiological Station. Concentration of nitrogen and phosphorus compounds was determined in accordance with State Standard Rules (GOST 4192–48 and GOST 18826–73) and expressed in $mg \cdot l^{-1}$.

Bacteriological analysis was conducted on 392 samples. As a measurement unit we used titre, which was taken as the minimal water volume in ml containing one microbic cell. The main aim was to define water contamination from *Escherichia coli*, enterococci and *Salmonella*. Analysis was done according to the methods of the Moscow Hygiene Institute. A water sample was considered uncontaminated when the *E. coli* titre was 1, the enterococci titre 10 and no traces of pathogenic microflora (Bey, 1980; Grigoryeva, 1975).

A bottom-sampler (area 0.025 m^2) was used for hydrobiological investigations. Bottom macroorganisms were preserved either in 4% formalin or 70% alcohol. For the laboratory studies on oligochaetes, whole mounts in glycerin or Canada balsam were used. Results were expressed in numbers per square meter on 86 samples.

Results

Data on water samples are summarized in Table 1.

In the Nezhegolyok River bacteriological contamination increased drastically during the veg-

Table 1. Water quality measurements for selected variables in two rivers.

Variable	Yerik R. Station 'A'	Yerik R. Station 'B'	Neyegolyok R.
Enterococci (titres)			
Spring	0.43	0.04	11.1
Summer	<0.01	0.04	0.4
Autumn	0.4	0.4	0.01
NH_4 (mg·l)			
Spring	0.08	0.19	0.05
Summer	0.20	0.10	0.03
Autumn	0.04	1.12	0.05
NO_3 (mg·l)			
Spring	2.38	7.53	1.56
Summer	2.66	2.33	1.38
Autumn	2.45	6.86	1.05
P compounds (mg·l)			
Spring	0.70	0.58	0.10
Summer	0.98	0.40	0.05
Autumn	1.28	0.56	0.11

etation period. In spring the water was clean according to the enterococci titre and contaminated according to the *E. coli* titre. In autumn the river water appeared heavily contaminated according to both titres and the pathogenic *Salmonella enteritides* was found. However, the variations in concentration of nitrogen and phosphorus forms were statistically insignificant and did not exceed permissible levels (Zhukinski, 1977; Bobyleva, 1980). The bottom fauna was represented by tubificids *Tubifex tubifex* and *Limnodrilus udekemianus*. Their numbers did not vary significantly with seasons either: in spring (1023 ind. $m^{-2)}$, in summer (1364 ind. $m^{-2)}$, and in autumn (1264 ind. $m^{-2)}$.

In the Yerik River environmental processes were more complex. In spring the water from station 'A' can be described as slightly contaminated in a bacteriological sense. Phosphorus compounds concentration equalled 0.70 mg l^{-1}. Bottom fauna was represented only by *Chironomus plumosus* larvae (1040 ind. m^{-2}). In the same period samples from the station 'B' showed higher bacteria counts while phosphorus concentration was slightly lower than at station 'A'. The bottom fauna also differed: in samples from station 'B' we found only the oligochaetes *Limnodrilus helveticus*, *L. udekemianus* and *T. tubifex*, with a significant preponderance of the first species (2610 ind. m^{-2}).

Our summer observation period was marked by high precipitation which confounded the differences between the two Yerik River stations. The water was heavily contaminated according to both the bacteriological titres at both stations, and at station 'A' the pathogenic *Salmonella virchow* was observed. Nitrogen compounds, though variable, did not exceed the permissible limits while phosphates exceeded the standard by 4–10 times (Table 1). Oligochaetes (*T. tubifex*, *Limnodrilus hoffmeisteri*, *L. udekemianus*, *L. helveticus*) gradually appeared at station 'A', in low numbers (up to 160 ind. m^{-2}). Station 'B', despite the small changes in the water quality, showed marked changes in the bottom fauna. In the first half of summer the zoobenthos was dominated by chironomids (the abundance of *C. plumosus* reaching 2800 ind. m^{-2}). The situation changed at the end of the summer, when the last chironomids emerged and the number of oligochaetes increased rapidly. At this time we found nearly 5000 ind. m^{-2} of *T. tubifex*.

In the autumn at station 'A' the water was cleanest based on the high *Escherichia coli* titre while the concentration of phosphates reached its maximum. Abundance of oligochaetes increased to 200 ind. m^{-2} of both *L. hoffmeisteri* and *L. udekemianus*, and 400 ind. m^{-2} of *T. tubifex*. At station 'B' the bacteriological condition of the river water remained unchanged, the concentration of ammonia nitrogen exceeded the standard twice while the concentration of phosphates reached the spring level. Two thousand ind. m^{-2} of *T. tubifex* and 760 ind. m^{-2} of *L. udekemianus* were found here in this period. However, after the appearance of leeches, the total abundance of all tubificids decreased to 400 ind. m^{-2}.

Discussion

It is well-known that 'self-purification' of water can be achieved by the activity of small organisms

(Pavlov, 1978; Grigyalis, 1963). Some aquatic species are capable of mineralizing large amounts of organic matter, e.g., *L. hoffmeisteri* or *L. udekemianus* can, under favourable conditions, ingest daily a sediment amount in excess of eight times their own weight (Poddubnaya, 1961).

The study of these two small rivers shows that tubificids occur here in considerable numbers and their activity limits water pollution from the by-products of agricultural fertilization. In comparatively stable environmental conditions, as in the Nezhegolyok River, the abundance of tubificids showed little seasonal change. In the case of stronger anthropogenic influences such as in the Yerik River, the intermittent organic contamination increases the amount of bacteria which is followed by a rise in tubificid abundance. However, this relationship can be hidden by the suppression of oligochaetes by their competitors or predators.

References

Bey, T.V. & G.M. Rachov, 1980. Znacheniye mikrobiologicheskikh pokazatelej pri prognozirovanii kachestva vody ustyevykh uchastkov rek. In Aktualnye voprosy okhrany okruzhayuschej sredy. Kiev: 53–59.

Bobyleva, O.A., 1980. Voprosy gosudarstvennogo sanitarnogo nadzora za vypolneniem meropriyatij po okhrane vodoyemov ot zagryazenij. In Aktualnye voprosy ochrany okruzhayuschej sredy. Kiev: 12–13.

Grigoryeva, P.V., 1975. Sanitarnaya bakteriologiya i virusologiya vodoyemov. Medicina, Moskva.

Grigyalis, A., 1963. Oligokhety vodoyemov Litovskoj SSR, ikh ekologiya i khozyajstvennoe znacheniye. Avtoreferat, kandidatskoj dissertatsii. Vilnius.

Pavlov, A.V., 1978. Laboratornye issledovaniya vneshnej sredy. Kiev.

Poddubnaya, T.L., 1961. Materialy po pitaniyu massovykh vidov tubificid Rybinskogo vodokhranilischa. Trudy instituta biologii vodokhranilisch, 4: 219–231.

Zhukinski, V.N. *et al.*, 1977. Nauchney osnovy controlya kachestva poverkhnostnykh vod po gidrbiologicheskim pokazatelyam. Gidrometeoizdat, Moskva: 43–53.

Hydrobiologia **278**: 133–138, 1994.
T.B. Reynoldson & K.A. Coates (eds), Aquatic Oligochaete Biology V.

Aquatic Oligochaeta of the Dnieper–Bug Estuary system

T.G. Moroz
Institute of Hydrobiology, Academy of Sciences of the Ukrainian SSR, Kherson Biological Station, Ukrainian SSR, 325016, Kherson, Maria Fortus, 87, Ukraine

Key words: Liman, salinity, distribution, abundance, species

Abstract

This paper describes the influence of different ecological factors on the distribution of Oligochaeta in parts of the Dnieper-Bug Estuary. The important environmental factors in the Dnieper-Bug Estuary were salinity, discharge, water quality, trophic condition and sediment characteristics. In the Lower Dnieper-Bug a correlation was found between abundance and number of oligochaete species and the concentration of ammonium nitrogen and mineral phosphorus in the water.

Introduction

The Oligochaeta are one of the most abundant groups of bottom macroinvertebrates in the Dnieper-Bug Estuary system, dominating in both frequency and occurrence.

The first descriptions of Oligochaeta in this area were by Grebnitsky (1874), Shmankevich (1873) and Ostroumov (1897). The complete list of Oligochaeta of the Dnieper-Bug Liman in 1937 was published by Jaroshenko (1948), listing 28 species but not describing their distribution in the Liman. Further investigations on species composition, biological and ecological characteristics of the Oligochaeta in this region were carried out by Fomenko (1971; 1972) Finogenova (1969; 1972a,b), Moroz (1977; 1983; 1987a,b) and Moroz & Grigoriev (1989).

Marine, brackish-water and fresh-water species are represented and their distribution and quantity in different areas are determined by different environmental factors. The Dnieper-Bug Estuary system consists of several separate water bodies with various hydrological and hydrochemical characteristics: the Dnieper-Bug Liman (the estuary proper); the Lower South Bug River; the main river channel, the delta and water bodies of the Lower Dnieper River (Fig. 1).

Materials and methods

This paper is based upon the analysis of 3518 quantitative samples taken from 1970 to 1990 on a network of stations. Samples were taken with a Peterson bottom-grab (0.025 m^2), seasonally (in winter, spring, summer, autumn) and monthly in selected areas (Fig. 1):

Stanislav shallows and Dnieper-Bug Liman – 1971–72
Delta and Lower Dnieper 1975, 1981–82.

Two samples were taken at each station, on each sampling occasion and sieved in the field through a 0.55 mm mesh. Samples were immediately preserved in 4% formalin. Worms were hand sorted in the laboratory and subsequently preserved in 70% alcohol.

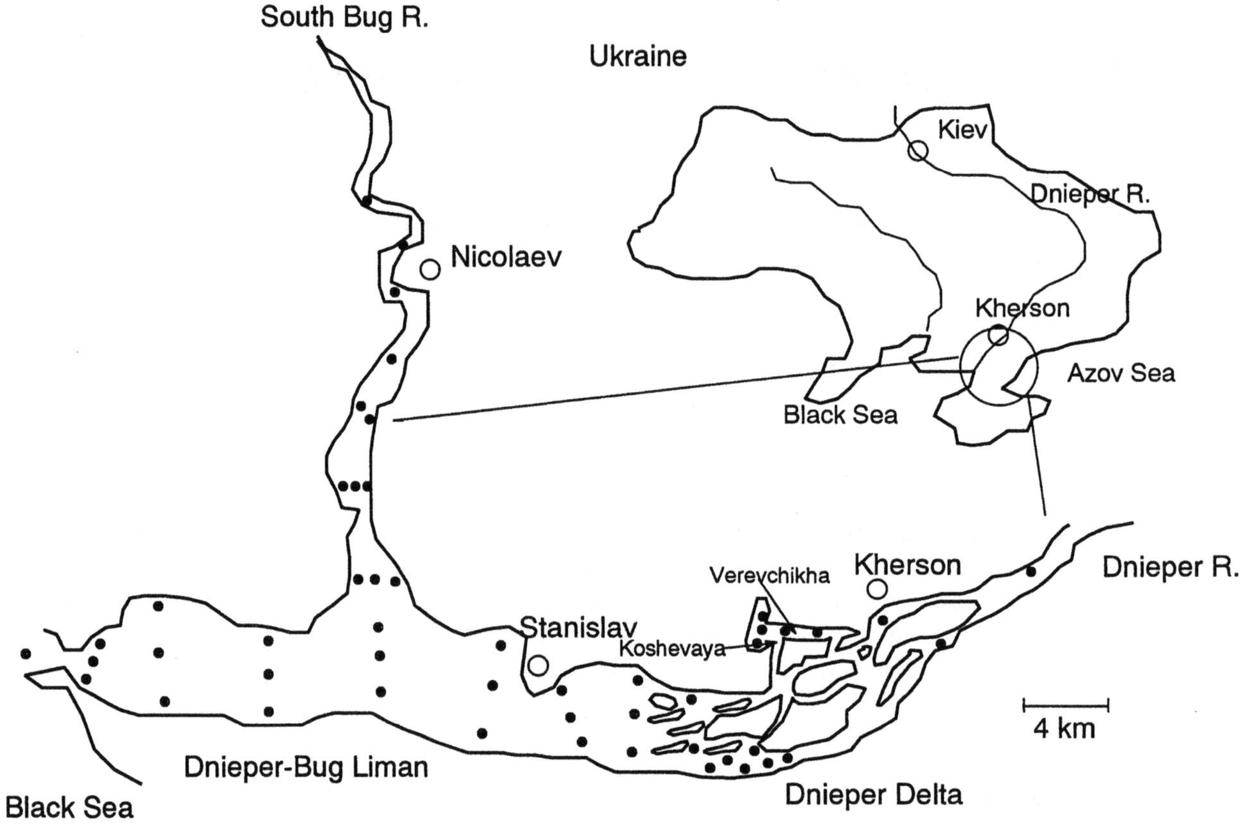

Fig. 1. Location of water bodies in the Dnieper-Bug Estuary and sampling sites (●).

Hydrochemical data were collected at the same time by Aleksandrova *et al.* (1985).

Results and discussion

Distribution of taxa

To date 89 species and sub-species of Oligochaeta have been described in the study area and the Tubificidae (35 species) and Naididae (25 species) dominate. Marine species have been found only in the Liman; brackish-water species predominate in the Lower South Bug and in the Liman. Freshwater species are most abundant in the Dnieper delta (Table 1). The distribution of marine species is mainly limited to the west part of Liman, which has a direct connection with the Black Sea. Among them *Heterochaeta costata* Claparede, 1863 and *Paranais litoralis* (Muller, 1784), on occasion *Clitellio arenarius* (Muller, 1776), predominate. Of the brackish-water species of marine origin *T. swirencowi* is mostly found in the estuarine areas. Farther in the Liman and especially in the Bug Liman occur *P. simplex, P. multisetosa, Paranais frici* Hrabe, 1941, as well as the Ponto-Caspian enchytraeids *Marionina mica* Finogenova, 1972, and *Marionina brevis* Finogenova, 1972. The euryhaline freshwater species

Table 1. Ecological composition of the Oligochaeta in the Dnieper-Bug estuary.

Number of species	Dnieper-Bug Liman	Lower River Channel	Dnieper Delta	Lower South Bug
Marine sp.	4	0	0	0
Brackish water sp.	18	9	3	9
Freshwater sp.	47	59	56	24
Total sp.	75	68	59	33

Table 2. Zoogeographical composition of the Oligochaeta in the Dnieper-Bug estuary.

Number of species	Dnieper-Bug Liman	Lower River Channel	Dnieper Delta	Lower South Bug
Cosmopolitan	22	21	21	6
Holartic	11	9	8	4
Palaeoarctic	20	30	27	15
Ponto-Caspian	11	7	3	7
Endemic	4	1	0	1

are distributed mainly in the deeper, freshwater areas of the Liman and in the river mouths.

Palaearctic and European species dominate in all water bodies, except the Liman, where cosmopolitan species are most common (Table 2). The Ponto-Caspian species are most fully represented in the Liman (11 species), among them three are endemic to the Black Sea basin. These are *Potamothrix danubialis* Hrabe, 1941, *Haber svirenkoi* Lastockin, 1937, and *Psammoryctides deserticola lastockini* Jaroshenko, 1948. The brackish-water species of marine origin *Paranais simplex* Hrabe, 1936, *Paranais multisetosa* Finogenova, 1972, and *Tubificoides swirencowi* Jaroshenko, 1948, are also endemic.

The Ponto-Caspian species are largely found on the Liman's freshwater shallows as well as in the lower reaches of the Dnieper and South Bug rivers.

Abundance of taxa

On average the oligochaetes comprise 80–100% of the benthic fauna and their annual abundance in different regions fluctuates from 1,000 to 15,000 ind. m^{-2}. Oligochaetes are most abundant in the eutrophic water bodies (the Koshevaya and Verevchikha branches) and in the channels between lakes. The lowest numbers of oligochaetes occur in the hypereutrophic and dystrophic lakes of the delta as well as mouths of the main branches of the Dnieper delta. Maximum densities were observed in the middle of the Liman (181,000 ind. m^{-2}) and in the Verevchikha branch (221,000 ind. m^{-2}) which receives sewage from the town of Kherson. In Verevchikha the dominant species were *Limnodrilus claparedeanus* Ratzel, 1868, and *Limnodrilus hoffmeisteri* Claparede, 1862, and in the centre of the Liman – *Potamothrix hammoniensis* (Michaelsen, 1901) and *Potamothrix heuscheri* (Bretscher, 1900). The maximum biomass was also found in the same regions, 120.1 and 112.2 g m^{-2}, respectively.

In the Dnieper-Bug Liman oligochaetes formed between 47 to 78% of the average abundance of total benthos. They were most abundant in the shallows. On the muds of the Liman *P. hammoniensis* and *P. heuscheri* dominated and on muddy sands *Potamothrix moldaviensis* Vejdovsky & Mrazek, 1903, *Isochaetides michaelseni* (Lastockin, 1936), *Potamothrix caspicus* (Lastockin, 1937) and sometimes *Psammoryctides barbatus* (Grube, 1861) were dominant. In the main channel of the Dnieper R. oligochaetes were most abundant on muddy sand mixed with shells, the dominant species included *P. moldaviensis, P. deserticola* (Grimm, 1876) and *Tubifex newaensis* (Michaelsen, 1902).

In the mouths of the main branches of the Dnieper R., *I. michaelseni* and *P. moldaviensis* also predominated. In the small branches overgrown with macrophytes and in the standing water bodies of the delta limnophilous *Psammoryctides albicola* (Michaelsen, 1901) and *P. hammoniensis* were found but in the strongly polluted areas of the delta *L. claparedeanus* and *L. hoffmeisteri* were most frequent.

Relationship with environmental factors

In different areas of the ecosystem the density and qualitative diversity of the oligochaetes were related to different factors: the nature of the sediment, the oxygen regime, and the rivers' inflow and water quality.

The majority of species inhabiting the Liman occur on a range of sediment types; nevertheless, there are species that are more or less restricted to certain types of sediment. *Marionina mica, I. michaelseni and P. deserticola lastockini* inhabit either sand or slightly silty-sand; *P. hammoniensis P. heuscheri, L. claparedeanus, L. hoffmeisteri and P. litoralis* occur on muds while *Stylaria lacustris* (Linnaeus, 1767), *P. albicola, Ophidonais serpentina* (Muller, 1773) and *Spirosperma ferox* Eisen, 1879, predominated among the macrovegetation.

The silty-sands are characterized by the greatest species diversity, and the pure sands and deep muds with the lowest maximum densities. Greatest densities (2,000–9,000 ind. m^{-2}) occurred in silty-sands of shallow areas and lower densities (700–4,000 ind. m^{-2}) were found in muds of the central areas of the eastern and central parts of the Liman. The abundance of oligochaetes is lowest in the western part (most saline), and averaged only 33–994 ind. m^{-2}. The greatest number of species and abundance of oligochaetes over the years of observation were found among the *Dreissena* shells on silty-sands of the Stranislav shallows in the eastern (freshwater) part of Liman. An increase in the abundance of oligochaetes on sandy sediments in brackish water was observed also by Pfannkuche (1977; 1980).

Salinity was used to discriminate between water masses. The brackish water masses were divided into the following salinity types: absolute (7.5–11.0 ‰), intermediate (2.5–5.0 ‰ – I subtype, 5.0–7.5 ‰ – II subtype) and relative (0.5–2.5 ‰); by Moroz & Gilman (1988) and were influential in determining the species composition of oligochaetes. The domain of marine species (*H. costata, P. litoralis*) was the zone of absolute brackish-water; the brackish-water species of marine and freshwater origin (*P. simplex, T. swirencowi, Tubificidae sp.* 1) inhabited the zone of intermediate brackish-water of the second subtype; the Ponto-Caspian (*M. mica, M. brevis, P. caspicus*) and euryhaline freshwater species (*P. hammoniensis, P. barbatus, I. michaelseni*) inhabited the intermediate brackish-water of the first subtype and the relative brackish water while the freshwater species (*T. newaensis, L. claparedeanus, L. hoffmeisteri, L. udekemianus*) occurred primarily in freshwater. The intrusion of a water mass of another type modified the distribution of the oligochaetes, as well as regulating the oxygen content of the near bottom water. The intrusion of the saltier marine water resulted in vertical stratification followed by a decline in the oxygen concentration of the near-bottom layers. This was observed in the Stanislav Shallows in August 1972 when the intrusion of brackish water resulted in a decline in oligichaete numbers from 5,340 to 620 ind. m^{-2} A similar observation was made by Bagge & Ilus (1973) in brackish waters in Finland. As a result, oligochaetes were most abundant on shallows where no stratification occurred.

The distribution of water masses is determined by the discharge patterns of the Dnieper and the South Bug. At high river flows the Liman is well flushed, has high oxygen concentrations and maintains a diverse oligochaete fauna, e.g., in the 1981 water-year the average abundance of oligochaetes in the Liman was 8,231 ind. m^{-2}, but in the 1984 water-year, when discharge was low, only 3,746 ind. m^{-2} were found. In addition, the annual discharge of the Dnieper is seasonal and monthly variation also influences the distribution of oligochaetes. In 1986 when discharge was reduced as a result of restricted outflow of water across the Kakhovka hydro-electric dam in August and September, a decline in abundance of oligochaetes occurred in the central part of Liman in the following month (from 1,619 ind. m^{-2} in July to 687 ind. m^{-2} in October).

Despite the importance of all other factors, both the total discharge of the Dnieper and its seasonal pattern are most important in determining the physical and chemical conditions for oligochaetes in the Dnieper-Bug Liman.

With regard to substrata the following relationships have been observed: on the pure and somewhat silty-sands of the Dnieper main bed the average abundance of macrobethic oligochaetes does not exceed 1,000 ind. m^{-2} with 2–3 species, on more silty-sands and muds the density increases 4,000–5,000 ind. m^{-2} with 4–5 species. A reduction in the current in the main channel resulted in similar numbers of oligochaetes in the littoral and central part of the river as well as a disappearance of rheophilous forms (*Propappus volki* Michaelsen, 1916) and dominance of limnophilous species.

In the delta of the Dnieper, current velocity and water exchange are the most important factors for oligochaetes. In weak currents (less than 1 cm s^{-1}) oligochaetes were rare (on the average from 5 to 78 ind. m^{-2}). With increased current (more than 10 cm s^{-1}) the abundance of oligochaetes averages 3,000 ind. m^{-2}, fluctuating from 1,381 to 2,995 ind. m^{-2}, and in the case of mean current (3–10 cm. s^{-1}) from 347 to 1,708 ind. m^{-2}).

Using statistical analysis of data obtained in the Lower Dnieper in 1981–1982, the relationship between the average abundance of oligochaetes (calculated on the basis of monthly samples on the permanent stations from March to November) and some hydrological and hydrochemical ingredients were investigated. Significant correlations were found between the abundance of oligochaetes and the oxygen concentration in the water ($r = -0.94 \pm 0.15$), and transparency ($r = -0.79 \pm 0.28$). The abundance of oligochaetes increases as oxygen concentration (31.3 to 90%) and transparency (0.5 to 2.7 m) decline. A close correlation was also found between the abundance of oligochaetes and the Shannon-Wiener diversity index ($r = -0.78 \pm 0.28$), the saprobic index of Pantle and Buck ($r = 0.95 \pm 0.14$), and with oxygen concentration ($r = -0.87 \pm 0.20$). The poorer the quality of water in the Lower Dnieper the higher the abundance of oligochaetes. As conditions deteriorate for the majority of bottom invertebrates, the abundance of oligochaetes species tolerant to organic pollution increases (Brinkhurst, 1965).

Similar relationships were established between the abundance of oligochaetes and, the concentration of ammonium nitrogen ($r = 0.97 \pm 0.08$) and mineral phosphorus ($r = 0.97 \pm 0.11$) in the water.

The species richness of oligochaetes and the occurrence of a number of the Ponto-Caspian species are primarily determined in the Lower Dnieper, by the water quality and the character of sediment. The higher the saprobic index the fewer the number of Ponto-Caspian species that occur in the habitat. There is a negative correlation between the number of Ponto-Caspian species and the saprobic index ($r = -0.88 \pm 0.27$). In the delta a positive correlation was found between the abundance of *P. hammoniensis* and the current velocity ($r = 0.79 \pm 0.14$).

Conclusion

The abundance and species richness of oligochaetes in the Dnieper-Bug estuary system depends: in the Dnieper-Bug Liman mainly on the type of water masses, the character of the sediment, the runoff of the Dnieper and its distribution during the year; in the main river-bed, branches and small channels of the Lower Dnieper on the quality and oxygen concentration of the water; in the delta water bodies on the current velocity and trophic state.

References

Aleksandrova, N.G., T.G. Moroz & V.S. Polischuk, 1985. Complex studies of water quality in lower reaches of the Dnieper. Vodnye resursy 6: 119–127. [in Russian]

Bagge, P. & E. Ilus, 1973. Distribution of benthic tubificids in Finnish coastal waters in relation to hydrography and pollution. Oikos, Supplementum 15: 214–225.

Brinkhurst, R.O., 1965. The biology of the Tubificidae with special reference to pollution. Proc. 3rd Seminar. Biological problems in water pollution, Cincinnati: 57–66.

Finogenova, N.P., 1969. Oligochaetes of the Ponto-Caspian basin (Lower reaches and estuaries of some rivers, the Black, Azov and Caspian Seas). Dissertation thesis, Leningrad, 17 pp. [in Russian]

Finogenova, N.P., 1972a. Oligochaetes in Ponto-Caspian brackish water basins. In G.M. Belyayvev *et al.* (eds), Aquatic Oligochaeta. Nauka, Moscow: 65–74. [in Russian]

Finogenova, N.P., 1972b. New species of Oligochaeta from the Dnieper-Bug Liman and the Black Sea with a revision of some species. Trudy Zool. Inst., Acad. Nauk SSSR, Leningrad 52: 97–116. [in Russian]

Fomenko, N.V., 1971. Distribution of oligochaetes in the Dnieper-Bug Liman in the summer of 1966. In A.V. Topachevskij *et al.* (eds), The Dnieper-Bug Liman. Naukova dumka, Kiev: 316–338. [in Ukrainian]

Fomenko, N.V., 1972. On ecological groups of oligochaetes on the Dnieper river. In G.M. Belyaev *et al.* (eds)., Aquatic oligochaete. Nauka, Moscow: 94–106. [in Russian]

Grebnitsky, N.A., 1874. Materials for fauna of Novorossian region. To fauna of the open limans. Zapiski Novoross. Obsch. Estestvoisp. 2: 267–270. [in Russian]

Jaroshenko, M.F., 1948. Oligochaeta of the Dnieper-Bug Liman. Nauchny Zapiski Modlavskoj nauchno-issl. Bazy Acad. Nauk SSSR 1: 57–59. [in Russian]

Moroz, T.G., 1977. Oligochaeta from the river mouths of the North-Western Black Sea. Hydrobiol. J. 4: 20–25. [in Russian]

Moroz, T.G., 1983. Biological and ecological peculiarities of Oligochaeta of the genus *Limnodrilus*. In B.E. Kurashvili (ed.), Aquatic Oligochaeta. Metsniereba, Tbilisi: 71–75. [in Russian]

Moroz, T.G., 1987a. Oligochaetes in brackish water bodies of the South Ukraine. In O.L. Kachalova & E.A. Parele (eds), Aquatic Oligochaeta. Riga: 96–100. [in Russian]

Moroz, T.G., 1987b. Oligochaetes from limans of the Black Sea. Vestnik zoologii, Kiev: 3–7. [in Russian]

Moroz, T.G. & V.L. Gilman, 1988. Hydrobiologically based typification of the brackish water. Hydrobiol. J. 5: 64–67. [in Russian]

Moroz, T.G. & B.F. Grigoriev, 1989. Macrozoobenthos. In U.P. Zaitsev (ed.), The Dnieper-Bug Estuary Ecosystem. Naukova dumka, Kiev: 160–165. [in Russian].

Ostroumov, A.A., 1897. On hydrobiological investigations in the Estuaries of the South Russian rivers. Bull. Acad. Imper. Sci. Sanct-Petersburg, 1896: 343–362. [in Russian]

Pfannkuche, O., 1977. Okologische und systematische Untersuchungen an naidomorphen Oligochaeten brackischer und limnischer Biotype. Dissert. Univ. Hamburg, 138 pp.

Pfannkunche, O., 1980. Aquatic Oligochaeta (Annelida) from the littoral zone of the Limfjord (Denmark). Verh. naturwiss. ver. Hamburg 23: 249–254.

Shamankevich, V., 1873. On the invertebrate animals of limans near Odessa. Zapiski Novoross. Obsch. Estestvoisp. 2: 273–341. [in Russian]

Hydrobiologia **278**: 139–149, 1994.
T.B. Reynoldson & K.A. Coates (eds), Aquatic Oligochaete Biology V.

On the distribution patterns and population dynamics of sublittoral and profundal oligochaeta fauna from Lake Banyoles (Catalonia, NE Spain)

M. Rieradevall & M. Real
Dept. d'Ecologia, Univ. Barcelona, Diagonal 645, 08028-Barcelona, Catalonia, Spain

Key words: Oligochaeta, *Potamothrix heuscheri*, Lake Banyoles, oxygen depletion, sulphide content, cestodes

Abstract

Oligochaeta assemblages from Lake Banyoles were composed of one species of Naididae and eight species of Tubificidae, *Potamothrix heuscheri* (8–1350 ind. m^{-2}) being the most abundant and widely distributed species, followed by *Limnodrilus hoffmeisteri* (25–858 ind. m^{-2}) and *Psammoryctides barbatus* (12–792 ind. m^{-2}). The remaining species were *Potamothrix hammoniensis, Potamothrix bavaricus, Limnodrilus claparedeanus, Branchiura sowerbyi, Aulodrilus pigueti* and *Dero digitata*, all of which occurred at lower densities and with restricted distributions.

Regardless of sampling depth, maximum densities of oligochaetes were found in winter (up to 5,142 ind. m^{-2} in December) and secondarily in June. Immature Tubificidae with hair setae and mature *P. heuscheri* were present and dominant throughout the year. Cestode parasites were recorded infesting *L. hoffmeisteri* and *P. heuscheri* populations.

Oligochaeta densities increased with depth, but low oxygen concentrations (less than 1 mg l^{-1}) during a long period (4 months) in some basins of the lake acted as a key factor in reducing the density of worms and the species richness to one species, *Potamothrix heuscheri*.

Introduction

The oligochaete fauna have an important role in all sublittoral and profundal lake and reservoir habitats with soft sediments. In lentic habitats of the Iberian Peninsula, oligochaetes have been studied from the point of view of geographical distribution in the profundal areas of reservoirs (Martinez-Ansemil & Prat, 1984). Later studies have examined relationships between profundal oligochaetes and physical and chemical characteristics of water and sediments in the same environments (Prat *et al.*, 1991; Real & Prat, 1991, in press; Real *et al.*, 1992).

Oligochaetes are known to be resistant to low oxygen conditions in the sediments (Gnaiger & Staudigl, 1987; Reynoldson, 1987). Species adapted to long anoxic periods are usually considered as indicators of eutrophic status (Howmiller & Scott, 1977; Lang & Lang-Dobler, 1980; Milbrink, 1980; Särkkä, 1987; Uzunov *et al.*, 1988). Lake Banyoles can be considered as an oligotrophic lake from the algal productivity and nutrient content data (Planas, 1973), but the profundal areas do not reflect such status. Meromictic conditions in some areas of the lake, high temperatures and high levels of sulphide in the water (Guerrero *et al.*, 1978; Garcia-Gil *et al.*, 1991; Rieradevall, 1991) make the availability of oxygen a key factor in understanding the distribution and dynamics of the benthic community in Lake Banyoles.

This paper deals with the depth distribution of oligochaetes and its relationship with some environmental factors such as the duration of the anoxic period. We also present some data on the population dynamics of some species and variations in the abundance of the total worm community. The sampling program was conducted in the context of a more extensive study on the benthic fauna of the sublittoral and profundal areas of Lake Banyoles (Rieradevall, 1991). The general results of that study and the details of different taxonomic groups are or will be published elsewhere (Rieradevall, in press; Rieradevall & Prat, 1989, 1991; Rieradevall & Busquets, 1990; Rieradevall & Gil, in press; Prat & Rieradevall, in press; Roca & Rieradevall, in prep.; and Prat & Rieradevall, in prep.).

Study site

Lake Banyoles is situated at 42° 7′ N, 2° 45′ E and is 175 m above sea level. Area and maximum length are 1.18 km^2 and 2,130 m, respectively. It is the largest of a series of small lakes forming a complex karstic system, where the main incoming of water is the vertical entrance from the bottom of the lake. These springs maintain a very fine clay material in suspension in several deep parts of the lake. Mean depth is 15 m and the maximum depth is 35–40 m above which is a layer of suspended sediment.

Secchi disk depth ranged between 0.9 and 6.8 m in the period studied (August 1986–December 1987). Bottom temperatures ranged between 8 and 18 °C in the profundal and 8–25 °C in the sublittoral (Rieradevall, 1991). The water was rich in calcium and sulphate (> 1 g l^{-1}) as result of the dissolution of calcareous and gypsiferous sequences. This fact maintains low $P\text{-}PO_4$ concentrations in water (0.05–0.08 μg at l^{-1}) (Planas, 1973). Conductivity and pH of surface water was around 1,300 μS cm^{-2} and 7–8.15 respectively.

Lake Banyoles has six main basins and, to date, fourteen bottom springs have been described (Moreno-Amich & Garcia-Berthou, 1989). Each basin presents different limnological characteristics of morphology and volume, which induce different duration of the anoxia period (Rieradevall, 1991; Rieradevall & Prat, 1991). It is a monomictic lake, stratified between April to October. In basins III and IV, the presence of a chemocline induces anoxic conditions. In basin III, the chemocline remains at 20 m depth all the year. There, meio and macrobenthic organisms were not found (Rieradevall, 1991, in press; Rieradevall & Prat, *loc. cit.*). In basin IV, the chemocline fluctuates between 13 to 17 m, which means that benthic organisms have to adapt their life cycles to four months of anoxia in the deep parts of the basin (Prat & Rieradevall, in prep.). In basin I and II, there is only one month of summer anoxia. Information about basin V, which is very small, and basin VI are not available.

Methods

Of the six basins, two (basins I and IV) were selected for study, which represent two different oxygen environments with varying duration of anoxia (1 and 4 months respectively). Temperature and oxygen content of water close to the sediment were monitored. Benthos of soft sediments in the sublittoral and the profundal areas of these basins were studied at sites of 5, 12 and 20 m depth in basin I and 7 and 13 m depth in basin IV. Each site was sampled monthly, from December 1986 to December 1987 with a modified (400 cm^2) Ekman grab (three replicates per site). Samples were filtered through 150 μm mesh and benthic organisms were sorted alive with the flotation method (Anderson, 1959). An aliquot of the remaining material was also sorted in order to test the efficiency of the method (Rieradevall, 1991). The material was fixed with formalin (4%) until counting under a stereoscopic microscope at 10 times magnification, and stored in 70% ethanol in plastic vials. For the oligochaete study, we selected one replicates at each sampling station per sampling event. For the identification of the oligochaete species, groups of about 5 worms of similar size were mounted in glycerin + lactic acid medium (1:1) (Lafont, 1987). The mounted material was kept at 80 °C on a heating plate to clear before being identified (Lafont, 1983). One replicate for each month was used for species identification, from two sampling stations: 20 m in basin I and 13 m in basin IV. For the rest of sampling stations, Oligochaeta fauna of a representative sample per season (December, April, June and October) were identified.

Results

Oligochaetes were well represented in the sublittoral and profundal habitats with soft sediments in Lake Banyoles. In Fig. 1, we summarize the relative abundance (%) of oligochaetes with respect to the total macrobenthos community, and the mean densities of worms over a thirteen month period at each sampling station.

Oligochaetes averaged up to the 64% of the total macrobenthos (mean densities being 1,303 ind. m^{-2}) at 20 m in basin I. However, their abundance decreased with decreasing depth, accounting for 27% (1,126 ind. m^{-2}) at 12 m and 8% (886 ind. m^{-2}) of the community at 5 m depth in the same basin. In contrast, in basin IV, mean densities and percentage of oligochaetes decreased with increasing depth; with 10% (567 ind. m^{-2}) at 7 m and 0.45% (11 ind. m^{-2}) at 13 m. In basin III, at 20 m depth, no worm or other fauna, except *Chaoborus flavicans*, was found (Rieradevall, 1991; Rieradevall & Prat, 1991). These different patterns are likely related to oxygen availability and sulphide concentrations, as we discuss below.

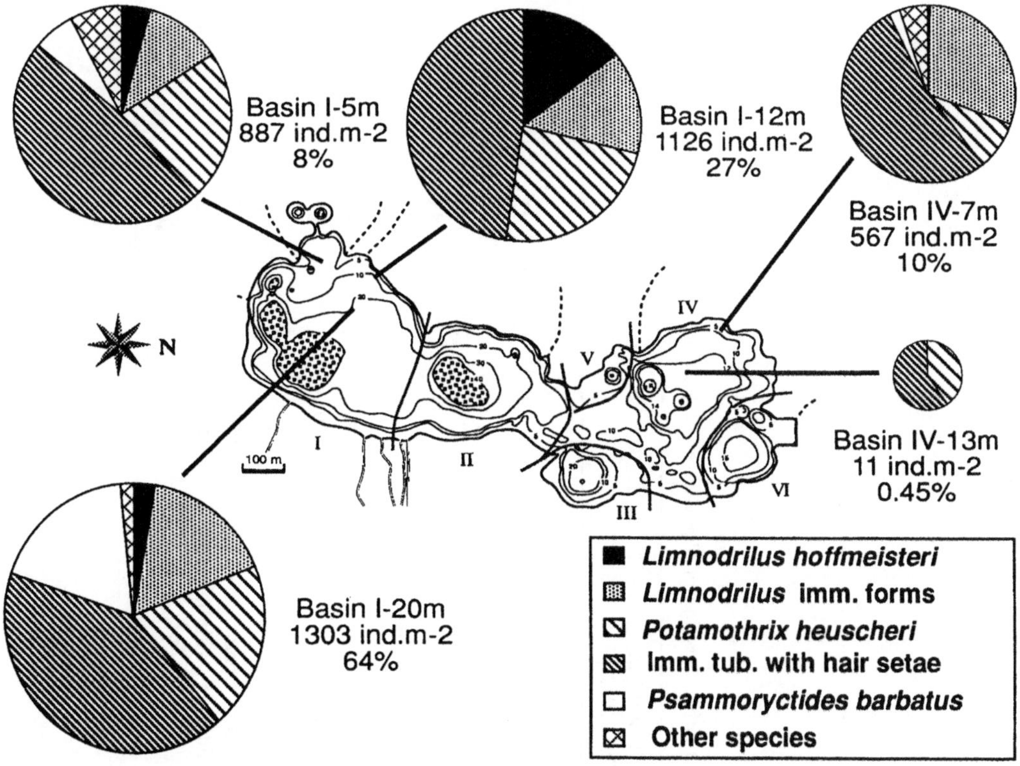

Fig. 1. Distribution and relative abundances of the Oligochaeta species in Lake Banyoles. Mean densities of oligochaetes (ind m^{-2}) based on a thirteen month survey and percentage of total oligochaetes in the macrobenthos. Strong shaded area shows the permanent suspended sediments. The basins are indicated in Roman numbers.

Table 1. Monthly variation of oligochaete taxa densities (ind. m^{-2}) found in Basin I at 20 m depth of Lake Banyoles.

Basin I – 20 m depth Month: 1986–1987	D	J	F	M	Ap	My	Jn	Jl	Aug	S	O	N	D
F. NAIDIDAE													
Dero digitata		120	43						12				
F. TUBIFICIDAE													
Aulodrilus pigueti		120	149										
Branchiura sowerbyi			43	59									
Limnodrilus claparedeanus			21	59									
L. hoffmeisteri			149	59	31		136			25	43		
Limnodrilus immature forms	570	336	191			12	546	67	6	99	43		
Potamothrix bavaricus	79		64				30						
P. hammoniensis						12							
P. heuscheri	825	360	276		94		136		12	172	365	19	167
Psammoryctides barbatus	570	792	617	471	314	12	212			197	214	249	37
Immature tubificids with hair setae	2281	1897	1913	118	219	347	456		28	99	43	115	195
TOTAL	4325	3625	3466	766	658	383	1516	67	58	592	708	383	399

Oligochaeta assemblage composition

Sublittoral and profundal oligochaete assemblages in Lake Banyoles were composed of eight species of tubificids and one naidid (Table 1). Tubificids with hair setae were most abundant (average 50%) at all the sampling stations (Fig. 1) and also over the year (Table 1).

Potamothrix heuscheri (Bretscher) (8–1,350 adult ind. m^{-2}) was the most abundant and widely distributed species, and represented between 9 and 40% of the worm community, being the only species present at 13 m in basin IV. This species did not show a clear variation with depth in basin I, but its relative abundance increased with depth in basin IV.

Limnodrilus hoffmeisteri Claparède (25–858 adult ind. m^{-2}) was also present at all the sampling stations (1–15% of the total oligochaete assemblage). If we consider immature *Limnodrilus* forms plus *L. hoffmeisteri* all together, no clear difference appears in their densities among stations and depths. However, mature individuals were more common in some locations of sublittoral and the upper part of the profundal of the lake (Fig. 1).

Other species of the genera *Potamothrix* and *Limnodrilus* appeared only in basin I at 20 m, one of the two most extensively studied stations, but at low densities. *Limnodrilus claparedeanus* Ratzel was present in February and March with 21–59 adult ind. m^{-2} (Table 1). At this station we found *Potamothrix bavaricus* (Öschmann) in December, February and June, with densities of 30–79 adult ind. m^{-2}. Mature forms of *Potamothrix hammoniensis* (Michaelsen) appeared in May at very low densities (12 ind. m^{-2}).

Psammoryctides barbatus (Grube) (12–792 ind. m^{-2}) appeared mainly in the profundal zone in basin I (20 m) where it represents 18% of oligochaetes, and occurred less frequently (2–6%) in the sublittoral (5–7 m). It was absent at 12 m (Fig. 1).

The remaining species occurred at lower densities and less frequently (Table 1 and 2). The small tubificid *Aulodrilus pigueti* Kowaleski appeared only sporadically at 7 and 20 m depths. Its low densities (120–153 ind. m^{-2}) are of little importance if we consider the lake as a whole (0.7%). *Branchiura sowerbyi* Beddard was present at 20 m depth in basin I in winter (43–59 ind. m^{-2}) but in the other seasons it was found at shallow stations (5 m depth). Finally, the naidid *Dero digitata* (Müller) usually was present in the sublittoral zone (1–7%), and also in the profundal (at 20 m in basin I).

Population dynamics

In Figs 2 and 3 we show the variation in densities of total oligochaetes and cocoons at the five

Table 2. Seasonal variation (December, April, June and October) of densities (ind m^{-2}) of oligochaeta taxa found in Basin I at 12 and 5 m depth and in Basin IV at 7 m depth in Lake Banyoles.

	Basin I – 12 m				Basin I – 5 m				Basin IV – 7 m			
Month: 1986–1987	D	Ap	Jn	O	D	Ap	Jn	O	D	Ap	Jn	O
Dero digitata						108	292	19			67	
Aulodrilus pigueti									153			
Branchiura sowerbyi						18	42	19				
Limnodrilus hoffmeisteri	69		858	42	287						27	
Limnodrilus immature forms	749	108		83	575	90	42	19	1265	60	54	
Potamothrix heuscheri	1350	42		167	575	343	479	57	364		54	
Psammoryctides barbatus					287	18		75		90		
Immature tubificids with hair setae	2974	117		42	575	814	1603	94	1092		1306	100
TOTAL	5142	267	858	334	2299	1391	2458	283	2874	150	1508	100

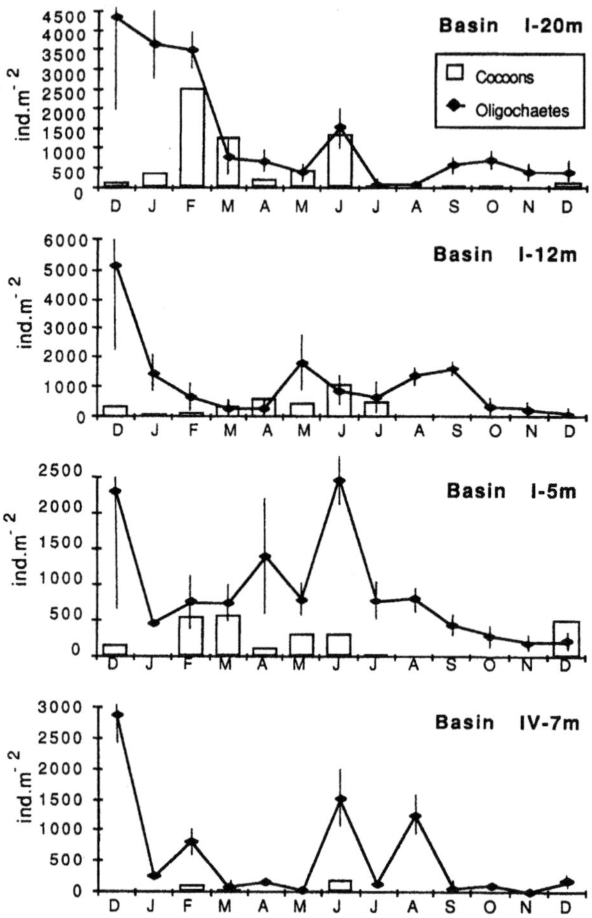

Fig. 2. Monthly variation of densities of total oligochaetes (ind m^{-2}) and the cocoons (No m^{-2}) in the stations studied in basin I (at 20, 12 and 5 m depth) and at 7 m in basin IV in Lake Banyoles (Dec. 1986–Dec. 1987). Standard error bars are indicated. Notice that different scales are used in the vertical axis.

stations throughout the period December 1986–December 1987. Two periods of density maxima occur at all the sampling stations. The first is in winter, with maximum values in December, and the second in spring with maximum values mainly in June. At the profundal stations (20 and 12 m depth) in basin I, which are anoxic for one month (October), both stations showed similar trends. At the stations closer to the shore (5 m in basin I and 7 m in basin IV) the June maximum is enhanced. Cocoons also appear in the late winter and or late spring at all stations.

The proportion of mature individuals over the thirteen month period was monitored at two profundal stations, at 20 m in basin I and 13 m in basin IV. In basin I mature individuals were found throughout the year, except in July, when the population was very scarce (Fig. 3a). The greatest number were found in winter, and secondarily in June and October. In general, this pattern reflects the population dynamics of *P. heuscheri*, which was the dominant mature species except in spring (Fig. 3b). Mature individuals of the rest of the species were more sporadic. Mature forms of *P. bavaricus* were found only in winter, while *P. hammoniensis* matured in spring. Mature *P. barbatus* appeared only in April in the profundal (basin I – 20 m) as well as in the sublittoral (basin IV – 7 m), suggesting a short and synchronized breeding period (Fig. 3b). On the other hand, mature forms of *L. hoffmeisteri* were observed mainly from late winter to late spring and secondarily in autumn; while *L. claparedeanus* matured in February–March.

The density of immature forms increased in winter months, and showed a secondary peak in June. Peak numbers of immature tubificids with hair setae and those of tubificids without hair setae were very similar. Maximum number of cocoons were laid in February, March and July, and in these months plus May, and September–October, developed embryos were observed inside the cocoons (Fig. 3a).

At 13 m in basin IV, a very different situation was observed (Fig. 3c). The oligochaete population was restricted from January to June, due to the long anoxic period (less than 1 mg l^{-1} over four months) and the high sulphide content (0.04–1 mM, Garcia-Gil *et al.*, 1991) in bottom waters, from July to November. Both immature and mature individuals of *Potamothrix heuscheri* appeared first in January, probably as a result of the migration from other parts of the lake. Mature forms disappeared in March and May. Cocoons were laid from March on, and greater numbers were found in June. Embryo development inside the cocoons was observed from May to June. However, because of severe environmental conditions which reduced the worm population to zero, further observations were not made (Fig. 3c).

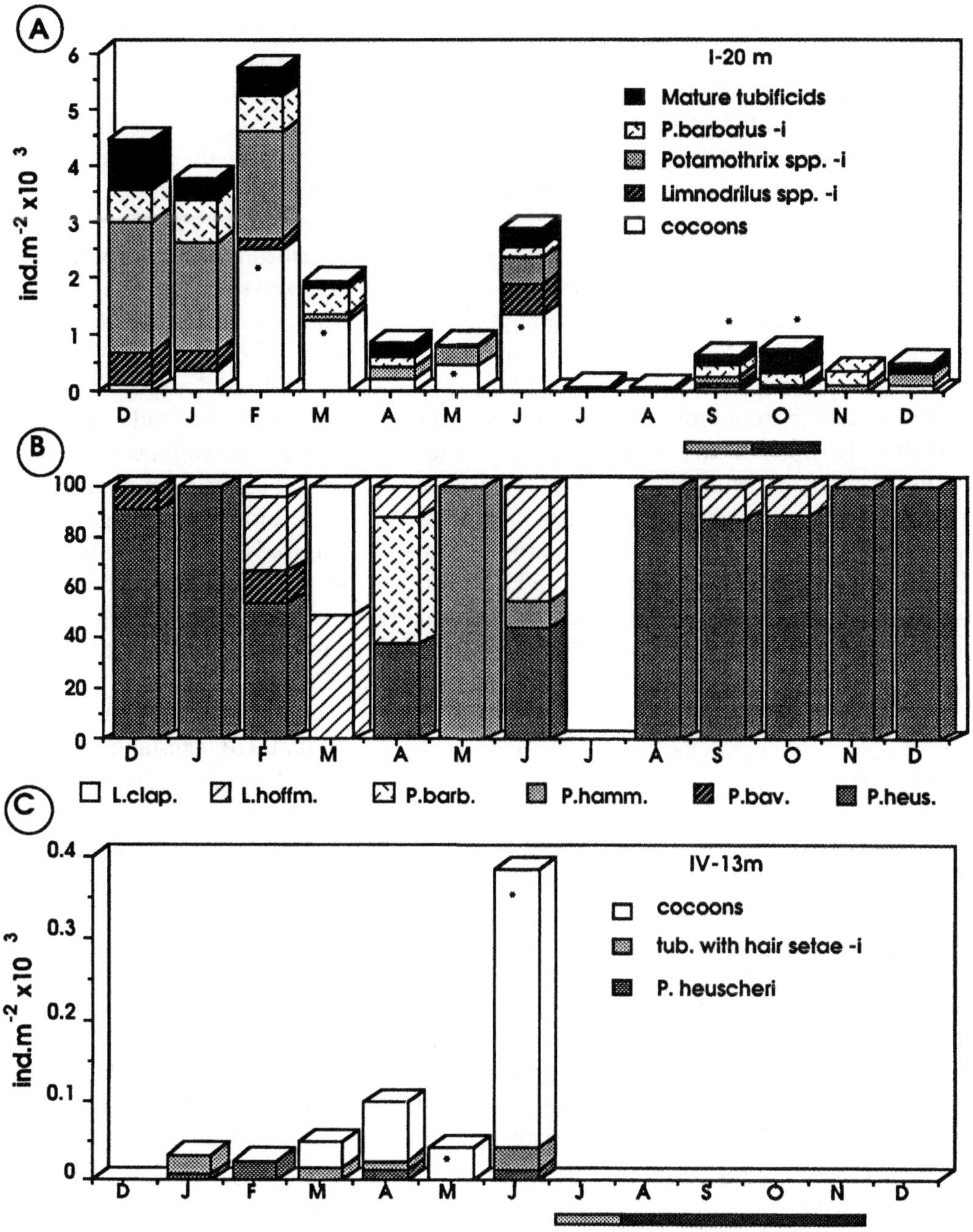

Fig. 3. (a) Monthly changes in the abundance of immature and mature oligochaetes (ind m^{-2}) and cocoons (No m^{-2}) at 20 m depth in basin I. The symbol * denotes the presence of embryos inside the cocoons. i = immature worms. Under the month axis, lighter shaded area indicates less than 3 mg l^{-1} of oxygen content while darker shaded area indicates less than 1 mg l^{-1}. (b) Monthly changes in the percentage of mature species of oligochaetes at 20 m depth in basin I. Abbreviations: *L. clap.* = *L. claparedeanus*; *L. hoffm.* = *L. hoffmeisteri*; *P. barb.* = *P. barbatus*; *P. hamm.* = *P. hammoniensis*; *P. bav.* = *P. bavaricus*; *P. heus.* = *P. heuscheri*. (c) Monthly changes in the abundance of immature and mature oligochaetes (ind m^{-2}) and cocoons (No m^{-2}) at 13 m depth in basin IV. Immature tubificids with hair setae could be assumed to be *Potamothrix heuscheri* at this sampling site. The symbol * denotes the presence of embryos inside the cocoons. i = immature worms. Under the month axis, lighter shaded area indicates less than 3 mg l^{-1} oxygen content while darker shaded area indicates less than 1 mg l^{-1}. Abbreviations: *L. clap.* = *L. claparedeanus*; *L. hoffm.* = *L. hoffmeisteri*; *P. barb.* = *P. barbatus*; *P. hamm.* = *P. hammoniensis*; *P. bav.* = *P. bavaricus*; *P. heus.* = *P. heuscheri*.

Tapeworm parasites in Oligochaeta

In oligochaete populations, tapeworm infested individuals can usually be found as intermediate or terminal hosts. In Lake Banyoles two species of worms were found to be infested by the tubificid-specific cestoda *Archigetes* cf. *sieboldi*. Mature and immature individuals of *L. hoffmeisteri* were the most affected. At 20 m depth in basin I, in February, 14% of the adult *L. hoffmeisteri* population was infested and in September 100% of adults and 25% of juveniles were infested. At 12 m, 33% of individuals in June and 67% in October had parasites. In April, in the sublittoral, 20% of juveniles contained cestodes. *Archigetes* cf. *sieboldi* also infested the *P. heuscheri* sublittoral population, 11% of adults and 12% of the immature Tubificidae with hair setae were infested. On two occasions (February and September) free gravid procercoids of *A.* cf. *sieboldi* were found at 20 m depth.

Discussion

Lake Banyoles has a moderate density and species richness of oligochaetes compared to other European lakes (Reynoldson, 1990; Jonasson, 1984; Lang, 1985; Lang, 1989; Moss & Timms, 1989; Bazzanti, 1981; Bazzanti, 1983). Higher numbers of species in the profundal parts of the lake, especially at 20 m in basin I, may be due to a higher numbers of samples being identified. Nevertheless, it is noteworthy that the same trend is revealed by other macroinvertebrate groups, mainly chironomids (Rieradevall, 1991; Rieradevall & Prat, 1991).

Oligochaete densities decrease with increasing proportion of coarse particles (Rieradevall, *op. cit.*, in press; Rieradevall & Prat, *op. cit.*), which is also related to depth. Although particle size would allow higher worm densities at 13 m depth in basin IV, oxygen depletion and sulphide content act as a limiting factor. These conditions reduced the abundance and species richness from seven to a single species in Lake Banyoles. These results agree with Newrkla & Wijegoonawardana (1987), who made similar observations in Mondsee (Austria) of a reduction in the number of Oligochaeta species from seven to two.

The dominance of *Potamothrix heuscheri* throughout the year and at all the sampling stations is remarkable. This species, together with *P. bavaricus* and *P. barbatus* were found at low densities in non eutrophic reservoirs in the alkaline zone of the Iberian Peninsula (Martinez-Ansemil & Prat, 1984; Real & Prat, 1991). Prat *et al.* (1991) and Real (1993) found *P. heuscheri* in the profundal areas of reservoirs with low oxygen and high sulphide concentrations. In these studies, *L. hoffmeisteri* and *Tubifex tubifex* appeared as the most abundant and frequent mature worms in all the reservoirs sampled (around one hundred). In Lake Banyoles, *L. hoffmeisteri* is the second species in percentage but no trace of the cosmopolitan *T. tubifex* was found; this could be due to its low competitive capability (Brinkhurst, 1969).

Other authors such as Lafont (1987) in Lake Léman (Switzerland) and Bazzanti & Lafont (1985) in Lake Nemi (Italy) have found such dominance of *P. heuscheri*. This species seems to be more important in South European lakes, although it has not been found in karstic Balkanic or Aegean lakes (Stankovic *et al.*, 1971). In these lakes, especially in Lake Ohrida, a great number of endemic species were recorded. Only two species are coincident with Lake Banyoles assemblages, *Psammoryctides barbatus* and *Potamothrix hammoniensis*, in sublacustrine springs of Lake Skadar (Jacobi, 1978).

Low oxygen concentration (less than 1 mg l^{-1}) and a long period of anoxia (more than 4 months) acted as key factors in reducing the species richness and the density of oligochaetes. Three different situations could be described according to these factors: a) one month of oxygen depletion in basin I (at 12 and 20 m depth) allowed the settlement of diverse oligochaete assemblages with high densities; b) four months of anoxia at 13 m in basin IV, combined with the presence of sulphide, reduced the Oligochaeta fauna to one species, *Potamothrix heuscheri*, with very low densities; and c) the extreme situation, in basin III at

20 m depth, a permanent meromixis throughout the year, did not allow the survival of any macroinvertebrates, except for *Chaoborus flavicans*, a planktonic Diptera.

P. heuscheri can exploit a wide range of environmental and trophic conditions. The species is found in oligotrophic, mesotrophic and also highly eutrophic or polluted waters (Bazzanti & Lafont, 1985). The most interesting feature of this species is that it constitutes monospecific populations in anoxic conditions with presence of sulphide. Similar situations have been described in lakes, especially from the Mediterranean area, such as Lake Tiberiades (Israel), Petichet (French Alps) and Lake Nemi (Italy) (Bazzanti & Lafont, 1985; Bazzanti & Seminara, 1987). The latter is extremely eutrophic with an oxygen deficit in the profundal area most of the year. *P. heuscheri* is the dominant species in the sediment of Lake Nemi, and at the maximum depth (31 m) this is the only species found, at very low densities (100 ind. m^{-2}). A similar situation can be described in basin IV at 13 m, where *P. heuscheri* is the only species, although it does not constitute a well formed population.

High numbers of predators such as *Chaoborus* and the chironomid *Procladius*, at 13 m in basin IV (Rieradevall & Prat, 1991), could also affect the establishment of worm communities when the re-oxygenation occurs.

The two peaks in density of *Limnodrilus* immature forms (June and December) at 20 m depth, are similar to that observed by Lafont (1987) in Lake Léman, where intensive hatching occurred in spring and winter. Lafont (1987) stated that the life cycle of this species took 2 to 3 years, but each worm could reproduce twice a year, and at least 3 times during its life.

P. barbatus as well as the less abundant species, *L. claparedeanus, P. hammoniensis* and *P. bavaricus*, showed shorter and well defined breeding periods. The opposite strategy was used by the more ubiquitous species, *P. heuscheri*, whose mature individuals were present all year long and, consequently, the recruitment of young worms occurred throughout the year. The same life history pattern for *P. heuscheri* has been described in Lake Léman (Lafont, 1987). In Lake Banyoles, two peaks of maximum density of oligochaetes can be defined: one in December, in the shallow depths, and throughout the winter months at 20 m depth in basin I; and the other in June at all the sampling stations. Lafont (*op. cit.*) recorded only a single peak of tubificids with hair setae (*Potamothrix* spp. + *T. tubifex*) in June. A more extensive period with relatively high densities of Oligochaeta may be related to higher temperatures in Lake Banyoles (8–25 °C in the sublittoral and 8–18 °C in the profundal).

Archigetes sieboldi has been cited as a parasite tapeworm of several tubificid species, mainly *L. hoffmeisteri, L. claparedeanus, P. hammoniensis* and *P. barbatus* (Timm, 1970) among the oligochaete species present in Lake Banyoles. Generally, infested worms account for a low proportion of the oligochaete population. Timm (*op. cit.*) reported maximum and average values of 22.6% and 10% or less of infested specimens, respectively. Kennedy (1965) found 0.5–5.6% of infested *L. hoffmeisteri* by the tapeworm *Archigetes limnodrili*. Under experimental conditions, higher values were reported: 31–63% individuals were infested by *Archigetes iowensis* (Calentine, 1964: in Kennedy, 1965). According to our data, Lake Banyoles populations of tubificids present locally high values of parasitic infestation but on average, the percentage of infestation is within recorded ranges.

In summary, the oligochaete community in Lake Banyoles has a richer composition and higher density in the deeper parts of the lake, where sediments are finer. In general, there are two peaks in oligochaete density, the maximum in winter (December) and in spring (June). At the sublittoral stations greater fluctuations in population numbers can be seen. The persistence of a chemocline at the profundal areas of certain basins in Lake Banyoles leads to anoxic conditions during long periods (4 months). As a result, the oligochaete community is strongly reduced and only *Potamothrix heuscheri* can be found in such an environment, which has been observed in other Mediterranean lakes.

Acknowledgements

Funds for the sampling programm were provided by the CAYCIT project PR84-0747 and a personal grant of 'Caixa de Barcelona' to M. Rieradevall. Our thanks are due to Enrique Martínez-Ansemil who introduced M. Real to the study of the Oligochaeta and who reviewed the identification of some specimens. We are also indebted to Narcís Prat for his useful comments and to two anonymous reviewers for their suggestions.

References

Anderson, R.O., 1959. A modified flotation technique for sorting bottom fauna samples. Limnol. Oceanogr. 4: 223–225.

Bazzanti, M. 1981. Survey of the macrobenthic community in an area of Lake Bracciano (Central Italy). Boll. zool. ital. 48: 295–303.

Bazzanti, M., 1983. Composition and diversity of the profundal macrobenthic community in the polluted Lake Nemi (Central Italy), 1979–1980. Acta Oecologica, Oecol. Applic. 4(3): 211–220.

Bazzanti, M. & M. Lafont, 1985. Variabilité des soies spermathécales de *Potamothrix heuscheri* (Bretscher) (Oligochaeta, Tubificidae) dans le lac de Nemi et remarques sur la relation entre cette espèce et le degré de trophie des lacs. Ann. Limnol. 21(2):107–115.

Bazzanti, M. & M. Seminara, 1987. Profundal macrobenthos structure as a measure of long-term environmental stress in a polluted lake. Wat. Air Soil Pollut. 33: 435–442.

Brinkhurst, R. O., 1969. Changes in the benthos in Lakes Erie and Ontario. Bull. Buffalo Soc. Nat. Sci. 25: 45–65.

Garcia-Gil, L. J., L. Sala-Genoher & C. A. Abella, 1991. Dynamics of phototrophic bacteria in relation to iron cycle in a meromictic basin of Lake Banyoles. Verh. int. Ver. Limnol. 24: 1016–1019.

Gnaiger, E. & I. Staudigl, 1987. Aerobic metabolism and physiological responses of aquatic oligochaetes to environmental anoxia. Heat dissipation, oxygen consumption, feeding and defaecation. Physiol. Zool. 60: 659–677.

Guerrero, R., C. Abellà & M.R. Miracle, 1978. Spatial and temporal distribution of bacteria in a meromictic karstic lake basin: relationships with the physicochemical parameters and zooplankton. Verh. int. Ver. Limnol. 20: 2264–2271.

Howmiller, R. P. & M.A. Scott, 1977. An environmental index based on the relative abundance of oligochaete species. Water Poll. Cont. Fe. 49: 809–815.

Jacobi, G.Z., 1978. Zoobenthos from sublacustrine springs in Lake Skadar, Crna Gora, Yugoslavia. Verh. int. Ver. Limnol. 20: 1067–1077.

Jonasson, P. M., 1984. Decline of zoobenthos through five decades of eutrophication in Lake Esrom. Verh. int. Ver. Limnol. 22: 800–804.

Kennedy, C. R., 1965. The life history of *Archigetes limnodrili* (Yamaguti) (Cestoda: Caryophyllaeidae) and its development in the invertebrate host. Parasitology 55: 427–437.

Lafont, M., 1983. Annélides Oligochétes. Bull. Soc. linn. Lyon 4: 108–135.

Lafont, M., 1987. Production of Tubificidae in the littoral zone of Lake Léman near Thonon-les-Bains: a methodological approach. Hydrobiologia 155: 179–187.

Lang, C., 1985. Eutrophication of Lake Geneva indicated by the oligochaete communities of the profundal. Hydrobiologia 126: 237–243.

Lang, C., 1989. Eutrophication of Lake Neuchatel indicated by the oligochaete communities. Hydrobiologia 174: 57–65.

Lang, C. & B. Lang-Dobler, 1980. Structure of tubificid and lumbriculid worm communities, and three indices of trophy based upon these communities, as descriptors of eutrophication level in lake Geneva (Switzerland). In R. O. Brinkhurst & D. G. Cook (eds), Aquatic Oligochaete Biology. Plenum Press, New York: 457–470.

Martinez-Ansemil, E. & N. Prat, 1984. Oligochaeta from profundal zones of Spanish reservoirs. Hydrobiologia 115: 223–230.

Milbrink, G., 1980. Oligochaete communities in pollution biology: the European situation with special reference to lakes in Scandinavia. In R.O. Brinkhurst & D. G. Cook (eds), Aquatic Oligochaete Biology. Plenum Press, New York: 433–455.

Moreno-Amich, R. & E. Garcia-Berthou, 1989. A new bathymetric map based on echo-sounding and morphometrical characterization of the Lake of Banyoles (NE Spain). Hydrobiologia 185: 83–90.

Moss, B. & M. Timms, 1989. Predation, sediment stability and food availability as determinants of the benthic invertebrate fauna in two shallow lakes. Hydrobiologia 185: 249–257.

Newrkla, P. & N. Wijegoonawardana, 1987. Vertical distribution and abundance of benthic invertebrates in profundal sediments of Mondsee, with special reference to oligochaetes. Hydrobiologia 155: 227–234.

Planas, M.D., 1973. Composición, ciclo y productividad del fitoplancton del lago de Banyoles. Oecol. Aquatica 1: 3–106.

Prat, N., F. Sanz & E. Martinez-Ansemil, 1991. El bentos profundo y litoral de una cadena de tres embalses españoles del río Guadiana (SW España). Limnética 7: 133–152.

Prat, N & M. Rieradevall, In press. Life cycle and production of *Cladopelma virescens* (Dipt: Chironomidae) in Lake Banyoles. Neth. J. Aquat. Ecol. 26 (2-4).

Real, M., 1993. Ecology of profundal zoobenthos in spanish reservoirs. Ph.D. Thesis Dissertation. Univ. of Barcelona (in Catalan).

Real, M. & N. Prat, 1991. Changes in the benthos of five Spanish reservoirs in the last 15 years. Verh. int. Ver. Limnol. 24: 1377–1381.

Real, M., F. Sabater & J. A. Morgui, 1992. Significant physiographic disturbances in the Ebro basin reservoirs (NE Spain) reflected by Oligochaeta size spectra. Hydrobiologia 235/236: 363–374.

Real, M. & N. Prat, In press. Factors influencing the distribution of chironomids and oligochaetes in profundal areas of spanish reservoirs. Neth. J. Aquat. Ecol. 26(2–4).

Reynoldson, T. B., 1987. The role of environmental factors in the ecology of tubificid oligochaetes – an experimental study. Holarct. Ecol. 10: 241–248.

Reynoldson, T. B., 1990. Distribution patterns of oligochaetes in the English Lake District. Arch. Hydrobiol. 118(3): 303–339.

Rieradevall. M., 1991. Ecology and production of benthos in Lake Banyoles. Ph.D. Thesis Dissertation. Univ. of Barcelona, 225 p. (in Catalan with some chapters in English).

Rieradevall, M. In press. Species composition and depth distribution of meiobenthos in Lake Banyoles. Verh. int. Ver. Limnol. 25.

Rieradevall, M. & J.M. Busquets, 1990. Bryozoans of Banyoles lake. Limnetica 6: 137– 146.

Rieradevall, M & M. J. Gil, In press. Distribution, density and specific composition of water mites in the sublittoral of Lake Banyoles (NE Spain). Ann. Limnol.

Rieradevall, M. & N. Prat, 1989. Chironomidae of profundal samples of Banyoles lake (NE Spain). Acta Biologica Debrecina, Oecol. Hung. 3: 267–274.

Rieradevall, M. & N. Prat, 1991. Benthic fauna of Banyoles lake (NE Spain). Verh. int. Ver. Limnol. 24: 1020–1023.

Särkkä, J., 1987. The occurrence of oligochaetes in lake chains receiving pulp mill waste and their relation to eutrophication on the trophic scale. Hydrobiologia 155: 259–266.

Stankovic, S. 1951. Le peuplement benthique des lacs Égéens. Verh. int. Ver. Limnol. 11: 367–388.

Stankovic, S., M. Tocko & J. Sapkarev, 1971. Différentiation écologique du peuplement benthique du lac d'Ohrid. Arch. Hydrobiol. 68(2): 163–203.

Timm, T., 1970. On the fauna of the Estonian Oligochaeta. Pedobiologia 10: 52–78.

Uzunov, J., V. Koesel & V. Sladecek, 1988. Indicator value of freshwater Oligochaeta. Acta hydrochim. hydrobiol. 16(2): 173–186.

Hydrobiologia **278**: 151–156, 1994.
T.B. Reynoldson & K.A. Coates (eds), Aquatic Oligochaete Biology V.

Depth distribution of oligochaetes in Lake Baikal (Siberia – Russia)

Patrick Martin, Boudewijn Goddeeris & Koen Martens
Royal Belgian Institute of Natural Sciences, rue Vautier 29, B-1040 Bruxelles, Belgium

Key words: aquatic oligochaetes, Lake Baikal, bathymetric distribution, vertical distribution

Abstract

In the course of a preliminary sampling program, oligochaetes were collected along two transects in soft sediments in Lake Baikal. The number of oligochaetes present in the samples was counted, without distinguishing between species. The results suggest an exponential decrease in number of individuals (N) relative to depth (11,165 N m^{-2} at 21 m, 265 N m^{-2} at 1200 m). Most oligochaetes were found in the top 7 cm of sediment. The orange colour of the sediments suggests a high oxygen availability, even at the greatest water depths.

Résumé

Au cours d'un programme d'échantillonnage préliminaire, les oligochètes du lac Baïkal ont été récoltés dans le sédiment mou prélevé le long de deux transects. Les oligochètes présents dans les échantillons ont simplement été dénombrés, sans identification d'espèces. Les résultats suggèrent une diminution exponentielle du nombre d'individus en fonction de la profondeur du lac (11165 N m^{-2} à 21 m, 265 N m^{-2} à 1200 m). La plupart des oligochètes ont été trouvés dans les 7 premiers cm de la couche supérieure du sédiment. La couleur orangée du sédiment suggère une grande disponibilité en oxygène, même aux profondeurs les plus grandes.

Introduction

Lake Baikal, in the Great Baikal Rift of eastern Siberia (53° N, 108° E), is the largest lake in the northern hemisphere (Kozhov, 1963). The "Pearl of Siberia" is nearly 650 km long and between 30 and 70 km wide. It is also the deepest of all lakes, the greatest depth being 1,637 m (Stewart, 1990a). With a total volume of fresh water of c. 23,000 km^3 (Kozhov, 1963), Lake Baikal contains about one fifth of the total surface freshwater supply in the world (Galazii, 1990). It is one of the most fascinating lakes on earth, mainly because of its age (20–25 million, possibly even 50 million years) and because of the fact that, unlike most other deep lakes, the water circulation carries oxygen to its deepest point (Weiss *et al.*, 1991; Maddox, 1989). The comparable Lake Tanganyika, for example, is much younger (2–6 millions years; Coulter, 1991) and virtually anoxic below 200 metres. Due to its unique ecological characteristics, the life in Lake Baikal is the richest in terms of both biomass and the number of recorded species, with numerous taxa being endemic.

For the Oligochaeta alone, the rate of endemism is nearly 90% (Kozhov, 1963). For the Tubificidae described before 1984, 25 species in 10 genera were reported from Lake Baikal, 20 of which were considered endemic (Brinkhurst, 1984). In a more recent publication, Snimschikova (1987) enumerated a total of 104 oligochaete species in 31 genera and seven families (Aeolosomatidae and Branchiobdellidae included) and this in the northern Baikal only.

Finally, in spite of local pollution from industry (mainly cellulose plants) and agriculture, the water of the lake remains extraordinarily pure. To date Lake Baikal has been spared the heavy development that blights North America's Great Lakes (Stewart, 1990b). Due to its characteristics, Lake Baikal is an exceptional ecosystem. Recently, various initiatives were taken to stimulate international research on different aspects of the Lake Baikal environment. As a result, a new research centre was founded, the Baikal International Centre for Ecological Research (BICER). A scientific collaboration was established between the Freshwater Biology Section at the Royal Belgian Institute of Natural Sciences and the Limnological Institute in Irkutsk, and research projects were defined. One research topic is focused on the distribution of zoobenthos in the lake (bathymetric, vertical and horizontal). This short note gives the preliminary results for Oligochaeta.

Material and methods

During a three week visit to Irkutsk in August 1990, a preliminary program was successfully carried out in the southern basin of the lake. Samples were taken with a modified Reineck box corer (surface 1000 cm^2), a device normally used for oceanographic work (Eleftheriou & Holme, 1984). Two transects and one isolated station at 1200 m were sampled on soft sediments: south of the Selenga Delta (20–80 m) and on the Posolskaya Bank (50–400 m) (Fig. 1). Four tubes (diameter 6.3 cm, surface 37.7 cm^2) were pushed into the sediment raised with the box-corer, and these are considered replicate samples. As the surface of the box corer was relatively large, it is here assumed that using this type of replicate, rather than taking each core from a separate Reineck sample (which was logistically impossible, due to time constraints) will introduce only limited bias in the analyses. One core was used for grain-size analysis, the other three for faunal analysis. These latter cores were subsequently divided with an extruder into six vertical slices (0–1 cm; 1–2 cm; 2–4 cm; 4–7 cm; 7–15 cm and 15–21 cm) and preserved in 10% formalin. The slices were later screened in the laboratory on a mesh size of 250 μm and the organisms then sorted.

In this preliminary work, only whole specimens of Oligochaeta were counted without distinguishing between species. The bathymetric distribution was obtained from the combination of the two transects investigated. The densities of the vertical distribution were obtained from the mean densities of the different stations.

Results

The bathymetric distribution of oligochaetes (Fig. 2) suggests an exponential decrease of the number of individuals in relation to lake depth (max. – min.: 11,165 N m^{-2} at 21 m, 265 N m^{-2} at 1,200 m). Smallest oligochaetes were found at the deepest station. Incidentally, it is interesting to note that the remarkable freshwater polychaete *Manayunkia baicalensis* Nusbaume, 1901 is occasionally encountered between 21 m and 45 m depth, near the shoreline.

Densities of oligochaetes within the sediment (Fig. 3) seem also to decrease exponentially from the 1–2 cm layer to the 15–21 cm layer. Most oligochaetes (97%) were found in the top 7 cm of the sediment, which appeared to be oxidized as suggested from the sediments orange colour. This indicate high oxygen availability even at the great-

Fig. 1. Map showing the location of Lake Baikal in the northern hemisphere and of the two transects (B) in the southern basin (C) (Posolskaya Bank, B1; south of the Selenga delta, B2). Depth isopleths are in metres (B).

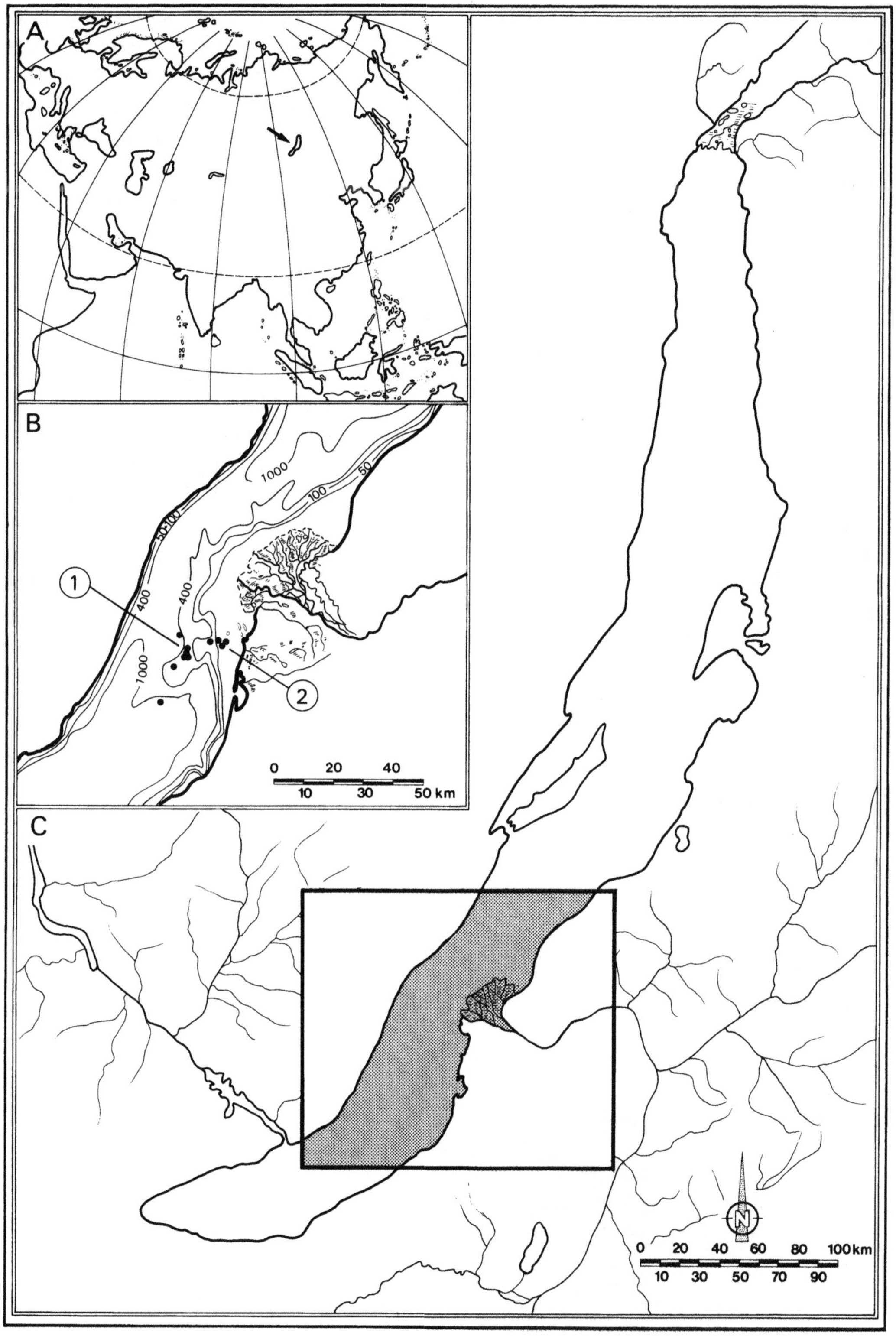
A
B
1000
100
50
50/100
400
400
1000
1
2
0 20 40
10 30 50 km
C
N
0 20 40 60 80 100km
10 30 50 70 90

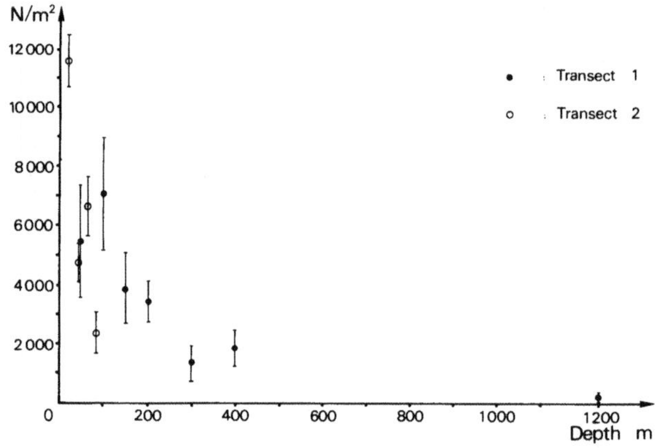

Fig. 2. Bathymetric distribution of oligochaetes. Densities are expressed in N $m^{-2} \pm$ s.e.

est depths within the lake. The deepest sediment layer (15–21 cm) was only inhabited by ?*Rhynchelmis brachycephala* (Michaelsen, 1901) (Lumbriculidae), the largest oligochaete of Lake Baikal. *Rhynchelmis brachycephala* places the largest part of its body in the deeper anoxic sulfide layer.

Significant interaction in the 2 – way ANOVA (Table 1) indicates that the main factors (bathymetric depth, sediment depth) are not linearly related and in some way may interact. This means that one cannot simply interpret bathymetric or sediment depth changes.

Discussion

The bathymetric distribution of species is one of the oldest problems in the study of lakes and is an extension of the earliest debates about the origin of organisms found to inhabit lakes (Brinkhurst, 1974). Is there a specially evolved fauna in relation to lake depth or not? With regard to Lake Baikal, the question is of great interest because of the particularly high rate of endemism in its waters. On the other hand, the

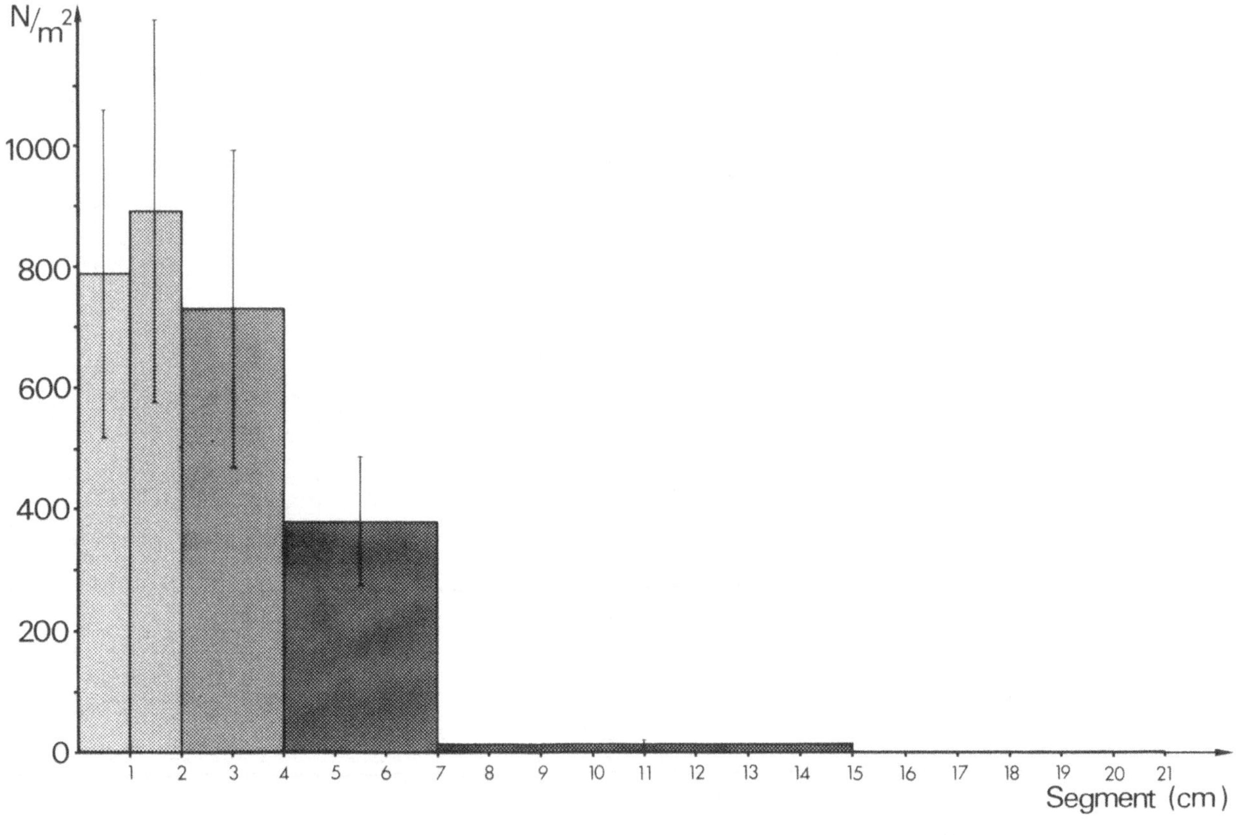

Fig. 3. Vertical distribution of oligochaetes. Densities are expressed in N $m^{-2} \pm$ s.e.

Table 1. Results of 2-way ANOVA's on ln (N + 1) for 11 depths and 5 segments; for 10 depths (−1200 m left out) and 4 segments (7–15 cm left out) and again for 10 depths (−21 m left out) and 4 segments (7–15 cm left out).

Source	F (v1, v2)	v1	*P*
11 depths, 5 segments (v2 = 164)			
Depth	17.673	10	***
Segments	39.829	4	***
Depth × segm.	6.283	40	***
10 depths, 4 segments (v2 = 119) (−1200 m)			
Depth	13.660	9	***
Segments	4.959	3	**
Depth × segm.	5.680	27	***
10 depths, 4 segments (v2 = 84) (−21 m)			
Depth	10.022	9	***
Segments	3.647	3	**
Depth × segm.	2.784	21	***

Levels of significance: *** $P \leq 0.001$; ** $P \leq 0.01$.
v1 = degrees of freedom amongst groups
v2 = degrees of freedom within groups.

vertical distribution of species in the sediment, i.e. the depth to which species can penetrate the sediment, is of particular importance for understanding their coexistence in a habitat.

Our preliminary results suggest:

(1) an exponential decrease of densities of oligochaetes with lake depth;
(2) an exponential decrease of densities of oligochaetes in relation to the sediment depth;
(3) a significant interaction between bathymetric and sediment depths with regard to the densities of oligochaetes; and
(4) a correlation between the greatest densities of oligochaetes in the sediment and its oxidized layer (i.e. the oxygen availability).

Due to its depth, Lake Baikal is characterized by a particular deep-water renewal ("ventilation"), depending on the decrease in the temperature of maximum density of freshwater with increasing pressure, and on wind action at the surface layer (Weiss *et al.*, 1991). As a result, deeper layers are always oxygenated and nutrients in the surface layer are recycled about four times before being removed to deep waters. The strongly oligotrophic nature of Lake Baikal is therefore a possible explanation of the very low densities of oligochaetes in the deepest zones.

While there are few studies on the macrohabitat distribution of oligochaetes (i.e. physical-chemical variables of water or plant associations; e.g. Verdonschot, 1984, 1987), their microdistribution in the sediment is somewhat better documented (Brinkhurst & Kennedy, 1965; Särkkä & Paasivirta, 1972; Milbrink, 1973; Kirchner, 1975; Newrkla & Wijegoonawardana, 1987; Healy, 1987). The vertical distribution of oligochaetes in the sediments may be the result of various factors (see Milbrink, 1973 and Newrkla & Wijegoonawardana, 1987): shelter from predators, nutritional preferences, specific rate of maturity, characteristics of the bottom substrate, tolerance to anoxia (Fisher & Beeton, 1975; Gnaiger *et al.*, 1987).

The second and the fourth factors, above, are probably related and suggest that, amongst the factors mentioned, oxygen availability is a key factor in the distribution of oligochaetes in the sediments. Kirchner (1975) showed that the vertical extent of the oxidized material had a very definite effect on the vertical distribution of the benthic fauna, which was reflected in increased dispersal or spreading out of the biomass in the community (1975: 429).

The burrowing activity of oligochaetes results in oxygen penetration in the sediment far below the oxic layer (oxygen is present in the immediate vicinity of animal burrows; Revsbech & Jørgensen, 1986). It can be assumed that oligochaetes penetrate deeper into the sediments not only because the oligotrophic condition of Lake Baikal stimulates this behaviour (search for food), but also because the high oxygen content at the water-sediment interface at all depths makes a deep penetration easier. Finally, it is possible to find oligochaetes such as *R. brachycephala* far below the oxidized layer, but this is not surprising because of their body length of up to 185 mm (Chekanovskaya, 1962) which enables them to reach the oxic layer even from such much greater depths in the sediments.

The statistical analyses indicate a significant interaction between bathymetric and sediment depths on the densities of oligochaetes. The direct or indirect nature of this interaction remains unknown. While a direct influence of depth on the densities of oligochaetes in the sediment is unlikely, an indirect influence related to the nature of the sediments (Brinkhurst, 1974) is easier to conceive. It is difficult, however, to substantiate here because of the observed, relative, homogeneity of the sediment between the stations investigated.

Due to the great depth of Lake Baikal, nutrients are recycled several times as they sink to the bottom, resulting in a nutritive depletion of the sediment in relation to depth. The interaction between bathymetric and sediment depths on the densities of oligochaetes could, therefore, be the result of this nutritive depletion if the feeding activity of these animals has an effect on their vertical distribution.

References

Brinkhurst, R. O., 1974. The benthos of lakes. Macmillan Press Ltd, London and Basingstoke, 190 pp.

Brinkhurst, R. O., 1984. A revision of the Tubificidae and Lycodrilidae (Annelida, Oligochaeta) known from Lake Baikal. Can. J. Zool. 62 (3): 494–509.

Brinkhurst, R. O. & C. R. Kennedy, 1965. Studies on the biology of the Tubificidae (Annelida, Oligochaeta) in a polluted stream. J. anim. Ecol. 34: 429–443.

Chekanovskaya, O. V., 1962. Aquatic Oligochaeta of the USSR (Vodnye Maloshchetinkovye Chervy Fauny SSSR), Academy of Sciences of USSR, Institute of Zoology, Transl. of Fauna SSSR Novaya Seriya, 78, I. Kohli 1981, 513 pp.

Coulter, G. W., 1991. Lake Tanganyika and its life. Natural History Museum Publications, Oxford Univ. Press, London, 354 pp.

Eleftheriou A. & N. A. Holme, 1984. Macrofauna Techniques. In Holme N. A. & A. D. McIntyre (eds), Methods for the Study of Marine Benthos. Second edition. Blackwell Scientific Publications. Oxford and London: 140–216.

Fisher, J. A. & A. M. Beeton, 1975. The effect of dissolved oxygen on the burrowing behaviour of *Limnodrilus hoffmeisteri* (Oligochaeta). Hydrobiologia 47: 273–290.

Galazii, G., 1990. Le lac Baïkal en sursis. La Recherche 221: 628–637.

Gnaiger, E., Kaufmann, R. & I. Staudigl, 1987. Physiological reactions of aquatic oligochaetes to environmental anoxia. Hydrobiologia 155: 155.

Healy, B., 1987. The depth distribution of Oligochaeta in an Irish quaking marsh. Hydrobiologia 155: 235–247.

Kirchner, W. B., 1975. The effect of oxidized material on the vertical distribution of freshwater benthic fauna. Freshwat. Biol. 5: 423–429.

Kozhov, M., 1963. Lake Baikal and its life. In W. W. Weisbach & P. van Oye (eds), Monographiae Biologicae, XI. Junk, The Hague, 344 pp.

Maddox, J., 1989. Baikal centre takes step forward. Nature 341: 481.

Milbrink, G., 1973. On the vertical distribution of oligochaetes in lake sediments. Inst. Freshwater Res. Drottningholm 53: 34–50.

Newrkla, P. & N. Wijegoonawardana, 1987. Vertical distribution and abundance of benthic invertebrates in profundal sediments of Mondsee, with special reference to oligochaetes. Hydrobiologia 155: 227–234.

Revsbech, N. P. & B. B. Jørgensen, 1986. Microelectrodes: Their use in microbial ecology. In K. C. Marshall (ed.), Advances in Microbial Ecology. Plenum Press, New York, 9: 293–352.

Särkkä, J. & L. Paasivirta, 1972. Vertical distribution and abundance of the macro- and meiofauna in the profundal sediments of Lake Päijänne, Finland. Ann. Zool. Fennici 9: 1–9.

Snimschikova, L. N., 1987. Oligokhety severnovo Baikala. Red. A. A. Linevitch, Fauna Baikala. Akademiya Nauk SSSR. Sibirskoe otdelenie. Limnologitcheskii Institut, 104 pp. [Oligochaetes of Northern Baikal. In A. A. Linevitch (ed.), The fauna of Baikal. Academy of Sciences of USSR, Siberian Department, Limnological Institute (in Russian)].

Stewart, J. M., 1990a. Baikal's hidden depths. New Scientist, 23 June: 42–46.

Stewart, J. M., 1990b. The great lake is in great peril. New Scientist, 30 June: 58–62.

Weiss, R. F., E. C. Carmack & V. M. Koropalov, 1991. Deep-water renewal and biological production in Lake Baikal. Nature 349: 665–669.

Verdonschot, P. F. M., 1984. The distribution of aquatic oligochaetes in the fenland area of N.W. Overijssel (The Netherlands). Hydrobiologia 115: 215–222.

Verdonschot, P. F. M., 1987. Aquatic oligochaetes in ditches. Hydrobiologia 155: 283–292.

Hydrobiologia **278**: 157–163, 1994.
T.B. Reynoldson & K.A. Coates (eds), Aquatic Oligochaete Biology V.

Ultrastructural and histochemical comparison in two haplotaxids

Magda de Eguileor, Giulio Lanzavecchia & Roberto Valvassori
Dipartimento di Biologia, Universita' degli Studi di Milano, Via Celoria 26, I-20133 Milano, Italy

Key words: haplotaxids, muscles, histochemistry, ultrastructure

Abstract

Pelodrilus leruthi Hrabe, 1958 and *Haplotaxis gordioides* Hartman, 1821 have been investigated from a histochemical point of view. A different distribution and number of fast and slow fibres has been detected in the two species. *Pelodrilus leruthi* specimens living in different caves, show some differences in the body wall muscle thickness and in the structure of the ventral nerve cord. A correlation with their different life styles is proposed.

Introduction

Haplotaxis gordioides Hartmann, 1821 and *Pelodrilus leruthi* Hrabe, 1958 (Haplotaxidae) show significant differences in the organization of their body wall (de Eguileor *et al.*, 1990). The longitudinal muscle fibres in *H. gordioides* have been described as cylindrical pseudoround circomyarian, due to the discontinuity of the contractile ring and to the connection between the central cytoplasmic axis and the cytoplasmic peripheral zone. In contrast, the longitudinal muscle fibres in *P. leruthi* belong to the flat circomyarian type (de Eguileor *et al.*, 1990).

Little is known about the habitat of these animals (Juberthie & Mestrov, 1965; Juberthie, 1967; Delay, 1970; Timm, 1980; Omodeo, 1985) and there are no data about general metabolic patterns and histochemical properties of muscle layers forming the body wall.

Materials and methods

Haplotaxis gordioides was collected in the periphyton of natural springs near Milan. Seven specimens were examined. *Pelodrilus leruthi* was collected during winter from geographically separated caves in the Pyrénées. Some animals were living in permanent ponds with their anterior part buried in a clayey sediment and the posterior portion constantly moving in water. These animals are indicated as 'Estelas' from the nickname of the collection site. Some animals were living completely buried in wet clay on the wall of a cave which was only periodically flooded. These animals are indicated as 'Bas Nistos' from the name of the nearby village. Eight worms from 'Estelas' and 12 'Bas Nistos' worms were examined.

Electron microscopy

For routine transmission electron microscopy, whole animals were cut in small pieces, fixed according to Karnowsky (1965) and embedded in an Epon 812 – araldite mixture. Sections obtained with an Ultrotome V LKB and stained with an Ultrastainer LKB were observed with a Jeol 100 B electron microscope.

Semi-thin sections

Cross sections (1 μm thick) of worms obtained with an Ultrotome V LKB were stained with cristal violet and basic fuchsin, according to Moore *et al.*, (1960).

Histochemistry

Frozen cross sections of all specimens were cut at 10 μm thickness in a Reichert-Jung Frigocut 2800, mounted on glass slides and stained, according to Dubowitz & Brooke (1973), utilizing a histoenzymatic kit [BIO-OPTICA] for: succinic dehydrogenase (SDH), NADH-tetrazoliun reductase (NADH-Tr), cytochrome-oxidase (COx) as linked to mitochondrial activity; periodic acid Shiff (PAS), for the presence of glycogen; oil-Red-O (ORO), to detect lipids; non-specific esteases, including acid lipases and cholinesterase activities; alkaline phosphatase, indicating secretory and regenerative processes. Controls, mainly involving the omission of substrate, show a negative result for each case examined. Sections were stained with either haematoxylin-eosin or Gomori trichrome for histological examination.

Results

Haplotaxis gordioides

The body wall of *H. gordioides*, as previously described by de Eguileor *et al.* (1990), is composed of a thick pluristratified cuticle, an epidermis, an outer layer of circular fibres and an inner layer of longitudinal fibres, both obliquely striated (Figs 1, 15, 16 & 17). The longitudinal muscle cells belong to the pseudoround circomyarian type. The circular muscle layer is rather thin and discontinuous. Between the circular and longitudinal layer small spaces are just visible. These spaces are delimited by epithelial cells (Fig. 15). A ventral nerve cord characterized by giant fibres is also visible.

Histochemical results: SDH (Fig. 3) and COX (Fig. 4): the two positive reactions indicate an oxidative activity. However, only a

few fibres in the outer portion of the longitudinal muscle layer show a strong positive reaction (arrowheads). PAS: only the inner longitudinal fibres are not stained (Fig. 7). Non-specific esterase (Fig. 6) and alkaline phosphatase activities are present on the entire surface of all fibres (Fig. 5). The massive presence of lipids (Fig. 2, arrowhead) is indicated by the ORO reaction.

Pelodrilus leruthi

The body wall organization of *P. leruthi* is typical of oligochaetes (Figs 8 & 9), with a thick cuticle, an epidermis, and circular and longitudinal muscle layers. Between the two muscle layers there is a basal lamina (Fig. 18). All muscle cells belong to the flat circomyarian type. They are closely packed and linked to one another by desmosomes (Fig. 19). In the longitudinal layer, two types of muscle cells are characterized by different thicknesses of the myosin filaments (Fig. 18). 'Bas Nistos' animals, burrowing in clayey sediments, have a thicker circular muscle layer (Fig. 20) than 'Estelas' animals, living with the anterior part of the body buried in the substratum and the posterior part moving in the overlying water (Fig. 22).

The organization of the ventral nerve cord shows substantial differences in specimens of *P. leruthi* collected in different sites. In 'Estelas' specimens the three giant fibres are visible only in the caudal end, whereas they are not easily seen in the anterior portion constantly buried in the substratum (Figs 23, 24 & 25). In 'Bas Nistos' specimens the three fibres are absent throughout (Fig. 21).

Histochemical results: SDH (Fig. 10) and NADH-T (Fig. 11): the two positive reactions indicate an oxidative activity; the outer layers of the longitudinal fibres and the circular ones show highly oxidated enzyme activities (Fig. 11, arrowheads). PAS: the outer layer of longitudinal fibres is weakly stained (Fig. 13). Non-specific esterase: the entire surface of all fibres reacts positively. In

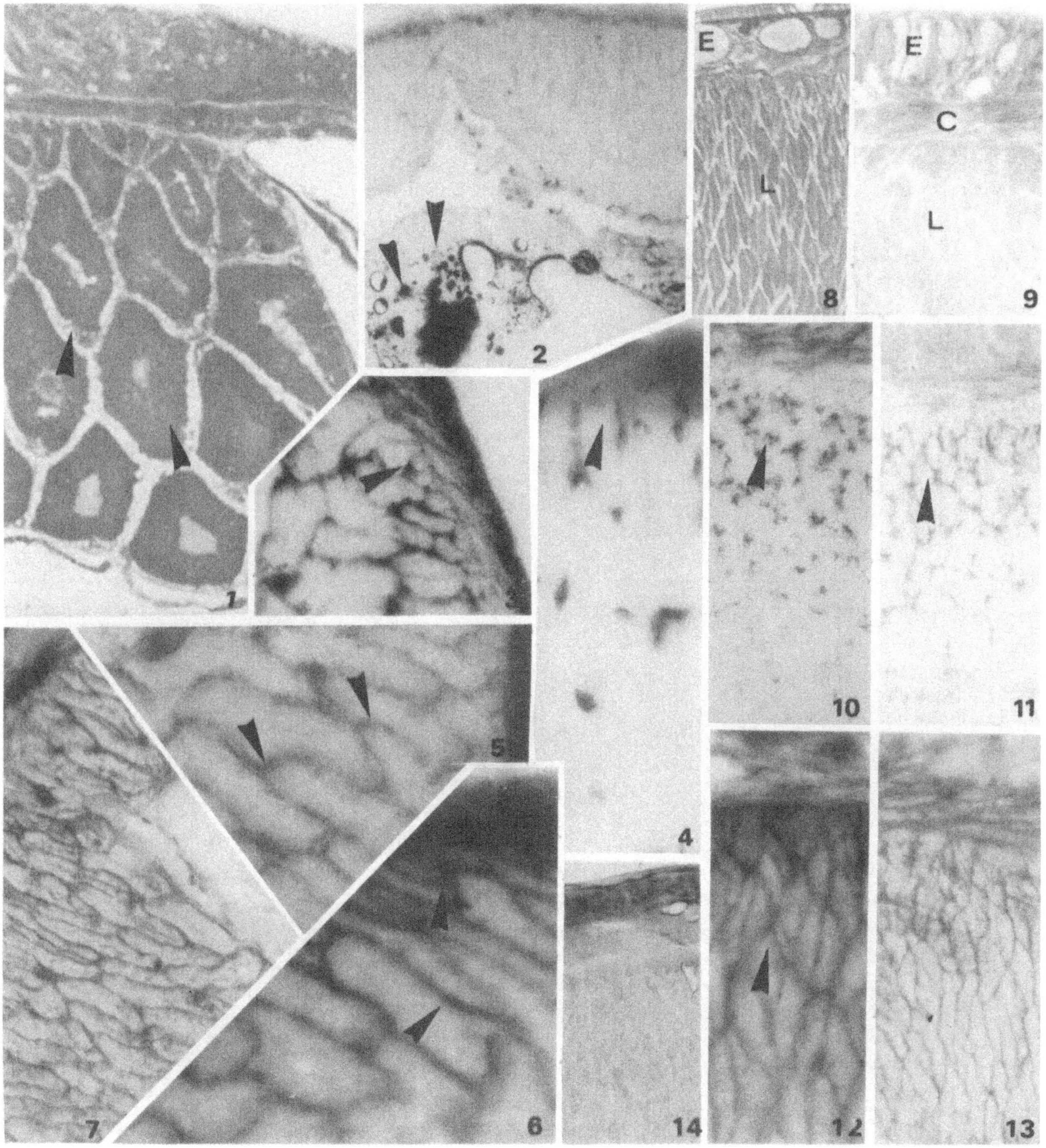

Figs 1–7. Cross sections of *Haplotaxis gordioides*; (1) Semi-thin section stained with cristal violet and basic fuchsin. The longitudinal pseudo circomyarian fibres are visible (arrowheads). (2) Oil-red-O (ORO) staining of a frozen section. A positive reaction is shown (arrowheads). (3) SDH and (4) COX: A positive reaction is characteristic of slow tonic fibres (arrowheads). (5) Alkaline phosphatase and (6) Non-specific esterase positive reactions (arrowheads). (7) PAS.

Figs 8–13. Cross sections of *Pelodrilus leruthi;* (8) Semi-thin section stained with cristal violet and basic fuchsin. The circular muscle layer (C) and the longitudinal ones (L) are visible under the epidermis (E). (9) Frozen section stained with Gomori-trichrome. (10) SDH. (11) NADH-T. (12) Non-specifric esterase. (13) PAS. The fibres of longitudinal layers are not stained. (14) Alkaline phosphatase negative reaction.

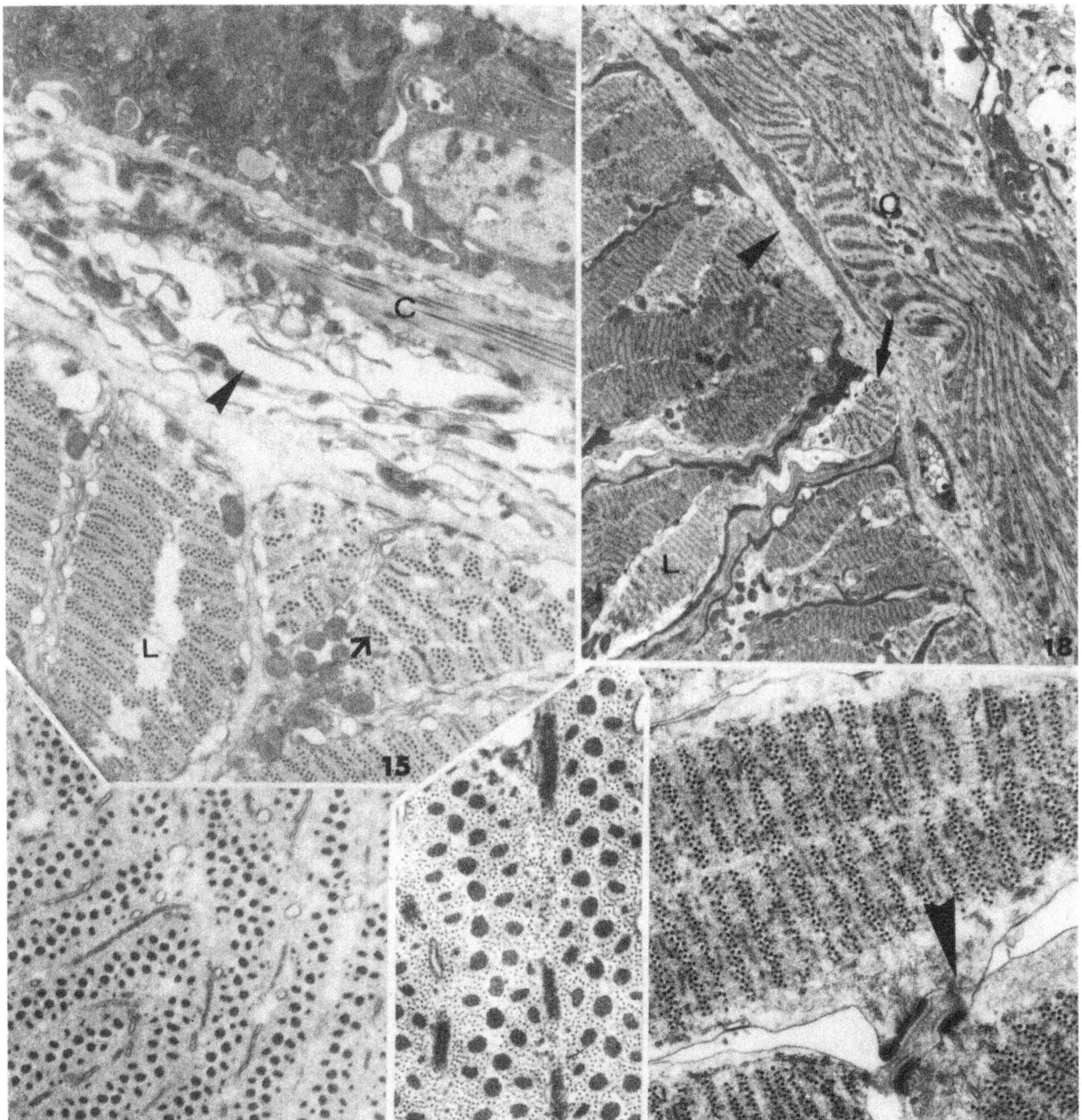

Fig. 15. Haplotaxis gordioides in cross section. In the outer portion of the longitudinal muscle layer (L) some triangular fibres with thicker myosin filaments are visible (arrow). Between the circular (C) and the longitudinal muscle layers (L), spaces delimited by a thin epithelium are sometimes visible (arrowhead). × 9000

Figs 16 & 17. Both the circular and the longitudinal muscle cells show a similar arrangement of the contractile material in a cross section. (16) × 23000; (17) × 42000

Fig. 18. Pelodrilus leruthi in a cross section. Numerous longitudinal fibres (L) with thicker myosin filaments are visible (arrow). The circular (C) and the longitudinal muscle layers are separated by a basal membrane (arrowhead). × 6000

Fig. 19. Detail of Fig. 18. The longitudinal muscle fibres are connected by mechanical junctions (arrowhead). × 18000

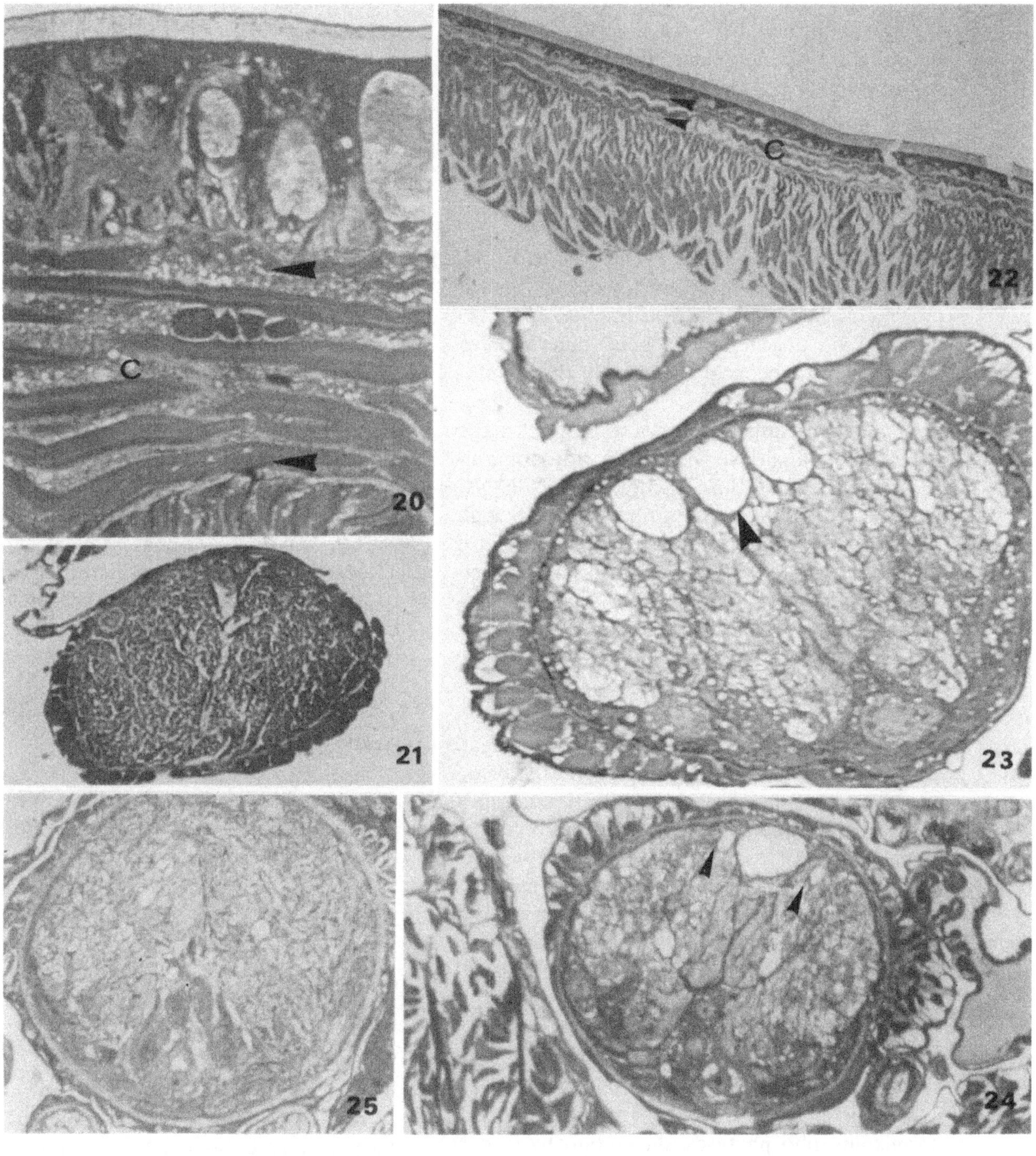

Figs 20–25. Pelodrilus leruthi in cross sections. (20) 'Bas Nistos' and (22) 'Estelas' specimens stained with cristal violet and basic fuchsin. The circular muscle layer (C, delimited by arrowheads) is thicker in Bas Nistos specimens and the ventral nerve cord (21) has no giant fibres. In 'Estelas' specimens the three giant fibres of the nerve cord are easily identifiable in the anterior region of the body (23, arrowhead indicates the middle giant nerve fibre). The giant fibres progressively reduce in the nerve cord from the intermediate region to the posterior part of the animal (24 arrowheads & 25).

particular, the outer longitudinal layer reacts more strongly than the inner ones (Fig. 12). Alkaline-phosphatase: the muscular layers did not show any reaction (Fig. 14). ORO: no lipid concentration was found.

Discussion

Haplotaxis gordioides and *P. leruthi* are clearly distinguishable from one another not only from a morphological point of view (Figs 1, 8, 15 & 18) but also by a general metabolic difference as shown by histochemical reactions of the longitudinal muscle layer.

In *P. leruthi* a large number of small fibres in the outer longitudinal layer behave as slow tonic fibres as indicated by their reaction to SDH, COX, NADH-T and PAS. Slow fibres show a high positivity for SDH, COX and NADH-T and a low activity for PAS. Fast fibres show a low positivity for SDH, COX and NADH-T and a strong reaction for PAS. In *P. leruthi,* as in all microdriles, few muscle fibres are comparable to the slow type from a morphological and histochemical point of view (e.g. myosin filaments with a larger diameter, Fig. 18). On the other hand, in these haplotaxids, as in all megadriles, several fibres are histochemically slow even though they would be considered fast from a morphological point of view (D'Haese & Carlhoff, 1987; de Eguileor *et al.*, 1989). In *H. gordioides* some scattered longitudinal fibres, with a thicker myosin filament, show similar histochemical reactions. These fibres lie right beneath the circular muscles (Figs 3 & 4).

Another substantial difference between the two species examined is that *P. leruthi* showed no fat reserves, whereas the body of *H. gordioides* is packed with lipids. The strong positive reaction for alkaline phosphatases, shown only by *H. gordioides* cannot yet be explained. A positive reaction for alkaline phosphates is generally linked to a secretory activity, or to regeneration and active exchange processes.

The variations in muscle ultrastructure observed in *P. leruthi* specimens collected from widely differing environments seem to be related to the different needs of the animals. Specimens living completely buried in dense clayey mud, burrowing in a very compact substratum, probably need a stronger circular muscle system than those undulating most of their body in water above the substratum.

Giant nerve fibres are typically found in oligochaetes that depend upon a rapid contraction of the body length as an escape strategy. In species, such as *Branchiura sowerbyi* Beddard, 1892, where the posterior region of the body extends into the water and performs rhythmical swaying motions, rapid transmissions of stimuli are essential for a specimen to survive (Drewes & Zoran, 1989).

In *P. leruthi* from 'Estelas' such giant fibres are visible in the extended posterior portion of the body. The mud-dwelling 'Bas Nistos' population of *P. leruthi* does not possess giant nerve fibres. The presence of these fibres in the 'Estelas' population probably indicates that predators do visit the ponds and so the ability to rapidly contract the longitudinal body muscles is retained.

Acknowledgements

This work has been supported by CNR and MPI grants. The authors are very grateful to Professor Omodeo for helpful information.

References

de Eguileor, M., S. Daniel, F. Cotelli, R. Valvassori & G. Lanzavecchia, 1989. Histochemical analysis of oligochaete body wall. Hydrobiologia 180: 99–107.

de Eguileor, M., R. Valvassori, G. Lanzavecchia & S. Giorgi, 1990. Body wall muscles in the haplotaxids *Haplotaxis gordioides* and *Pelodrilus leruthi* (Annelida, Oligochaeta). Zoomorphology 110: 27–36.

Delay, B., 1970. Données sur l'écologie, l'anatomie et la biologie d'*Haplotaxis (Pelodrilus) leruthi* (Hrabe) (Oligochète, Haplotaxidae). Ann. Speleol. 25: fasc. 3: 621–649.

D'Haese, J. & D. Carlhoff, 1987. Localization and histochemical characterization of myosin isoforms in earthworm body wall muscle. J. Comp. Physiol. 157: 171–179.

Drewes, C. D. & M. J. Zoran, 1989. Neurobehavioral specializations for respiratory movements and rapid escape from predators in posterior segments of the tubificid *Branchiura sowerbyi*. Hydrobiologia 180: 65–71.

Dubowitz, V. & M. H. Brooke, 1973. Muscle biopsy: a modern approach. W. B. Saunders, London, pp. 177–181.

Juberthie, C., 1967. Progres récènts sur la biologie des Oligochètes souterrains *Haplotaxidae* et *Lumbricidae*. Spelunca Mémoires 5: 267–272.

Juberthie, C. & M. Mestrov, 1965. Sur les Oligochètes terrestres des sediments argileux des grottes. Ann. Speleol. 20: 210–233.

Karnowsky, M. J., 1965. A formaldehyde-glutaraldehyde fixative of high osmolarity for use in electron microscopy. J. Cell Biol., 27: 137A.

Moore, R. D., V. Mumaw & M. D. Schoenberg, 1960. Optical microscopy of ultrathin tissue sections. J. Ultrastruct. Res. 4: 113–116.

Omodeo, P., 1985. Some new species of Haplotaxidae (Oligochaeta) from Guinea and remarks on the history of the family. Hydrobiologia 155: 1–13.

Timm, T., 1980. Distribution of aquatic Oligochaetes. In R. O. Brinkhurst & D. G. Cook (eds), Aquatic Oligochaete Biology. Plenum Press, New York: 55–77.

Hydrobiologia **278**: 165–178, 1994.
T.B. Reynoldson & K.A. Coates (eds), Aquatic Oligochaete Biology V.

An ultrastructural overview of tubificid spermatozoa

Marco Ferraguti[1], Daniela Ruprecht[1], Christer Erséus[2] & Olav Giere[3]
[1]*Department of Biology, Università degli studi di Milano, 26 Via Celoria, I-20133, Milano, Italy;*
[2]*Department of Zoology, University of Göteborg, and (present address) Department of Invertebrate Zoology, Swedish Museum of Natural History, Box 50007, S-10405 Stockholm, Sweden;*
[3]*II Zoological Institute and Museum, Martin-Luther-King Platz 3, D-2000, Hamburg, Germany*

Key words: spermatozoa, Tubificidae, oligochaete ultrastructure, taxonomy, phylogeny

Abstract

The spermatozoa of seven species of tubificid oligochaete were compared to those already described in order to supply spermatozoal characters for systematic work on Tubificidae. The spermatozoa of *Clitellio arenarius* (subfamily Tubificinae) resemble those of the two other known members of the subfamily (*Tubifex tubifex* and *Tubificoides amplivasatus*) in being of two different types and in showing the same morphology of acrosomes and nuclei. Among Phallodrilinae, the three gutless species *Inanidrilus bulbosus, Inanidrilus leukodermatus* and *Olavius planus* share with *Bathydrilus formosus* the shape of the nucleus and various characters of the acrosome, whereas *Thalassodrilus prostatus* is particular in having the acrosome vesicle external to the tube and the longest middle piece recorded in oligochaetes. *Monopylephorus limosus* and *Rhizodrilus russus* (Rhyacodrilinae) differ widely in acrosomal and nuclear characters: *M. limosus* has a twisted nucleus, whereas *R. russus* has an apically flanged nucleus with a conspicuous apical concavity and an endonuclear canal. The data published on spermatozoa of the Limnodriloidinae are reviewed.

Introduction

Ultrastructure of spermatozoa is considered a useful tool for phylogenetic studies (Jamieson, 1987). A cladistic analysis of sperm characters in oligochaetes, using single species as representatives for each family, has given results in agreement with phylogenetic patterns based on conventional characters (Jamieson *et al.*, 1987). However, few studies have used sperm characters to study phylogeny at lower taxonomic levels (some examples are: Dallai & Mazzini, 1990; Baccetti & Daccordi, 1988; Giusti *et al.*, 1990), partly because the physiological pressures due to variations in reproductive biology of the particular species have been considered to exceed the phylogenetic changes (Ferraguti & Gelder, 1991).

Taxonomic information about the family Tubificidae has increased significantly in the last years with the addition of a number of primarily marine genera. This group is now the most speciose aquatic oligochaete family (at least 600 species known to date). Five subfamilies are currently recognized within Tubificidae: Tubificinae, Phallodrilinae, Telmatodrilinae, Limnodriloidinae, and Rhyacodrilinae, although the last-

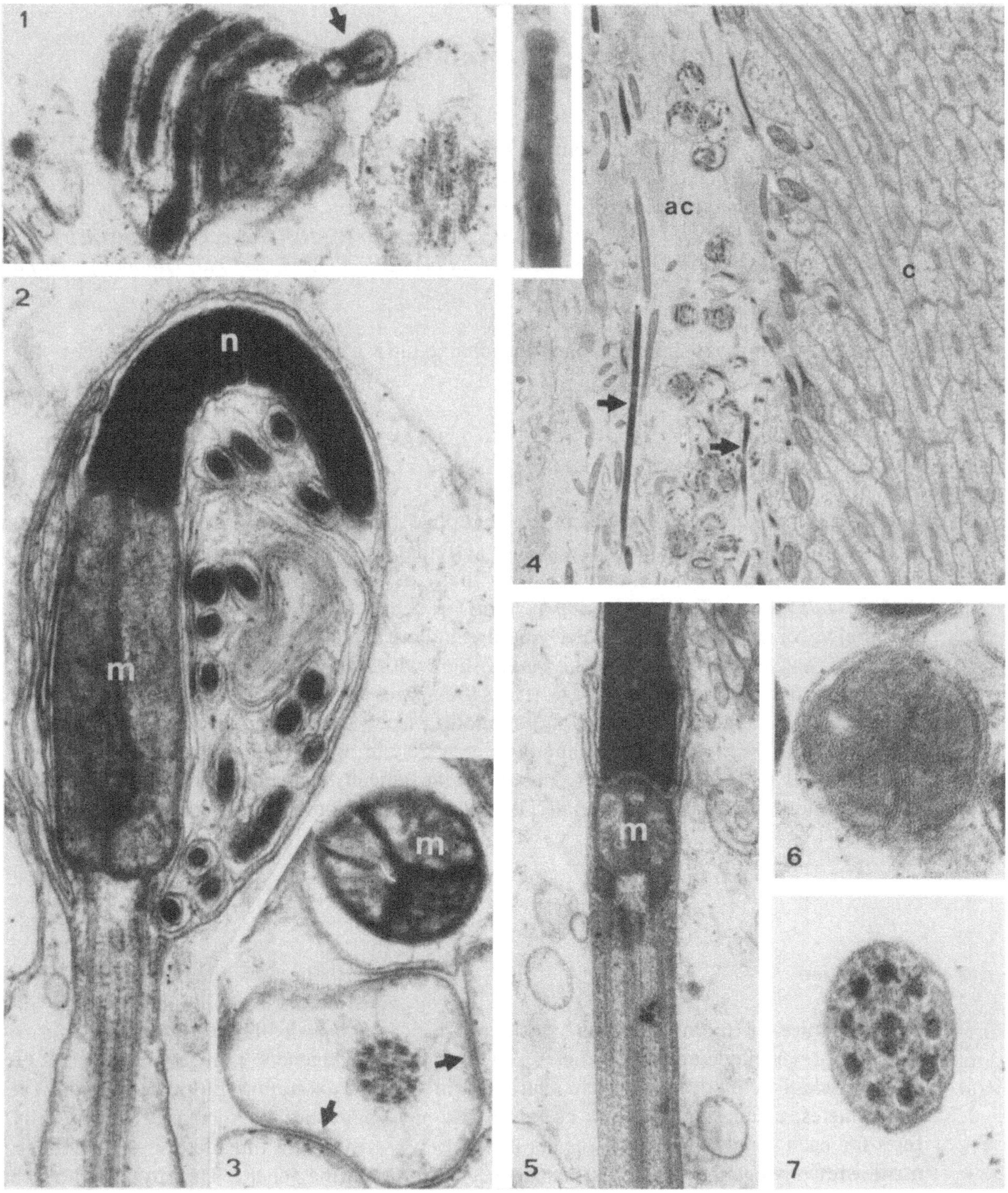

Figs 1–7. Clitellio arenarius. (1) Rolled-up nucleus of an atypical spermatozoon in the spermatozeugmata. The acrosome (arrow) is protruding from the skein. (× 55,000). (2) Longitudinal section of the mitochondria (m) and the proximal portion of the nucleus (n) of an atypical spermatozoon. Both are rolled-up and wrapped by the same cell membrane. (× 55,000). (3) Cross section of the three mitochondria (m) and the flagellum of an atypical spermatozoon. The plasma membrane is dilated around the axoneme and very close to the neighbouring membranes. No septate junctions are visible in the areas of contact (arrows). (× 50,000). (4)

Table 1. References on Tubificid species investigated.

Species	Spermiogenesis	Spermatozoa	Spermatozeugmata
TUBIFICINAE			
Tubifex tubifex	Ferraguti & Lanzavecchia (1971)	Braidotti & Ferraguti (1982)	Braidotti *et al.* (1980) Braidotti & Ferraguti (1982) Ferraguti *et al.* (1989)
Tubifex hattai	Jaana (1983)	–	Jaana (1982)
Limnodrilus hoffmeisteri	Block & Goodnight (1980)	Block & Goodnight (1980)	–
Tubificoides amplivasatus	–	this study	Ferraguti *et al.* (1989)
Clitellio arenarius	–	this study	Ferraguti *et al.* (1989)
Peloscolex (= *Guistadrilus*) *multisetosus*	Correll (1969)	–	–
LIMNODRILOIDINAE			
Kaketio (= *Thalassodrilides*) *ineri*	–	Ferraguti *et al.* (1989)	Ferraguti *et al.* (1989)
Marcusaedrilus tuber	–	–	Ferraguti *et al.* (1989)
Limnodriloides winckelmanni (= *australis*)	Jamieson & Daddow (1979) Jamieson *et al.* (1978)	Jamieson & Daddow (1979) Jamieson *et al.* (1978)	–
RHYACODRILINAE			
Rhizodrilus arthingtonae	Jamieson & Daddow (1979)	Jamieson *et al.* (1978)	–
Rhizodrilus russus	–	this study	this study
Monopylephorus limosus	–	this study	–
Branchiura sowerbyi	Casellato *et al.* (1987) Hirao (1973)	–	–
PHALLODRILINAE			
Bathydrilus formosus	–	Ferraguti *et al.* (1989)	Ferraguti *et al.* (1989)
Bathydrilus rhodei	Jamieson & Daddow (1979)	–	–
Thalassodrilus prostatus	–	this study	–
Inanidrilus bulbosus	–	this study	–
Inanidrilus leukodermatus	–	this study	–
Olavius planus	–	this study	–

mentioned is probably paraphyletic (Erséus, 1990a). Erséus (*op. cit.*) suggests that the Naididae are likely to be cladistic members of the Tubificidae *sensu lato* on the basis of conventional (light microscopical) characters. However, phylogenetic relationships of the different subfamilies are still largely unresolved.

The aim of this paper is to add spermatological characters to the conventional ones (cf. Erséus, 1990a) to facilitate future assessments of the phylogeny of the Tubificidae (Erséus & Ferraguti, in prep.). Knowledge of tubificid sperm morphology has been confined to a few species unrepresentative of the different subfamilies and the data are scattered in many papers of the last twenty years (Table 1).

Tubificid spermatozoa share the general characters of those in the other oligochaetes (Fig. 28):

Longitudinal section of a spermatozeugma showing part of the axial cylinder (ac) containing the fertilizing spermatozoa (arrows), and the cortex (c) formed by the atypical spermatozoa. Inset: detail of the acrosome of a fertilizing sperm. (× 4,750; inset: × 42,000). (5) Longitudinal section of the middle piece of a fertilizing spermatozoon. Note the reduced size of mitochondria (m) compared to those of the atypical spermatozoon in Fig. 2 (× 53,000). (6) Cross section of the five mitochondria of a fertilizing spermatozoon. (× 100,000). (7) Cross section of the proximal portion of the fertilizing sperm flagellum (× 100,000).

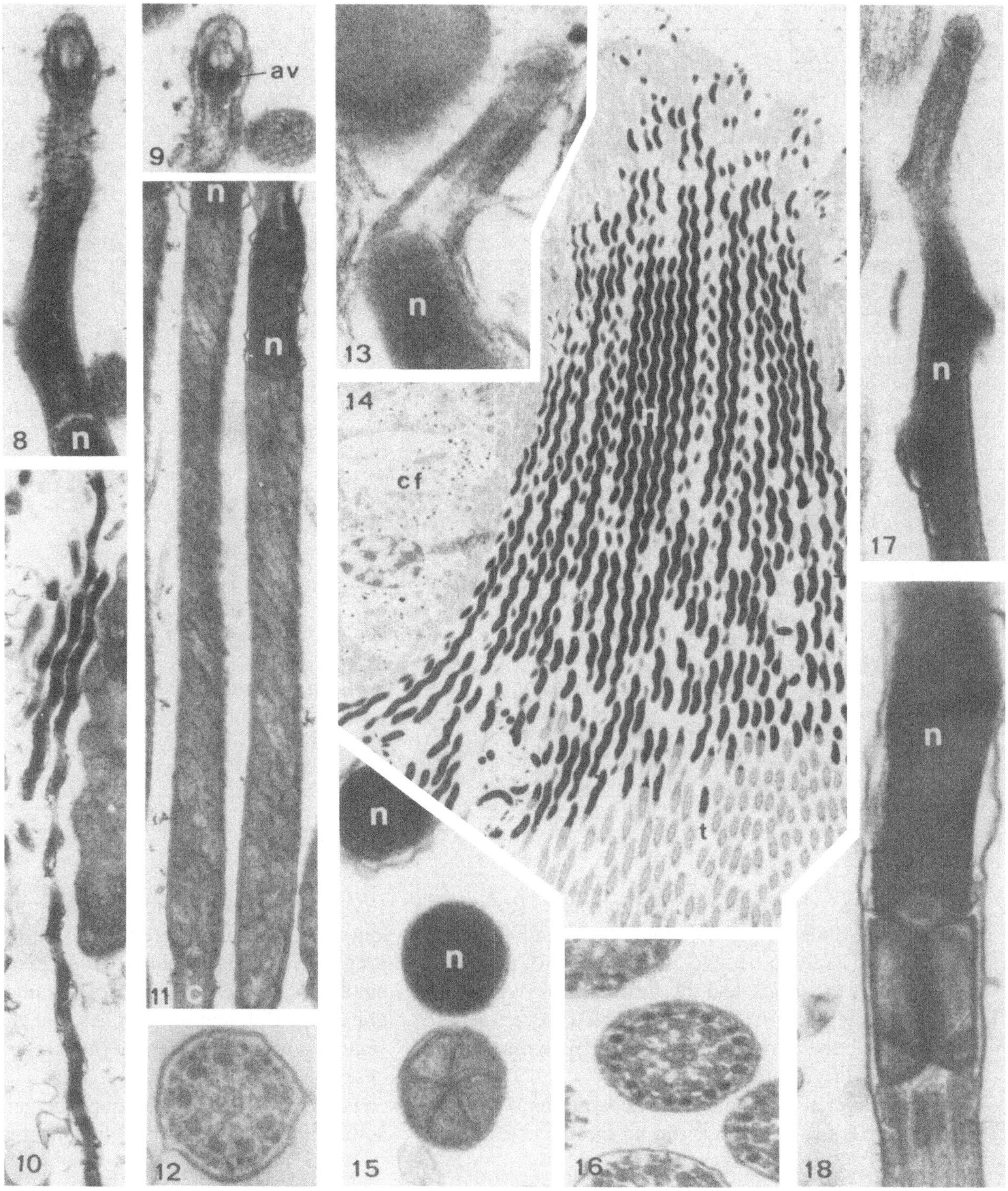

Figs 8–12. Thalassodrilus prostatus. (8) Longitudinal section of the acrosome showing its twisted shape. n = nucleus. (× 67,500). (9) Detail of the acrosomal tip. The acrosome vesicle (av) is external to the tube. (× 67,500). (10) Longitudinal section of some nuclei. (× 10,500). (11) Longitudinal section of the extremely elongated mitochondria. n = nuclei; c = centriolar remnant (× 25,000). (12) Cross section of the main portion of the flagellum. (× 75,000).

Figs 13–16. Olavius planus. (13). Longitudinally sectioned acrosome. n = nucleus. (× 98,000). (14) Ciliated funnel collecting mature an acrosome with an acrosome tube containing a part of the acrosome vesicle which in turn surrounds an acrosome rod; a secondary acrosome tube contained within the tube and starting from the base of the vesicle; a long and thin nucleus with different shapes, a variable number of cylindrical-sector-shaped mitochondria interposed between the nucleus and the flagellum; a flagellum starting with a centriolar remnant containing a basal cylinder; the central apparatus of the remaining part of the axoneme being modified by the presence of two tetragon fibers or a prominent central sheath; and beta glycogen granules surrounding the axoneme (Ferraguti & Lanzavecchia, 1971; Jamieson & Daddow, 1979). Variations of all these characters will be discussed in this paper which describes the spermatozoa of seven species belonging to three different subfamilies. The data will be compared with those for tubificid sperm types already described in the literature.

Materials and methods

Specimens of *Clitellio arenarius* (Müller, 1776) (Tubificinae), and *Thalassodrilus prostatus* (Knöllner, 1935) (previously placed in *Phallodrilus*, see Erséus, 1992) (Phallodrilinae) were collected from marine intertidal habitats near Göteborg (Sweden), *Monopylephorus limosus* (Hatai, 1898) (Rhyacodrilinae) from the sediments of River Lambro in Milano, Italy, and *Rhizodrilus russus* (Erséus, 1990) (Rhyacodrilinae) from an intertidal mudflat off mangroves at Mai Po, Deep Bay, New Territories, Hong Kong. Material of *Inanidrilus bulbosus* Erséus, 1979 (Phallodrilinae) was collected from calcareous sands off Crystal River, Florida, U.S.A.; *Inanidrilus leukodermatus* (Giere, 1977) and *Olavius planus* (Erséus, 1979) (Phallodrilinae) from similar sediments in Bermuda.

A variety of fixatives was used in the present study: *M. limosus* was fixed in a paraformaldehyde-glutaraldehyde mixture in a saturated solution of picric acid; *C. arenarius*, and *T. prostatus*, in phosphate buffered 3% glutaraldehyde; *R. russus* in cacodylate buffered 3% glutaraldehyde with 10% sucrose added; the gutless species (*Inanidrilus* and *Olavius*) in 2.5–5% glutaraldehyde buffered with either phosphate or cacodylate. After washing in the corresponding buffer and postfixation in similarly buffered 1% osmium tetroxide, the worms were dehydrated in a graded ethanol series and embedded in Spurr's resin (gutless species, *M. limosus*, and *R. russus*) or Epon-Araldite mixture (*C. arenarius*, and *T. prostatus*). Sections of *C. arenarius, M. limosus, R. russus*, and *T. prostatus* were cut with LKB ultratomes III, IV, and V, stained following Daddow (1983) and observed under a Jeol 100 SX electron microscope. The gutless phallodrilines were cut with a Reichert OmU2 ultramicrotome, stained with uranyl acetate and lead citrate and observed under a Zeiss EM9 – S2 electron microscope.

Observations

Subfamily Tubificinae

Clitellio arenarius (Figs 1–7 & 29). Two types of spermatozoa are produced in *Clitellio*: eupyrene (fertilizing) ones and oligopyrene, sterile ones (Fig. 29). In the spermathecae they aggregate to form rod-like spermatozeugmata (Fig. 4) similar to those described in *Tubifex* (Braidotti & Ferraguti, 1982): the fertilizing spermatozoa lie parallel in the center of the spermatozeugma forming the axial cylinder, whereas the oligopyrene spermatozoa are partly linked to one another to form a sheath (cortex) surrounding the eupyrene ones. Unlike other known models of tubificine spermatozeugmata, there are no evident junctions be-

spermatozoa. cf = ciliated funnel tissue; n = nuclei; t = tails. (× 4,400). (15) Cross sectioned nuclei (n) and mitochondria. (× 63,000). (16) Cross sectioned flagellum. (× 91,000).

Figs 17–18. Inanidrilus leukodermatus. (17) Apex of the nucleus (n) and acrosome longitudinally sectioned. (× 63,000). (18) Base of the nucleus (n) and middle piece. (× 63,000).

tween the tails of the oligopyrene sperm (Fig. 3) and the nucleus of each non-fertilizing spermatozoon is wound into a skein-like ball (Figs 1 & 2) (Ferraguti & Ruprecht, 1992).

The shape of the two kinds of mature spermatozoa clearly differs. The acrosome of the eupyrenes is 0.4 μm long and shows a deep withdrawal of the acrosome vesicle surrounding the rod and a short secondary tube (Fig. 4, insert). The nuclei of these cells are long and thin, witha twisted apical part and a straight basal portion (Figs 4 & 5). The middle piece contains five (sometimes four) short and small mitochondria forming a spheroidal structure (Figs 5 & 6) and the flagellum, which lacks a proper centriole, has a portion with a prominent central sheath and contains glycogen granules (Fig. 7).

The oligopyrene spermatozoa are much shorter. The acrosome, 0.2 μm long, has no rod, no secondary tube, and its acrosome vesicle lies completely external to the tube (Fig. 1). The thread-like nucleus is extremely thin (less than 50 nm at its distal end) and never condenses completely. The middle piece contains three (rarely four) mitochondria, their overall volume being eight times that of the mitochondria in the fertilizing spermatozoa (Figs 2 & 3). The tail starts proximally with a centriolar area, followed by a conventional axoneme surrounded by a swollen plasma membrane (Figs 2 & 3). The glycogen of the flagellum is in the gamma form.

Subfamily Phallodrilinae

Thalassodrilus prostatus (Figs 8–12 & 30). The acrosome of *T. prostatus* has a long twisted tube (Fig. 8). The vesicle is completely external to the

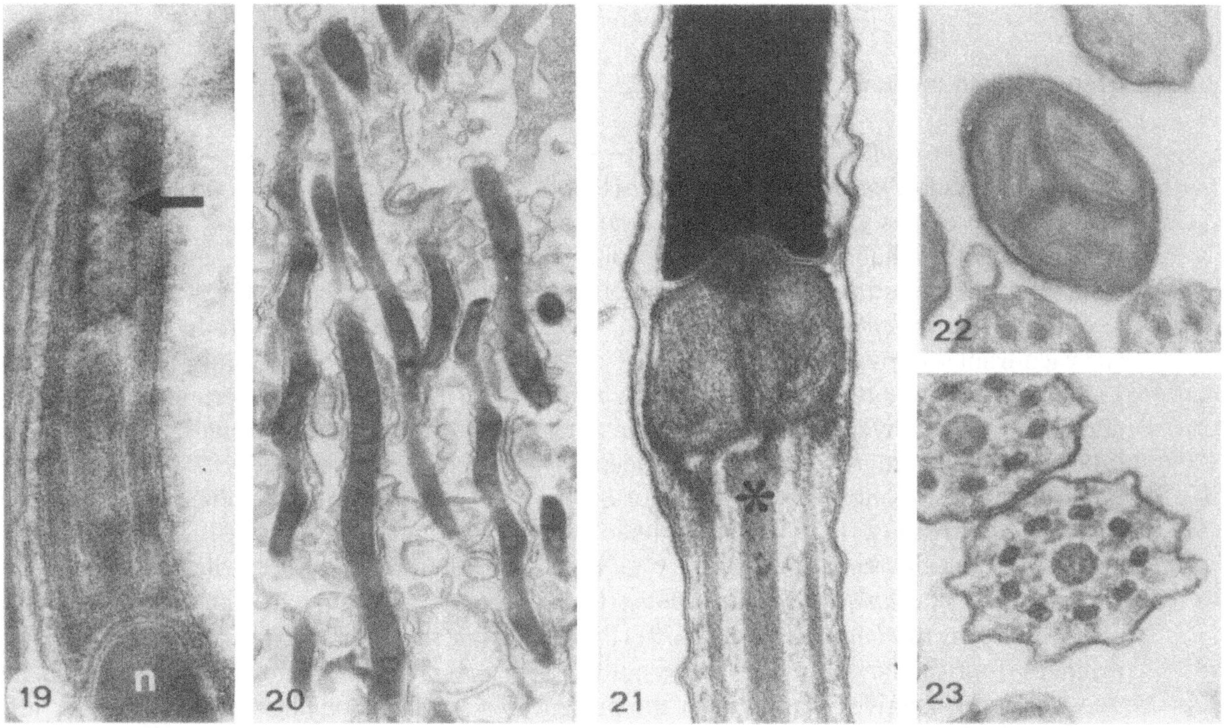

Figs. 19–23. Monopylephorus limosus. (19) Longitudinally sectioned acrosome. Dense material is present inside the acrosome vesicle (arrow). The acrosome vesicle is deeply withdrawn reaching the base of the tube. n = nucleus. (× 120,000). (20) Longitudinally sectioned nuclei. (× 19,500). (21) Longitudinal section of the middle piece. The basal cylinder of the flagellum (asterisk) nearly reaches the mitochondria. (× 82,500). (22) Cross section of the three mitochondria. (× 82,500). (23) Cross section of flagella showing the prominent central sheath. (× 82,500).

tube and surrounds a small rod. A short, barely visible, secondary tube starts from the base of the vesicle (Fig. 9). The nucleus is twisted-column-shaped (Fig. 10) and shows oblique striations here interpreted as the imprints of the manchette. The five, twisted, mitochondria represent the longest (up to 5.2 μm) middle piece known among oligochaetes (Fig. 11). The axoneme has a proximal portion with prominent central sheath (Fig. 12) and a distal one with tetragon fibers.

The gutless species (Figs 13–18). The overall shape of the spermatozoa in *I. bulbosus, I. leukodermatus* and *O. planus* is remarkably similar, thus the three species will be described together. The acrosomes are less than 0.4 μm long and their main axis forms an obtuse angle with the nuclear main axis as the acrosomes continue the pitch of the nuclear corkscrew. Withdrawal of the vesicle is deep, but the rod is short. A short secondary tube is visible only in *Olavius* (Figs 13 & 17). In

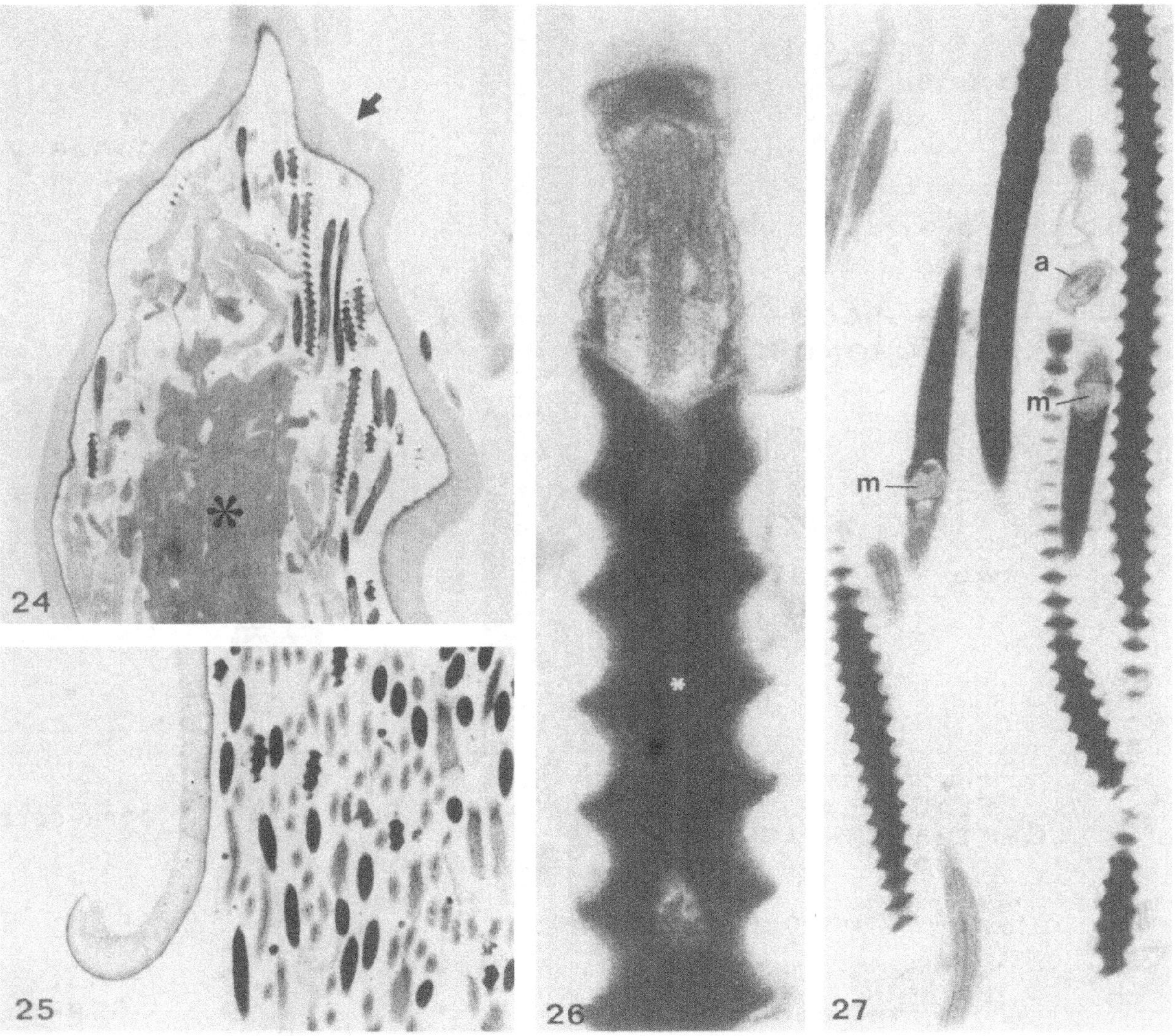

Figs 24–27. Rhizodrilus russus. (24) Portion of a spermatozeugma showing the dense capsule (arrow) surrounding the spermatozoa and internal masses of dense material (asterisk). (× 5,000). (25) The dense capsule ends like a rim of a hat (× 6,000). (26) Anterior extremity of the spermatozoon. An endonuclear canal (asterisk) deeply penetrates the distal part of the flanged nucleus ending with an enlarged portion. The acrosome tube had a variable thickness. (× 65,000). (27) At low magnification the nuclei show their straight base and the flanged anterior portion. a = acrosome; m = mitochondria. (× 1,800).

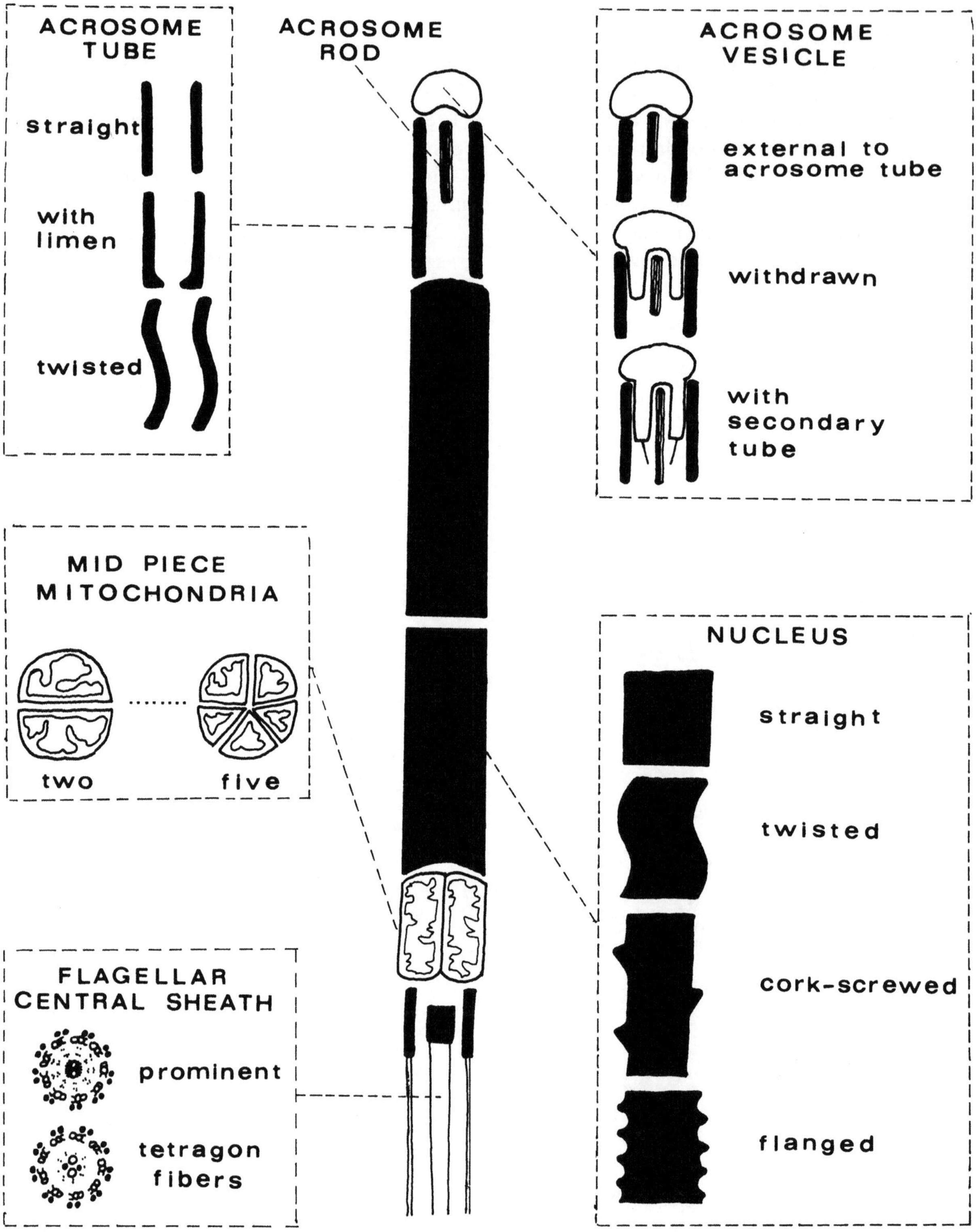
ACROSOME TUBE
straight
with limen
twisted
ACROSOME ROD
ACROSOME VESICLE
external to acrosome tube
withdrawn
with secondary tube
MID PIECE MITOCHONDRIA
two
five
NUCLEUS
straight
twisted
cork-screwed
flanged
FLAGELLAR CENTRAL SHEATH
prominent
tetragon fibers

all three species the nuclei are basally twisted-column-shaped (Fig. 14) and apically cork-screwed (Fig. 17). Both *I. bulbosus* and *I. leukodermatus* show the imprint of the manchette microtubules on the nucleus (Fig. 18). The middle pieces, formed by five mitochondria, are 0.5 μm long in *I. leukodermatus*, 0.28 μm in *I. bulbosus*, and 0.34 μm in *O. planus* (Fig. 15). The flagella show a tetragon fiber arrangement (Fig. 16).

Subfamily Rhyacodrilinae

Monopylephorus limosus (Figs 19–23 & 31). The acrosome is 0.63 μm long and has an acrosome tube slightly bent on one side with the vesicle completely withdrawn into it. The tube shows a varying thickness, being thicker at the proximal end, where it forms a conspicuous limen (*sensu* Jamieson, 1978). The rod is short (0.25 μm) and is located at the base of the tube; a secondary tube is lacking (Fig. 19). The nucleus is twisted along its whole length (the wavelength increasing from the distal to the proximal extremity) (Fig. 20) and is followed by a short (0.35 μm) middle piece composed of three mitochondria (Figs 21 & 22). The flagellum has a prominent central sheath (Fig. 23).

Rhizodrilus russus (Figs 24–27 & 31). In this species the spermatozoa in the spermathecae are, at least in part, grouped in bundles hooded at one end by a capsule. The capsule is an amorphous matrix holding together the anterior portion of the sperm bundle (Figs 24 & 25). The spermatozoa have a complex acrosome, 0.84 μm long, with the acrosome tube varying in thickness in the different portions. The acrosome vesicle, withdrawn into the tube, surrounds the rod and its base is characteristically moulded like the jaws of a forceps. The acrosomal rod is as long as the entire acrosome tube, but the secondary tube is extremely short (if present at all) (Fig. 26). The anterior end of the nucleus shows a V-shaped concavity continuing into a 1 μm long endonuclear canal (Fig. 26). The distal half of the nucleus is flanged (Figs 26 & 27), whereas the basal portion is straight (Fig. 27). Four or five mitochondria form the short (0.25 μm) middle piece (Fig. 27). A flagellum follows with a prominent central sheath.

Discussion

Subfamily Tubificinae (Fig. 29)

The spermatozoa of *C. arenarius* are of two different types, as in *Tubifex* and *Tubificoides* (Ferraguti *et al.*, 1989). The differences between oligopyrene and eupyrene spermatozoa are consistent in the three species. In the oligopyrene, atypical sperm, the acrosomes are "empty", the nuclei are short (much shorter in *Tubifex* than in *Clitellio*), the mitochondria (two in *Tubifex* and *Tubificoides*, three in *Clitellio*) are larger, the flagella have a dilated membrane, and the glycogen is in gamma form. Various characters seem to be similar among the spermatozoa of the three species belonging to the two types: in the eupyrene spermatozoa of *Tubifex* and *Clitellio* acrosomes are nearly identical (we have no information for *Tubificoides*); nuclei and flagella have the same shape, whereas the mitochondria are two in *Tubifex* and four or five in *Clitellio*. It may be relevant that in *Clitellio* and *Tubificoides* (unpublished), but not in *Tubifex*, the number of mitochondria varies between the eupyrene and the oligopyrene spermatozoa. It may be concluded that, from the spermatological point of view, Tubificinae is a very coherent subfamily.

Subfamily Phallodrilinae (Fig. 30)

The only character shared by the spermatozoa of all four phallodriline species examined in this paper and the other phallodriline, *Bathydrilus for-*

Fig. 28. An overview of character variation in tubificid spermatozoa. In the center is a longitudinally sectioned, generalized tubificid spermatozoon. Characters depicted are: shape of the acrosome tube; position of the acrosome vesicle; nuclear shape; number of midpiece mitochondria; structure of the central apparatus of the flagellum. For simplicity, the plasma membrane is not drawn.

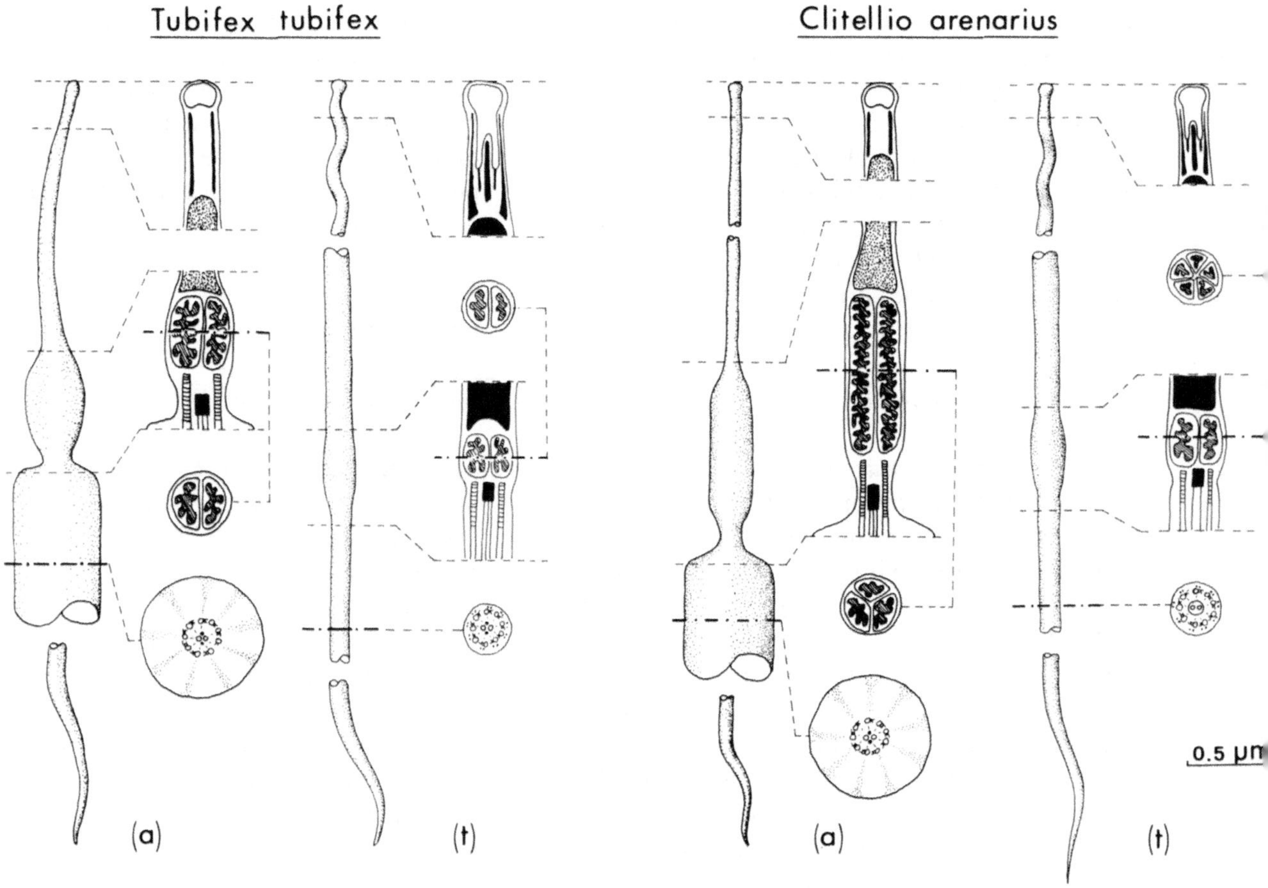

Fig. 29. Species belonging to the subfamily Tubificinae produce two kinds of spermatozoa: atypical (a) and typical, fertilizing ones (t).

mosus Erséus, 1986 studied previously (Ferraguti *et al.*, 1989), is the possession of five midpiece mitochondria, but this is likely to be a plesiomorphic number. The shape of the nuclei is a possible synapomorphy: a basally twisted and apically corkscrewed nucleus occurs in the gutless species, in *B. formosus* and, probably, in *Bathydrilus rhodei* (Jamieson, 1977) (Jamieson & Daddow, 1979; the species was then regarded as a member of *Phallodrilus*), but in *T. prostatus* the nucleus is twisted along its whole length and lacks the apical corkscrew. In *Inanidrilus*, *Thalassodrilus*, and *Bathydrilus* the nucleus of mature spermatozoa shows the imprint of the manchette microtubules. The acrosomes are similar in *Bathydrilus*, *Inanidrilus* and *Olavius*, being eccentric relative to the nuclear axis. This condition may be due to the shortness of the acrosomes, continuing the corkscrew shape of the apical portion of the nuclei. In *Thalassodrilus* the acrosome tube is twisted. So far, the spermatozoon of *Thalassodrilus* seems to be unique within the subfamily in having the acrosome vesicle completely external to the tube, a character that has been considered plesiomorphic for Oligochaeta as a whole (Jamieson *et al.*, 1987). On the other hand, the middle piece of *T. prostatus* is the longest recorded in the class, an apomorphic condition.

Subfamily Rhyacodrilinae (Fig. 31)

The sperm characters of the members of the subfamily are highly variable. The only significant character shared by the spermatozoa of *M. limo-*

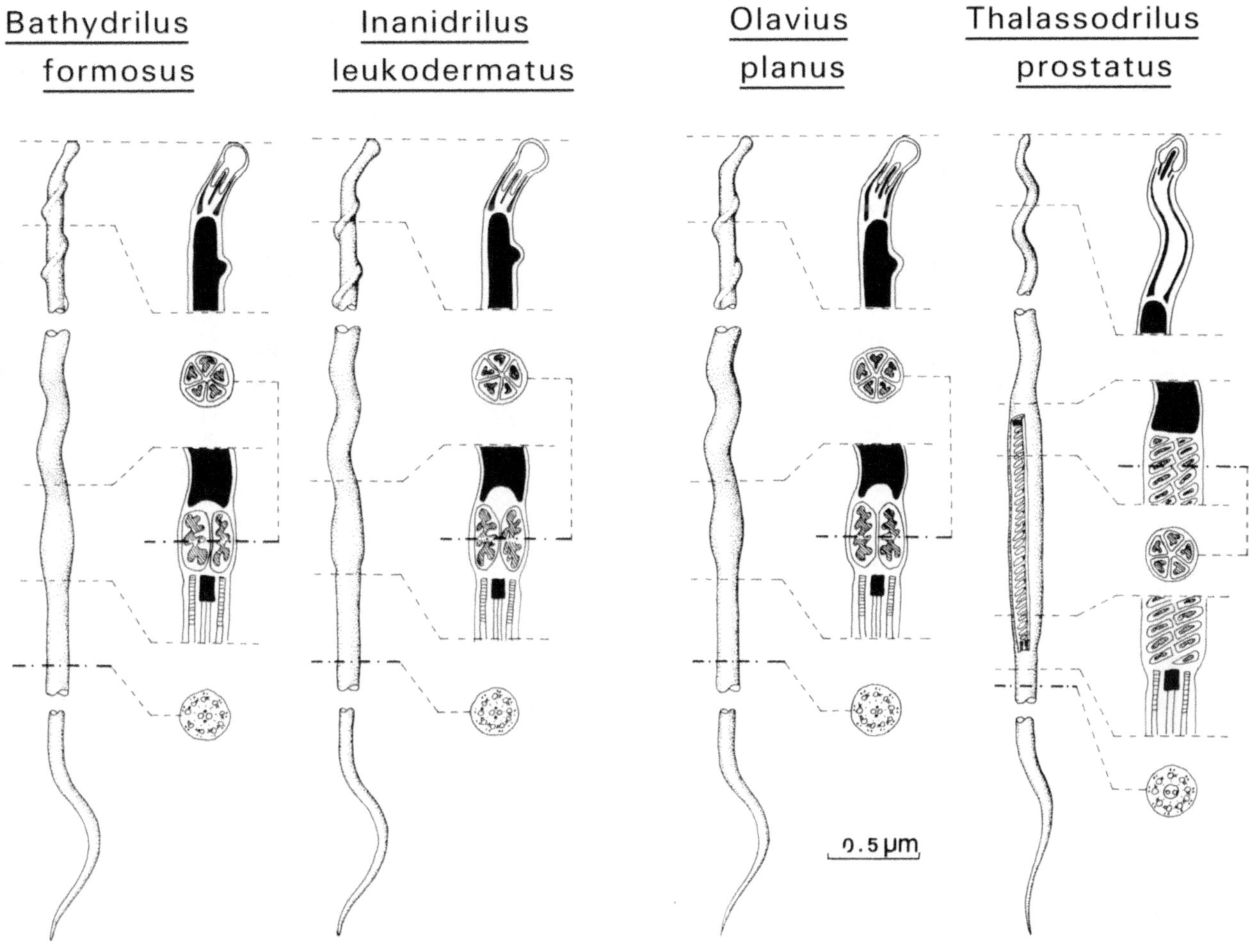

Fig. 30. Sperm models of four phallodriline species.

sus and *R. russus* is the presence of a prominent central sheath in the flagellum, which may be a synapomorphic feature for clitellates, as suggested by Ferraguti (1984). This would be a plesiomorphy for Rhyacodrilinae. The number of mitochondria, the nuclear shape, and the main features of the acrosome differ greatly. Furthermore, the nucleus of *Rhizodrilus* sperm is unique among tubificids by being extremely similar in its general shape (basally straight and apically flanged) to that of enchytraeid spermatozoa (Webster & Richards, 1977). The apical concavity has been reported only for branchiobdellids and, possibly, some leeches (Ferraguti & Gelder, 1991). This feature is, however, present also in the lumbriculid *Kincaidiana* (Ferraguti, unpublished). An endonuclear canal is present, as far as we know, only in *Rhizodrilus* among Tubificidae. However, Jamieson & Daddow (1992) reported the presence of an endonuclear canal in the eudrilid *Eudrilus eugeniae* Kinberg, 1867. It is noteworthy that the sperm of *Rhyacodrilus arthingtonae* Jamieson, 1978 [a species placed in the genus *Rhizodrilus* by Baker & Brinkhurst (1981)] differs in morphology (see Jamieson & Daddow, 1979) from *Monopylephorus* and *R. russus*. *Rhizodrilus arthingtonae* appears more plesiomorphic, having a primitive acrosome with a short secondary tube, a short withdrawal of the acrosome vesicle and a straight nucleus.

Monopylephorus limosus is devoid of a secondary acrosome tube, as are the phallodrilines *Inanidrilus* and *Bathydrilus*, but it has a spiral nucleus like *Thalassodrilus*.

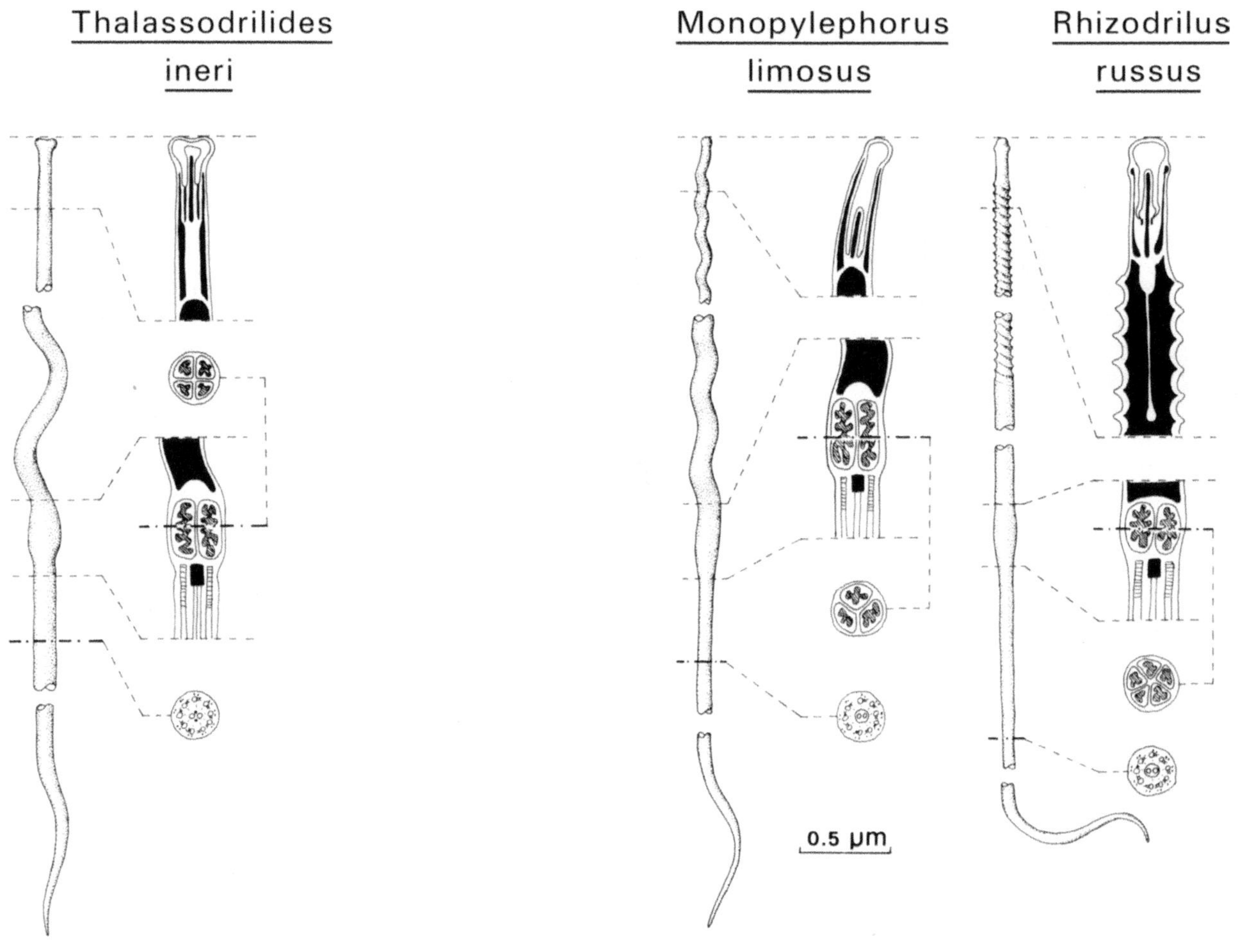

Fig. 31. Sperm models of one limnodriloidine (left) and two rhyacodriline (right) species.

Subfamily Limnodriloidinae (Fig. 31)

Only two species of this subfamily have been investigated [*Limnodriloides winkelmanni* Michaelsen, 1914 by Jamieson & Daddow, 1979, (now identified as *L. australis* Erséus, 1982; Erséus, 1982), and *Kaketio ineri* Righi & Kanner, 1979 (now placed in *Thalassodrilides* by Erséus, 1990b) by Ferraguti *et al.*, 1989]. It is therefore difficult to draw any conclusions about the general appearance of the limnodriloidine spermatozoa. Furthermore, it is accepted that *T. ineri* is a highly apomorphic form of the subfamily (Righi & Kanner, 1979; Ferraguti *et al.*, 1989). However, in both species, the acrosome tube has two portions of distinctly different thicknesses, a detail unknown from elsewhere in the family. In *Limnodriloides* the acrosome rod nearly reaches the nucleus, whereas in *Thalassodrilides* it terminates where the tube wall increases in thickness, at a distance from the nucleus. Characters shared by the two species are the nuclear shape (basally cork-screwed and apically straight), the four mitochondria in the middle piece and the tetragon fibers in the flagellum.

Conclusions

The spermatozoa of Tubificidae show a considerable variation of nearly all morphological characters (Fig. 28). Compared to the other oligochaete spermatozoa, some of the sperm types show inferred plesiomorphic characters, like the acrosome vesicles completely external to the tube, up to now known only in enchytraeids (Webster & Richards, 1977). Other species have apomorphic characters, as the endonuclear canal of *R. russus* and the extreme length of mitochondria in *T. prostatus*.

Our results indicate a coherence of sperm characters within Tubificinae; within Phallodrilinae there is a good agreement of the gutless species sperm morphology with that of *Bathydrilus*, whereas *Thalassodrilus* seems to have a more isolated position. At present, it is more difficult to see links between the sperm characters and the taxonomic positions within Rhyacodrilinae and Limnodriloidinae. Studies of additional sperm types within these two subfamilies are needed for any conclusive statement.

References

Baccetti, B. & M. Daccordi, 1988. Sperm structure and phylogeny of the Chrysomelidae. In P. Jolivet, E. Petitpierre & T. H. Hsiao (eds), Biology of Chrysomelidae. Kluwer Academic Publishers, Dordrecht: 357–378.

Baker, H. R. & R. O. Brinkhurst, 1981. A revision of the genus *Monopylephorus* and redefinition of the subfamilies Rhyacodrilinae and Branchiurinae (Tubificidae: Oligochaeta). Can. J. Zool. 59: 939–965.

Block, E. M. & C. J. Goodnight, 1980. Spermatogenesis in *Limnodrilus hoffmeisteri* (Annelida, Tubificidae): a morphological study of the development of two sperm types. Trans. am. micros. Soc. 99: 369–384.

Braidotti, P. & M. Ferraguti, 1982. Two sperm types in the spermatozeugma of *Tubifex tubifex* (Annelida, Oligochaeta). J. Morph. 171: 123–136.

Braidotti, P., M. Ferraguti & T. Fleming, 1980. Cell junctions between spermatozoa flagella within spermatozeugmata of *Tubifex tubifex* (Annelida, Oligochaeta). J. ultrastr. Res. 73: 299–309.

Casellato, S., G. Martinucci & E. Zoja, 1987. Ultrastructural features of gametogenesis during the life cycle in *Branchiura sowerbyi* Beddard (Oligochaeta, Tubificidae). Hydrobiologia 155: 145–154.

Correl, P. J., 1969. Spermatogenesis in *Peloscolex multisetosus* (Oligochaeta, Tubificidae). MSc Thesis, University of Wisconsin, Milwaukee, 98 pp.

Daddow, L. Y. M., 1983. A double lead stain method for enhancing contrast of ultrathin sections in electron microscopy: a modified multiple staining technique. J. Microsc. 129: 147–153.

Dallai, R. & M. Mazzini, 1990. The spermatozoon of the gall-midge Oligotrophidi (Diptera, Cecidomyiidae). Boll. Zool. 56: 13–27.

Erséus, C., 1982. Taxonomic revision of the marine genus *Limnodriloides* (Oligochaeta, Tubificidae). Verh. naturw. Wer. Hamb. (N.F.). 25: 207–277.

Erséus, C., 1990a. Cladistic analysis of the subfamilies within the Tubificidae (Oligochaeta). Zool. Scr. 19: 57–63.

Erséus, C., 1990b. The marine Tubificidae (Oligochaeta) of the barrier reef ecosystems at Carrie Bow Cay, Belize and other parts of the Caribbean Sea, with descriptions of twenty-seven new species and revision of *Heterodrilus, Thalassodrilides* and *Smithsonidrilus*. Zool. Scr. 19: 243–303.

Erséus, C., 1992. A generic revision of the Phallodrilinae (Oligochaeta, Tubificidae). Zool. Scr. 21: 5–48.

Ferraguti, M., 1984. The comparative ultrastructure of sperm flagella central sheath in Clitellata reveals a new autapomorphy of the group. Zool. Scr. 13: 201–207.

Ferraguti, M. & S. R. Gelder, 1991. The comparative ultrastructure of spermatozoa from five branchiobdellidans (Annelida, Clitellata). Can. J. Zool. 69: 1945–1956.

Ferraguti, M. & G. Lanzavecchia, 1971. The morphogenetic effects of microtubules. J. submicr. Cytol. 3: 121–137.

Ferraguti, M. & D. Ruprecht, 1992. The double sperm line in the tubificid *Clitellio arenarius* (Annelida, Oligochaeta). Boll. Zool. 59: 349–362.

Ferraguti, M., G. Grassi & C. Erséus, 1989. Different models of tubificid spermatozeugmata. Hydrobiologia 180: 73–82.

Giusti, F., G. Manganelli & M. G. Selmi, 1990. Phylogenetic implications of spermatozoon fine structure in the Onchidiidae (Mollusca: Gastropoda). Lavori della Societa' Italiana di Malacologia 23: 327–347.

Hirao, J., 1973. Spermatogenesis in a freshwater Oligochaete, *Branchiura sowerbyi* Beddard. Bull. Fac. Med. Wakayama Univ. 3: 55–62.

Jaana, H., 1982. The ultrastructure of the epithelial lining of the male genital tract and its role in the spermatozeugmata formation in *Tubifex hattai* Nomura (Annelida, Oligochaeta). Zool. Anz. 209: 159–176.

Jaana, H., 1983. The fine structural study of spermiogenesis in the freshwater oligochaete, *Tubifex hattai*, with a note on the histone transition in the spermatid nucleus. J. Fac. Sci. Hokkaido Univ. Ser., VI, Zool. 23: 163–184.

Jamieson, B. G. M., 1978. A comparison of spermiogenesis and spermatozoal ultrastructure in megascolecid and lumbricid earthworms (Oligochaeta: Annelida). Aust. J. Zool. 26: 225–240.

Jamieson, B. G. M., 1987. The ultrastructure and phylogeny of insect spermatozoa. Cambridge University Press, Cambridge, England, 320 pp.

Jamieson B. G. M. & L. Daddow, 1979. An ultrastructural study of microtubules and the acrosome in spermiogenesis of Tubificidae (Oligochaeta). J. ultrastr. Res. 67: 209–224.

Jamieson, B. G. M. & L. Daddow, 1992. The spermatozoon of *Eudrilus eugeniae* (Oligochaeta, Annelida): the first record of an endonuclear canal in oligochaete spermatozoa. J. Submicr. Cytol. Pathol. 24: 323–333.

Jamieson, B. G. M., L. Daddow & J. D. Bennet, 1978. Ultrastructure of the tubificid acrosome (Annelida, Oligochaeta). Zool. Scr. 7: 115–118.

Jamieson, B. G. M., C. Erséus & M. Ferraguti, 1987. Parsimony analysis of the phylogeny of some Oligochaeta (Annelida) using spermatozoal ultrastructure. Cladistics 3: 145–155.

Righi, G. & E. Kanner, 1979. Marine Oligochaeta (Tubificidae and Enchytraeidae) from the Caribbean Sea. Stud. Fauna Curacao and Caribbean Isl. 58: 44–68.

Webster, P. M. & K. S. Richards, 1977. Spermiogenesis in the enchytraeid *Lumbricillus rivalis* (Oligochaeta, Annelida). J. ultrastr. Res. 61: 62–77.

Hydrobiologia **278**: 179–188, 1994.
T.B. Reynoldson & K.A. Coates (eds), Aquatic Oligochaete Biology V.

Body wall muscles in oligochaetes

Giulio Lanzavecchia, Roberto Valvassori & Magda de Eguileor
Dipartimento di Biologia, Universita' degli Studi di Milano, Via Celoria 26, I-20133 milano, Italy

Key words: oligochaetes, microdriles, megadriles, muscles, ultrastructure

Abstract

Longitudinal body wall muscles in tubificids, enchytraeids, lumbriculids, haplotaxids and lumbricids are always histologically and ultrastructurally distinguishable from one another. Some structural peculiarities, as those typical of lumbriculids, may be used as taxonomic tools. The longitudinal body wall muscles in all oligochaetes belong to slow and fast contracting fibre types. Two possible phylogenetic patterns for the formation of slow fibres are discussed.

Introduction

Motility is one of the fundamental factors governing animal evolution, since it is linked to the ability to adapt to different enviromental conditions (Clark, 1964). Nevertheless, the organization of muscle fibres has not been regarded as important in the interpretation of animal phylogeny (Mettam, 1967). As Fauchald remarks (1974), though not pointing specifically at muscles, phylogenetic schemes based on modifications to a single organ system invariably show internal inconsistency and have proved unsatisfactory. However, we believe that most of the difficulties we encounter when using muscle fibres as taxonomic tools are due to the fact that knowledge of the structural and ultrastructural organization of these fibres in invertebrates, especially in soft-bodied animals, is fragmentary and brief (Jamieson, 1981; Lanzavecchia *et al.*, 1988; Sawyer, 1986; Bird & Bird, 1991).

We should consider that, when speaking of muscular systems, we are not referring to a single character (one gene, one protein) but to a very high number of characters interacting with one another, in a largely unknown way (Lewin, 1984). These form complex character systems, with precise morphologies, performing precise functions. Thus, the overall organization of musculature, frequently used as a taxonomic marker (see, for example, Lorenzen, 1985), may be split into many elementary characters. The characters include: (1) the large number of proteins making up the contractile machine, (2) the proteins defining the Z elements and sarcotubular system, (3) the spatial relations among the various elements within one fibre, (4) the interactions established among the various fibres, and (5) the morphological variability of the fibres within the different layers or sectors.

The muscular system can express itself only through its interaction with the skeletal system (in annelids, with the hydroskeleton) and with the nervous system.

In this work we intend to use the data concerning the relationships between different fibres and their fine structure, as they have been described in numerous appearances for taxonomic and phylogenetic purposes, and to compare the conclusions obtained with those derived from traditional systematics. This analysis takes into account only data pertaining to longitudinal muscle fibres of the body wall of some microdriles and megadriles.

Materials and methods

For routine transmission electron microscopy small pieces of animals were fixed according to Karnowsky (1965) and embedded in an Epon 812-araldite mixture. Thin sections obtained with an LKB Ultrotome V were observed with a Jeol 100 B electron microscope. Semi-thin sections (1 μm) were observed on a light microscope after staining with crystal violet and basic fuchsin (Moore *et al.*, 1960).

Terminology

The contractile material of an obliquely-striated fibre, in a cross section, appears subdivided into a series of regions or fields with the same sequence of bands normally seen in longitudinal sections of cross-striated muscle fibres. These fields are normally indicated as 'sarcomeres'.

Results

The longitudinal muscular fibres of the body wall of some oligochaete species belonging to different families are illustrated (Scheme 1), with particular reference to the Lumbriculidae. Characters of the muscles considered in this paper are: dimensions of fibres in cross section; width of 'sarcomeres'; distribution of Z rods and sarcotubules; thin:thick filament ratio; diameter of thick filaments (larger filaments have greater length). According to Rosenbluth (1968: 257), "In a fibre with long myofilaments more actin-myosin cross-links will be in parallel than in a fibre with short myofilaments and, other things being equal, the maximum force developed by the respective fibre will be directly proportional to myofilament length and the velocity will be inversely proportional". In *Glycera* (bloodworm) there are some fibres, besides the predominant type, with thick filaments that are markedly long and wide. Bloodworm musculature is thus composed of a mixture of fibre types which may correspond to the fast and slow fibres of cross-striated muscles (Rosenbluth, 1968). Following the results of previous studies (de Eguileor *et al.*, 1988, 1989), fibres with a myosin filament ranging from 25 to 30 nm in diameter and with a low ratio beteen myosin and actin filaments (around 7–8) are considered to be 'fast', whereas fibres with a myosin filament exceeding 50 nm in diameter and with a ratio between myosin and actin filaments over 12 are considered to be 'slow'.

Tubificidae

Species examined: *Tubifex tubifex* (Müller, 1774), *Limnodrilus udekemianus* Claparède, 1862, *Monopylephorus* sp., *Peloscolex* sp., *Branchiura sowerbyi* Beddard, 1892. In all the species examined the organization of the body wall is constant (Scheme 1) as far as the characters taken into account in this work are concerned. In cross section, the main axis of the fast longitudinal muscles of the flat circomyarian type is perpendicular to the body surface (radial arrangement) (Fig. 1; Scheme 1:A). The contractile region is more or less oval, with a mean ratio of 4.5 between the diameters. The contractile layer measured from the periphery to the middle line splitting the two regions along a reversed obliquity reaches 1.2 μm. Myosin filaments have a small diameter (30 nm) and a thick:thin filament ratio of 7–8. Towards the coelomic cavity, the cytoplasmic region of the fibres forms pouches in which the nuclei are located.

Among the fast fibres, towards the circular muscle layer, there are small triangular or irregular fibres, with very thick myosin filaments (up to 60 nm in diameter) (Fig. 7; Scheme 1:A) and a high ratio (15–16) between actin and myosin filaments (Lanzavecchia, 1971). These fibres have a general morphology that can be ascribed to the flat circomyarian type.

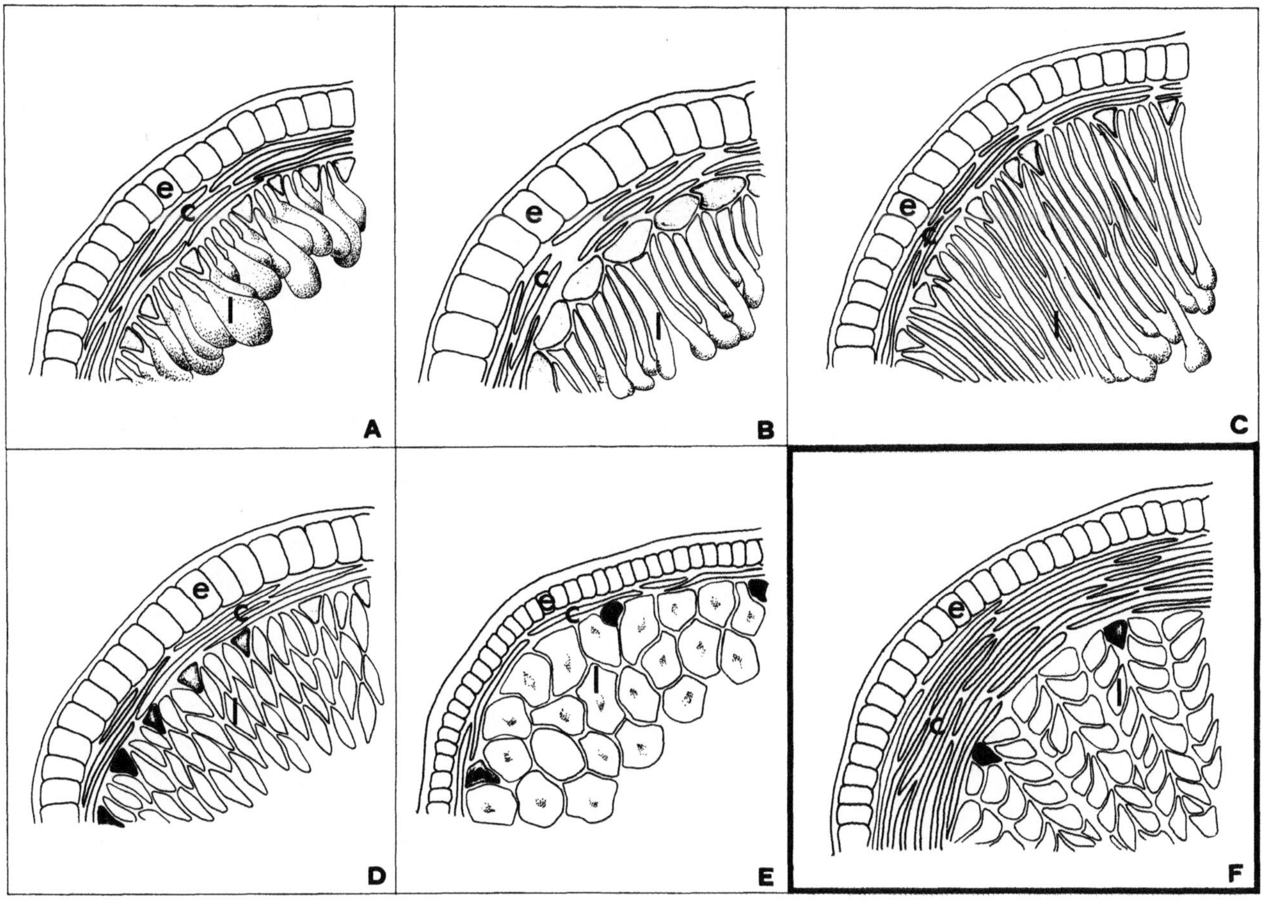

Scheme 1: Representation of the body wall organization in: A: tubificids; B: enchytraeids; C: lumbriculids; D–E: haplotaxids; F: lumbricids. Under the epidermis (e) and the circular muscle layer (c) is drawn the longitudinal layer (l). Dark shading: longitudinal fibres characterized by large myosin filaments. Light shading: longitudinal fibres with smaller myosin filaments.

All longitudinal fibres, that are separated by rather wide spaces, are not connected by junctions and are not surrounded by a basal lamina.

Enchytraeidae

Species examined: *Enchytraeus minutus* Nielsen and Christensen, 1961, *Enchytraeus albidus* Henle, 1837, *Lumbricillus lineatus* (Müller, 1774), *Lumbricillus mirabilis* Tynen, 1969. For data pertaining to the muscles of this family and to morphological modulations existing in the different species, see: de Eguileor *et al.*, 1989; Lanzavecchia *et al.*, 1989; and Valvassori *et al.*, 1989.

Lumbriculidae

Species examined: *Rynchelmis limosella* Hoffmeister, 1843, *Lumbriculus variegatus* (Müller, 1774), *Bythonomus lemani* (Grübe, 1879), *Kincaidiana* sp., *Phagodrilus* sp.. The fast longitudinal fibres of the body wall of *R. limosella* have already been desribed thoroughly (Lanzavecchia *et al.*, 1987; de Eguileor *et al.*, 1989). Observations on other species in the family have confirmed such descriptions. These fibres are one of the most constant and characteristic structures of the whole family. In cross section they look like long, thin ribbons perpendicular to the body surface (Fig. 3). The contractile layer is about 1 μm thick and is

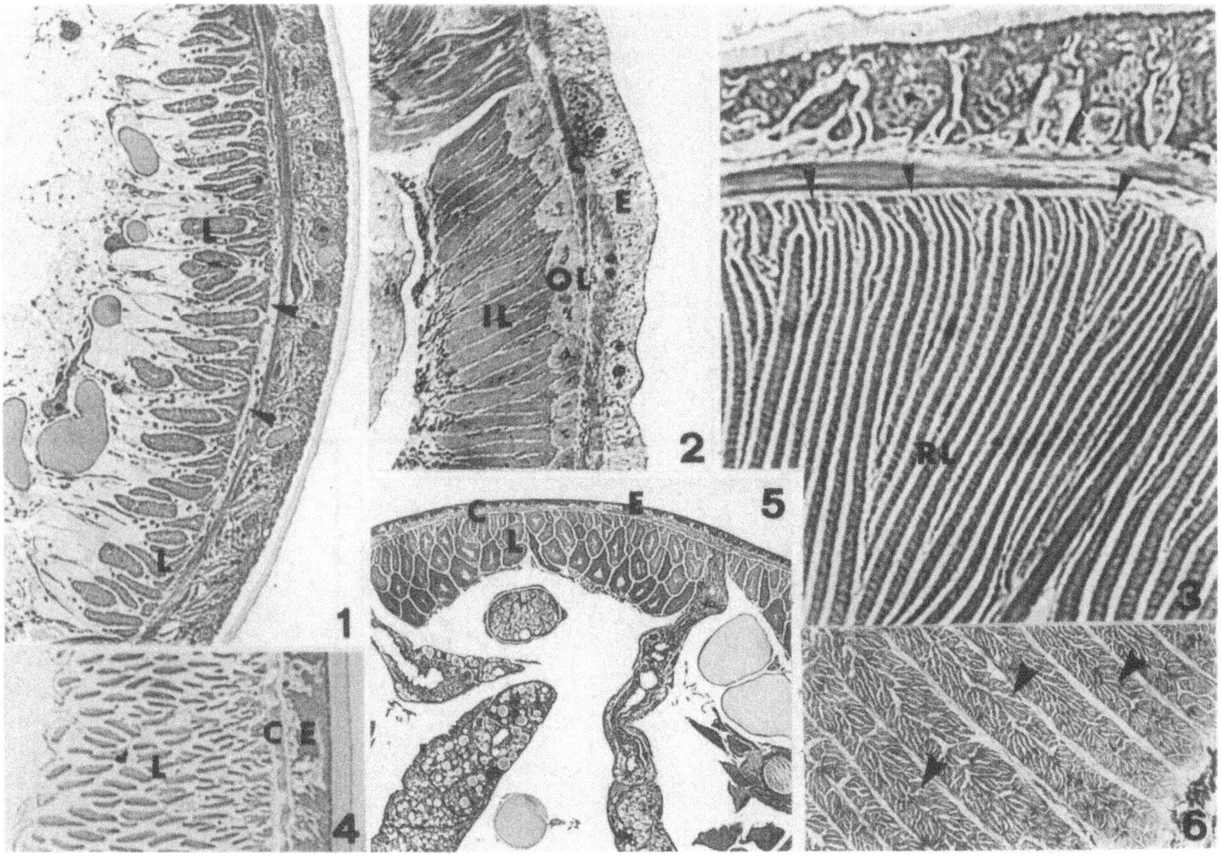

Fig. 1. Tubifex tubifex. Light micrograph of cross sectioned body wall. In the longitudinal muscle layer (L), among flattened circomyarian fibres, triangular ones are visible (arrowheads) × 1000.

Fig. 2. Enchytraeus sp. Light micrograph of cross sectioned body wall. The different morphology of longitudinal muscle fibres are clearly visible. Under the epidermis (E) and circular muscle layer (C), the outer longitudinal fibres (OL) and the inner longitudinal muscles (IL) are visible × 850.

Fig. 3. Bythonomus lemani. Light micrograph of cross sectioned body wall. Among ribbonlike longitudinal fibres (RL), triangular ones are recognizable (arrowheads) × 2500.

Fig. 4 & 5. Haplotaxids body wall in cross sections. 4. *Pelodrilus leruthi* × 450: 5. *Haplotaxis gordioides* × 200 (E = epidermis; C = circular muscles; L = longitudinal muscles).

Fig. 6. Eiseniella tetraedra. Typical pinnate fibres (arrowheads) are visible in a cross section of the body wall × 250.

characterized by a great number of very wide subsarcolemmal cisternae and by sarcotubules and Z elements, often running uninterrupted along the whole thickness of the fibre (Fig. 15). In this layer there are myosin filaments with a rectangular cross section, a unique feature among annelids and a constant in the family (Lanzavecchia *et al.*, 1987) (Figs 14, 15 & 17). The perimeter of these

Fig. 7. Limnodrilus udekemianus. In a cross section of the body wall, two types of longitudinal fibres are visible, characterized by myosin filaments with different thickness. The fibre on the left shows large myosin filaments (arrows) while the right one shows smaller myosin filaments (arrowhead). × 35000.

Fig. 8. Enchytraeus albidus. Under the epidermis (E) and circular muscles (C) the triangular longitudinal fibres (TL) are distinguishable from the other longitudinal fibres (L). × 5500.

Fig. 9. Detail of Fig. 8, showing the difference in myosin filament diameters (arrowheads). × 30000.

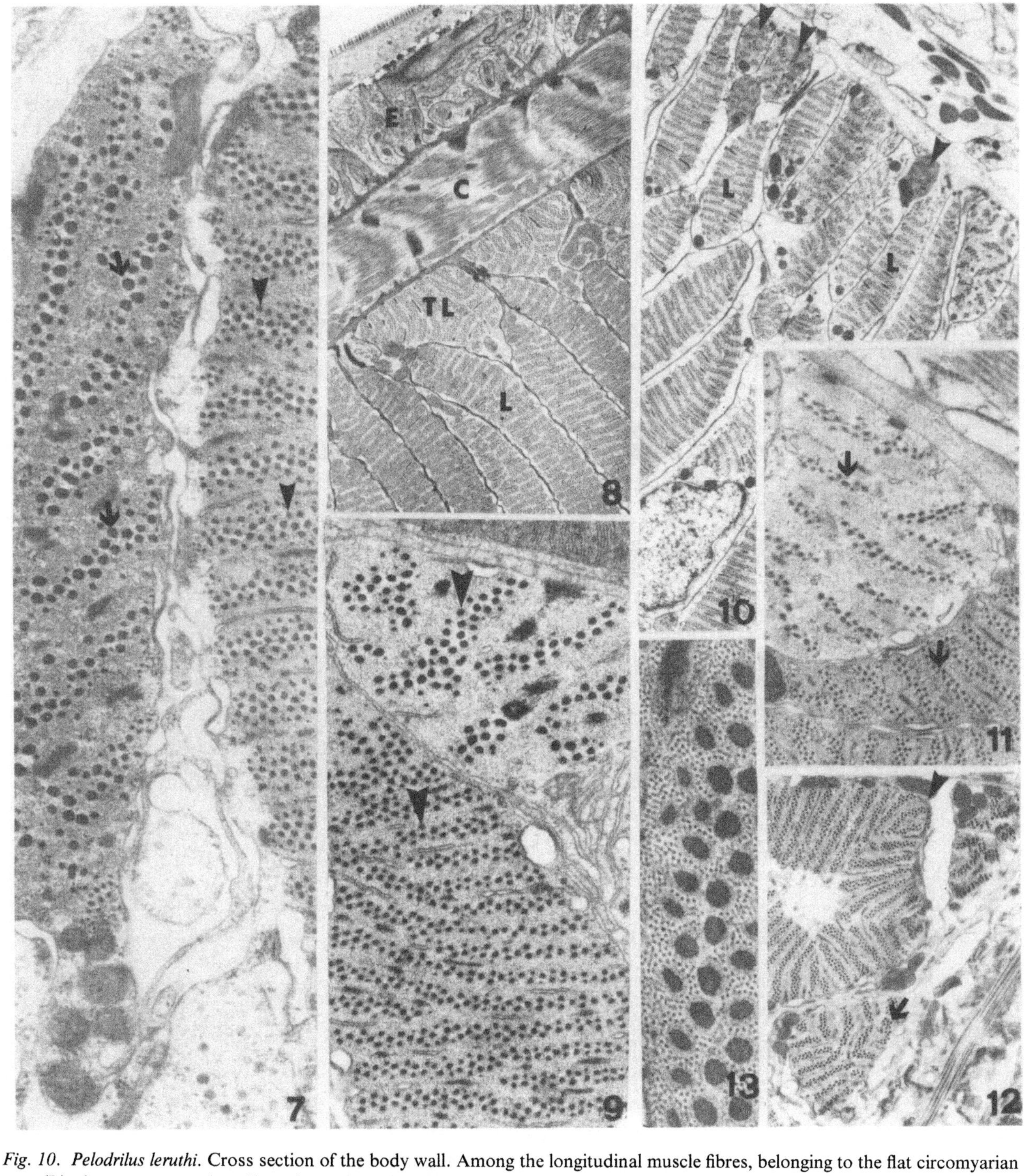

Fig. 10. Pelodrilus leruthi. Cross section of the body wall. Among the longitudinal muscle fibres, belonging to the flat circomyarian type (L), there are some slow fibres (arrowheads) × 8000.

Fig. 11. Detail of Fig. 10, showing the different dimensions of myosin filaments (arrows) × 16000.

Fig. 12. Haplotaxis gordioides in cross section. The longitudinal muscle fibres are round circomyarian (arrowhead), or triangular (arrow) × 6800.

Fig. 13. Detail of Fig. 12, showing the arrangement of thick and thin filaments in a 'sarcomere' × 80000.

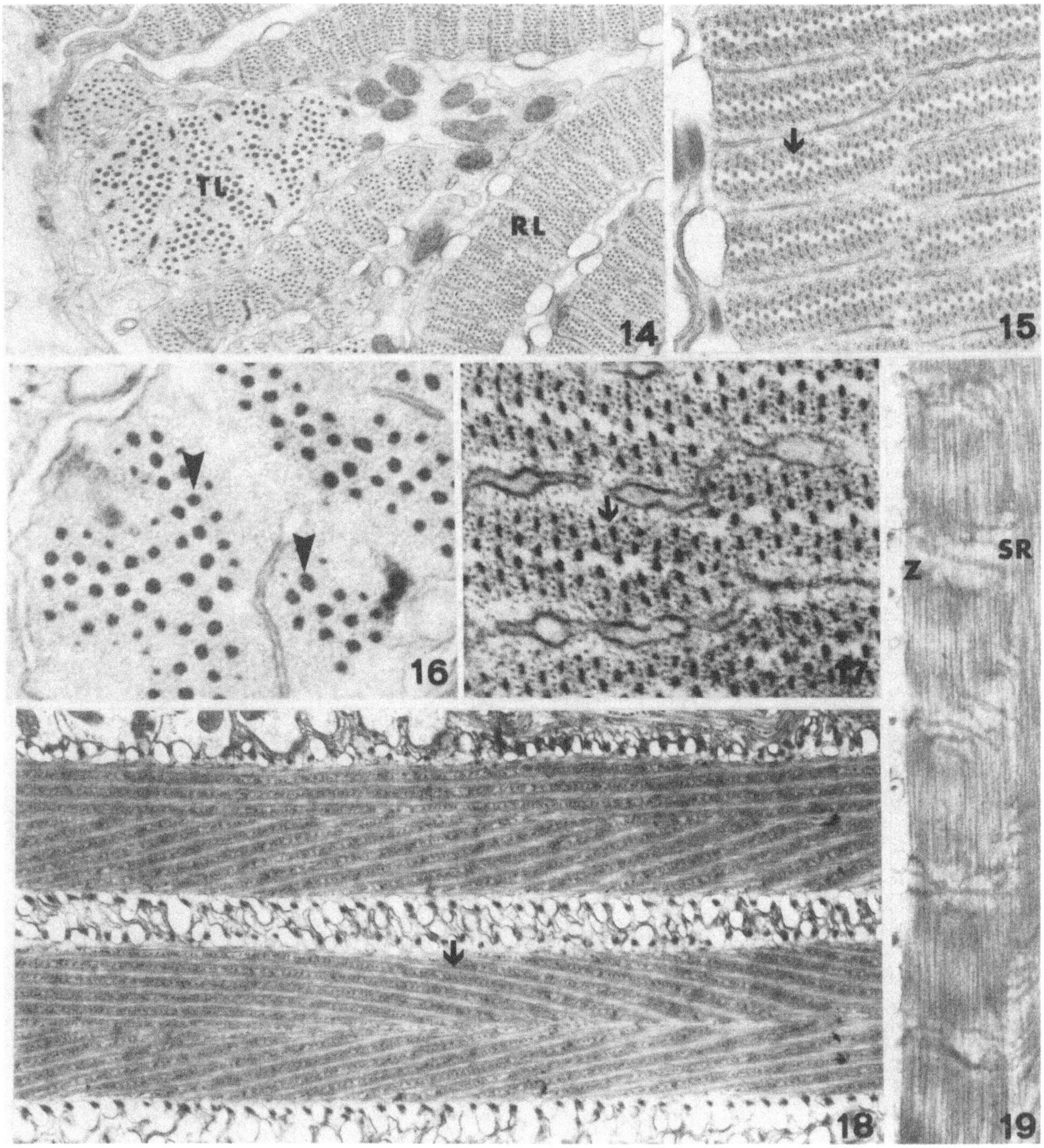

Figs 14–17. The body wall longitudinal muscle fibres of lumbriculids are ribbon like (RL) and characterized by thick filaments with a rectangular profile in a cross section (arrows). The slow fibres are triangular in shape (TL) and their myosin filaments have a round profile in cross section (arrowheads). 14 & 16: *Lumbriculus variegatus*. 15 & 17: *Kincaidana* sp. (14) × 13000; (15) × 27000; (16) × 35000; (17) × 50000.

Figs 18 & 19. *Kincaidiana* sp. The ribbonshaped longitudinal muscle fibre shows a characteristic morphology in a longitudinal section. 18. Section parallel to the XZ plane, with the typical 'chevron' pattern (arrow). × 7000. 19. Section parallel to YZ plane. The Z elements (Z) and the sarcotubules (SR) can easily be distinguished × 14000.

filaments measured in cross section is the same as the perimeter of a round filament; the diameter is 25 nm. Each of these myosin filaments is surrounded by an elliptic crown of 10–12 actin filaments, whereas, the overall ratio between actin and myosin filaments in the longitudinal muscle layer is around 7 (Lanzavecchia & de Eguileor, 1976).

The cytoplasmic region containing the mitochondria and the nucleus always faces the coelomic cavity (Scheme 1:C). These fibres show very peculiar morphologies (Figs 18 & 19) when longitudinally sectioned, following different planes (Lanzavecchia *et al.*, 1987).

Slow fibres are few. They have a triangular or irregular shape and are intermingled among fast fibres in the proximity of the circular muscle layer (Fig. 14; Scheme 1:C). Based on different diameters of the myosin filaments, either 40 or 60 nm, these fibres can be subdivided into two different categories. The myosin filaments are perfectly circular in a cross section (Figs 14 & 16).

Haplotaxidae

Species examined: *Haplotaxis gordioides* (Hartmann, 1821), *Pelodrilus leruthi* (Hrabe, 1958). For data referring to the two species of this family see: de Eguileor *et al.*, 1990, 1994. It is important to underline that these two species show two completely different conditions in terms of organization of body wall musculature and ultrastructural morphology of the fibres (Figs 4, 5, 10, 11, 12 & 13; Scheme 1:D,E).

Megadriles

Species examined: *Lumbricus terrestris* (Linnaeus, 1758), *Eiseniella tetraedra* Savigny, 1826, *Eisenia foetida* Savigny, 1826. Among megadriles, only the Lumbricidae, with its well known 'caissons' arrangement of the longitudinal muscles, has been examined (Fig. 6; Scheme 1:F).

From a morphological viewpoint, all the longitudinal muscles should be considered as being fast since the myosin filament has a small diameter. However, histochemical observations (d'Haese & Carlhoff, 1987; de Eguileor *et al.*, 1989) have demonstrated that the most external longitudinal fibres in lumbricids, although morphologically similar to the others, show biochemical features pertaining to slow fibres and their myosin is of the slow type.

Thorough observations also demonstrate that small and very rare fibres with thicker myosin filaments are present in the longitudinal muscular layer next to the layer of circular fibres.

The data discussed are summarized in Table 1.

Table 1

	Tubificidae		Enchytraeidae		Lumbriculidae		Haplotaxidae				Lumbricidae
							H. gordioides		*P. leruthi*		
	long. f.	Δ long. f.	long. f.	outer long.	long. f.	Δ long. f.	long. f.	outer long.	long. f.	outer long.	long. f.
Dimension of fiber in cross section (μ)	30 x 4	0.1 x 1.6	11 x 3	5 x 1.3	60 x 1.5	2.4 x 1.8	13 x 8.2	2.2 x 2.3	1.5 x 2.5	2.5 x 1.5	15 x 3
Width of "sarcomere" (μ)	0.3	0.4	0.26	0.36	0.3	0.4	0.2	0.3	0.25	0.45	0.3
Diameter of thick filament (nm)	30	60	25-35	40	25	40-60	45	55	35	60	37
thin/thick filament ratio	8	15	9	11	7	15	15	17	10	15	9

Discussion

The ultrastructural organization of muscular fibres may be useful in identifying a taxon. Lumbriculids, as well as enchytraeids, tubificids, haplotaxids and lumbricids, that is all oligochaetes which have been exhaustively studied under these terms, can be recognized simply by the ultrastructure of their muscle fibres. However, among haplotaxids completely different muscular organizations, at structural (Figs 4 & 5), ultrastructural (Figs 10, 11, 12 & 13) and histochemical levels (de Eguileor *et al.*, 1989, 1990, 1994) are found. In line with what can be derived from other characters (Brinkhurst, 1988), by studying the muscular system of the body wall we can draw the conclusion that this family should be subdivided into different taxa.

On the basis of the data reported up to now, we can state that in oligochaetes longitudinal muscles never make up a single layer, since there are always two types of fibres: the external ones being of the slow type, the internal ones of the fast type. Between circular and longitudinal muscles, usually interposed at the periphery of the longitudinal fibres, there are always some fibres which can be defined as slow both from the morphological and the histochemical point of view. The number of slow fibres varies according to the different families (Figs 7, 8, 10, 12 & 14; Scheme 1:A–F).

The most characteristic and apparently clearest situation is the one to be found in enchytraeids, with circomyarian or triangular slow longitudinal fibres forming an almost uninterrupted layer which is easily recognized (Figs 8 & 9; Scheme 1:B). In the tubificids, lumbriculids and haplotaxids as well, there are some fibres which are typically slow from the morphological point of view, although they are variable in number and are intermingled among fast fibres (Scheme 1:A, C, D, E).

Recent observations on lumbricids have demonstrated the existence of small and very rare longitudinal fibres with the same morphology and the same topographical arrangement found in most microdriles. The presence of small fibres with myosin filaments having a large diameter, intermingled among fast longitudinal fibres is not limited to oligochaetes. Similar fibres have also been described in some polychaetes: Glyceridae (Rosenbluth, 1968), Spionidae and Nephtyidae (Lanzavecchia *et al.*, 1988) and in Sipunculida (de Eguileor & Valvassori, 1977). They are always located in the most external layer of the large, longitudinal muscle bundles. A similar situation was observed in myzostomids (de Eguileor, unpublished data), whereas it is more difficult to interpret the images of the body wall of echiuroids. In the dwarf male of *Bonellia* sp., Schuchert & Rieger (1990) describe a discontinuous layer of small muscular fibres, right under the circular muscles, with myosin filaments having a large diameter, and interpret them as diagonal fibres.

This seems to lead to the conclusion that the presence of superficial slow longitudinal fibres should be considered as a plesiomorphism typical of oligochaetes and perhaps of annelids; however, in all the families studied it is possible to detect specific apomorphies related to this character.

There is a clear-cut difference between micro- and megadriles in terms of slow and fast muscles. In all microdriles slow myosin and large myosin filaments are associated but this is not the case in lumbricids. Histochemical analysis (D'Haese & Carlhoff, 1987) shows that a layer of peripheral longitudinal fibres presents slow myosin, even though these fibres have the same morphology as that of the other deeper fibres (de Eguileor *et al.*, 1989).

The very small number of fibres with large myosin filaments can in no way be held responsible for the whole slow-myosin ATPase activity of the external longitudinal layer. This is more likely to reflect the ancestral situation (Scheme 1:F). It is interesting to note that a similar morphological and histochemical feature has been described in leeches (Arcidiacono, 1977; Rowlerson, 1991), where the longitudinal muscles are organized in fields formed by series of circomyarian fibres. The fibres in the most external layer contain slow-type myosin (Rowlerson, 1991) but the myosin filaments have a small diameter, like those of the inner series, which however have fast-type myosin. External longitudinal fibres can be morphologically distinguished from the internal ones since their contractile layer is much thinner and their cytoplasmic axis is very rich in mitochondria. It is interesting to compare these data with those pertaining to white (with fast myosin) and red (with slow myosin) muscle fibres in fishes. Red fibres are involved in slow swimming and have the same ultrastructural morphology as the white ones in terms of contractile machinery (Nag, 1971), however, they show a remarkable difference when considering the T-system morphology, the sarcoplasmic reticulum, the terminal cisternae, and, in particular, the large number of mitochondria.

The observations on oligochaetes (and in particular on lumbricids) and on hirudineans suggest that in clitellates there are two basic forms of slow muscles: 1 – fibres with slow myosin and myofilaments as those observed in fast muscles. 2 – fibres with slow myosin and myofilaments larger in caliber and longer than the former. The second condition has been adopted by microdriles for all slow fibres (circular and outer longitudinal fibres); lumbricids have employed this system only to build circular fibres, but used the first option to build longitudinal slow fibres.

Further observations on other annelid taxa are needed to confirm the reliability of this type of analysis. Up to now it seems reasonable to conclude that each family is characterized by peculiar morphologies of the muscular wall and it is sometimes possible to find apomorphies of the muscle wall for different genera in a family.

Acknowledgements

This work was supported by C.N.R. and M.P.I. grants. The authors are very grateful to Tiziana Benedetti for graphical assistance.

References

Arcidiacono, G., 1977. Osservazioni comparative sulla muscolatura faringea degli Irudinei. I. La proboscide di *Helobdella stagnalis* Blanchard. Accad. Naz. Lincei 63: 103–112.

Bird, A. S. & J. Bird, 1991. The structure of Nematodes. Academic Press, London, 126 pp.

Brinkhurst, R., 1988. A taxonomic analysis of the Haplotaxidae (Annelida, Oligochaeta). Can. J. Zool. 66: 2243–2252.

Clark, R. B., 1964. Dynamics in Metazoan evolution. The origin of the coelom and segments. Clarendon Press, Oxford, 313 pp.

de Eguileor, M. & R. Valvassori, 1977. Studies on the helical and paramyosinic muscles. VII. Fine structure of body wall muscles in *Sipunculus nudus*. J. Submicr. Cytol. 9: 363–372.

de Eguileor, M., G. Lanzavecchia & R. Valvassori, 1994. Ultrastructural and histochemical comparison in two haplotaxids. Hydrobiologia 278 (Dev. Hydrobiol. 95): 157–163.

de Eguileor, M., G. Lanzavecchia, R. Valvassori & P. Lanzavecchia jr., 1987. Unusual model of lumbriculids' helical muscles: comparison with body wall muscles in other microdriles. Hydrobiologia 155: 135–144.

de Eguileor, M., R. Valvassori, G. Lanzavecchia & S. Giorgi, 1990. Body wall muscles in the haplotaxids *Haplotaxis gordioides* and *Pelodrilus leruthi* (Annelida, Oligochaeta). Zoomorphology 110: 27–36.

de Eguileor, M., F. Cotelli, R. Valvassori, M. Brivio & L. Di Lernia, 1988. Functional significance of intermediate filament meshwork in annelid helical muscles. J. Ultrastruct. Mol. Struct. Res. 100: 183–193.

de Eguileor, M., S. Daniel, F. Cotelli, R. Valvassori & G. Lanzavecchia, 1989. Histochemical analysis of oligochaete body wall. Hydrobiologia 180: 99–107.

D'Haese, J. & D. Carlhoff, 1987. Localization and histochemical characterization of myosin isoforms in earthworm body wall muscle. J. Comp. Physiol. 157: 171–179.

Fauchald, K., 1974. Polychaete phylogeny: a problem in protostome evolution. Syst. Zool. 23: 493–506.

Jamieson, B. G. M., 1981. The ultrastructure of the Oligochaeta. Academic Press, London, 462 pp.

Karnowsky, H., 1965. A formaldehyde-glutaraldehyde fixative of high osmolarity for use in electron microscopy. J. Cell Biol. 27: 137A.

Lanzavecchia, G., 1971. Studi sulla muscolatura elicoidale e paramiosinica. IV. La muscolatura longitudinale e circolare della parete corporea degli Anellidi Tubificidi. Accad. Naz. Lincei 50: 50–56.

Lanzavecchia, G. & M. de Eguileor, 1976. Studies on the helical and paramyosinic muscles. V. Ultrastructural morphology and contraction speed of muscular fibres of *Erpodella octoculata* and *Erpobdella testacea* (Annelida, Hirudinea). J. submicr. Cytol. 8: 69–88.

Lanzavecchia, G., M. de Eguileor & R. Valvassori, 1985. Superelongation in helical muscles of leeches. J. Muscle Res. Cell Motility 6: 569–584.

Lanzavecchia, G., M. de Eguileor & R. Valvassori, 1988. Muscles. In W. Westheide & C. O. Hermans (eds), The ultrastructure of Polychaeta. G. Fisher Verlag, Stuttgart. Microfauna Marina 4: 71–88.

Lanzavecchia, G., M. de Eguileor, R. Valvassori & P. Lanzavecchia jr., 1987. Analysis and reconstruction of unusual obliquely striated fibres in lumbriculids (Annelida, Oligochaeta). J. Musc. Res. Cell Motility 8: 209–219.

Lanzavecchia, G., M. de Eguileor, R. Valvassori, L. Di Lernia & C. Cambiaso, 1989. Morphogenesis of body wall muscle fibres in *Enchytraeus minutus*. Hydrobiologia 180: 91–97.

Lewin, R., 1984. Why is development so illogical? Science 224: 1327–1329.

Lorenzen, S., 1985. Phylogenetic aspects of pseudocoelomate evolution. In S. Conway Morris, J. D. George, R. Gibson & H. M. Platt (eds), The origins and relationships of lower invertebrates. Clarendon Press, Oxford: 210–223.

Mettam, C., 1967. Segmental musculature and parapodial movement of *Nereis diversicolor* and *Nephtys hombergi* (Annelida: Polychaeta). J. Zool., Lond. 153: 245–275.

Moore, R. D., V. Mumaw & M. D. Shonberg, 1960. Optical microscopy of ultrathin tissue sections. J. Ultrastruct. Res. 4: 113–116.

Nag, A. C., 1971. Ultrastructure and adenosine triphosphatase activity of red and white muscle fibres of the caudal region of a fish, *Salmo gairdneri*. J. Cell Biol. 55: 42–57.

Rosenbluth, J., 1968. Obliquely striated muscle. IV. Sarcoplasmic reticulum, contractile apparatus, and endomysium of the body muscle of a polychaete, *Glycera*, in relation to its speed. J. Cell Biol. 36: 245–259.

Rowlerson, A. M. & S. E. Blackshaw, 1991. Fibre types in leech body wall muscle. J. exp. Biol. 157: 299–311.

Sawyer, R. T., 1986. Leech biology and behaviour. Vol. I – Anatomy, physiology and behaviour. Clarendon Press, Oxford: 223–280.

Schuchert, P. & R. M. Rieger, 1990. Ultrastructural observations on the dwarf male of *Bonellia viridis* (Echiura). Acta zool. 71: 5–16.

Valvassori, R., M. de Eguileor, G. Lanzavecchia & G. Scari', 1989. Body wall organization in enchytraeids. Hydrobiologia 180: 83–89.

Hydrobiologia **278**: 189–199, 1994.
T.B. Reynoldson & K.A. Coates (eds), Aquatic Oligochaete Biology V.

Comparative body wall musculature and muscle fibre ultrastructure in branchiobdellidans (Annelida: Clitellata), and their phylogenetic significance

Roberto Valvassori [1], Magda de Eguileor [1], Giulio Lanzavecchia [1] & Stuart R. Gelder [2]
[1] *Department of Biology, Universita' degli Studi di Milano, Via Celoria 26, I – 20133 Milano, Italy;*
[2] *Department of Biology, University of Maine at Presque Isle, 181 Main Street, Presque Isle, Maine 04769–2888, USA*

Key words: branchiobdellidans, muscles, ultrastructure, phylogeny

Abstract

The body wall muscles in five species of branchiobdellidans are all arranged in the oligochaete pattern and the muscle fibres are obliquely striated. The structure of the circular muscle fibres do vary to some degree. The longitudinal muscle fibres in *Ankyrodrilus legaeus, Branchiobdella kozarovi*, and *Xironogiton instabilis* all are round circomyarian and thus double-obliquely striated. These species represent three of the four genera composing the family Branchiobdellidae. Although *Bdellodrilus illuminatus* and *Cambarincola fallax*, from the families Bdellodrilidae and Cambarincolidae, respectively, also possess a few round circomyarian fibres, most are 'polyplatymyarian' comparable to single-obliquely striated fibres. A similar division of branchiobdellidan families is obtained based on the number of anterior nephridial pores. The muscular structure in the branchiobdellidans shows both similarities and differences with the leeches and the lumbriculid oligochaetes. One phylogenetic explanation for this is that the branchiobdellidans separated from the common clitellate ancestor before the oligochaetes and leeches became recognizable taxa.

Introduction

The systematic position of branchiobdellidans in the Clitellata is controversial; in particular, the relationships of branchiobdellidans with the lumbriculid oligochaetes and leeches (Holt, 1986; Sawyer, 1986; Brinkurst & Gelder, 1989, 1991) are debated.

The anatomical characters employed in assessing the phylogeny of soft bodied animals usually exclude the structure of the body wall muscle system, and the ultrastructure of the muscle fibres. This is because it is not realized that muscle cells can, and have, evolved different morphologies. Such an evolution may have been associated with an organism successfully adapting its method of movements to different environments (Lanzavecchia *et al.*, 1994). By accepting the premise that differences in muscular structures do arise during evolution, then a detailed comparative analysis of muscle organization and cell structure would be expected to provide insight into the phylogeny of a taxon (Lanzavecchia *et al.*, 1988; de Eguileor *et al.*, 1990).

Information on the histology and ultrastructure of branchiobdellidan muscles is limited to representatives from two of the 21 genera in the taxon. Specifically, the histology of the muscular system in *Xironogiton instabilis* (Moore, 1893) was studied by Franzen (1962) and in *Branchiobdella parasita* Henle, 1835 by Schmidt (1903); the ultrastructure of *Branchiobdella pentodonta* Whitman, 1882 was described by Farnesi & Vagnetti (1972), Ferraguti & Lanzavecchia (1977), and de Eguileor & Ferraguti (1980). From this scant information, the muscle cells from the two genera do appear to have discernibly different structures. This is not unexpected if the premise stated above is accepted. By extending this line of reasoning, closely related species and genera will share more derived structural similarities in their muscle cells and organization than distantly related ones.

This investigation was designed to study the ultrastructure of the body wall muscles in representatives from four branchiobdellidan genera to compare with existing information on *B. pentodonta*. The four species used were *X. instabilis*, *Bdellodrilus illuminatus* (Moore, 1893), *Cambarincola fallax* Hoffman, 1963, and *Ankyrodrilus legaeus* Holt, 1965.

Branchiobdella kozarovi Subchev, 1978 was included in the study to ascertain an indication of intrageneric variation in this genus.

Materials and methods

Specimens of *X. instabilis, B. illuminatus*, and *C. fallax* were collected from Mantle Lake stream,

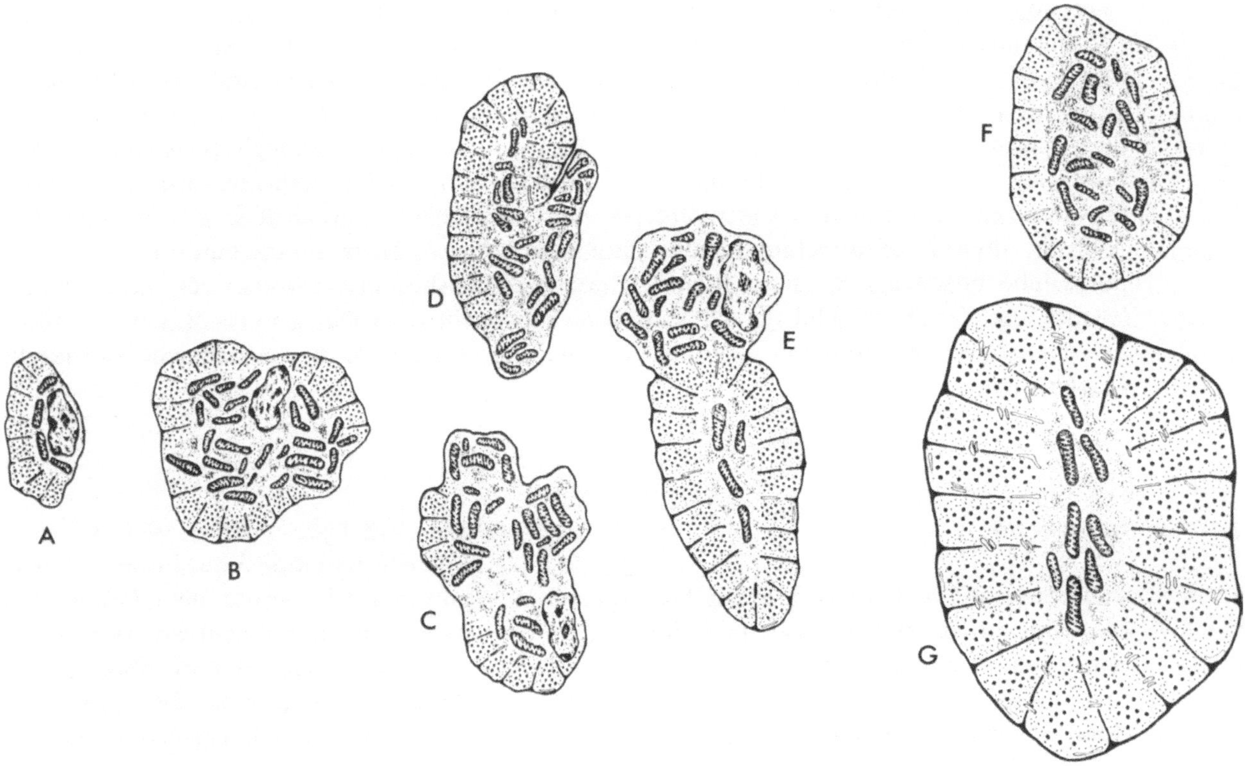

Scheme 1. Schematic drawing of the different structural types of the longitudinal body wall muscles in branchiobdellidans. A: platymyarian fibre (all Annelids). B and C: polyplatymyarian fibres, with a discontinuous contractile ring present in some branchiobdellidans (Bdellodrilidae and Cambarincolidae). D: c-shaped fibre (branchiobdellidans and oligochaete embryos). E: pseudo-circomyarian fibre, with a cytoplasmic bulge containing nucleus and mitochondria (oligochaetes and branchiobdellidans). F and G: round-circomyarian fibres, with contractile layers of different widths (hirudineans and branchiobdellidans). In all fibres but that shown in G there are only peripheral diads.

Presque Isle, Maine, U.S.A. (Gelder & Smith, 1987), *A. legaeus* from the Harpeth River 12 km SE of Franklin, Tennessee, U.S.A. and *B. kozarovi* from fish ponds near Razdelma, district of Varna, Bulgaria (Gelder *et al.*, 1994). For routine transmission electron microscopy the animals were anaesthetized with a saturated solution of Mephenesin (3-o-Toloxy-1, 2-propandiol, SIGMA) for 30 minutes, and fixed according to Karnowsky (1965). They were embedded in an Epon 812-araldite mixture, sectioned with an LKB Ultrotome V and observed with a Jeol 100 B electron microscope.

Terminology

The contractile material of an obliquely-striated fibre, into a cross section, appears subdivided into a series of regions or fields with the same sequence of bands normally seen in longitudinal sections of cross-striated muscle fibres. According to the terminology proposed by Rosenbluth (1965), these regions are commonly indicated as 'sarcomeres' (Lanzavecchia *et al.*, 1985)

Results

The body wall muscles in the five species examined are arranged as a circular and a longitudinal layer consistent with those found in oligochaetes. The contractile material in these muscle cells is organized as in all obliquely striated fibre (Lanzavecchia, 1977).

Circular layer

The circular layer appears in a cross section of an organism as a discontinuous ring composed of small, isolated muscle fibres. These fibres lie against the epidermis, and they are often partially surrounded by epidermal cells (Fig. 1). The morphology of the muscle fibres in the five species examined varies. The muscle fibres, in *B. kozarovi* are round circomyarian (Scheme 1F, G) and similar to those previously described in *B. pentodonta* by de Eguileor & Ferraguti (1980). In the other four species the contractile material is arranged in a flat or C-shaped layer (Scheme 1A, D) oriented towards the epidermis, with the cytoplasmic material always facing towards the longitudinal muscle layer. The cytoplasm sometimes forms bulges and these contain numerous mitochondria (Fig. 2).

Longitudinal layer

Due to the great heterogeneity of the muscle cell structure in this layer, the species examined are described separately.

Bdellodrilus illuminatus – The longitudinal muscle fibres are normally separated from the body cavity and from the outer circular and epidermal layers by a very thin, roughly fenestrated epithelium (Figs 3 & 4). Where the circular muscle layer is interrupted, the epithelium is in close contact with the epidermis. In cross section, the longitudinal muscle fibres are circular, elliptical or polygonal in shape and from 1.5 to 30 μm in diameter (Figs 3 & 5). The contractile layer is peripheral measuring from 0.1 to 5.0 μm in thickness. It is arranged around the central cytoplasmic core in a complete ring, C-shaped, U-shaped or in a series of isolated plates of irregular shape and variable length (Figs 3 & 5). Inside the cytoplasmic region mitochondria and lipid droplets are present and thus, due to the discontinuity of the contractile layer, they are also numerous in the subsarcolemma. The fibre membrane forms numerous folds which are either parallel to the fibre surface, or penetrate deeply into the fibre, giving rise to structures comparable to T-tubules (Figs 6 & 7). The sarcoplasmic reticulum is poorly developed inside the contractile layer; few cisternae are close to the sarcolemmal folds forming diads (Fig. 7, arrow).

The longitudinal fibres are surrounded by a loose connective tissue and the sarcolemma of each muscle cell is separated from the connective tissue by a well-defined basal lamina (Fig. 8). The basal lamina sometimes follows the large infoldings of the cell surface, but more frequently it does not (Figs 6 & 8).

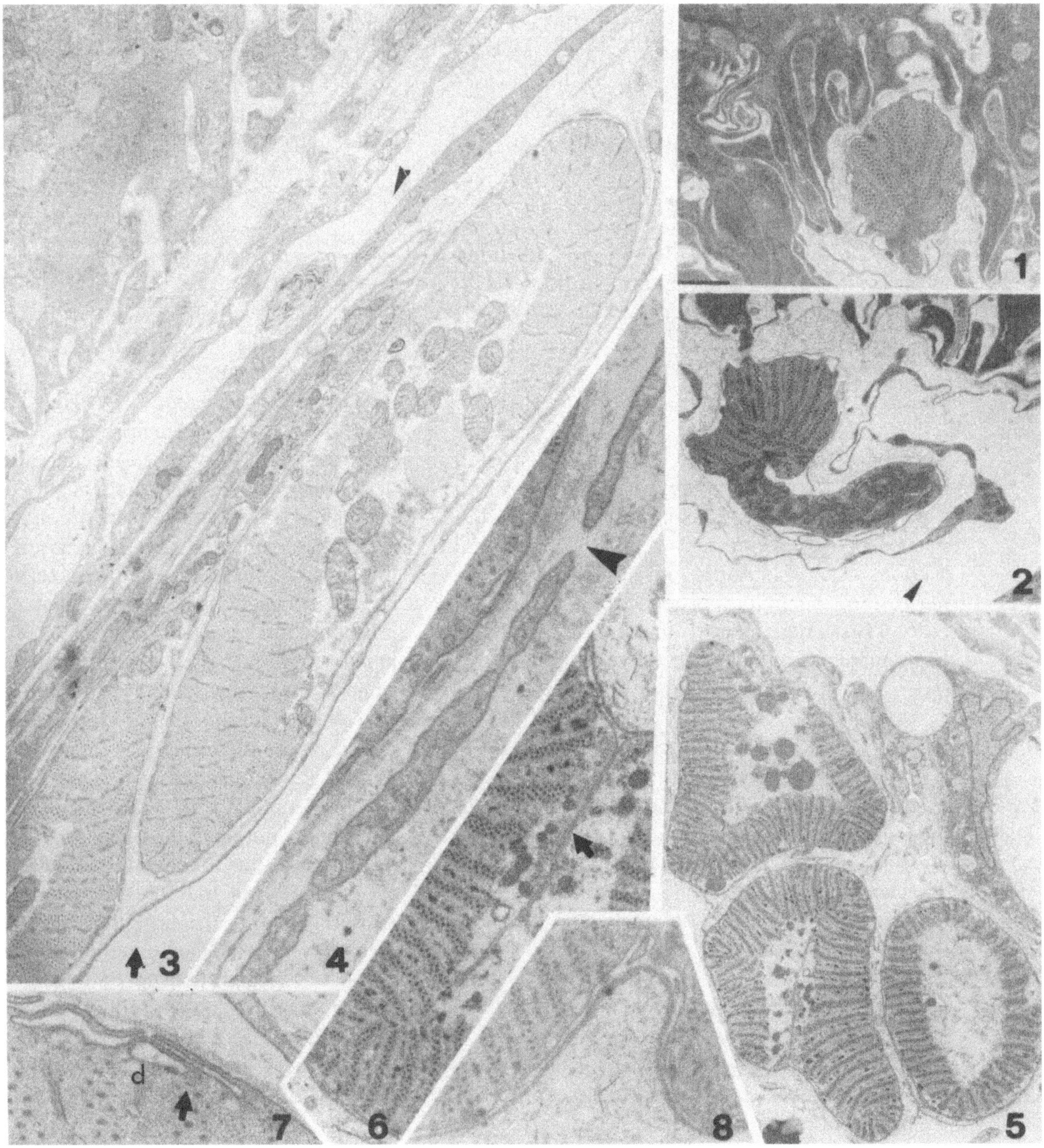

Fig. 1. The small and isolated circular muscle fibres are partially surrounded by epithelial cells. *Cambarincola fallax* × 11,000.
Fig. 2. Circular muscle fibre: the sarcoplasm forms buds in which mitochondria are numerous. The intermuscular space is evident (arrowhead). *Cambarincola fallax* × 9,000.
Figs 3–8. Bdellodrilus illuminatus.
Fig. 3. The longitudinal muscle fibres are separated from the body cavity (arrow) and from the intermuscular space (arrowhead) by a thin epithelium × 8,500.
Fig. 4. Detail of Fig. 3, showing the discontinuity (arrowhead) of the epithelium that surrounds the longitudinal muscle layer. × 37,000.

Cambarincola fallax – The body wall muscle system is much more complicated than in *B. illuminatus*. Under the circular layer, a complicated series of muscle cells can be identified (Figs 9 & 10). The shape of the fibres is heterogeneous, with the contractile layer forming a continuous ring or being subdivided into fields of different extent and thickness (Fig. 11). Some fibres are characterized by lateral extensions where the direction of the myofilaments forms an angle with respect to the main fibre axis (Fig. 9). This condition is clearly identifiable both in cross- and longitudinal sections of the organism. Due to the uneven distribution of the contractile region, the sarcoplasm is localized either in the core, or at the periphery of the cell where it forms buds of different sizes (Figs 9 & 12). The thickness of the contractile layer ranges from 0.1 to 6.5 μm. The 'sarcomeres' are defined by Z elements and sarcoplasmic reticulum cisternae. Numerous diads are present inside the contractile layer, but not necessarily oriented parallel to the long axis of the 'sarcomeres' (Figs 13 & 14).

A thin, incomplete epithelium is usually identifiable, separating the longitudinal fibres from the body cavity and from the outer circular muscle and epidermal layers (Fig. 14). Some of the muscle fibres in the longitudinal layer are anchored to the epidermal cells by hemidesmosomes and to the cuticle by large bundles of tonofilaments (Fig. 15).

Ankyrodrilus legaeus, Xironogiton instabilis and *Branchiobdella kozarovi* – The contractile layer in the longitudinal fibres forms a peripheral ring around the sarcoplasmic core (Fig. 16), although occasionally it appears interrupted (Fig. 17). The nucleus and mitochondria are located in the central core, with a few small mitochondria scattered among the contractile material or in the subsarcolemma. The thickness of the contractile layer ranges from 0.5 to 7 μm in *A. legaeus* and *X. instabilis*, and from 1 to 5 μm in *B. kozarovi*. As a result, the diameter of the cytoplasmic core varies considerably (Fig. 16). The 'sarcomeres' are regular and radially arranged in the fibres where the contractile layer is thin (Fig. 19). However, when the thickness of the contractile layer increases, the arrangement of the 'sarcomeres' is less clearly discernable due to the large number of irregularly distributed and randomly oriented diads inside the contractile ring (Fig. 20).

A few small longitudinal fibres, scattered beneath the circular layer, and characterized by thick myosin filaments measuring 60 nm at their widest part are present in *A. legaeus* and *X. instabilis* (Fig. 18). The longitudinal muscle layer is usually separated from the body cavity by a thin epithelium. Occasionally a narrow space, bounded by a similar thin epithelium, is visible between the longitudinal muscle layer and the outer circular muscle and epidermal layers (Figs 16 & 18).

Discussion

The body wall musculature of the five species of branchiobdellidans analyzed here and previously is composed of a circular outer layer and a longitudinal inner one. Some fibres in the latter layer are characterized by a variable orientation of the contractile material. This condition is particularly evident in *C. fallax*, where the branches from the longitudinal muscle fibres run towards the epidermis. In spite of this peculiarity, the over-all organization of the body wall muscular system in this taxon is consistent with the plan found in the oligochaetes and, with slight variations, in polychaetes. In contrast, leeches are characterized by the presence of two layers of fibres interposed between the circular and the longitudinal layers. The fibres of the intermediate layers are oriented, with respect to the longitudinal axis of the body, at an angle depending on the state of contraction of the animal.

Fig. 5. The shape of the longitudinal muscle fibres and the thickness of the contractile layer appear heterogeneous in a cross section ×4,500.

Figs 6 & 7. Details showing the different arrangement of the T-tubules (arrows) and a diad (d). (6) ×14,000 (7) ×37,000.

Fig. 8. Detail of the basal lamina that surround each muscle fibre ×25,000.

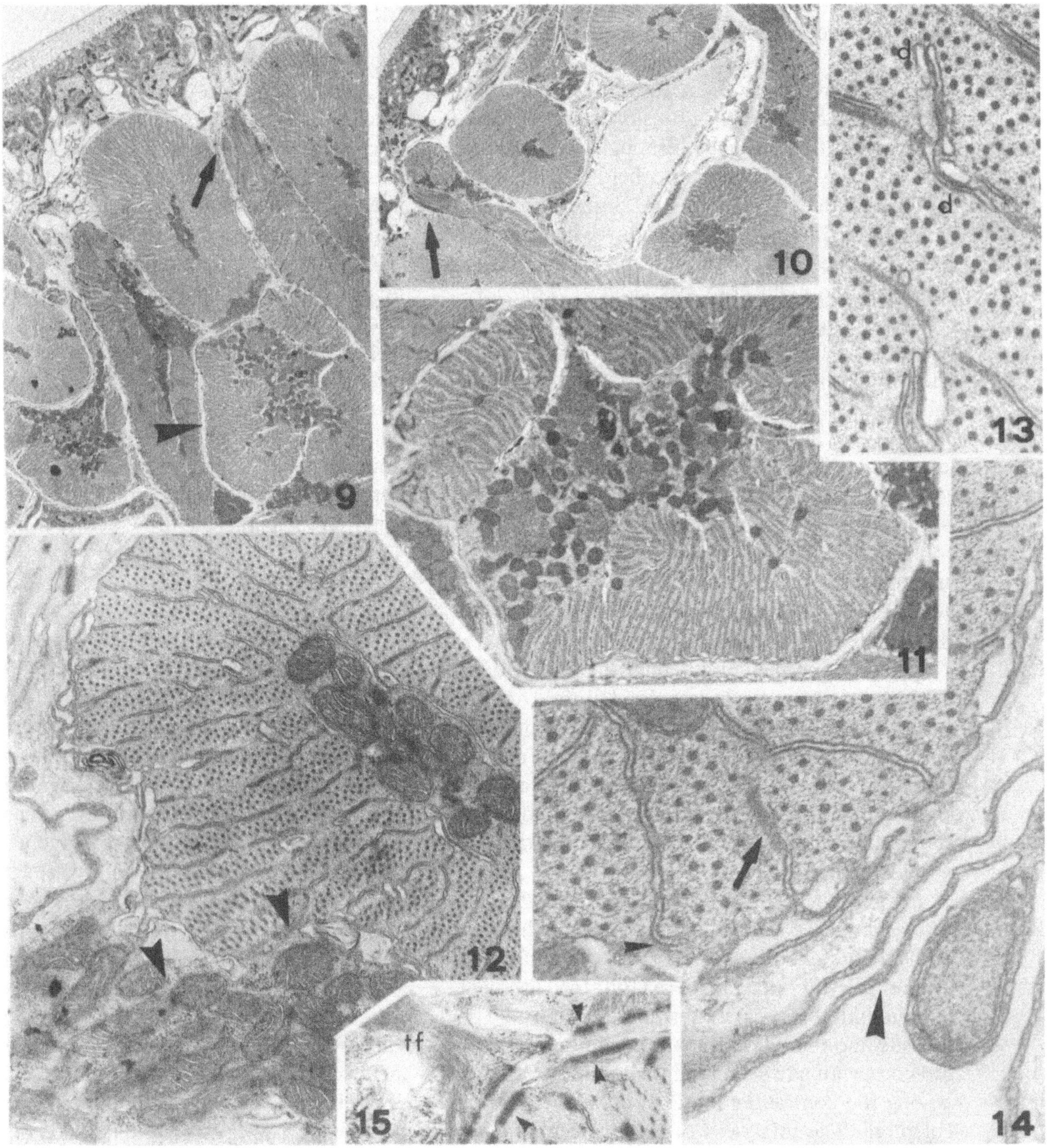

Figs 9–20. Cambarincola fallax.

Figs 9 & 10. Body wall in cross sections. The musculature is composed of different fibre types. The contractile ring sometimes encircles the sarcoplasm but frequently it is subdivided into peripheral fields (arrowhead). Lateral extensions of muscle fibres run towards the epidermis (arrows). (9) × 2,200 (10) × 1,500.

Fig. 11. Detail of Fig. 9 in which the subdivision of the contractile material into separated peripheral fields is clearly visible × 5,500.

Fig. 12. The sarcoplasm is sometimes localized into peripheral extensions filled with mitochondria (arrowheads) × 17,500.

Fig. 13. Detail of the contractile material of a cross sectioned longitudinal muscle fibre where some diads (d) are visible × 37,000.

Intermuscular space

In the branchiobdellidans, the circular layer is composed of separate fibres that are partially surrounded by the epidermis. This condition was reported in *B. illuminatus* by Moore (1895), and he noted spaces between the longitudinal and circular muscle layers, which he named 'inter-muscular'. The small inter-muscular spaces are lined by a thin epithelium, similar to that separating the longitudinal layer from the body cavity. The spaces and their lining are also present in the other four species. It has not been possible to determine whether these spaces are isolated units or coelomic diverticula.

Moore (1895) proposed that the inter-muscular spaces in *B. illuminatus* would enable the muscle layers to easily move over each other during contraction. However, *B. illuminatus* leads a sedentary life style where such an advantage would be of no conceivable use. The spaces are smaller in the other species examined, and it is these species that are active. Therefore, the hypothesis proposed by Moore (1895) appears unsupported. Although the function of the spaces remains a mystery, it is possible to offer an explanation for the differences in the sizes observed in these branchiobdellidans.

If the presence of small spaces in the active species is considered a plesiomorphic state, an increase in space size may be a result of the reduction in muscle fibre in *B. illuminatus*, which has a sedentary life style in the protected branchial chambers of the host. If the spaces are indeed parts of coelomic diverticula, such a consequence would be logical. Similar inter-muscular spaces have never been described in other annelids, even though a thin, and ill-defined separation between the circular and the longitudinal layers can sometimes be observed in *Haplotaxis gordioides* Hartman, 1821 (de Eguileor *et al.*, 1994).

T tubules and sarcoplasmic reticulum

The arrangement of the contractile material is typical of an obliquely striated fibre. In a cross section, the 'sarcomeres' appear to be separated by sarcotubules and Z elements radially arranged. In *B. illuminatus* these structures are poorly developed, although they are very common in the four active species. The active species also have a great number of diads distributed throughout the contractile layer. The level of development of the sarcoplasmic reticulum cisternae and the abundance of diads are commonly related to the efficiency of the regulation of calcium ion concentration inside the contractile material. Obliquely striated fibres are normally characterized by a reduced thickness of the contractile layer and usually by the absence of a T-system and diads (de Eguileor *et al.*, 1989). The coupling between excitation and contraction is limited to the few sarcoplasmic reticulum cisternae connected with the sarcolemma and thus called the subsarcolemmal cisternae. However, a true T-system coupled with sarcotubules has been described only in *B. pentodonta* (de Eguileor & Ferraguti, 1980).

Structures like the T-system that are capable of conducting the membrane depolarization, that determines the release of calcium ions to the center of the fibre, have been related to the thickness of the contractile layer. In the longitudinal muscle fibres, this thickness reaches 5 μm in *B. kozarovi*, 6.5 μm in *C. fallax*, and 7 μm in *X. instabilis* and *A. legaeus*. These values are close to the one reported for *B. pentodonta* by de Eguileor & Ferraguti (1980), where a mean thickness of 2.5 μm was measured. In *B. illuminatus* the contractile layer of the longitudinal fibres reaches a maximum thickness of 5 μm. Consequently, the absence of diads and the limited extension of both the sarcolemmal infoldings and the sarcoplasmic reticulum, indicate that the organism should be

Fig. 14. Sarcotubules (small arrowhead) and Z elements (arrow) separate the 'sarcomeres'. The thin epithelium separating the longitudinal muscle fibres from the inter-muscular space and from the body cavity is clearly visible (large arrowhead) × 50,000.
Fig. 15. The longitudinal muscle fibres are mechanically linked to the cuticle through hemidesmosomes (arrowheads) and tonofilament bundles (tf) × 17,500.

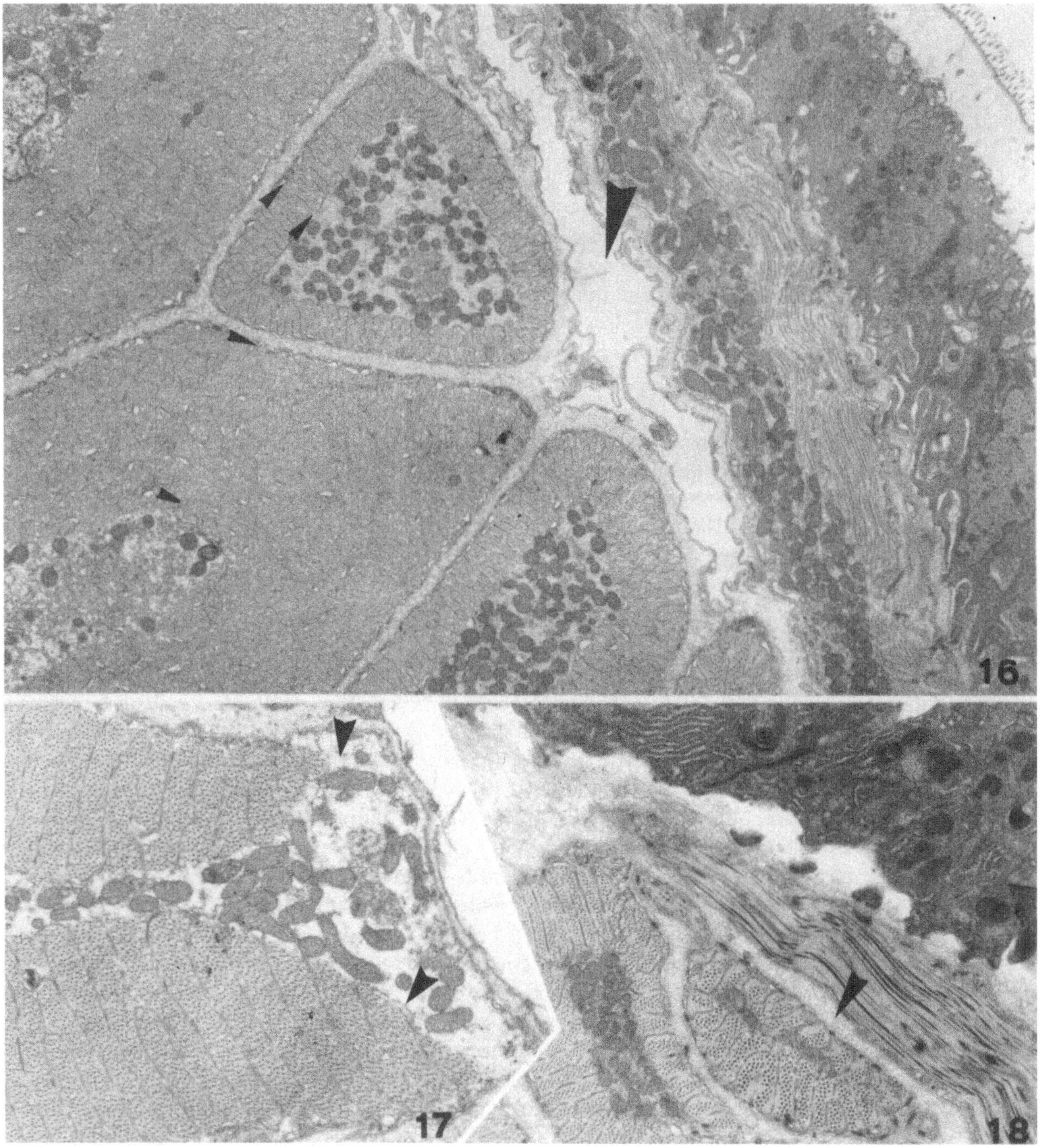

Fig. 16. Ankirodrilus legeaus in cross section. The body wall is composed of round circomyarian muscle fibres. The thickness of the contractile ring is variable (small arrowheads). A narrow intermuscular space separates the longitudinal muscle layer from the circular one (large arrowhead) × 6,000.

Fig. 17. Branchiobdella kozarovi in cross section. The contractile peripheral ring of the longitudinal muscle fibres sometimes may be interrupted (arrowheads) × 10,000.

Fig. 18. Xironogiton instabilis in cross section. Some longitudinal muscle fibres are characterized by thick myosin filaments (arrowhead) × 9,000.

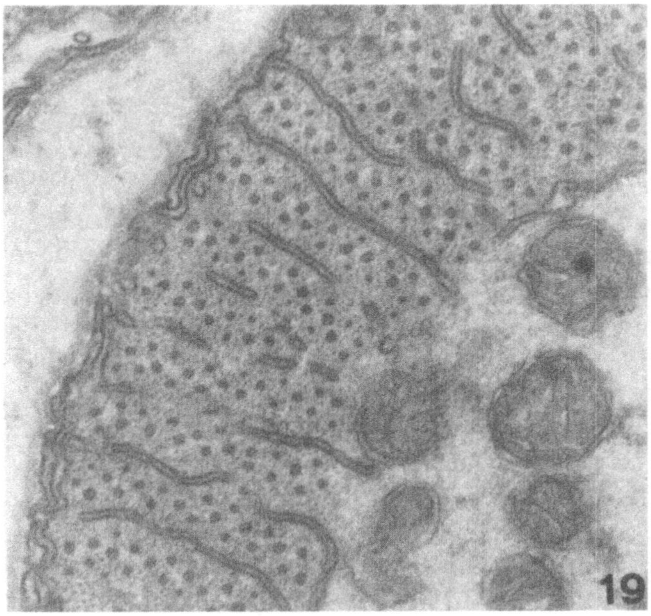

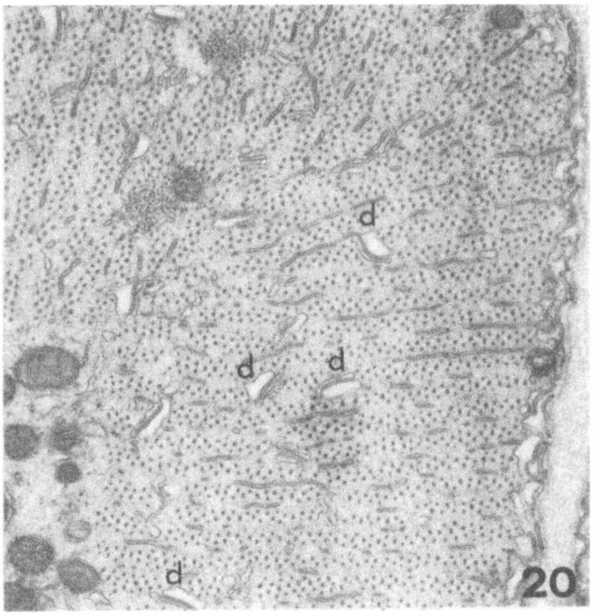

Fig. 19. Ankirodrilus legaeus in cross section. The 'sarcomeres' are well defined and radially arranged in the fibres with a thin contractile layer ×45,000.

Fig. 20. A. legaeus in cross section. The diads (d) are numerous and the radial arrangement of the 'sarcomeres' is less clearly identifiable in the fibres with a thicker contractile layer ×20,000.

expected to have poor motility. This is indeed the case with *B. illuminatus* which exhibits a sedentary life style. It is noteworthy that the arrangement and number of diads seen in a cross section of a fibre from the active species is identical to that observed in *B. pentodonta* (de Eguileor & Ferraguti, 1980).

Phylogenetic considerations

In the branchiobdellidans several types of muscle fibres can be recognized based on the arrangement of the contractile layer with respect to the sarcoplasm (Scheme 1). In *X. instabilis*, *A. legaeus* and *B. kozarovi*, the round circomyarian type is the only fibre type present in the longitudinal layers (Scheme 1F, G). The contractile material completely surrounds the cytoplasmic core and thus these fibres belong to the helical or double-obliquely striated type; these are also found in leeches (Lanzavecchia, 1977). The round circomyarian condition is seldom present in *B. illuminatus* or *C. fallax*. Instead, the peripheral contractile layer is normally incomplete, forming a C-shaped or U-shaped arrangement (Scheme 1D). As analyzed by Valvassori and de Eguileor (1991) these fibres have to be functionally considered as single-obliquely striated. As is clearly evident in Figs 9 & 11, the contractile layer is further subdivided into peripheral portions by the intervention of sarcoplasm, so that each contractile portion can be considered structurally and functionally an independent unit. This peculiar arrangement may be referred to as a 'poly-platymyarian' (Scheme 1B, C) condition.

Both circular and longitudinal muscle layers, in all the species of branchiobdellidans reported, are characterized by a well-developed connective tissue layer interposed between the muscle cells and by a well defined basal lamina surrounding each fibre. This condition is typical of leeches (Lanzavecchia, 1977; Lanzavecchia *et al.*, 1985) and has also been described in *Haplotaxis gordioides* (de Eguileor *et al.*, 1990), Sipunculida (de Eguileor & Valvassori, 1977), Priapulida (Candia Carnevali & Ferraguti, 1979) and in all annelids that do not possess the flat circomyarian fibre type (de Eguileor *et al.*, 1989).

The branchiobdellidans species, *A. legaeus, B. kozarovi*, *B. pentodonta*, and *X. instabilis* can be grouped together because they all possess round circomyarian fibres in the longitudinal layer (Scheme 1F, G). It is important to note that these species represent three of the four genera composing the family Branchiobdellidae. The other two species, *B. illuminatus* and *C. fallax*, both possess 'polyplatymyarian' fibres (Scheme 1B, C) and belong to the families Bdellodrilidae and Cambarincolidae, respectively. The separation of these branchiobdellidans into two groups is identical to that obtained using the presence of one or two anterior nephridial pore(s) (Holt, 1986: 682, Table 1). Such a positive correlation as this, between an accepted anatomical character and the muscle structure, certainly warrants further investigations. Applying these data to the tentative phylogeny of the class Branchiobdellae as proposed by Gelder and Brinkhurst (1990), the fibre type of Branchiobdellidae is considered plesiomorphic and the others would then be apomorphic for this character.

The clitellate taxa considered to be phylogenetically closest to the branchiobdellidans are either the lumbriculid oligochaetes or the hirudineans (Brinkhurst & Gelder, 1991; Sawyer, 1986). No attempt will be made here to discuss this topic in detail, but rather to use selected, salient information for comparison with the results obtained in the present study.

The branchiobdellidans and the hirudineans are considered similar in body form, presence of a posterior attachment organ and method of locomotion. Although the two taxa both have round circomyarian fibres (Scheme 1F, G), this fibre type is restricted in the branchiobdellidans to members of the plesiomorphic family Branchiobdellidae. Further similarities are not found in the two, since intermediate muscle layers present in the body wall of the hirudineans are absent in branchiobdellidans. This difference can be added to a number of other characters that were used to distinguish the branchiobdellidans from the hirudineans by Holt (1989).

The same arrangement of reproductive organs, specifically two pairs of male ducts converging into the single, median atrium in the second male segment (GII), and the absence of a second pair of ovaries in the second female segment (GIV) (Brinkhurst & Gelder, 1989) is found in both the branchiobdellidans and lumbriculids. These and other anatomical characteristics were used by Brinkhurst and Gelder (1989) to show that the two taxa were more likely to have a close common ancestor, than the branchiobdellidans and hirudineans. The present study shows that the branchiobdellidans and lumbriculids both share a similar oligochaetous organization of the body wall musculature. However, these two taxa differ in that the branchiobdellidans have round circomyarian muscle fibres, and the lumbriculids, as the majority of oligochaetes, have flat circomyarian fibres.

The integration of muscle structure with other anatomical characters currently used in the phylogeny of the Clitellata clearly shows both similarities and differences between the branchiobdellidans, hirudineans, lumbriculids. The possibility of utilizing morphological data concerning the organization of muscle fibres in phylogenetic analyses has been considered in Lanzavecchia *et al.* (1988). Unfortunately, the different conditions for each of the above mentioned groups are not consistent with the recognized diagnostic characters for the taxa. In contrast, a phylogenetic assessment of the available information on muscle structure alone suggests that the branchiobdellidans separated from the common clitellate stock before the oligochaetes and leeches became recognizable taxa.

This study simply adds new information to the data-base on anatomical and ultrastructural characters from which a phylogeny of the Branchiobdellae, and its position within the Clitellata will ultimately be constructed. The fact that four species, from three out of the four genera in the Branchiobdellidae share a similar muscle morphology, compared to two species from other families, is a fact few would have believed prior to this investigation. Further studies are necessary to determine whether similar traits can be identified in the muscle structure of the other branchiobdellidan families.

Acknowledgements

This work was supported by C.N.R. grant CT 89.02920.04.115.20804. Additional support to S.R. Gelder was provided by Faculty Development and Mini-Grant funds from the University of Maine at Presque Isle, U.S.A., and by Dr M.A. Subchev together with the Institute of Zoology, Bulgarian Academy of Sciences, Bulgaria. The authors are grateful to Tiziana Benedetti for graphical assistance.

References

Brinkhurst, R. O. & S. R. Gelder, 1989. Did the lumbriculids provide the ancestors of the branchiobdellidans, acanthobdellidans and leeches? Hydrobiologia 180: 7–15.

Brinkhurst, R. O. & S. R. Gelder, 1991. Chapter 12. Annelida: Oligochaeta and Branchiobdellida. In J. H. Thorpe & F. Covitch (eds), Ecology and Classification of North American Freshwater Invertebrates. Academic Press, New York: 401–435.

Candia Carnevali, M. D. & M. Ferraguti, 1979. Structure and ultrastructure of muscle in the priapulid *Halicriptus spinulosus*; functional and phylogenetic remarks. J. mar. bio. Ass. 59: 734–744.

de Eguileor, M. & M. Ferraguti, 1980. Architecture of the T-system in helical paramyosin muscles of an annelid (Branchiobdella). Tissue & Cell 12: 739–747.

de Eguileor, M. & R. Valvassori, 1977. Studies on the helical and paramyosinic muscles. VII. Fine structure of body wall muscles in *Sipunculus nudus*. J. Submicr. Cytol. 9: 363–372.

de Eguileor, M., G. Lanzavecchia & R. Valvassori, 1994. Ultrastructural and histochemical comparison in two haplotaxids. Hydrobiologia 278 (Dev. Hydrobiol. 95): 157–163

de Eguileor, M., R. Valvassori, G. Lanzavecchia & L. Di Lernia, 1989. T-system in muscles of microdriles. Hydrobiologia 180: 109–114.

de Eguileor, M., R. Valvassori, G. Lanzavecchia & S. Giorgi, 1990. Body wall muscles in the haplotaxids *Haplotaxis gordioides* and *Pelodrilus leruthi* (Annelida, Oligochaeta). Zoomorphology 110: 27–36.

Farnesi, R. M. & D. Vagnetti, 1972. Morphological investigations on the parietal musculature in *Branchiobdella pentodonta* Whitman at the optical and ultrastructural levels. Boll. Zool. 39: 343–350.

Ferraguti, M. & G. Lanzavecchia, 1977. Comparative electron microscopic studies of muscles and sperm cells in *Branchiobdella pentodonta* Whitman and *Bythonomus lemani* Grube (Annelida: Clitellata). Zoomorphologie 88: 19–36.

Franzen, A., 1962. Notes on the morphology and histology of *Xironogiton instabila* (Moore, 1893) (Fam. Branchiobdellidae) with special reference to the muscle cells. Zool. Bidr. Upps. 35: 369–383

Gelder, S. R. & R. O. Brinkhurst, 1990. An assessment of the phylogeny of the Branchiobdellida (Annelida: Clitellata) using PAUP. Can. J. Zool. 68: 1318–1326.

Gelder, S. R., M. Ferraguti & M. A. Subchev. 1994. A description of spermatozoan ultrastructure and some anatomical characters in *Branchiobdella kozarovi* Subchev, 1978 (Annelida: Clitellata), and review of the spermatozoan morphology within the genus. Hydrobiologia 278 (Dev. Hydrobiol. 95): 17–26.

Gelder, S. R. & R. C. Smith, 1987. Distribution of branchiobdellids (Annelida: Clitellata) in northern Maine, U.S.A. Trans. Am. microsc. Soc. 106: 85–88.

Holt, P. C., 1986. Newly established families of the order Branchiobdellida (Annelida: Clitellata) with a synopsis of the genera. Proc. biol. Soc. Wash. 99: 676–702.

Holt, P. C., 1989. Comments on the classification of the Clitellata. Hydrobiologia 180: 7–15.

Karnowsky, M. J., 1965. A formaldehyde-glutaraldehyde fixative of high osmolality for use in electron microscopy. J. Cell Biol. 27: 137A.

Lanzavecchia, G., 1977. Morphological modulations in helical muscles (Aschelminthes and Annelida). Int. Rev. Cytol. 51: 133–186.

Lanzavecchia, G., M. de Eguileor & R. Valvassori, 1985. Superelongation in helical muscles of leeches. J. Muscle Res. Cell Motility 6: 569–584.

Lanzavecchia, G., M. de Eguileor & R. Valvassori, 1988. Muscles. In W. Westheide & C. O. Hermans (eds), The ultrastructure of Polychaeta. G. Fisher Verlag, Stuttgart, New York. Microfauna Marina 4: 71–88.

Lanzavecchia, G., R. Valvassori & M. de Eguileor, 1994. Body wall muscles in oligochaetes. Hydrobiologia (Dev. Hydrobiol. 95): 179–188.

Moore, J. P., 1895. The anatomy of *Bdellodrilus illuminatus* an american discodrilid. J. Morph. 10: 497–540.

Rosenbluth, J., 1965. Ultrastructural organization of obliquely striated muscle fibres in *Ascaris lumbricoides*. J. Cell Biol. 25: 495–515.

Sawyer, R. T., 1986. Leech Biology and Behaviour. Clarendon Press, Oxford, 1065 pp.

Schmidt, F., 1903. Die Muskulatur von *Branchiobdella parasita*. Z. wiss. Zool. 75: 596–705.

Valvassori, R. & M. de Eguileor, 1991. Helical muscles: models and adaptations. In G. Lanzavecchia & R. Valvassori (eds), Form and function in Zoology. Mucchi, Modena. Selected Symposia and Monographs U.Z.I. 5: 141–162.

Hydrobiologia **278**: 201–212, 1994.
T.B. Reynoldson & K.A. Coates (eds), Aquatic Oligochaete Biology V.

Response of tidal freshwater macrobenthos to sediment disturbance

Robert J. Diaz
Virginia Institute of Marine Science, School of Marine Science, College of William and Mary, Gloucester Point, VA 23062, USA

Key words: disturbance, dredging effects, fluid mud, macrobenthos, tidal freshwater, Tubificidae

Abstract

The macrobenthic fauna in the tidal freshwater James River, Virginia, USA, exhibited a high degree of resilience and limited temporal response to sediment disturbance caused by large quantities of low bulk density (< 1.3 g cm^{-3}) fluid mud. The fluid mud was produced by hydraulic dredge deepening of a ship channel with disposal of dredged sediments onto a nearby shoal. The response of tidal freshwater communities to fluid mud was limited to mainly quantitative changes in abundance of dominant taxa (*Limnodrilus* spp., *L. hoffmeisteri*, *Ilyodrilus templetoni*, *Corbicula fluminea*, *Coelotanypus scapularis*) and was directly related to the thickness of fluid mud layers. Disturbance effects were short lived and for tubificids most obvious in areas with > 0.3 m fluid mud. In areas that received < 0.3 m fluid mud, acute effects were limited to chironomids and small (< 10 mm) *C. fluminea*. The fauna colonizing the areas disturbed by fluid mud was the same as that inhabiting the shoal prior to disturbance. There was no indication of a successional sequence, as reported for other freshwater and marine habitats. Three weeks after the disturbance ended, all but a few insect taxa had recolonized. Changes in community structure from fluid mud disturbance were slight with total taxa best characterizing the disturbance. The insensitivity of community structure measures reflects the high resiliency of macrobenthic communities to physical stresses in tidal freshwater systems.

Introduction

The disturbance response of tidal freshwater fauna has received little attention despite extensive tidal freshwater systems around the world that serve as important nursery grounds for many anadromous and catadromous fish, and feeding grounds for wading and shore birds (Pfannkuche *et al.*, 1975; Odum *et al.*, 1984; Setzler-Hamilton, 1987; Grant & Olney, 1991). Population and industrial centers also tend to locate on tidal freshwater rivers placing various pollutant stresses on the fauna (Pfannkuche, 1981; Diaz, 1989).

Tidal freshwater systems are transitional areas between riverine and estuarine habitats and possess unique features. They are depositional environments where the bulk of the suspended load carried by nontidal rivers settles. This settlement is due in part to reduced currents caused by tides, and increased cross-sectional area as the river-estuary proceeds to the sea (Nichols, 1979). Physical stress is a prominent feature of tidal freshwater areas, particularly stress due to sedimentation and sediment instability. These stresses reduce habitat heterogeneity, leaving most bottom habitats as unconsolidated mud.

Dredging to maintain shipping channels and other habitat disturbing activities are common in tidal freshwater (Nichols, 1979; Howarth *et al.*, 1991).

One of the most striking features of tidal freshwater habitats is the low number of occupant species compared to the number of occupant species in both nontidal freshwater and brackish estuarine habitats (Pfannkuche *et al.*, 1975; Lopez, 1988; Diaz, 1989). The most comparable freshwater habitats to tidal freshwater, in terms of species diversity and composition, are large lakes similar to Lake Eire (Brinkhurst, 1974; McCall & Soster, 1990). Tidal freshwater areas tend to have low species diversity because of the harsh physical environment and lack of habitat heterogeneity. While species diversity is low, the fauna of tidal freshwater areas are important to energy flow in estuarine and marine systems (van den Broek, 1978; Diaz, 1980). Major faunal components are tubificid oligochaetes and chironomid insect larvae as well as mollusc and other insect and crustacean groups.

A channel dredging project in the James River, Virginia, U.S.A., involving a hydraulic dredge and open water unconfined disposal of muddy sediments allowed the evaluation of the short-term response of tidal freshwater macrobenthos to sediment disturbance. Large volumes of fluid mud (Inglis & Allen, 1957) were produced by this project. Fluid mud is defined as dense suspension of sediment with concentrations of 10 to 480 $g\,l^{-1}$ and bulk densities <1.3 $g\,cm^{-3}$ (Nichols *et al.*, 1978).

Site description

The James River is a major tributary of Chesapeake Bay that has a long history of dredging activity (Nichols, 1979). The study site is a shoal area 2.4 to 3.6 m deep off of Buckler's Point (Fig. 1). It is located near the middle of the tidal freshwater James River, 50 km from the fall line at Richmond and 43 km upstream from the average position of measurable salinity at Swann's Point. The river's physiography in the Buckler's Point area is characterized by a single 7.5 m deep and 90 m wide dredged navigation channel, with broad shoal areas on either side. The study shoal, on the north side of the river, is a partly protected meander bend and in the past has been used as a dredged sediment disposal area. Bottom topography is a result of these past disposal events and tidal reworking of the sediments (Nichols, 1979). Tides in the area are semidiurnal with a mean range of 0.7 m. Salinity is normally 0.0‰, but under drought conditions can reach as high as 5‰.

Municipal and industrial wastes are discharged into the river at Hopewell (15 km upstream from Buckler's Point) and Richmond. Water quality below Richmond and Hopewell is poor because the magnitude of biological oxygen demand (BOD) in the river is greater than the river's assimilative capacity. In the area of Windmill point, water quality is improved and macrobenthic community structure reaches its greatest complexity for the tidal freshwater James River (Diaz, 1989).

Material and methods

Prior to dredged sediment disposal, in July 1976, macrobenthic sampling sites were located in the most probable path of fluid mud dispersal. A 1350 m transect was sampled from the center of the disposal site down river parallel to the channel. Another transect extended from the center of the disposal site to the edge of the channel. Stations were scattered along each transect and not kept at fixed locations in an attempt to follow the varying thickness of fluid mud (Fig. 2).

Macrobenthic sampling sites were stratified *a posteriori* by thickness of fluid mud as determined by Nichols *et al.* (1978) using a nuclear transmission density probe and visual fluff measurements (Fig. 2). Four strata were chosen corresponding to fluid mud thicknesses of <0.1, 0.1 to 0.3, 0.3 to 0.9, and 0.9 to 1.6 m above the natural bottom contour. All samples located within each fluid mud stratum were averaged to evaluate the effects of fluid mud disturbance. A reference site, considered to be 0.0 m fluid mud, was located 1650 m upriver from the disposal site, out of the influ-

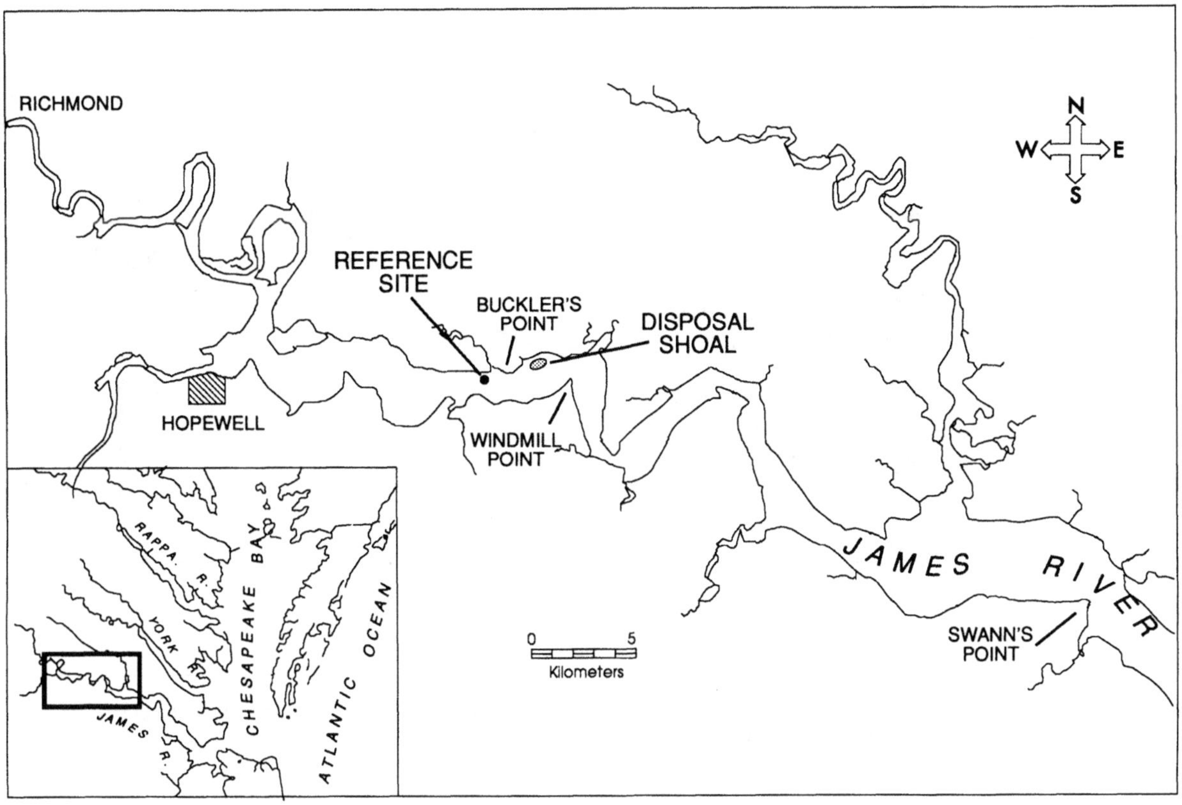

Fig. 1. Location of study site in the tidal James River.

ence of the disposal operation (Fig. 1). Samples were collected as follows: on 1 July 1976 (11 days predisposal) nine stations (P1 through P9); on 26 July (one day after disposal ended) ten stations (D0–D9); on 13 August (3 weeks after disposal) eight stations (W1–W8); and on 20 October (3 months after disposal) eight stations (M1–M8) were sampled (Fig. 2). Stations were located by triangulating on fixed survey poles. All data are contained in Diaz & Boesch (1977).

Macrobenthos were collected using a 0.05 m^2 Ponar grab. At each station three grabs were taken. After removing 40 to 50 g of sediment for grain size and organic content, each grab was placed in a plastic bag, returned to the lab, and sieved through a 500 μm screen. Sieved residue was relaxed for half an hour in a 1% propylene phenoxetol solution and fixed in 5 to 10% buffered formalin containing a vital stain (phloxine B). After sorting under a dissecting microscope, all organisms were identified and enumerated to the lowest practical taxon.

Percent sand, silt, and clay was determined by wet sieving and pipette analysis following procedures of Folk (1968). Sediment descriptions refer to the Udden-Wentworth classification described in Folk (1968). Organic content of the sediment was determined by incineration of approximately 5 g dry sediment at 550 °C for 4 hours.

Species diversity, and species richness and evenness, was calculated using the Shannon H′ Index (Pielou, 1975). Taxa abundance data were variable and not normally distributed, therefore trends in response to fluid mud through time were evaluated using the Goodman-Kurskal gamma coefficient, an ordinal measure of association that evaluates concordant and discordant pairs (Agresti, 1990). Gamma ranges from −1 to +1 and is similar in interpretation to correlation coefficients. Gamma was calculated using only

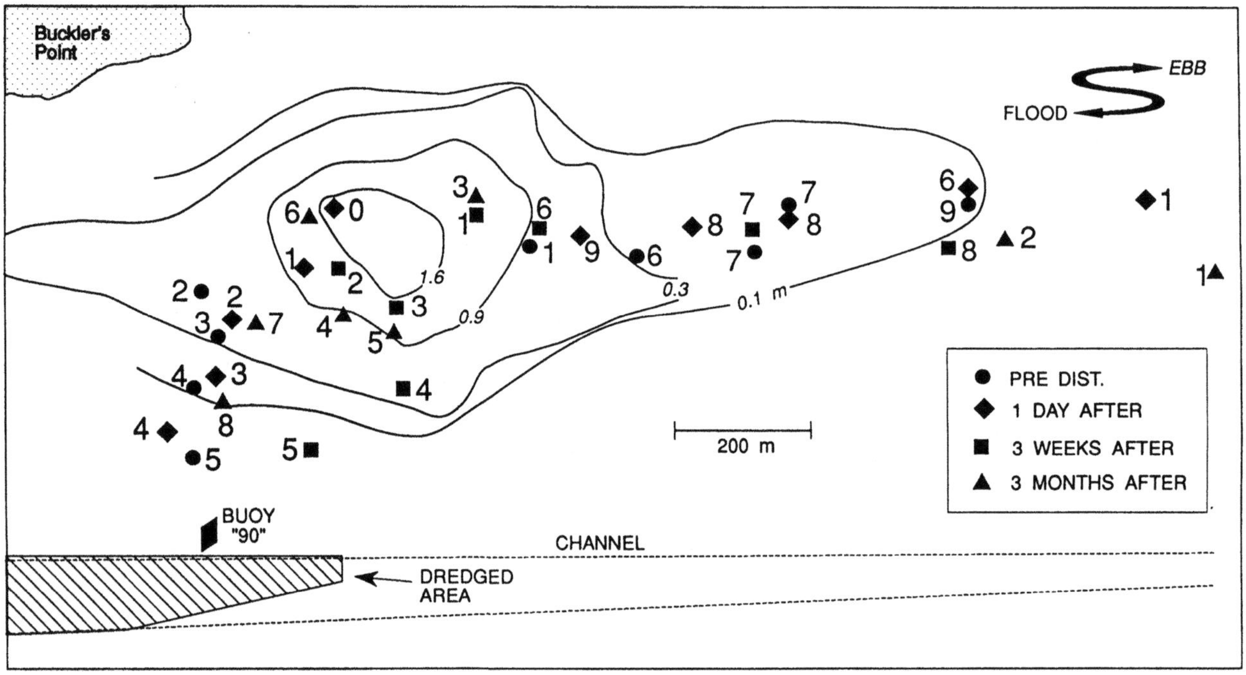

Fig. 2. Sampling sites on the disposal shoal, contours indicate thickness of fluid mud above the natural bottom on July 20 (Nichols *et al.*, 1978). Depth on the shoal before disposal ranged from 2.4 to 3.2 m.

postdisposal stations (D, W, M) because predisposal samples (P) were not disturbed by fluid mud. Gamma was considered significantly different from zero if the 95% confidence interval did not include zero.

Results

Fluid mud spread

A hydraulic cutterhead dredge started maintenance dredging on 11 July 1976 and ended on 25 July. A total of 168200 m^3 of sediment was removed from the channel and pumped via pipe line to the nearby disposal shoal. In order to pump the dredged sediments large volumes of water were added, producing large quantities of low density fluid mud. Approximately 676000 m^2 was covered by more than 0.1 m and about 49000 m^2 by more than 0.5 m of fluid mud. The maximum depth of fluid mud was 2.1 m recorded the day after the operation ended. Two months later, on 21 September, the maximum depth of fluid mud was 2.0 m (Nichols *et al.*, 1978).

Sediment grain size and organic content

Prior to disposal, sediments on the shoal ranged from clayey silt to silty sand. Stations P2, P5, and P6 had the most coarse sediment with over 30% very fine sand. One day after the completion of disposal the sediments were finer grained, with clayey-silt and silt predominating over the shoal. After three months sediments were still finer grained and were predominantly silt (Table 1). There was also a decrease in average clay content between the 26 July (16.9%) and August and October samples (5.5%). In general, sediments became finer grained and more homogeneous through time with no apparent pattern related to the depth of fluid mud (Table 1). Sediments at the reference site were clayey-silt to silt throughout the study. One day after disposal organic content ranged from 10.0 to 12.3% at both reference

Table 1. Sediments parameters averaged by fluid mud strata. All data are percent.

	Sampling period			
	Pre-	Post-disturbance		
Depth of fluid mud (m)	P	1 day D	3 weeks W	3 months M
Sand				
0.0	5.4	7.0	10.7	0.0
0.0–0.1	30.3	8.2	7.6	0.0
0.1–0.3	19.2	8.2	8.4	35.6
0.3–0.9	26.3	15.8	13.6	14.9
0.9–1.6	*	8.6	9.5	10.0
Silt				
0.0	73.6	63.7	82.3	90.8
0.0–0.1	54.4	74.6	85.6	94.0
0.1–0.3	68.2	71.6	87.1	57.4
0.3–0.9	61.4	69.4	81.9	79.7
0.9–1.6	*	79.2	85.0	83.2
Clay				
0.0	21.0	29.3	7.0	9.2
0.0–0.1	15.3	17.2	6.9	5.9
0.1–0.3	12.6	20.2	4.5	6.9
0.3–0.9	12.3	14.8	4.5	5.4
0.9–1.6	*	12.2	5.5	6.7

* No data.

and disposal shoal stations. Most of the organic matter appeared to be bits of refractory detritus, with 0.6 to 3.5% of the total organic matter retained on a 63 μm sieve.

Temperature and dissolved oxygen:

Bottom water temperature in the study area was 26 °C in July, 25 °C in August, and 14 °C in October. Predisposal bottom dissolved oxygen concentration over the reference area was 7.4 mg l^{-1} (94% saturation) and averaged 6.7 mg l^{-1} (85% saturation) over the disposal shoal. One day after disposal ended there was a horizontal gradient of dissolved oxygen with lowest values of 3.2 mg l^{-1} (40%) in the area that received the most fluid mud and 4.7 (59%) in the <0.1 m fluid mud areas. At the reference area, oxygen was 5.7 mg l^{-1} (72%). In August, dissolved oxygen was at 7.8 (96% saturation) over the reference area and 6.1 to 7.1 mg l^{-1} (75 to 88%) over the disposal shoal.

Fauna

Tubificidae were the most numerous taxa, comprising 75% of all individual collected. The genus *Limnodrilus*, which was represented by two species, comprised 89% of all the oligochaetes. Chironomidae was the most species rich group, represented by 14 taxa, which were 9.5% of the individuals collected. Larvae of the genus *Coelotanypus* comprised 62% of all chironomids. Bivalves had five taxa representing 11% of the individuals, with *Corbicula fluminea* being 95% of all bivalves. Over 99% of the *C. fluminea* were small (<10 mm length). Combined, all tubificids, chironomids, and bivalves were 99% of all individuals and 27 taxa. The five taxa that occurred at over 90% of the stations were considered dominant. The next 9 taxa were considered common (25 to 60% of stations) and the last 27 uncommon (Table 2).

Limnodrilus spp., the most common taxa in the tidal freshwater James River (Diaz, 1989), typified the response of the dominant taxa. It was consistently present, but variable, at the reference site and disposal shoal, averaging 1443 ± 910 m^{-2} ($\pm$ standard deviation) before disposal and 639 ± 479, 924 ± 619, and 1004 ± 953 m^{-2} in the three sample periods after disposal. One day after disposal was completed, *Limnodrilus* spp. declined in areas that received >0.3 m of fluid mud to 410 ± 203 m^{-2}. At station D0, where fluid mud thickness was >1.6 m, *Limnodrilus* spp. density was reduced to 7 ± 12 m^{-2} one day after disposal. Three weeks later, *Limnodrilus* spp. had increased to near predisposal densities in all fluid mud strata. Three months later, they were still near predisposal densities, with highest density in the 0.3 to 0.9 m stratum (Fig. 3). The Goodman-Kurskal gamma coefficient was significantly positive, 0.22 ± 0.02 ($\pm 95\%$ confidence interval based on asymptotic standard error), indicating a

Table 2. Taxa occurrence totaled for all stations and sampling dates (39 = 100%).

Taxon	Occurrence
Turbellaria	
Hydrolimax grisea Haldeman	1
Naidae	
Nais spp.	2
Tubificidae	
Limnodrilus spp.	39
(Immature w/o hair seta)	
Limnodrilus hoffmeisteri Claparède	37
Limnodrilus cervix Brinkhurst	2
Ilyodrilus templetoni Southern	38
Isochaetides freyi Brinkhurst	10
Quistadrilus multisetosus Smith	11
Tubifex sp.	2
Branchiura sowerbyi Beddard	7
Hirudinea	
Helobdella elongata (Castel)	6
Chironomidae	
Coelotanypus scapularis (Loew)	36
Coelotanypus concinnus (Coquillett)	1
Procladius bellus (Loew)	23
Harnischia spp.	23
Cryptochironomus spp.	18
Polypedilum spp.	14
Chironomus spp.	8
Dicrotendipes nervosus (Staeger)	5
Ablebesmyia sp. E Roback	2
Tanypus neopunctipennis Meigen	2
Stictochironomus devinctus (Say)	1
Psectrocladius sp.	1
Chironomid sp. 1	1
Chironomid sp. 2	2
Other Insects	
Chaoborus punctipennis (Say)	19
Odonata	3
Cyclorrahapha larvae	3
Hexagenia mingo Walsh	2
Coleopteran larvae	1
Tricopteran A	1
Tricopteran B	1
Amphipoda	
Gammarus fasciatus Say	7
Mollusca	
Valvata sincera Say	1
Ferrissia sp.	1
Gastropod	1
Corbicula fluminea (Mueller)	38
Rangia cuneata Sowerby	3
Sphaerium (Musculium) transversum Say	10
Pisidium sp.	4
Elliptio complanata (Lightfoot)	1

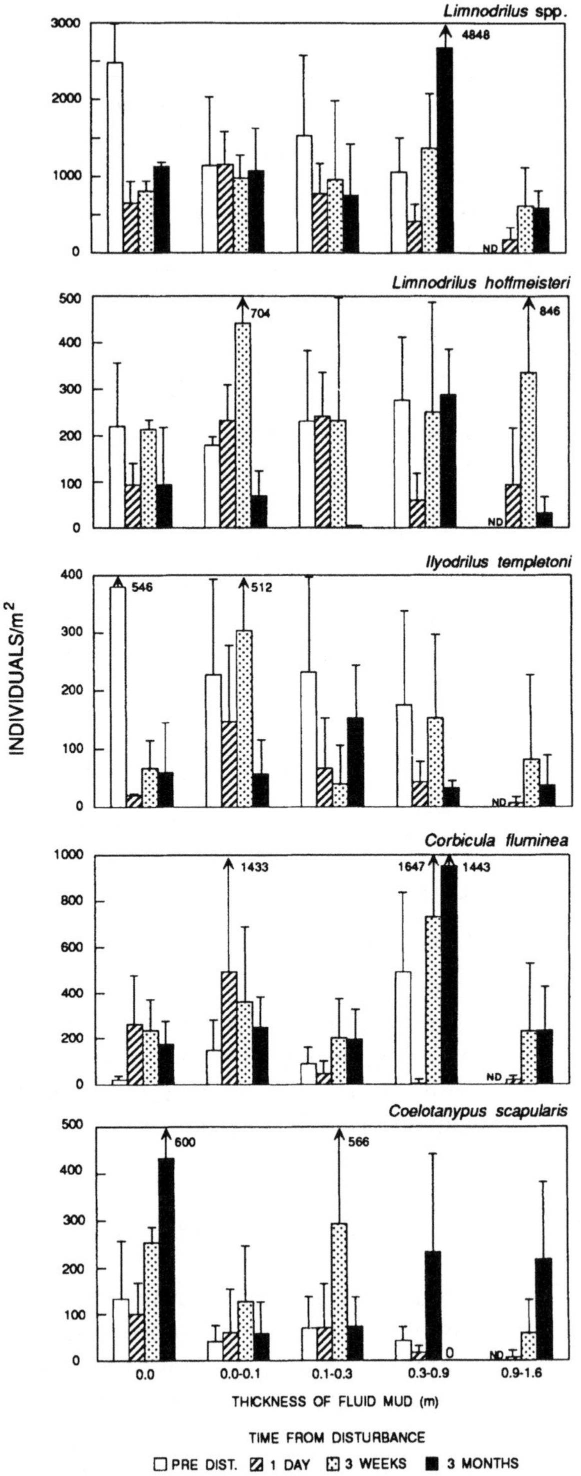

Fig. 3. Abundance of dominant taxa through time by fluid mud strata. Error bars are one standard deviation.

tendency for *Limnodrilus* spp. abundance to be higher through time at deeper fluid mud strata (> 0.3 m).

Predisposal densities of *Limnodrilus hoffmeisteri* averaged 239 ± 135 m^{-2}. One day after disposal, densities dropped at the reference site ($93 \pm 50 m^{-2}$) and in areas that were covered with 0.3 m fluid mud (71 ± 83 m^{-2}). In areas that had < 0.3 m of fluid mud, densities remained near predisposal levels (219 ± 91 m^{-2}). Three weeks later average densities were high at all strata, indicating recolonization or increased maturation of immature *Limnodrilus* spp. (Fig. 3). A low but significant Goodman-Kurskal gamma coefficient (0.12 ± 0.04) indicated that there was a tendency for the abundance of *L. hoffmeisteri* to increase through time at deeper fluid mud strata.

The response of *Ilyodrilus templetoni* was similar to *L. hoffmeisteri* with high predisposal densities (averaging 234 ± 179 m^{-2}), and lowest densities one day after disposal in the reference area (20 ± 0 m^{-2}) and in areas receiving > 0.9 m fluid mud (6 ± 11 m^{-2}). Three weeks after disposal, *I. templetoni* increased in abundance, but did not reach predisposal levels at the 0.1 to 0.3 m stratum (Fig. 3). The Goodman-Kurskal gamma coefficient was low but significantly positive, 0.09 ± 0.06, indicating a weak tendency for *I. templetoni* abundance to increase through time at deeper fluid mud strata.

Predisposal densities of small (< 10 mm length) *Corbicula fluminea*, the Asiatic clam introduced to the James River in the late 1960's (Diaz, 1974), were variable averaging 208 ± 271 m^{-2}. One day after disposal average density declined in areas that received > 0.3 m of fluid mud to 9 ± 14 m^{-2}. At the reference site (260 ± 225 m^{-2}) and areas that received < 0.1 m of fluid mud (493 ± 940 m^{-2}) densities increased one day after disposal, but were variable. Three weeks after disposal, densities increased in all areas and remained relatively constant in October, even in areas with 0.3 m of fluid mud (Fig. 3). The Goodman-Kurskal gamma coefficient was significantly positive, 0.47 ± 0.03, indicating small *C. fluminea* abundances to increase through time at deeper fluid mud strata.

Larvae of *Coelotanypus scapularis*, the most abundant chironomid in the tidal James River, were variable and averaged 66 ± 66 m^{-2} during predisposal sampling. One day after disposal, *C. scapularis* were not much affected and only declined in areas with > 0.9 m fluid mud (Table 3). Three weeks later, average densities increased in all areas. By October, average abundance was highly variable (Fig. 3). A negative and marginally significant Goodman-Kurskal gamma coefficient (-0.064 ± 0.059) indicated that there was a tendency for *C. scapularis* abundance to decline through time at deeper of fluid mud strata.

Abundances of the remaining taxa were too low to be certain about their responses to fluid mud, but their general trends are presented in Table 3. Predisposal densities of *Chaoborus punctipennis* larvae were low and variable. One day after disposal, it occurred in all strata. Fluid mud did not seem to discourage the settlement of *C. punctipennis*, which is epibenthic and migrates vertically based on light intensity (Swift & Forward, 1988). Densities three weeks and three months later were similar but patchy (Table 3). Densities of the larval chironomids *Harnischia* spp., *Procladius bellus*, and *Cryptochironomus* spp. were low, but their response to fluid mud appeared to be similar to *C. scapularis*, except that *Harnischia* spp. was reduced in areas that received 0.1 m fluid mud (Table 3) and *Procladius bellus* did not recolonize after three months. Reference site densities of *P. bellus* were high through July and August and seasonally declined in October. At the disposal site *Cryptochironomus* spp. densities were low up to three weeks after disposal. By October *Cryptochironomus* spp. had colonized to predisturbance densities.

Quistadrilus multisetosus, *Helobdella elongata*, and *Rangia cuneata* did not occur at the reference site but were consistently present on the disposal shoal. The occurrence of *Isochaetides freyi, Branchiura sowerbyi*, large (> 10 mm length) *Corbicula fluminea*, *Gammarus fasciatus*, *Sphaerium transversum* and *Pisidium* sp. were all too low to determine any response to fluid mud disturbance (Table 3).

Table 3. Abundance of common taxa (± one standard deviation) relative to fluid mud thickness and time from disturbance.

Fluid mud (m)	P Predisposal	D 1 day	W 3 weeks	M 3 months	P Predisposal	D 1 day	W 3 weeks	M 3 months
Limnodrilus spp.					*Coelotanypus scapularis*			
0.0	2480 ± 399	646 ± 320	806 ± 110	1126 ± 30	133 ± 122	100 ± 72	280 ± 35	433 ± 166
0.0–0.1	1140 ± 880	1153 ± 451	976 ± 302	1070 ± 552	40 ± 35	60 ± 100	86 ± 124	56 ± 70
0.1–0.3	1529 ± 1054	768 ± 418	960 ± 1041	746 ± 702	68 ± 68	71 ± 100	293 ± 273	73 ± 64
0.3–0.9	1055 ± 450	410 ± 203	1373 ± 747	2680 ± 2168	48 ± 41	33 ± 31	233 ± 221	0
0.9–1.6	*	93 ± 140	617 ± 550	588 ± 288	*	6 ± 12	57 ± 76	218 ± 178
Limnodrilus hoffmeisteri					*Harnischia* sp.			
0.0	233 ± 136	93 ± 50	240 ± 20	93 ± 128	13 ± 23	13 ± 23	66 ± 30	33 ± 41
0.0–0.1	180 ± 20	233 ± 81	440 ± 264	70 ± 58]	26 ± 30	13 ± 23	40 ± 52	6 ± 11
0.1–0.3	229 ± 149	211 ± 98	233 ± 277	0	16 ± 22	6 ± 11	33 ± 30	0
0.3–0.9	277 ± 138	60 ± 66	250 ± 243	286 ± 100	22 ± 32	0	73 ± 46	0
0.9–1.6	*	93 ± 128	335 ± 511	31 ± 37	*	0	6 ± 10	42 ± 36
Ilyodrilus templetoni					*Procladius bellus*			
0.0	433 ± 113	20 ± 0	66 ± 50	60 ± 87	66 ± 57	60 ± 52	46 ± 46	13 ± 23
0.0–0.1	226 ± 170	146 ± 137	303 ± 209	56 ± 63	46 ± 41	0	13 ± 11	6 ± 11
0.1–0.3	230 ± 186	66 ± 87	40 ± 69	153 ± 94	64 ± 49	20 ± 20	0	0
0.3–0.9	175 ± 163	43 ± 34	153 ± 145	33 ± 11	30 ± 37	6 ± 10	6 ± 10	6 ± 11
0.9–1.6	*	6 ± 11	82 ± 148	36 ± 59	*	6 ± 11	6 ± 11	13 ± 23
Isochaetides freyi					*Cryptochironomus* spp.			
0.0	26 ± 23	46 ± 30	6 ± 11	0	6 ± 11	40 ± 69	0	6 ± 11
0.0–0.1	0	53 ± 41	0	0	0	0	6 ± 11	6 ± 10
0.1–0.3	6 ± 11	0	0	0	16 ± 16	0	0	13 ± 23
0.3–0.9	46 ± 11	0	0	20 ± 20	20 ± 17	0	0	40 ± 34
0.9–1.6	*	0	6 ± 10	6 ± 11	*	0	0	15 ± 21
Quistadrilus multisetosus					*Polypedilum* spp.			
0.0	0	0	0	0	0	0	6 ± 11	20 ± 20
0.0–0.1	0	20 ± 25	26 ± 37	40 ± 52	20 ± 0	0	6 ± 11	13 ± 23
0.1–0.3	55 ± 137	0	0	0	13 ± 17	26 ± 46	0	66 ± 61
0.3–0.9	6 ± 11	0	46 ± 46	0	17 ± 25	0	0	13 ± 11
0.9–1.6	*	0	20 ± 20	0	*	0	0	0
Corbicula fluminea (< 10 mm length**)					*Chaoborus punctipennis*			
0.0	20 ± 20	260 ± 225	233 ± 127	173 ± 90	0	13 ± 11	6 ± 11	20 ± 34
0.0–0.1	146 ± 130	493 ± 940	360 ± 325	246 ± 149	6 ± 11	6 ± 11	6 ± 11	20 ± 0
0.1–0.3	88 ± 83	45 ± 51	200 ± 174	193 ± 133	6 ± 11	15 ± 17	0	0
0.3–0.9	491 ± 342	6 ± 11	730 ± 912	953 ± 490	0	13 ± 23	20 ± 20	0
0.9–1.6	*	10 ± 16	228 ± 304	231 ± 213	*	6 ± 10	0	31 ± 24

* No data.

** < 1% of all individuals were 10 mm.

Community structure

Diversity at all stations was moderate to low, as is typical of the tidal freshwater James River (Diaz, 1989). The range of diversity was 1.27 to 2.56 bits/individual (Table 4). Average diversity by stratum fluctuated slightly through time. Lowest diversity occurred one day after disposal in areas with 0.1 to 0.9 m of fluid mud. An examination of other community parameters indicated that diversity by itself presented a limited view of community structure. Diversity at station D0, receiving > 1.6 m fluid mud, was low (1.50) but near the average for all stations one day after disposal (1.69 ± 0.39 standard deviation). This area had the highest evenness (0.95) for the study and its species richness of 1.44 was only slightly below average (1.70 ± 0.56), but there were only 4 individuals representing 3 taxa in the three grabs taken at this station. At the reference site on the one day after disposal, diversity (2.56) and richness (3.01) were above average with 515 individuals representing 17 taxa in the 3 grabs. Thus, diversity and its components, evenness and richness, were of little value in interpreting effects of fluid mud.

The average number of taxa per 0.15 m^2 and total macrofauna per m^2 did exhibit patterns related to depth of fluid mud. There was a decrease in the number of taxa one day after disposal in areas that received > 0.1 m fluid mud. Three weeks later, the entire study area had recovery of taxa and after three months the average number of taxa was near predisposal levels. Total macrofaunal density decreased one day after disposal in areas receiving > 0.3 m fluid mud, three weeks later density was near predisposal levels. In October, three months later, density declined in all strata, likely from seasonal changes with the onset of Fall (Table 4).

Table 4. Community structure parameters (H′, individuals m^{-2}, taxa 0.15 m^{-2}) averaged by fluid mud strata.

	Sampling period			
	Pre-	Post-disturbance		
Depth of fluid mud (m)	P	1 day D	3 weeks W	3 months M
Diversity H′				
0.0	1.54	2.56	2.46	2.15
0.0–0.1	1.91	2.08	2.42	1.90
0.1–0.3	1.76	1.54	1.88	1.81
0.3–0.9	2.19	1.27	1.92	1.72
0.9–1.6	*	1.59	1.92	2.04
Individuals m^{-2}				
0.0	3433	1353	1780	2060
0.0–0.1	1847	2410	2500	1640
0.1–0.3	2303	1230	1760	1253
0.3–0.9	2256	557	2843	4460
0.9–1.6	*	170	1344	1207
Taxa 0.15 m^{-2}				
0.0	12	17	13	11
0.0–0.1	10	14	11	16
0.1–0.3	11	8	6	7
0.3–0.9	14	7	8	11
0.9–1.6	*	6	8	10

Discussion

Recovery from disturbance

The disposal of fine grained sediments as fluid mud (low sediment bulk densities between 10 to 480 g l^{-1} (Nichols *et al.*, 1978)) caused reductions in macrobenthos that were related to depth of fluid mud. Disturbance effects were short lived and most obvious in areas with > 0.3 m fluid mud (Table 5). In areas that received < 0.3 m fluid mud, acute effects were limited to chironomids and small *Corbicula fluminea*. Tubificids were un-

Table 5. Fluid mud strata that effected mortality of dominant taxa and time to recovery.

Taxa	Fluid mud thickness	Recovery time
Limnodrilus spp.	0.3–0.9 m	3 w
Limnodrilus hoffmeisteri	0.3–0.9	3
Ilyodrilus templetoni	0.1–0.3	3
Coelotanypus scapularis	0.9–1.6	3
Corbicula fluminea	0.1–0.3	3

affected at these thinner layers of fluid mud and through time became slightly more abundant than at the reference site.

The biological disturbance associated with the fluid mud was short lived because of the eurytopic tolerance of the fauna that inhabits tidal freshwater. Tubificidae are well adapted to life in within soft unstable sediments. They are subsurface deposit feeders that can loose contact with the water-sediment-interface for long periods of time. When environmental conditions become unsuitable, the tubificid *Limnodrilus* spp. can undergo subsurface migration in search of a more suitable habitat (Fisher & Beeton, 1975). Many chironomid larvae, in contrast, have respiratory structures (gills) whose function may be impaired by the fluid mud. Fluid mud also presents support problems for denser organisms like large bivalves.

The species that colonized after the disturbance were the same as those inhabiting the shoal prior to fluid mud disposal. There did not appear to be any succession of functional or adaptive types, as seen in other disturbed freshwater (Soster & McCall, 1990; McCall & Soster, 1990) and marine (Pearson & Rosenberg, 1978) systems. The disturbance successional model proposed by Soster & McCall (1990) for Lake Erie macrobenthos does not fit the tidal freshwater James River. Basically, there is a lack of species that posses the classic profile of early succession colonizers (rapid colonization with highly elevated populations that quickly crash, Grassle & Grassle (1974)).

The resilient nature of tidal freshwater fauna facilitated the rapid colonization of areas disturbed by fluid mud. Three weeks after the disposal ended, all but a few of the more sensitive insect taxa had recovered, mainly through migration of individuals from the surrounding areas or up through the fluid mud. A similar mode of colonization was found by Soster & McCall (1990) for *Limnodrilus* spp., *Ilyodrilus templetoni*, *Coelotanypus* sp., and *Procladius* sp. entering defaunated sediment trays placed on the bottom of Western Lake Erie. However, Soster & McCall (1990) found these two tubificids to be late colonizers, requiring months to build populations, and the two chironomids to be intermediate colonizers. The more lotic nature of tidal freshwater may promote colonization, as has been seen in other freshwater habitats (Tevesz, 1985).

In general, community structure indices in tidal freshwater are low (Diaz, 1989) and are not a good measure of faunal response. The number of taxa per unit area was most instructive in defining the disturbance response of fauna to fluid mud. The day after the disposal ended total taxa was lowered by >0.1 m of fluid mud. It took three months for total taxa to recover (Table 4). Diversity was a poor reflection of community response to sediment disturbance suggesting the danger of evaluation environmental effects with only diversity indices (Diaz, 1992).

It appears that the high level of physical disturbance in tidal freshwater systems leads to an underutilization of benthic resources, both space and food. The successful invasion of North American tidal freshwater systems by the Asiatic clam, *C. fluminea*, in the late 1960's has not displaced tidal freshwater species, but has successfully utilized a portion of benthic resources that were not being used by other species. The degree to which benthic resource in tidal freshwater are underutilized is not known.

General effects of fluid mud

Fluid mud presents some unique biological problems that are not associated with turbidity or burial by more consolidated sediments. Turbidity, either natural or man-made, tends to be short-lived with conflicting reports on its environmental effects (Jean & Pine, 1975; Morton, 1977). The initial effects of increased turbidity are interference with respiration and food collection. Depending on the stamina of the species exposed and duration of exposure to excess turbidity, the outcome could range from minor irritation or death for nonmotile forms unable to escape, to benefits for motile forms that enter turbidity in search of food or protection.

Fluid mud could conceptually be regarded as intermediate between turbidity and burial by more consolidated sediments. Unlike turbidity, which moves with local currents, fluid mud movement is controlled by gravity and tidal action, and has sediment-induced buoyancy effects which inhibit vertical mixing (Wolanski *et al.*, 1988). Consolidated sediments, which have concentrations greater than 175 g l^{-1}, do not flow. Fluid mud starts to form at 10 g l^{-1} and will flow up to concentrations of 175 g l^{-1}, after which consolidation and settlement begin (Nichols *et al.*, 1978).

Nichols *et al.* (1978) found fluid mud produced from the dredging in the area of Buckler's Point to be persistent and consolidation to occur slowly (Table 6). After 332 days they found the average rate of consolidation to be 0.18 ± 0.06 cm d^{-1} (mean $\pm$ sd). Fluid mud thus poses a more formidable threat to respiration and feeding of macrobenthos than does turbidity. Because of its low bulk density and ability to flow, fluid mud spreads over broad areas and provides little physical support to macrobenthos. The day after disposal ended there was a 10 to 20 cm layer of $^{-3}$ bulk density sediment in areas that had 0.1 to 0.3 m of fluid mud. In areas with > 0.9 m of fluid mud this low density sediment layer was 60 to 70 cm thick Nichols *et al.* (1978). The main disturbance mechanism for fluid mud is to separate organisms from the water-sediment interface, upon which they depend for respiration.

Frequent resuspension of surface sediments in tidal rivers, from tidal and wind energy, is an integral part of the disturbance regime of these habitats. On settling, the resuspended sediments may produce thin layers with densities in the fluid mud range, but they do not persist. Disposal of dredged sediments usually results in more dramatic mobilization of surface and subsurface sediments over smaller areas (Nichols, 1979). Thus, while instability of fine-grained sediments is natural or indigenous to tidal rivers, thick fluid mud layers represent a foreign or exotic disturbance regime eliciting a response from the fauna.

Table 6. Consolidation rates for fluid mud placed on the disposal shoal, from Nichols *et al.* (1978).

Days after disposal	Consolidation rate (mean $\pm$ SD)
0– 17	1.14 ± 0.41 cm d^{-1}
17– 40	0.48 ± 0.20
40–275	0.08 ± 0.03
0–332	0.18 ± 0.06

In tidal freshwater systems interactions between physical disturbance regimes and species are such that species richness is kept low and does not reach levels that would allow for a broader spectrum of responses to physical disturbance (Pickett & White, 1985; Soster & McCall, 1990). The response of tidal freshwater communities to fluid mud was limited to mainly quantitative changes in abundance, a response consistent with other systems that possess a high degree of resilience (Denslow, 1985). Even though the dominant species (*Limnodrilus hoffmeisteri, Ilyodrilus templetoni, Corbicula fluminea, Coelotanypus scapularis*) are eurytopic and colonized quickly they did not exhibit large increases in numbers, a key characteristic of an opportunistic life history (sensu Grassle & Grassle, 1974).

Acknowledgements

I would like to acknowledge J. Hauer, L. Stone, and C. Scott for field collection and processing of samples, and P. McCall, R. Seitz, and R. O. Brinkhurst for their review that greatly improved this manuscript. This study was part of the Dredged Material Research Program, U. S. Army Corps of Engineers, Waterways Experiment Station, Vicksburg, MS contract number DACW39–75–C–0121. This is contribution number 1774 of the Virginia Institute of Marine Science.

References

Agresti, A., 1990. Categorical data analysis. John Wiley & Sons, N.Y., 558 pp.
Brinkhurst, R. O., 1974. The benthos of lakes. St. Martin's Press, N.Y., 190 pp.
Denslow, J. S., 1985. Disturbance-mediated coexistence of species. In S. T. A. Pickett & P. S. White (eds), The Ecology of Natural Disturbance and Patch Dynamics. Academic Press, San Diego: 307–323.

Diaz, R. J., 1974. Asiatic clam, *Corbicula manilensis* (Philippi), in the tidal James River, Virginia. Chesapeake Sci. 15: 118–120.

Diaz, R. J. 1980. Ecology of tidal freshwater and estuarine Tubificidae (Oligochaeta). In R. O. Brinkhurst & D. Cook (eds), Aquatic Oligochaete Biology. Plenum Press, New York: 319–330.

Diaz, R. J., 1989. Pollution and tidal benthic communities of the James River Estuary, Virginia. Hydrobiologia 180: 195–211.

Diaz, R. J., 1992. Ecosystem assessment using estuarine and marine benthic community structure. In G. A. Burton (ed.), Sediment Toxicity Assessment, Lewis Publishers, Boca Raton: 67–85.

Diaz, R. J. & D. F. Boesch, 1977. Impact of fluid mud dredged material on benthic communities of the tidal James River, Virginia. Tech. Rpt. TR–D–77–45, U.S. Army Corps Engin., Waterways Exp. Stat., Vicksburg, Miss., 57 pp.

Fisher, J. A. & A. M. Beeton, 1975. The effect of dissolved oxygen on the burrowing behavior of *Limnodrilus hoffmeisteri* (Oligochaeta). Hydrobiologia 42: 273–290.

Folk, R. L., 1968. Petrology of sedimentary rocks. Hemphill's, Austin, 170 pp.

Grant, G. C. & J. E. Olney, 1991. Distribution of striped bass *Morone saxatilis* (Walbaum) eggs and larvae in major Virginia rivers. Fish. Bull. 89: 187–193.

Grassle, J. F. & J. P. Grassle, 1974. Opportunistic life histories and genetic systems in marine benthic polychaetes. J. Mar. Res. 32: 253–284.

Howarth, R. W., J. R. Fruci & D. Sherman, 1991. Inputs of sediment and carbon to an estuarine ecosystem: influence of land use. Ecol. Appl. 1: 27–39.

Inglis, C. C. & F. H. Allen, 1957. The regimen of the Thames Estuary as affected by currents, salinities and river flow. Proc. Inst. Civil Engin. 7: 827–878.

Jean, G. S. & R. E. Pine, 1975. Environmental effects of dredging and spoil disposal. J. Water Poll. Cont. Fed. 47: 553–561.

Lopez, G. R., 1988. Comparative ecology of the macrofauna of freshwater and marine muds. Limnol. Oceanogr. 33: 946–962.

McCall, P. L. & F. M. Soster, 1990. Benthos response to disturbance in Western Lake Erie: regional faunal surveys. Can. J. Fish. Aquat. Sci. 47: 1996–2009.

Morton, J. W., 1977. Ecological effects of dredging and dredged spoil disposal: a literature review. U.S. Fish Wildl. Tech. Paper 94, 33 pp.

Nichols, M. M., 1979. The problem of misplaced sediment. In H. D. Palmer & M. G. Gross (eds), Ocean Dumping and Marine Pollution, Dowden, Hutchinson & Ross, Stroudsburg, Pa.: 147–161.

Nichols, M. M., G. S. Thompson & R. W. Fass, 1978. A field study of fluid mud dredged materials; its physical nature and disposal. Tech. Rpt. TR–D–78–40, U.S. Army Corps Engin., Waterways Exp. Stat., Vicksburg, Miss. (NTIS AD A058 952), 74 pp.

Odum, W. E., T. J. Smith, J. K. Hoover & C. C. McIvor, 1984. The ecology of tidal freshwater marshes of the United States east coast: a community profile. U.S. Fish Wildl. Serv. FWS/OBS–83/17. 177 pp.

Pearson, T. H. & R. Ronsenberg, 1978. Macrobenthic succession in relation to organic enrichment and pollution of the marine environment. Ocenaogr. Mar. Biol. Annu. Rev. 16: 229–311.

Picket, S. T. A. & P. S. White, 1985. Patch dynamics: a synthesis. In S. T. A. Pickett & P. S. White (eds), The Ecology of Natural Disturbance and Patch Dynamics. Academic Press, San Diego: 371–384.

Pfannkuche, O., 1981. Distribution, abundance and life cycles of aquatic Oligochaeta (Annelida) in a freshwater tidal flat of the Ele Estuary. Arch hydrobiol. Suppl. 43: 506–524.

Pfannkuche, O., H. Jelinek & E. Hartwig, 1975. Zur fauna eines susswasserwattes im Elbe-Aestuar. Arch. Hydrobiol. 76: 475–498.

Pielou, E. C. 1975. Ecological diversity. Wiley-Interscience, N.Y., 196 pp.

Setzler-Hamilton, E. M., 1987. Utilization of Chesapeake Bay by early life history stages of fishes. In S. K. Majumdar, L. W. Hall & H. M. Austin (eds), Contaminant Problems and Management of Living Chesapeake Bay Resources. Pennsylvania Academy of Science, Philadelphia: 159–164.

Soster, F. M. & P. L. McCall, 1990. Benthos response to disturbance in Western Lake Erie: field experiments. Can. J. Fish. Aquat. Sci. 47: 1970–1985.

Swift, M. C. & R. B. Forward, 1988. Absolute light intensity vs. rate of relative change in light intensity: The role of light in the vertical migration of *Chaoborus punctipennis* larvae. Bull. Mar. Sci. 43: 604–619.

Tevesz, M. J. S., 1985. Benthic colonization in fresh water: a synthesis. Kirtlandia 41: 3–14.

van den Broek, W. L. F., 1978. Dietary habits of fish populations in the Lower Medway Estuary. J. Fish. Biol. 13: 645–654.

Wolanski, E., J. Chappell, P. Ridd & R. Vertessy, 1988. Fluidization of mud in estuaries. J. Geophys. Res. 93: 2351–2361.

Hydrobiologia **278**: 213–222, 1994.
T.B. Reynoldson & K.A. Coates (eds), Aquatic Oligochaete Biology V.

Oligochaetes and water pollution in two deep Norwegian lakes

Goran Milbrink
Department of Zoology, Upsala University, Box 561, 751 22 Uppsala, Sweden

Key words: Oligochaeta, water pollution, indicators, Norway

Abstract

Analyses of the oligochaete fauna of two of the deepest lakes in Scandinavia – the Norwegian lakes Mjösa (450 m) and Tyrifjorden (295 m), revealed a totally different species composition in the deep profundal compared with the upper profundal – in contact with the nutrient-enriched epilimnion. In both lakes a pronounced thermal stratification develops in the summer, thus the epilimnion receiving gross organic pollution behaves differently from the profundal. The lakes are each effectively divided into two bodies of water with limited water exchange between them, i.e. one major oligotrophic body and one minor more nutrient-rich. Since the 1950s both lakes have been exposed to heavy pollution of various kinds. In Lake Mjösa in 1975 and 1976 unpleasant algal blooms of the blue-green alga *Oscillatoria bornetii* fa. *tenuis* occurred. Bottom samples obtained at the same time revealed that the deep central bottoms of the lake were totally dominated by oligotrophic oligochaete indicators, i.e. by *Stylodrilus heringianus* and *Spirosperma ferox*, while the fauna of the upper profundal in the vicinity of domestic and agricultural sewage outfalls, wood processing industries, etc. was dominated by *Limnodrilus hoffmeisteri* and *Tubifex tubifex* in great abundance, indicating enriched conditions. Several other species indicative of eutrophy, were absent, most of them belonging to the genus *Potamothrix*. A fairly similar situation exists in Lake Tyrifjorden, where, for instance, in the shallow bay of Steinsfjorden – heavily eutrophied by agricultural wastes – blooms of blue-green algae have caused problems from time to time. The same oligochaete communities as in Lake Mjösa distinguish the central oligotrophic bottoms from the regionally more enriched upper profundal. The likely reasons for an intact profundal oligochaete fauna are great volumes of oxygen-rich hypolimnic water of low temperature and a high bottom/lake surface area ratio.

Introduction

The deep fjord lakes Mjösa and Tyrifjorden are located near Oslo and other population centres in south-eastern Norway. Consequently they are of great importance as drinking-water supplies and as recreational areas and it has been in the public interest to keep these lakes clean. The lakes also have famous populations of large, fast-growing brown trout. Since 1950 both lakes have increasingly become the recipients of waste-water of various kinds seriously threatening the water quality.

Until about 1950 Lake Mjösa (the largest lake in Norway) was oligotrophic (Holtan, 1979a,b, 1990). However, untreated sewage from human settlements, from agriculture and various industries, particularly pulp mills, resulted in eutrophic conditions developing and the water quality rapidly deteriorated. In 1971 the Norwegian Institute for Water Research (NIVA) began long-term studies in Lake Mjösa (Holtan, 1979a, 1990) and in 1976 a 'Save Lake Mjösa Campaign' was proclaimed by the national authorities resulting in considerable reduction in pollution. Even though the lake has responded to this campaign it has not yet returned to normal (Holtan, 1990).

Eutrophication of Lake Tyrifjorden, which was historically oligotrophic, began in 1930 (Langeland, 1974). In 1983 the open lake could be described as oligotrophic, the north-western part as mesotrophic, while the north-eastern basin – Steinsfjorden – was clearly eutrophic (Berge, 1983). The major pollution sources were sewage from human settlements, agriculture and various industries and mercury contamination due to emissions from wood processing industries. A 'Save Lake Tyrifjorden Campaign' similar to that for Lake Mjösa including limnological studies was started in the late 1970s with NIVA as one of the organizers.

A comprehensive investigation of the bottom fauna of the two lakes was undertaken as part of the limnological programme to determine the extent to which the deep soft bottoms in 1975–80 had been affected by eutrophication of the surface waters? Analyses of the profundal oligochaete fauna were used despite the fact that several reliable indicator species of eutrophy have not yet, as far as is known, reached Norway, among them several so-called Ponto-Caspian species of the genus *Potamothrix* (cf. Milbrink, 1980), which have their origin in the south-east of Europe (Timm, 1980). These species have only recently reached eastern Sweden (Milbrink, 1980) and Central Europe (Lang, 1985). The species *Potamothrix hammoniensis* (Michaelsen) is presently considered to be the overall most characteristic species of eutrophic lowland-lakes in Europe (Milbrink, 1980), but with few findings in Norway, so far (see below). NIVA generously allowed the author access to oligochaete material collected in Lake Mjösa in 1975 and 1976 and in Lake Tyrifjorden in 1980 for analyses and evaluation.

Limnological background

The Mjösa and Tyrifjorden lake-basins and catchment areas

Lake Mjösa (area about 365 km^2, max. depth 449 m) and Lake Tyrifjorden (area about 135 km^2, max. depth 295 m) are,respectively, the largest and fifth largest lakes of Norway (Figs 1 & 2). The theoretical water retention time for Lake Mjösa is 5–6 years and for Lake Tyrifjorden is 2–3 years (Holtan *et al.*, 1979a; Berge, 1983). Characteristic hydrological and limnological data for the two lakes have been summarized in Table 1. The catchment areas of both lakes are large, 16,420 km^2 and 9,808 km^2, respectively (Fig. 1). Nearly 70% of the Lake Mjösa catchment area consists of galciated mountainous regions. Parts of two of the highest mountain chains of Norway lie within this area. The south-eastern portion of the watershed comprises one of the best agricultural areas of Norway and forms nearly 20% of the catchment area. There are approximately 50,000 people living in the forested valley system. About 150,000 people live in the immediate surroundings of Lake Mjösa (Holtan, 1990), most of them in the cities of Hamar, Gjövik and Lillehammer (Fig. 2). The lake is bordered by several wood processing, food and metallurgic industries. The River Gudbrandsdalslågen is the main tributary to Lake Mjösa contributing about 75% of the inflow. Since 1953 the river has been regulated for hydro-electric power production and 60% of the inflow occurs between June and August.

The Lake Tyrifjorden catchment area is adjacent to the Lake Mjösa basin to the east (Fig. 1). Two wide valleys with three rivers drain the area (Berge, 1983). To the east is the long and deep Lake Randsfjorden (Fig. 1) the outflow of which flows into Lake Tyrifjorden. Much of the lower part of this catchment area is densely populated agricultural land. About 90,000 people live in the whole area. Organic pollution from human settlements has been an ongoing problem in the lake. Other sources of pollution such as pulp mill and other industries are localized in the Hönefoss area in the lower reaches of River Storelva (Fig. 1).

Hydrological conditions and water chemistry

Lake Mjösa is regulated for hydro-electric production with an annual fluctuation in level of 3.6 m (1961). Water is withdrawn during the winter and the lake-basin is subsequently refilled during the yearly spring flood (Holtan, 1979a,b). Vernal circulation normally only mixes the upper 200 m of the lake and takes place between April and June, while autumnal circulation lasts from October to January or February. The very deep bottom-water never exceeds + 4.0 °C. As in many other deep lakes in the temperate zone water flow is complex and marked internal seiches occur to depths of 40 and 50 m and Coriolis forces affect the north-south flow of River Gudbrandsdalslågen through the entire lake. As a consequence, the central parts of the lake are often ice-free during the winter. During the stratification period the cooler river water often flows below the surficial water layers at depths corresponding to the river water temperature (Holtan *et al.*, 1979b).

The water of Lake Mjösa, like most surface water of Norway, has a low salt content with little buffering capacity. The pH is close to neutral, the mean water colour is 13 mg Pt/l and the mean organic content 8.8 mg $KMnO_4\,l^{-1}$ (Holtan, 1990). Conductivity increases from about 20 $\mu S\ cm^{-1}$ (20 °C) in the northern parts of the lake to about 40 $\mu S\ cm^{-1}$ (20 °C) in the

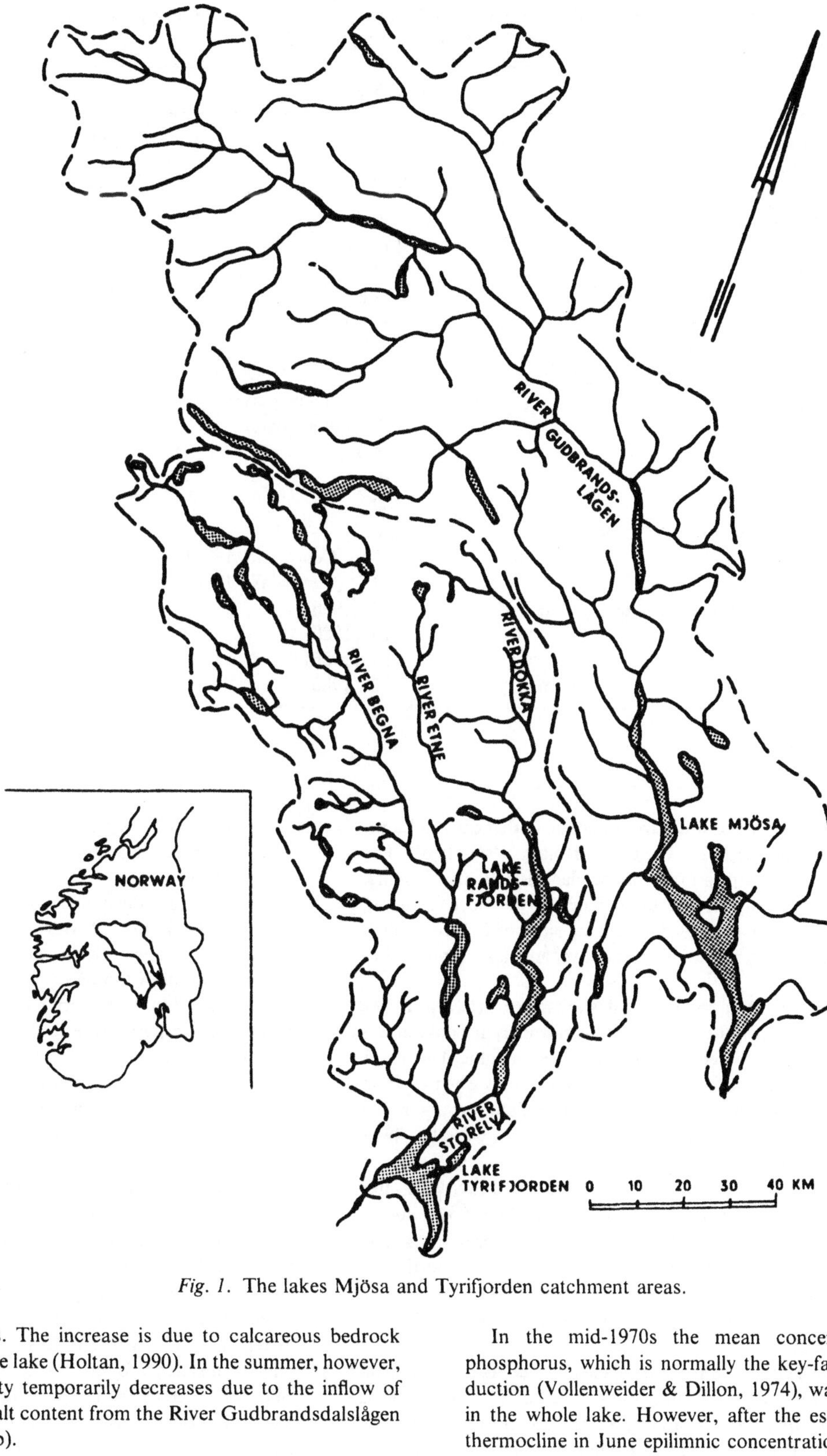

Fig. 1. The lakes Mjösa and Tyrifjorden catchment areas.

southern parts. The increase is due to calcareous bedrock surrounding the lake (Holtan, 1990). In the summer, however, the conductivity temporarily decreases due to the inflow of water of low salt content from the River Gudbrandsdalslågen (Holtan, 1979b).

In the mid-1970s the mean concentration of total-phosphorus, which is normally the key-factor for algal production (Vollenweider & Dillon, 1974), was about 10 $\mu g\ l^{-1}$ in the whole lake. However, after the establishment of the thermocline in June epilimnic concentrations, due to inflows

Table 1. Limnological back-ground data for Lakes Mjösa and Tyrifjorden (with Steinsfjorden Bay) in the late 1970's. Mainly from Holtan (1979a,b, 1981, 1990), Kjellberg (1982, 1986) and Berge (1983).

	L Mjösa	L Tyrifjorden	Steinsfjorden Bay
Max. depth, m	449	295	24
Mean depth, m	153	114	10
Area, km^2	365	135	14
Water retention time, years	6	2.7	
Catchment area, km^2	16,420	9,808	63.7
Mean water flow, $m^3\ s^{-1}$	250	170	
Amplitude of water regulation, m	3.6	1–2	
Mean summer pH-level	neutral	neutral	
Mean water colour, mg Pt l^{-1}	13	10	10
Mean organic content, mg $KMnO_4\ l^{-1}$	8.8	–	
Mean conductivity $\mu S\ cm^{-1}$	20 (northern part) 40 (soutern part)	32	80–90
Secchi transparency, m	6.0	5–6.5	4.5
Mean total, $\mu g\ l^{-1}$	10	7–9	10
Mean total-N, $\mu g\ l^{-1}$	400	200	
Yearly phosphorus load 1976, tons	400	70	
Chlorophyll a, $\mu g\ l^{-1}$	5.0 (Centr. lake 1976) 2.5 (Centr. lake 1980)	2.5–2.8	5
Primary productivity, June–Aug., g C $m^{-2}\ d^{-1}$	570 (1976) 110 (1980) 250 (1986)	200–300	400–600
g C $m^{-2}\ y^{-1}$	22 (1980)	36 (1981)	72 (1981)

of untreated sewage water from human settlements and agriculture, could reach 20–25 $\mu g\ l^{-1}$ (Holtan, 1981, 1990). The water of Lake Mjösa may thus be classified as mesotrophic. Mean concentrations of total-N generally fall close to 400 $\mu g\ l^{-1}$, and N:P-ratios exceed the critical value of 12 below which nuisance nitrogen-fixating cyanophycean algae tend to develop.

There are clear hydrological similarities between Lakes Mjösa and Tyrifjorden and there is normally no oxygen deficiency in the extreme profundal of either lake. The River Storelva, accounts for nearly 90% of the inflow. The two main tributaries as well as the lake itself are regulated (within a 1–2 m range). The lake is drained to the south-west by River Drammenelva.

Maximum water flow normally occurs in May–June after ice and snow melt in the mountains, but high flows due to precipitation can also take place in the summer months. The shallow, half-secluded basin of Steinsfjorden in the north-east (area about 14 km^2, maximum depth about 24 m) has a small water catchment area and limited water exchange with the rest of the lake.

Lake Tyrifjorden is a typical clear-water lake with a mean conductivity of about 30 $\mu S\ cm^{-1}$ at 20 °C and the pH is neutral. Mean Secchi transparencies are close to 6 m. The Steinsfjorden basin is more eutrophic and surrounded by more calcareous rocks. Its mean conductivity is between 80 and 90 $\mu S\ cm^{-1}$ at 20 °C and its mean Secchi transparency 4.5 m (Berge, 1983).

Phosphorus is the main factor limiting primary production in both lakes (Berge, 1983). Not surprisingly, values of total-phosphorus, chlorophyll a and algal biomass are closely linked. Total-phosphorus concentrations in the summer are about 7 $\mu g\ l^{-1}$ in the eastern basin of Lake Tyrifjorden, while in the western part, close to the inlet of River Storelva, the concentration is 9 $\mu g\ l^{-1}$ and in Steinsfjorden about 10 $\mu g\ l^{-1}$. Nitrogen is mainly found in the form of nitrates (about 200 $\mu g\ l^{-1}$) in Lake Tyrifjorden, but observations from the 1930s reveal that nitrate concentrations have doubled since then. Blue-green algal blooms have so far mostly been a problem in Steinsfjorden. High concentrations of nitrates here decrease from 100–150 $\mu g\ l^{-1}$ in the spring to about 10 $\mu g\ l^{-1}$ in the summer and blue-green algae increase at the same time. The N:P ratios are usually below 12, normally indicating that there are good opportunities for nitrogen-fixing blue-greens, but much of the nitrates are likely to be absorbed by extremely dense carpets of *Elodea canadensis*, which have attained nuisance proportions in Steinsfjorden (Berge, 1983) since the latter 1970s. Long-term studies have shown that the chemical properties of Lake Tyrifjorden had not changed significantly over a ten-year period (Berge, 1983).

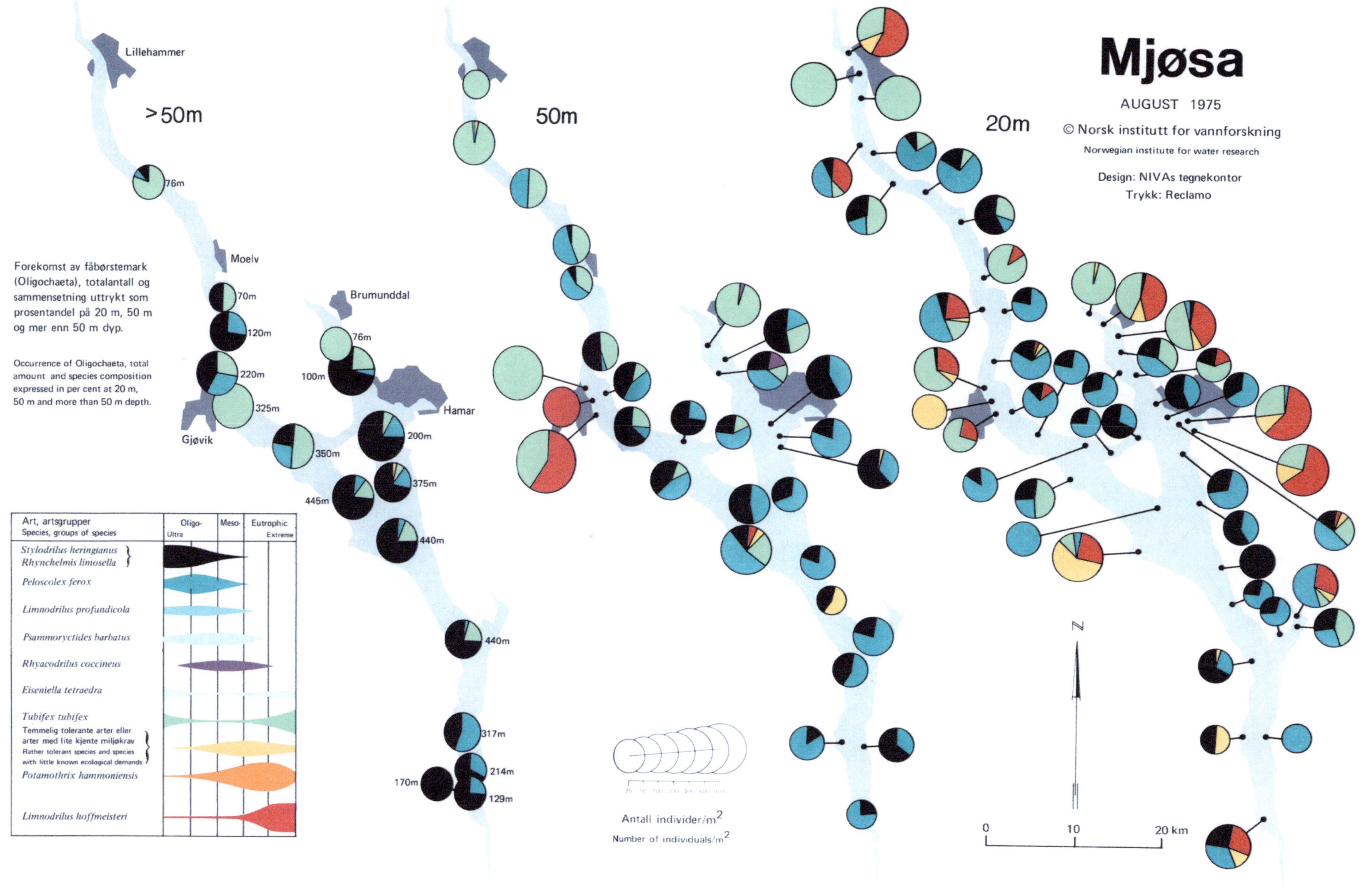
Mjøsa
AUGUST 1975
© Norsk institutt for vannforskning
Norwegian institute for water research
Design: NIVAs tegnekontor
Trykk: Reclamo
>50m
50m
20m
Lillehammer
Moelv
Brumunddal
Hamar
Gjøvik
76m
70m
120m
220m
325m
100m
350m
200m
375m
445m
440m
317m
214m
170m
129m
Forekomst av fåbørstemark (Oligochaeta), totalantall og sammensetning uttrykt som prosentandel på 20 m, 50 m og mer enn 50 m dyp.
Occurrence of Oligochaeta, total amount and species composition expressed in per cent at 20 m, 50 m and more than 50 m depth.
Art, artsgrupper
Species, groups of species
Oligo-
Ultra
Meso-
Eutrophic
Extreme
Stylodrilus heringianus
Rhynchelmis limosella
Peloscolex ferox
Limnodrilus profundicola
Psammoryctides barbatus
Rhyacodrilus coccineus
Eiseniella tetraedra
Tubifex tubifex
Temmelig tolerante arter eller arter med lite kjente miljøkrav
Rather tolerant species and species with little known ecological demands
Potamothrix hammoniensis
Limnodrilus hoffmeisteri
Antall individer/m2
Number of individuals/m2
N
0
10
20 km

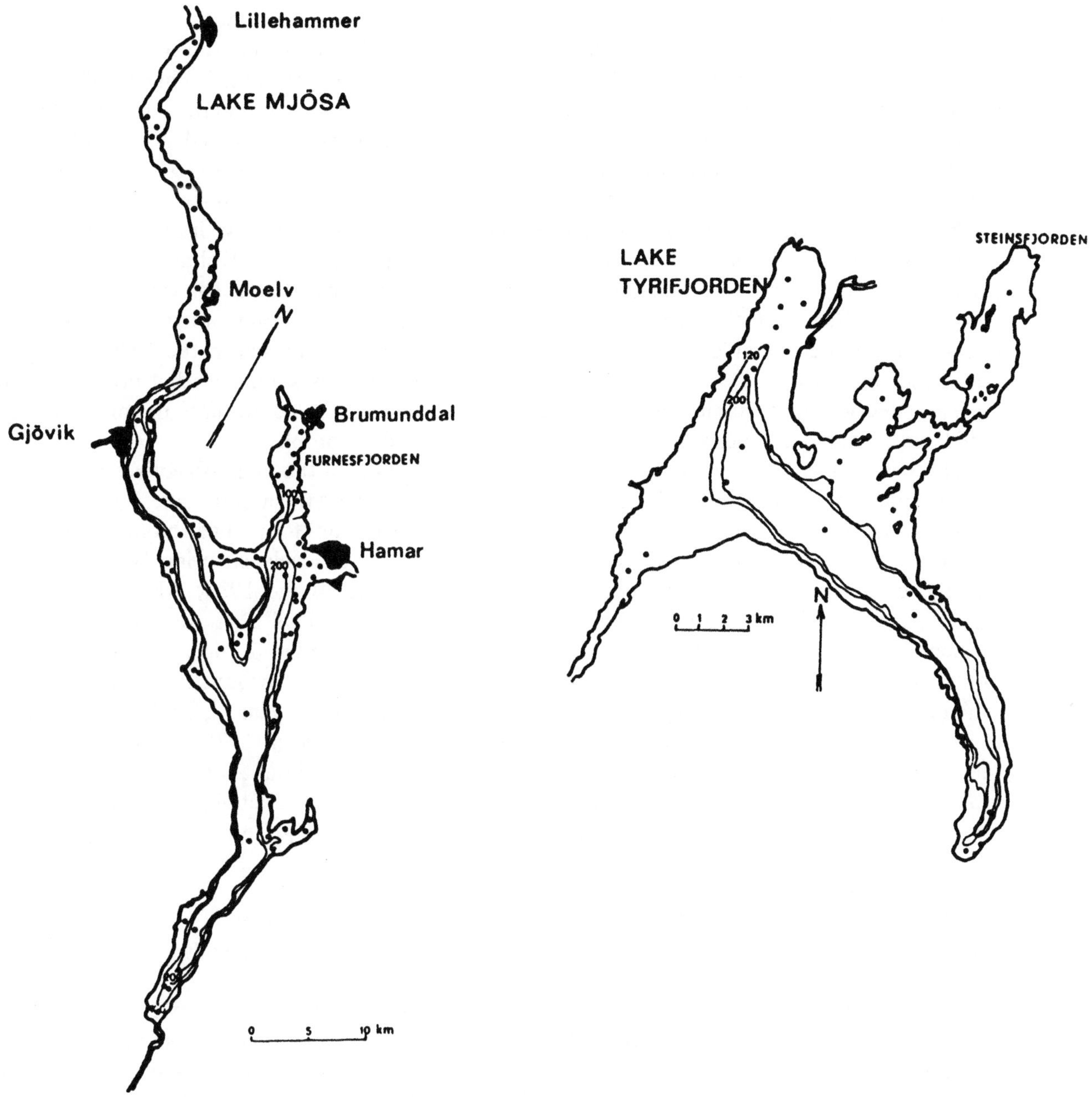

Fig. 2. Lakes Mjösa and Tyrifjorden with depth contours and station nets.

Eutrophication affecting the biology of Lake Mjösa and Lake Tyrifjorden – past situation

The main sources of pollution in Lake Mjösa before the 'Save Mjösa Campaign' were the wood processing industries at Gjövik, Lillehammer and Brumunddal (Fig. 2). Raw sewage outlets also discharged from these centres as well as other dense human settlements (e.g., Hamar). Discharge of waste from food processing industries and from agriculture occurred mainly in small rivers flowing into the central region of the lake and to Furnesfjorden bay in the north-east (Mr G. Kjellberg, pers. comm.).

Before 1930 Lake Mjösa was a clear, oligotrophic lake (Huitfeldt-Kaas, 1906, 1946; Ström,

1930; Lindström *et al.*, 1973). Diatoms such as *Asterionella formosa* Hass. were important phytoplankton species. After about 1930 this species and the bluegreen *Anabaena flos-aquae* Bréb. gradually increased in abundance and from about 1950 these species dominated the algal community, the sessile green alga *Ulothrix zonata* (Web. & Mohr) became abundant along the shores and the algal biomass increased from year to year (Holtan, 1990). In the 1960s the biomass increased further, sensitive algal species disappeared and under suitable conditions *Anabaena flos-aquae* bloomed.

This development culminated, in 1975 and 1976, in the sudden appearance of the bluegreen *Oscillatoria bornetti* fa. *tenuis* resulting in 'taste and odour' problems in the drinking water supply (Holtan, 1979a).

Most of the anthropogenic loadings to Lake Tyrifjorden are in the north-western part of the lake, through the river Storelva. Several wood processing industries have generated large banks of sedimented fibres and emissions of organic mercury (methylated mercury). Many small industries are localized in the Hönefoss-area and some 30,000 people live just to the north of the lake. Some of the best agricultural areas of Norway are to be found in the lower reaches of the Lake Tyrifjorden catchment area. Tourism increases the summer population of the lake, especially the eastern part. The Steinsfjorden basin has also been the recipient of considerable amounts of sewage and agricultural wastes (Berge, 1983).

Eutrophication of Lake Tyrifjorden due to external sources of phosphorus peaked in the 1950s and 1960s. At the end of the 1970s local authorities decided that the yearly load of external phosphorus should be reduced considerably to avoid future undue development of blue-green algae (Berge, 1983).

Remedial actions to save Lakes Mjösa and Tyrifjorden

By 1976 the adverse effects of eutrophication, particularly as it affected drinking water, prompted authorities to undertake 'Save the Lake' campaigns. Sewage plants for phosphorus removal were built, discharges of silage and stored manure were eliminated and phosphorus free detergent use encouraged. As a result the total yearly supply of phosphorus to Lake Mjösa was reduced by about 30%. Since 1976 mean Secchi disc transparencies have increased from 4 to 6 metres (Kjellberg, 1986; Holtan, 1990) and the bluegreen alga *O. bornetti* fa. *tenuis* has only occurred infrequently. Although phytoplankton biomass has slightly increased in the 1980s the species composition is similar to that observed in the 1950s and 1960s, i.e. to diatoms (primarily *A. formosa*) and cryptophyceans.

One of the main objectives of the 'Campaigns' in Lake Mjösa and Lake Tyrifjorden was to reduce planktonic algal biomass and productivity to 'safe' levels ensuring 'stable ecological conditions' (Berge, 1983). Algal biomass (as mg m^{-3} of chlorophyll a) is still slightly in excess of the recommended 2 mg m^{-3} (Table 1). The target for primary productivity of phytoplankton is 25 g C m^{-2} $year^{-1}$, peak values in Lake Mjösa of about 90 in 1975 and 1976 were actually reduced to about 22 in 1980. In Lake Tyrifjorden primary productivity remains above the target value at 36 g C m^{-2} $year^{-1}$.

The bottom fauna
Material and methods

Bottom fauna samples were taken in Lake Mjösa, by NIVA, in the late summer of 1975 and the early spring of 1976. A total of 93 stations were selected at three depth levels – 20 m, 50 m and depths over 50 m (Figs 2 & 3). At each station 5 random Ekman grab samples (area 240 cm^2) were taken. A similar strategy was used for sampling 31 stations in Lake Tyrifjorden in May 1980 (Figs 2 & 4). In both surveys each sample was sieved through a 0.5 mm sieve and the residues were preserved in 4% formalin. All oligochaetes – most of them sexually immature – were identified to species.

Oligochaetes as Indicators of Water Pollution in Lakes Mjösa and Tyrifjorden

The profundal bottom fauna of both lakes is characterized by oligochaete worms and chironomid larvae. Other fauna groups, such as pisiid mussels, are present to about 100 m depth in Lake Tyrifjorden and to about 200 m depth in Lake Mjösa. Three large, relict, crustacean species are common in Lake Mjösa. *Mysis relicta* Lovén is considered to be planktonic at great depths. *Pallasea quadrispinosa* (Sars) and *Gammaracanthus loricatus* (Sars) have been found down to about 170 m (Holtan, 1979b; Kjellberg, 1984).

The bottom fauna of the deep profundal of both lakes is relatively poor both in abundance (generally <500 specimen m^{-2}) and biomass (<1 g wet weight m^{-2}). However, in the fibre banks in outfalls from wood processing industries such as at Gjövik and Brumunddal in Lake Mjösa (Fig. 2), abundance of oligochaetes may exceed 10,000 specimens m^{-2} and biomass about 20 g wet weight m^{-2} (Holtan, 1979b). Stomach analyses of 20 whitefish (*Coregonus*) caught from the fibre banks outside Gjövik showed that each contained hundreds of *Limnodrilus hoffmeisteri* Claparéde as well as fibre material (Milbrink, unpubl.; courtesy of Dr P. Aass).

Figure 3 provides an overview (sector diagrams) of the abundance and composition of oligochaetes in Lake Mjösa in August 1975, at the time of a bluegreen algal bloom. Sector diagrams are presented separately for each depth strata. The sectors in the diagram represent percentage composition of the species at each station. The diameter of each circle is proportional to the abundance of oligochaetes in accordance with the traditional European 'trophic scale' (Liebmann, 1962).

Oligotrophic species (blue and black) such as *Stylodrilus heringianus* Claparéde and *Spirosperma* (*Peloscolex*) *ferox* (Eisen) dominate the deep stations indicating oligotrophic conditions in this region of the lake (Fig. 3 – left). Abundance was only slightly elevated in the vicinity of Hamar and Gjövik, otherwise there was no indication of eutrophication.

At the 50 m depth (Fig. 3 – centre), especially outside Gjövik and the inflow of River Lena (agricultural wastes, Mr G. Kjellberg, pers. inform.) the highly 'tolerant' species *Limnodrilus hoffmeisteri* appears and abundance increases. *Tubifex tubifex* (Müller) also responds to more local nutrient enrichment by increasing in abundance. In Furnesfjorden near Brumunddal the fairly tolerant species *Rhyacodrilus coccineus* (Vejdovsky) is indicative of local organic enrichment.

Finally at the 20-meter level (Fig. 3 – right) outside the cities, smaller settlements and at all main river inflows, the species composition changes to include more tolerant species; for instance *L. hoffmeisteri, L. udekemianus* Claperéde, *Ilyodrilus templetoni* (Southern), *Aulodrilus pluriseta* (Piguet) and *Tubifex ignotus* (Stolc), and the ecologically ambivalent species *Tubifex tubifex* (Milbrink, 1980). The last species dominates (largely monocultures) the northern most part Furnesfjorden indicating high organic load. Otherwise, as expected, diversity increases dramatically in the upper profundal – a combination of availability of nutrients and oxygen (cf. Milbrink, 1980). *Limnodrilus hoffmeisteri* and *T. tubifex* dominate the fibre banks in the northern portion of Furnesfjorden.

The sector diagram of Lake Tyrifjorden (Fig. 4) does not include estimates of abundance of oligochaetes. The species composition is similar to that in Lake Mjösa. In the organically most enriched parts of the lake in the north-west and in Steinsfjorden in the north-east abundance and biomass reach 2,000 specimens and 7 g wet weight m^{-2}, respectively, compared to values 500 specimens and 2 g wet weight m^{-2} in most of the lake (Kjellberg, 1982). The central areas of the lake, particularly the deep profundal and some shal-

Fig. 3. (To be found between pages 216 and 217.) Occurrence of oligochaeta in a station net in Lake Mjösa in August 1975. Abundance and percentage composition at three depth levels – 20 m, 50 m and > 50 m. The sectors in the diagram represent the percentage composition of the species at each station. The length of the radius of each circle is proportional to the abundance of oligochaetes in accordance with the scale below. Species occurrence in relation to a generalized trophic scale (to the left in the figure).

lower areas, are characterized by oligotrophic species (Fig. 4), i.e. by *Stylodrilus heringianus*, *Rhynchelmis limosella* Hoffmeister (rare) and *Spirosperma ferox* (see *P. ferox*). In the shallow zone in the north-west, the River Storelva is the main source of pollution, and in Steinsfjorden the oligochaete fauna is dominated by 'tolerant' species such as *L. hoffmeisteri*, *T. tubifex* and to some extent *A. pluriseta*. Locally enriched parts of the lake, for instance at Sylling in the south, are indicated by the occurrence of *L. hoffmeisteri*.

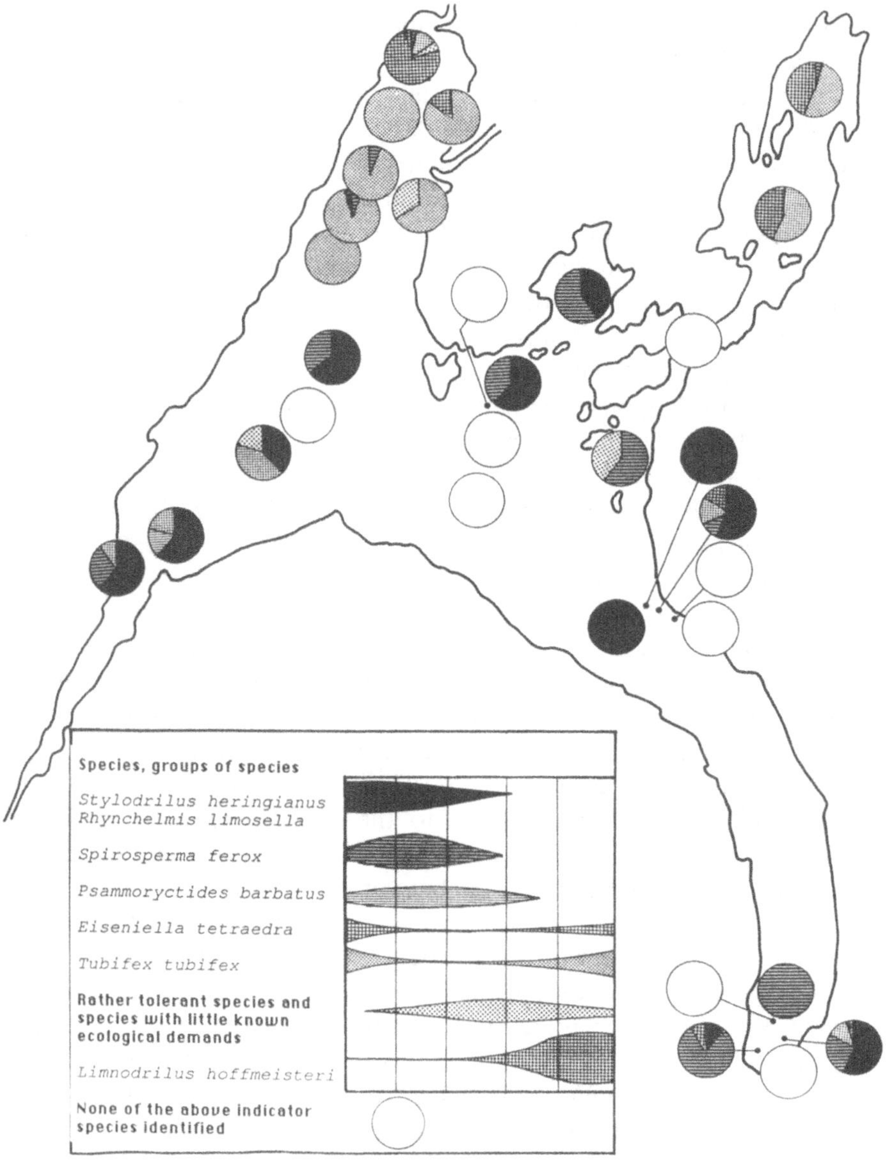

Fig. 4. Occurrence of oligochaetea in a station net in Lake Tyrifjorden in May 1980. Percentage composition in relation to a generalized trophic scale as in Fig. 3. Actual abundance values not included in the figure.

Discussion

The profundal benthic fauna of temperate lakes can usually be used as an indicator of lake trophic status and is particularly useful in detecting long term trends and responses to pollutant loading (Brinkhurst, 1966; Wielderholm, 1976; Milbrink, 1980).

Both the abundance and biomass of the benthic fauna of Lakes Mjösa and Tyrifjorden are dominated by oligochaetes and chironomids (80–100%) at depths greater than 50 m. In 1975 the deep bottom fauna were largely unaffected by eutrophication. The large crustaceans *P. quadrispinosa*, *G. loricatus* and *M. relicta* and the chironomid and oligochaete species found were all characteristic of oligotrophic conditions. Most being cold-stenothermous species with high oxygen demands typical of those found in lakes where the hypolimnetic waters are oxygen-saturated and with low organic content in the sediment. There is a particularly good correspondence between the oligochaete fauna and chironomid species indicative of oligotrophy. In both lakes the 'ultra-oligotrophic' chironomid *Heterotrissocladius subpilosus* (Kieff.) predominates in the central deep areas, together with 'oligotrophic' indicators such as *Paracladopelma obscura* Brundin, *Tanyarsus* spp. and *Microspectra* spp. *Chironomus anthracinus* Zett. dominates the organically enriched areas such as the area in the vicinity of Hamar and the inner portion of Furnesfjorden in Lake Mjösa, and the north west and Steinsfjorden in Lake Tyrifjorden. Other species indicative of mild eutrophication are *Phaenopsectra coracina* (Zett.) and *Stictochironomus rosenschoeldi* (Zett.) (Kjellberg, 1984).

The deep central bottoms of both Lakes Mjösa and Tyrifjorden are characterized by species that normally dominate the oligochaete fauna of deep oligotrophic lakes of northern Europe (Milbrink, 1980). The oligotrophic lumbriculid species *Rhynchelmis limosella* has only been found in Lake Tyrifjorden (at 100 m) and has not yet been found in Lake Mjösa. Comparatively few species indicative of eutrophic conditions have been found in Norway. Ponto-Caspian species of the genus *Potamothrix*, which are locally very common in Lake Mälaren, eastern Sweden, *P. vejdovskyi (Hrabe)*, *P. heuscheri* (Bretscher) and *P. moldaviensis* Vejdovsky & Mrazek and the more infrequently found *P. bedoti* (Piguet). However, a few records of *P. hammoniensis* from Norwegian waters exist today (Slorejn pers. inf., Milbrink, unpubl.).

Most of the main sources of pollution into Lakes Mjösa and Tyrifjorden would seem to have been identified by 'the oligochaete community method', and local sources such as outflows from silage and fish pens could also be identified. The precision is likely to increase with a more intensive station network and more frequent sampling. This in combination with chemical-physical data and available information on, for instance, the composition of the chironomid fauna, would provide a valuable bio-monitoring programme.

In conclusion Lake Mjösa behaves as two separate bodies of water, one epilimnic body which in the 1970s for six months of the year was markedly nutrient-enriched due to numerous surficial inflows of organic material and one nutrient-poor hypolimnic body. Such a clear separation in one lake is quite interesting. In 1975–76 the deep-profundal oligochaete fauna clearly reflected the situation prevailing in the hypolimnion and influences on this fauna from epilimnic water masses seem negligible. The oligochaete fauna of the upper profundal was characteristic of the eutrophied epilimnion. Thus, three decades of quite heavy eutrophication and pollution of both Lake Mjösa and Lake Tyrifjorden have brought about little or no change to the deep-profundal oligochaete fauna. The likely reasons for an intact profundal oligochaete fauna here are great volumes of oxygen-rich hypolimnic water of low temperature and high bottom/lake surface area ratios. It is expected that ongoing measures to reduce enrichment and pollution of these lakes will turn out to be particularly effective and it is likely that these lakes, due to natural dilution (Edmondson, 1970), will return to oligotrophy or mesotrophy within a limited period of time.

Acknowledgements

The author is most grateful to Mr Gösta Kjellberg, NIVA, without whose help this work would not have been possible.

References

Berge, D. (ed.), 1983. Tyrifjorden. Tyrifjordundersökelsen 1978–1981. Sammanfattande slutrapport. NIVA, 156 pp.

Brinkhurst, R.O., 1966. The Tubificidae (Oligochaeta) of polluted water. Verh. Int. Ver. Limnol. 16: 854–859.

Edmondson, T.E., 1970. Phosphorus, nitrogen and algae in Lake Washington after diversion of sewage. Science 169: 690–691.

Holtan, H., 1979a. The Lake Mjösa Story. Arch Hydrobiol. Beih. 13: 242–258.

Holtan, H. 1979b. Mjosprosjektet. Hovedrapport for 1971–76. NIVA-rapport nr. 0–69091 (in Norwegian).

Holtan, H., 1981. Eutrophication of Lake Mjösa and its recovery. Wat. Qual. Bull., Vol. 6. No. 4.

Holtan, H., 1990. Lake Mjösa, Recipient and Ecosystem. Limnologica (Berlin) 20: 285–289.

Huitfeldt-Kaas, H., 1906. Planktonundersögelser i Norske Vande. Christiania, 199 pp.

Huitfeldt-Kaas, H., 1946. The plankton in Mjösa. Nytt. Mag. Naturvid. 85: 161–221.

Kjellberg, G., 1982. Overvåking av Mjösa. Bakgrunnsdata, historikk og videreforing. Del A. NIVA-rapport nr. 54/82, 104 pp. (in Norwegian).

Kjellberg, G., 1984. Studier av bunnfaunan i Tyrifjorden og Steinsfjordens profundalområder, Mai 1980. NIVA-rapport nr. 28/84, 20 pp. (in Norwegian).

Kjellberg, G., 1986. Overvåking av Mjösa. Sammendrag, trender og kommentarer 1976–85. NIVA-rapport nr. 241/86, 80 pp. (in Norwegian).

Lang, C., 1985. Eutrophication of Lake Geneva indicated by the oligochaete communities of the profundal. Hydrobiologia 126: 237–243.

Langeland, A., 1974. Long-term changes in the plankton of Lake Tyrifjord, Norway. Norw. J. Zool. 22: 207–219.

Liebmann, H. 1962. Handbuch der Frischwasser und Abwasserbiologie. Band I, R. Oldenbourg, München, 588 pp.

Lindström, E.A., R. Skulberg & O.M. Skulberg, 1973. Observations on planktonic diatoms in the lake-river system Lake Mjösa-Lake Öyeren-River Glåma, Norway. Norw. J. Bot. 29: 183–195.

Milbrink, G., 1980. Oligochaete communities in Pollution Biology. The European Situation with special reference to lakes in Scandinavia. In R.O. Brinkhurst & D.G. Cook (eds), Aquatic Oligochaete Biology. Plenum Press: 433–455.

Ström, K.M., 1930. Limnological observations on Norwegian Lakes. Arch. Hydrobiol. 21: 97–124.

Timm, T., 1980. Distribution of aquatic oligochaetes. In R.O. Brinkhurst & D.G. Cook (eds), Aquatic Oligochaete Biology. Plenum Press: 35–79.

Vollenweider, R.A., 1976. Advances in defining critical loading levels for phosphorus in lake eutrophication. Mem. Ist. ital. Idrobiol. 33: 53–83.

Vollenweider, R.A. & P.J. Dillon, 1974. The application of the phosphorus loading concept to eutrophication research. Canada Centre for Inland Waters, publ. No. NRCC 13690, 42 pp.

Wiederholm, T., 1976. Chironomids as indicators of water quality in Swedish lakes. NLU Information 10, 17 pp.

Hydrobiologia **278**: 223–230, 1994.
T.B. Reynoldson & K.A. Coates (eds), Aquatic Oligochaete Biology V.

A field test of a sediment bioassay with the oligochaete worm *Tubifex tubifex* (Müller, 1774)

Trefor B. Reynoldson
National Water Research Institute, Environment Canada, CCIW, 867 Lakeshore Road, Burlington, Ont. L7R 4A6, Canada

Key words: Oligochaeta, sediments, toxicity, multivariate analysis, Great Lakes

Abstract

A sediment toxicity test using reproduction in the tubificid oligochaete *Tubifex tubifex* was tested in 11 geographic locations in the Laurentian Great Lakes where sediments have previously been documented as being contaminated. Samples were taken from both potentially contaminated and "clean" sites in each of the 11 areas. Using test endpoints such as survival and growth, results from 72 samples were classified using cluster and ordination techniques. Six response groups were identified and four groups of toxic sites determined. Toxic effects ranged from mortality of adult worms to reduced reproduction. Using correlation and discriminant analysis there was good evidence that high concentrations of metals such as copper and lead were responsible for the observed effects at some sites.

Introduction

Contaminated sediments have been identified as an important environmental management concern at many locations with a history of industrial development and have been identified as a major concern in 39 of 42 of the most contaminated areas in the Laurentian Great Lakes (Reynoldson *et al.*, 1988). Designation of a sediment as contaminated is typically based on chemical concentration. However, there has been a developing trend to use biological techniques including sediment bioassays as a key component of sediment assessments (Long & Chapman, 1985; IJC, 1987, 1988; Zarull & Reynoldson, 1992). In a previous paper the development of a laboratory sediment bioassay using the tubificid oligochaete *Tubifex tubifex* (Müller, 1774) was described (Reynoldson *et al.*, 1991). The test appeared to meet many of the criteria proposed by Giesy & Hoke (1989) for an ideal sediment bioassay, *i.e.*, the bioassay is quick to set up and dismantle, there is little maintenance required during the test, and it requires simple equipment and technology. The test is also ecologically relevant as it measures reproduction and was shown to be repeatable and sensitive.

The main objective of this work was to establish whether the *T. tubifex* reproductive assay could: (1) identify differences in field sediments, (2) discriminate contaminated from non-contaminated sites and (3) to determine which physical and chemical characteristics of field sediments are the most influential in affecting test variables and specifically whether the test could resolve normal variation in sediment characteristics from sediment contamination.

Methods and materials

Site selection

The International Joint Commission (IJC) has designated twelve Canadian locations as being Great Lakes Areas of Concern (IJC, 1987). Sampling sites were located in 11 of these areas. In each area a number of sites were selected based on proximity to likely sources of contamination. An attempt was made to locate sites at varying distances from suspected areas of contamination as well as to locate one site in a potentially "clean" location. Thus, the range in potential toxicity of the sampled sediments was maximized to test the response range of the assay. The actual numbers of sites sampled in each location is shown in Fig. 1.

Field methods

At each site, sediment for testing was taken with a Ponar sampler. The sediments were transferred to a sealed container and held under ice for transport to the laboratory. Samples were then stored at 4 °C until bioassays were performed. In addition, at each station a sediment core was sampled using a modified Benthos Corer (Benthos Inc.). The top 5 cm was homogenized and sub-sampled for analyses of major ions, metals, organics and particle size.

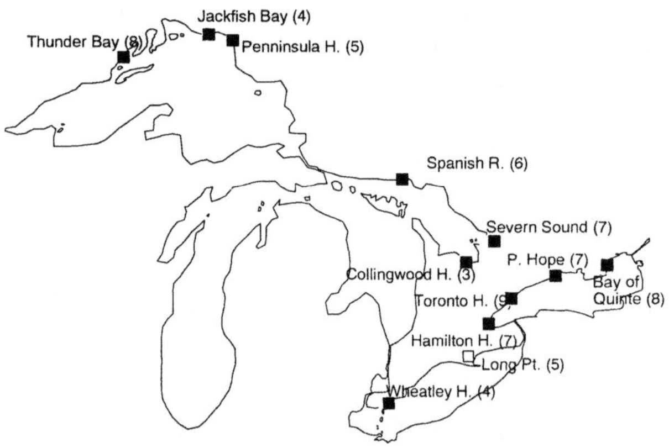

Fig. 1. Location of sampling areas in the Great Lakes (solid squares) and reference site (open square). The number of stations sampled in each area is indicated in parentheses.

Laboratory methods

Tubifex tubifex bioassays were conducted according to the methods described by Reynoldson *et al.* (1991). Sediment was sieved through 250 μm mesh before testing to remove the indigenous fauna. Five laboratory replicates were established for each sediment sample. Four sexually mature specimens of *T. tubifex* were placed in each replicate beaker and maintained in the dark at 22.5 °C $\pm$ 0.5 for 28 days. At the end of the bioassay the numbers of surviving adults, cocoons (empty and full) and small (pass through 500 μm mesh) and large (retained by 500 μm mesh) young were enumerated. The storage time for the sediments before testing ranged from 77 to 182 days. The longer storage period is more than the 168 days demonstrated to have no effect on results in our earlier paper (Reynoldson *et al.*, 1991).

Particle size was estimated on wet samples using the sedigraph method (Duncan & LaHaie, 1979) and data are expressed in terms of percentage sand, silt and clay. Major ion, metal and organic chemistry analyses were conducted on freeze dried samples. Concentrations of major elements and trace metals were determined by X-ray fluorescence spectrometry and the precision of the analysis ranged between 0.2 and 10% depending on the element or trace metal (Mudroch, 1985).

Statistical methods

The test produces several endpoints: number of surviving adults, number of empty (hatched) cocoons, number of full (unhatched) cocoons, number of small young and number of large young. These variables were used to classify the sites using cluster analysis and ordination. Clustering was done using an agglomerative polythetic method (UPGMA – unweighted pair group mean average) with the Bray Curtis association metric. The groups produced by cluster analysis were examined in ordination space to establish whether the classification groups were truly distinct. Ordination was done using semi-strong hybrid multidimensional scaling (SSH) which is superior to other techniques as it equally weights the input distances (Belbin, in press). Vector scores from SSH were correlated with environmental variables and the relationships between site groups and sediment characteristics were determined by multiple discriminant analysis (MDA). All analyses were done using the SYSTAT (Wilkinson, 1990), SAS and PATN packages.

Results

Sixty-seven sites were sampled from 11 areas of concern as well as five samples taken from a reference location (Long Point Marsh, L. Erie) which is used as the culture sediment.

Classification of sites

From the cluster analysis six groups of sites were identified (Fig. 2). The dendrogram shows that below this level small groups of sites begin to separate. The reference sites (LP1A–LP4A) are all located in Group 2, which is the largest group of 37 sites. Seven of the ten "clean" sites are also in Gp. 2, suggesting they were not contaminated, the "clean" site from the Bay of Quinte (BQ1) was classified as part of Gp. 1, and two other "clean" sites were classified in Gp. 4. The first separation of sites distinguishes the 10 sites comprising Groups 4, 5 and 6 from 62 sites forming Groups 1, 2 and 3. The next separation is the one site in Hamilton Harbour (HH42), forming Gp. 6, that is characterized by being the most toxic site tested, where adult survivorship was less than 25% (Fig. 3) and there was no reproduction. On this branch of the classification tree two other groups of sites can be distinguished. Three sites forming Gp. 5 are all from Port Hope Harbour (PH4, 5 and 7), these are also characterized by mortality of adult worms and as a group has the second lowest level of reproduction. Total cocoon production is only 25% of the reference sediment (Long Point) and the total number of young produced is only 10% that observed in the reference sediment. The third group of sites (Gp. 4) on this branch of the tree are from different locations; the Bay of Quinte, Port Hope and Toronto Harbour in Lake Ontario, the Spanish River area in Lake Huron and Thunder Bay and Jackfish Bay in Lake Superior. There is no apparent acute toxicity at these sites as adult survivorship was high (Fig. 3) however both cocoon and young production were low.

Three groups of sites are distinguished on the second branch of the tree (Fig. 2). First, a group (Gp. 3) of 11 sites from three areas, Hamilton Harbour (4 sites), Toronto Harbour (4 sites) and Thunder Bay (3 sites). This group shows no evidence of acute toxicity but the number of total young produced is lower than the sites forming Gp. 4. The last two groups are, Gp. 2 that is the largest single group of sites (37) which includes the reference sites (LP1A–LP4A) and most of the "clean" sites and Gp. 1, the second largest group (14 sites). There was little difference in number of surviving adults, total cocoons and total young in these two groups (Fig. 3).

The results of ordination show the six groups to be clearly separated with a strong single axis from the lower left of the ordination plot to the upper right (Fig. 4). The first separation of Groups 1, 2 and 3 from Groups 4, 5 and 6 occurs on the first vector that is best related to the number of small young (higher to the right) and empty cocoons (lower to left) (Table 1). The second vector, to which the number of full cocoons and large young contribute most (Table 1), separates all the groups except Gp. 4. In particular, the sites forming Groups 1, 2 and 3 are separated from each other on this axis and similarly Groups 4, 5 and 6 are separated. The reference sites were classified in Gp. 2 but have been marked separately in Fig. 2 and show very little variation in ordination space. Using only two vectors the stress level for the ordination is 0.0811 suggesting a satisfactory goodness of fit (Kruskal, 1962, 1964).

Relationship with sediment characteristics

The ordination scores are good descriptors of the structure of the data, as indicated by both the

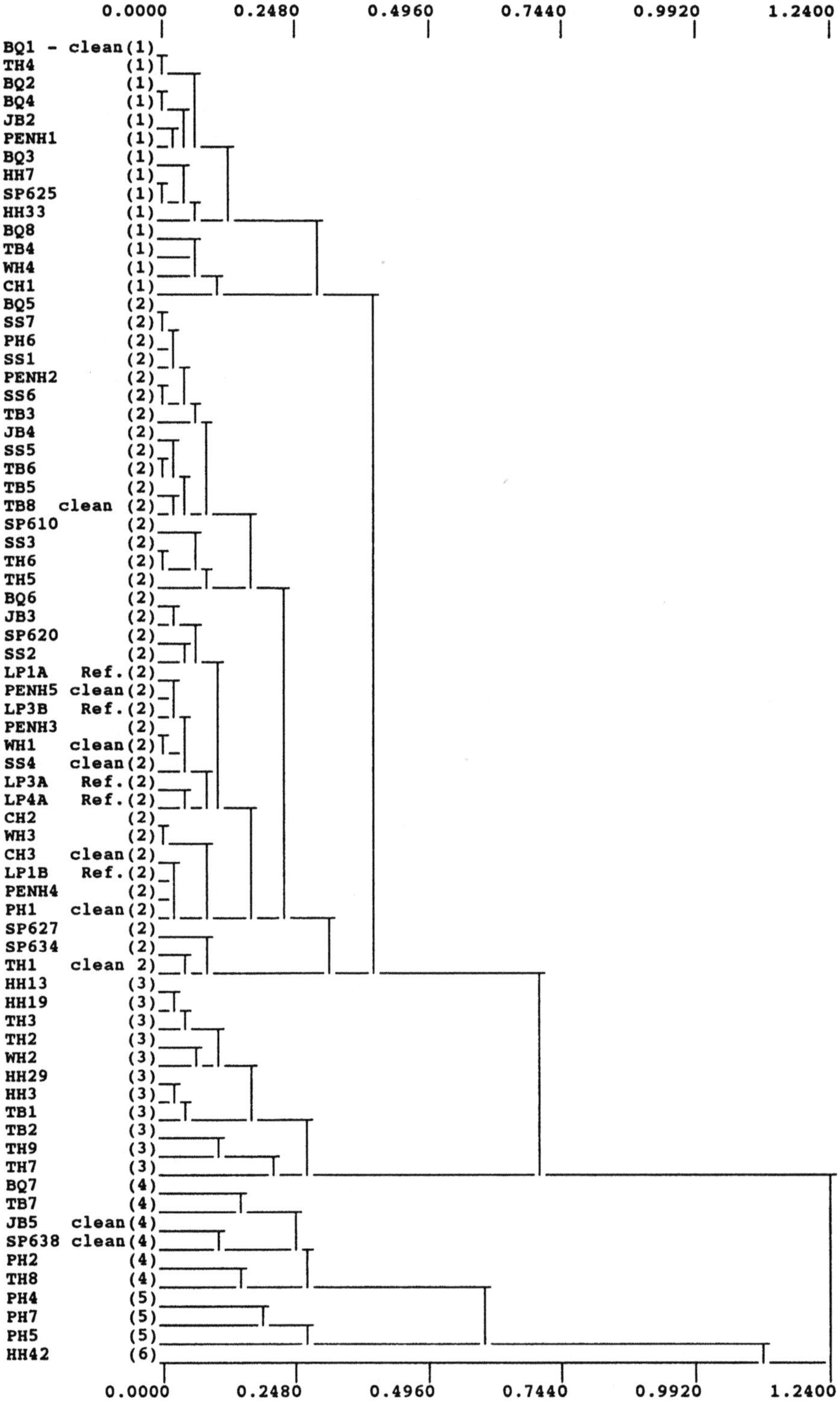

Fig. 2. Dendrogram of 72 sites using an unweighted pair group mean average clustering on five variables from *Tubifex* reproduction bioassay data.

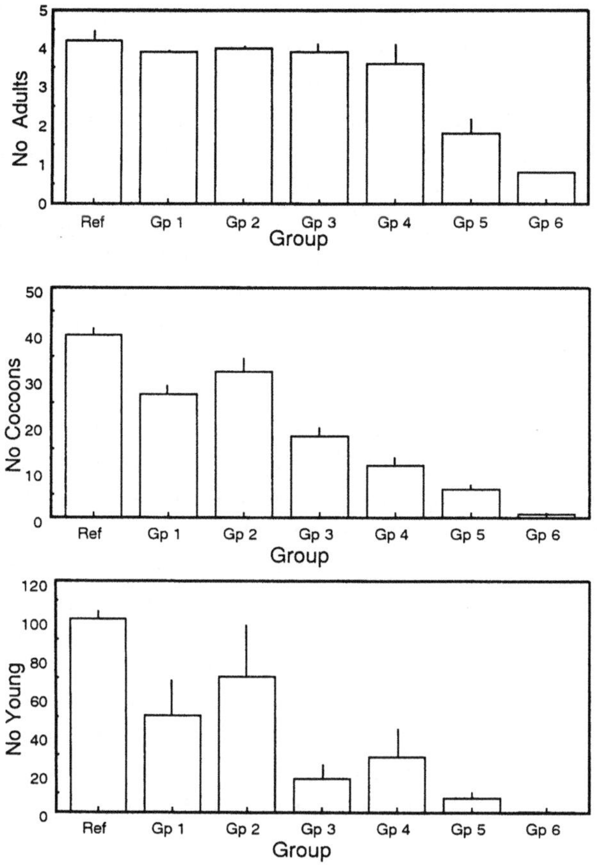

Fig. 3. Numbers of surviving adults, total cocoons and total young produced (mean ± S.D.) in each test response group from cluster analysis and at reference sites.

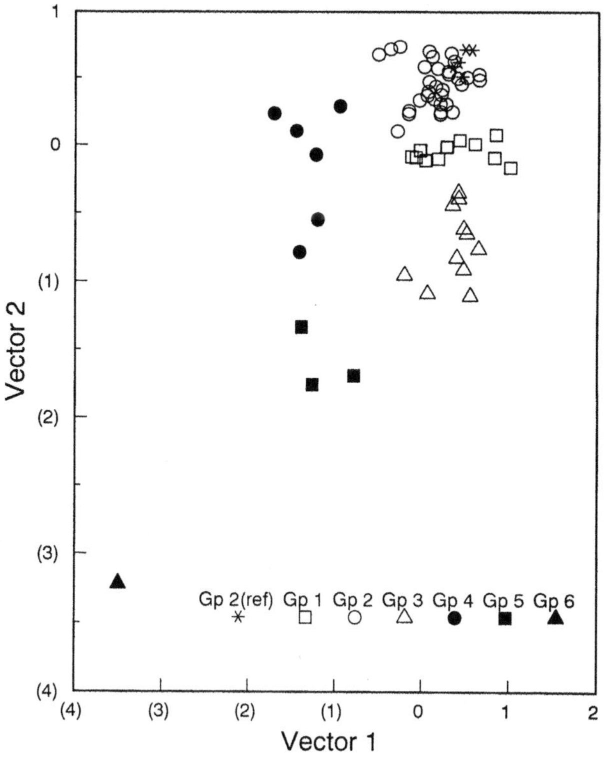

Fig. 4. Multidimensional scaling ordination plot using two vectors, from five variables produced from *Tubifex* reproduction bioassay; groups from cluster analysis and reference sites are represented by different symbols.

goodness of fit and the discrete nature of the site groupings in ordination space. Correlation analysis was used between the ordination vectors and 22 measured sediment variables to determine which sediment characteristics were best related with the structure of the data matrix from the bioassay variables. Seven sediment variables were correlated ($P < 0.001$) with the first vector (Table 2), which distinguished Gp. 6 from Groups 4 and 5 and Groups 1, 2 and 3 (Fig. 4). The only variable that was as strongly correlated with vector 2 scores was chromium concentration, three other variables were significant at a lower level ($P < 0.005$), concentration of manganese and zinc and percentage of sand. The concentrations of these variables were compared between the reference sites, Groups 1 and 2 with values for Groups 3 to 6 to indicate which if any contaminants are responsible for the observed acute and chronic effects. Sites in Gp. 3 (Table 3) show elevated concentrations of copper, chromium and zinc, as well as a slight increase in cadmium concentration and the percentage of sand is considerably lower. Sites in Gp. 4 show only an elevation in manganese concentration and

Table 1. Relationship between classification variables and ordination vectors.

Variable	Vector 1	Vector 2	Correlation
Adults	0.645	0.764	0.755
E. Cocoons	−0.789	−0.614	0.472
F. Cocoons	−0.146	0.989	0.898
S. Young	1.000	−0.007	0.886
L. Young	−0.026	−0.999	0.219

Table 2. Correlation coefficients between environmental variables and ordination vectors.

Variable	Vector 1	Variable	Vector 2
Cu	0.665	Cr	−0.470
Cr	0.577	Mn	0.437
Cd	0.568	Sand	−0.426
Co	0.528	Zn	−0.424
S	0.523	Co	−0.384
Fe	0.519		
OC's	0.465		

site 5 shows no increase in any of the correlated variables. The one site forming Gp. 6 has high concentrations of almost all the measured metals.

An alternate method of relating environmental characteristics with biological classification is the use of multiple discriminant analysis (MDA). Stepwise discriminant analysis was used initially, as an alternate to correlation analysis to identify which environmental variables best explained the variance within the biological structure of the data matrix. Five variables were identified by this analysis in rank order they were, concentration of lead, arsenic, cobalt, copper and phosphate. The values for these variables are also shown in Table 3, and show lead concentration to be higher in Groups 3, 5 and 6 than in the non-toxic sites (ref., Groups 1 and 2), arsenic to be higher in Gp. 5 and phosphate in Gp. 6.

Discriminant analysis was also used to determine the probability of sites being correctly classified to the biological site groups using the variables identified from correlation and stepwise analyses (Table 4). Only 52 sites had a complete data matrix for the variables identified by correlation analysis. These variables correctly predicted 58% of the sites. The poorest predictions were in Groups 1 and 2 and the variables were able to predict 22 of 34 sites correctly to one of these two groups. The more contaminated sites, Groups 4, 5 and 6, were well predicted, 5 of 7 being correct. The sites were almost as well predicted using only the five variables identified in the stepwise analysis. Overall 56% of the sites were correctly assigned. Again the poorest predictions were of sites in Gp. 1 where only 4 of 14 were correctly assigned, a further 3 were assigned as Gp. 2 and 5 Gp. 2 sites were assigned to Gp. 1. In total 65% of Gp. 1 and 2 sites were assigned to one of the two groups. Of the remaining sites over 70% were assigned correctly.

Discussion

The classification of sites by cluster analysis (Fig. 2) and confirmed by ordination (Fig. 4) ap-

Table 3. Values for selected environmental variables in site groups (expressed in $\mu g\ g^{-1}$ dry weight of sediment unless otherwise indicated).

Variable	Ref	Gp 1	Gp 2	Gp 3	Gp 4	Gp 5	Gp 6
Cu	12.1	60.9	44.6	85.5	46.8	50.6	154.0
Cr	41.5	97.6	84.6	167.5	89.4	48.6	483.5
Cd	0.4	1.5	1.1	2.1	1.0	1.2	2.0
Co	4.0	22.8	18.0	26.2	19.1	6.7	119.8
S ($mg\ g^{-1}$)	9100	4540	2029	3377	1834	2560	13267
Fe_2O_3 (%)	2.8	5.3	4.9	6.9	4.9	2.65	19.8
OC	–	0.020	0.008	0.014	0.007	0.017	0.008
MnO (%)	0.12	0.15	0.19	0.17	0.30	0.07	0.32
Sand (%)	14.5	21.6	36.8	3.8	37.2	25.5	0.0
Zn ($mg\ g^{-1}$)	96.6	393.8	169.1	1185.3	198.0	164.2	4385.4
Pb	25.8	88.2	53.2	197.1	80.5	193.8	704.1
As	1.9	10.3	9.3	10.5	8.4	49.2	14.5
P_2O_5 (%)	0.2	0.3	0.2	0.4	0.2	0.3	0.6

Table 4. Accuracy of biological site group predictions from two sets of environmental variables.

Biological grouping	Sites correctly predicted using variables from Correlation analysis (52 sites – 10 variables)	Sites correctly predicted using variables from stepwise MDA (72 sites)
Gp 1	4 of 11 (36%)	4 of 14 (29%)
Gp 2	14 of 24 (58%)	21 of 37 (57%)
Gp 3	7 of 10 (70%)	8 of 11 (73%)
Gp 4	1 of 3 (33%)	3 of 6 (50%)
Gp 5	3 of 3 (100%)	3 of 3 (100%)
Gp 6	1 of 1 (100%)	1 of 1 (100%)
Total	30 of 52 (58%)	40 of 72 (56%)

pears to be a good descriptor of the severity of sediment contaminant effects. Sites in Groups 6 then 5, respectively, are most toxic, both Groups 5 and 6 demonstrate acute toxicity with reduced survivorship of adults and severely impaired reproduction. Sites in Groups 3 and 4 show no indication of mortality but all demonstrate reduced reproduction in both cocoon and young production. From these data the *T. tubifex* reproduction bioassay is able to identify differences in field collected sediments and to discriminate both clean from contaminated sites and able to define severity of impact ranging from acute to chronic effects. It appears to be sensitive enough to identify contaminated sites but not so sensitive that spurious results are provided.

There appears to be a strong relationship between the test response groups and sediment chemistry in three of the four groups showing toxic effects. Metal concentrations appear to best explain the observed effects. Concentrations of copper, chromium, cadmium and cobalt are best correlated with the ordination vectors (Table 2) and lead and arsenic are first selected by stepwise discriminant analysis. Lead was markedly higher in three of the four groups of toxic sites, copper, chromium, cadmium and zinc were elevated in two of the four toxic groups. While it is not possible to determine whether one or all of the elements, either in combination or singly, were responsible for the observed effects this is strong circumstantial evidence for metal toxicity. It indicates that the observed toxicity is strongly correlated with those metals or that metals are good surrogates for general anthropogenic inputs.

However, the fact that only 56–58% of the sites were correctly assigned to the biological groups indicates other factors may be important in explaining the observed responses, particularly in the non-toxic sites that were less well-predicted. It is also important to note that correlation does not imply causality. The sites that were toxic had elevated concentrations of a range of substances any of which could be the agent of the observed effects, or a substance that was not analysed could be responsible. It is interesting that neither organochlorines or PCB congeners showed any relationship with the observed effects, which suggests this assay may be more sensitive to metal contaminants.

At one group of sites, Gp. 4, where the number of young, but particularly the numbers of cocoons were reduced none of the measured variables were elevated above the levels observed at the non-toxic sites (Groups 1 and 2). This suggests that a contaminant that was not measured may be causing the observed effect and demonstrates the importance of biological assessment as a chemical assessment would suggest these sites to be uncontaminated.

Acknowledgements

I would like to thank Larry Benner and other Technical Operations staff for their assistance in the collection of these samples. Also Michelle Burley and Sherri Thompson for their assistance in counting untold numbers of small *T. tubifex* in conducting the bioassays.

References

Belbin, L. In press. SSH: Semi-strong hybrid scaling. Classification Society.

Duncan, G. A. & G. G. LaHaie, 1979. Size analysis procedures used in the sedimentology laboratory. NWRI, CCIW, Hydraulics Div. Manual. September, 1979, *pp.

Giesy, J. P. & R. A. Hoke, 1989. Freshwater sediment toxicity bioassessment: rationale for test species selection and test design. J. Great Lakes Res. 15: 539–569.

International Joint Commission, 1987. Guidance on characterization of toxic substances problems in Areas of Concern in the Great Lakes basin. Report to the Great Lakes Water Quality Board. Windsor, Ontario, 179 pp.

International Joint Commission, 1988. Procedures for the assessment of contaminated sediment problems in the Great Lakes. Report to the Great Lakes Water Quality Board. Windsor, Ontario, 140 pp.

Long, E. R. & P. M. Chapman, 1985. A sediment quality triad: measures of sediment contamination, toxicity and infaunal community composition in Puget Sound. Mar. Poll. Bull. 16: 405–415.

Mudroch, A., 1985. Geochemistry of the Detroit River sediments. J. Great Lakes Res. 11: 193–200.

Reynoldson, T. B., A. Mudroch & C. J. Edwards, 1988. An overview of contaminated sediments in the Great Lakes with special reference to the international workshop held at Aberystwyth, Wales, U.K. Report to the Great Lakes Science Advisory Board. Windsor, Ontario, 41 pp.

Reynoldson, T. B., S. P. Thompson & J. L. Bamsey, 1991. A sediment bioassay using the tubificid oligochaete worm *Tubifex tubifex*. Env. Toxicol. Chem. 10: 1061–1072.

Wilkinson, L., 1990. SYSTAT: The system for statistics. Evanston, Illinois SYSTAT, Inc.

Zarull, M. A. & T. B. Reynoldson, 1992. A management strategy for contaminated sediment: Assessment and remediation. Water Poll. Res. J. Canada. 27: 871–882

Hydrobiologia **278**: 231–241, 1994.
T.B. Reynoldson & K.A. Coates (eds), Aquatic Oligochaete Biology V.

Lacustrine, profundal meiobenthic oligochaetes as indicators of trophy and organic loading

Jukka Särkkä
University of Jyväskylä, Department of Biology, POB 35, FIN-40351 Jyväskylä, Finland

Key words: Oligochaeta, Naididae, Aeolosomatidae, indicators, meiofauna, profundal, eutrophication

Abstract

Meiobenthic and macrobenthic oligochaetes were collected from two depths in Lake Päijänne and examined together with environmental data in which two environmental gradients can be seen. Numbers of *Piguetiella blanci, Aeolosoma quaternarium* and *Aeolosoma hemprichi* were positively correlated with oxygen saturation and negatively correlated with total phosphorus and COD suggesting that these species are indicators of oligotrophic conditions. Numbers of *Amphichaeta leydigii, Specaria josinae* and *Vejdovskyella comata* were positively correlated with total phosphorus which implies that these species prefer eutrophic conditions. The last three species and *Dero digitata* showed positive correlations with COD and organic sedimentation, which suggests that they also indicate organic loading. A decrease in aeolosomatid biomass was seen with increasing phosphorus and organic loading. The average body size of detritus-feeding naidids in the upper part of the profundal zone was reduced by organic loading, and more clearly eutrophication, which emphasizes the importance of including meiobenthic oligochaetes in pollution investigations. Of the naidid and aeolosomatid species, *Chaetogaster langi* preferred the greatest depths of the profundal zone while most of the other species were more abundant at 20 m rather than at the maximum depth. Clustering of the oligochaete taxa supported these observations and clustering of sampling stations based on oligochaete variables generally corresponded to clustering based on environmental data.

Introduction

The profundal zone of lakes is an environment where one of the most important environmental parameters, oxygen content, reaches its minimum. Sediment is transported there from shallower areas and there is ongoing deposition of dead and decaying phytoplankton and precipitating organic matter. Temperature fluctuates less than in shallower zones, for example, the annual fluctuation in the deepest parts of Lake Päijänne is 2 to 8 °C.

Macrobenthic chironomids were used as indicators of lake type (e.g. Brinkhurst, 1974) and macrobenthic oligochaetes (Milbrink, 1980; Wiederholm, 1980) have similarly been used.

Although marine meiofauna has become important in pollution investigations freshwater meiofauna is less well known. However, profundal species, especially naidids and aeolosomatids, have become increasingly well known (Särkkä, 1989).

Use of the lacustrine meiofauna in pollution investigations has a number of advantages. First, its potential response to pollution events is much faster than that of the macrofauna because generation times are usually months rather than years. Second, the small size and high densities of meiofauna enable the use of a small sample volume, so samples can be easily transported.

This paper extends the results of studies by Särkkä (1989, 1992), and examines the relationship between pollution variables, bathymetric differences and the occurrence of species.

Study area, material and methods

Lake Päijänne (Fig. 1) is the deepest and second largest lake in Finland (max depth 94 m, av. depth 17 m, area 1100 km^2, length 120 km). The lake is characterized by two environmental gradients originating from loadings by pulp and/or paper mills. The northern part of the lake receives treated urban effluent and waste water from a pulp mill located 40 km upstream. The central area of the lake also receives discharge from a pulp and paper mill. Station 1 may be considered somewhat semi-lotic, owing to the flow of water (about 150 m^3 s^{-1}) and narrowness of the lake.

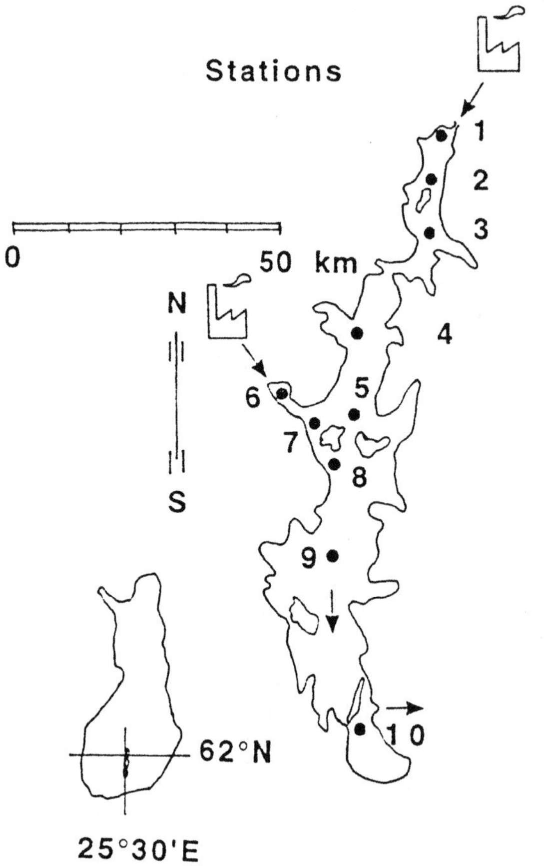

Fig. 1. Map of Lake Päijänne and the sampling stations (1–10)

Two depths (20 m and maximum) were sampled at each of ten stations. The top 5 cm of sediment was taken using a 15.9 cm^2 corer, five replicates were taken at each station. Samples were preserved immediately with 4% formaldehyde, washed through a 0.08 mm sieve and stained with rose bengal before picking. The environmental data are according to Särkkä (1989) and Maatela *et al.* (1990).

The untransformed data were analyzed by means of StatView 512 + (correlations) and SYSTAT (clustering) statistical packages. Cluster analysis was done using an average linkage with Pearson correlation as the distance metric. Average individual body size was counted by dividing the total biomass by the number of individuals.

Results

Selected environmental variables and quantitative data for oligochaetes are presented in Table 1. The effect of organic waste discharge can be seen in the northern part of the lake, in which there is, from station 1 to stations 4 and 5, a decreasing amount of total phosphorus and organic matter and an increasing amount of oxygen along the gradient. A second gradient is seen from stations 6 and 7 southwards to station 10; the oligotrophic character of the southern part of the lake can be seen in several variables. Spatial differences in oxygen content are less important at 20 m depth than at the maximum depths.

A total of 1,546 individuals of oligochaetes belonging to 20 taxa were identified (9 lumbriculids, 444 naidids, 392 tubificids, 700 aeolosomatids and 1 enchytraeid).

Table 2 presents average numbers of individuals for 20 taxa, of which most of the naidids and all the aeolosomatids are of meiobenthic size (for average sizes of the various species, see Särkkä,

Table 1. Environmental variables and oligochaete community characteristics at stations 1–10 in Lake Päijänne. Oxygen is expressed as % of saturation at maximum depth (1 m above the bottom) and at 20 m in March. COD = chemical oxygen demand in the epilimnion, measured as $KMnO_4$ consumption. Tot. P = total phosphorus (mg m^{-3}) in the epilimnion; Phyt. = average phytoplankton biomass (g m^{-3}). Org. sed. = recent sedimentation of organic matter (g m^{-2} a^{-1}). Org. Cl = organically bound chlorine (μg g^{-1} dw) in the sediment layer where it reaches its maximum (at maximum depth). Oligochaete data: naidid biomass excludes *Chaetogaster langi*, all biomass data expressed as dry weight per core sample of 15.9 cm^{-2}. Diversity is Shannon's index. Olig. spec. number = number of species of oligochaetes; Olig./Chir. ind. ratio = ratio of numbers of individuals of Oligochaeta: Chironomidae.

Maximum depth

Stat.	Oxyg. at max. depth	COD	Tot. P	Phyt.	Org. sedim.	Org. Cl	Olig. tot. abund.	Olig. tot. biom.	Naidid biom.	Aeol. biom.	Tubif. biom.	Olig. divers.	Olig. spec. number	Olig./Chir. ind. ratio
1	19	10.0	20	1.26	83	680	31.2	709	15.0	0	694	0.52	7	–
2	10	9.7	21	1.33	157	490	8.2	273	1.4	0	272	0.23	3	41.00
3	42	9.2	17	0.95	124	510	5.0	63.3	2.0	0	61	1.17	5	6.25
4	54	9.1	15	0.99	36	350	2.2	61.7	0.4	0	61	1.03	4	5.50
5	57	8.4	14	0.68	37	520	6.4	129	0.9	4.97	122	1.12	6	6.40
6	0	14.0	40	1.27	158	790	2.6	38.2	0.1	0	38	0.90	3	6.50
7	9	10.0	24	1.40	80	1330	10.0	0	2.5	0	236	0.85	6	–
8	73	8.0	14	0.44	44	390	17.2	71.8	6.0	0.39	61	1.04	9	7.17
9	71	7.7	10	0.44	33	130	6.0	23.2	4.2	3.88	14	1.38	8	3.33
10	86	7.4	8	0.38	39	70	7.4	35.5	6.2	0.45	27	1.48	10	37.00

Depth of 20 m

Stat.	Oxyg. at 20 m	Olig. tot. abund.	Olig. tot. biom.	Naidid biom..	Aeol. biom.	Tubif. biom.	Olig. divers.	Olig. spec. number	Olig./Chir. ind. ratio
1	82	22.2	193	36.9	4.8	151	1.10	7	55.5
2	78	12.8	140	21.3	0	119	1.32	8	64.0
3	82	1.0	4.2	2.6	1.5	0	1.14	5	0.71
4	92	1.0	7.4	5.4	2.0	0	1.22	6	1.00
5	85	6.2	60.4	13.9	8.8	37	0.74	9	5.17
6	0	17.2	411	7.6	0	403	1.04	8	86.00
7	80	2.0	18.8	4.6	0.1	14	1.89	7	2.00
8	85	5.0	16.0	3.2	2.4	9	1.36	9	1.79
9	85	4.0	11.3	0	1.0	9	1.24	6	3.33
10	91	3.6	17.7	4.2	3.6	9	1.10	8	3.60

1989). Some taxa are known as macrobenthic species (lumbriculids and tubificids).

Discussion

Indicator value of quantitative data

Certain requirements should be met if a selected taxonomic group, such as the aquatic oligochaetes, is to be used as an indicator (Cranston, 1990):

1. the group should be naturally exposed to the environmental parameters to be monitored,
2. the group must be functionally important in the ecosystem,
3. the group must be ubiquitous and/or widely distributed within the area,

Table 2. Numbers of individuals of Oligochaeta per 10 cm^2 at maximum depths and at 20 m at 10 stations in Lake Päijänne.

Stations	1	2	3	4	5	6	7	8	9	10
Maximum depths										
Lumbriculidae										
Lamprodrilus isoporus Michaelsen	–	–	–	–	–	–	–	–	–	0.13
Stylodrilus heringianus Claparede	–	–	–	–	–	–	–	–	–	–
Naididae										
Chaetogaster langi Bretscher	0.25	–	0.13	0.13	1.64	0.63	3.62	7.42	2.89	2.39
C. setosus Svetlov	–	–	–	–	–	0.13	0.16	–	–	–
Amphichaeta leydigii Tauber	5.91	–	–	–	–	–	–	0.38	–	–
Specaria josinae (Vejdovsky)	0.25	0.25	–	–	–	–	0.16	–	–	0.25
Uncinais uncinata (Oersted)	–	0.13	–	–	–	–	–	0.13	0.25	0.25
Slavina appendiculata d'Udekem	0.13	–	–	–	–	–	–	–	–	–
Vejdovskiella comata (Vejdovsky)	0.38	–	–	0.13	–	–	–	–	–	–
Piguetilla blanci (Piguet)	–	–	–	0.13	–	–	–	0.38	0.13	0.13
Dero digitata (Müller)	–	–	1.64	–	0.13	–	0.16	–	–	–
Tubificidae										
Tubifex tubifex (Müller)	12.83	4.91	1.01	0.88	2.14	0.88	5.35	1.01	–	0.13
Limnodrilus hoffmeisteri Claparede	–	–	–	–	–	–	–	–	0.13	–
Psammoryctides barbatus (Grube)	–	–	–	–	–	–	–	–	–	0.13
Spirosperma ferox Eisen	–	–	–	0.25	–	–	0.16	–	–	0.25
Aulodrilus pluriseta Piguet	–	0.13	0.13	–	0.13	–	–	0.13	0.13	–
Aeolosomatidae										
Aeolosoma hemprichi Ehrenberg	–	–	–	–	0.25	–	–	0.88	4.91	0.88
A. quaternarium Ehrenberg	–	–	–	–	6.16	–	–	0.13	3.02	0.25
Rheomorpha neiswestnovae (Last.)	–	–	–	–	–	–	–	0.13	0.25	–
Depth of 20 m										
Lumbriculidae										
Lamprodrilus isoporus Michaelsen	–	–	–	–	–	–	–	–	–	0.38
Stylodrilus heringianus Claparede	–	–	0.50	–	–	–	–	–	0.13	–
Naididae										
Chaetogaster langi Bretscher	0.13	0.13	0.13	0.13	1.76	0.25	0.16	2.52	2.26	1.38
C. setosus Svetlov	0.13	–	–	–	0.13	0.13	–	0.13	–	–
Amphichaeta leydigii Tauber	0.13	0.13	0.13	0.13	–	1.01	–	–	–	–
Specaria josinae (Vejdovsky)	1.89	1.51	–	–	0.25	0.38	0.16	–	–	–
Uncinais uncinata (Oersted)	–	–	–	0.13	–	–	–	0.13	–	0.25
Slavina appendiculata d'Udekem	–	–	–	–	–	–	–	–	–	–
Vejdovskiella comata (Vejdovsky)	6.54	1.26	–	–	–	0.50	0.31	0.13	–	–
Piguetilla blanci (Piguet)	–	0.13	0.25	0.38	1.13	–	–	0.13	–	0.13
Dero digitata (Müller)	–	0.50	–	–	–	0.13	0.31	–	–	–
Tubificidae										
Tubifex tubifex (Müller)	4.03	4.15	–	–	0.50	8.55	0.31	0.13	–	–
Limnodrilus hoffmeisteri Claparede	1.26	–	–	–	–	0.50	–	–	–	–
Psammoryctides barbatus (Grube)	–	–	–	–	–	–	–	–	–	–
Spirosperma ferox Eisen	–	–	–	–	–	–	–	–	0.13	0.13
Aulodrilus pluriseta Piguet	–	–	–	–	–	–	–	–	–	–
Enchytraeidae	–	–	–	–	0.13	–	–	–	–	–
Aeolosomatidae										
Aeolosoma hemprichi Ehrenberg	16.73	–	0.38	2.14	26.92	–	–	5.03	2.14	9.18
A. quaternarium Ehrenberg	–	–	1.76	1.64	1.26	–	0.16	1.01	0.38	1.01
Rheomorpha neiswestnovae (Last.)	–	–	0.25	0.13	0.25	–	0.31	0.50	0.25	0.63

Table 3. Spearman correlations between Oligochaeta community characteristics and environmental variables. For significant Spearman correlations ($p < 0.05$), the sign of the correlation and p value are given. % = percentages of total variance explained by a Pearson correlation > 35%. f = frequency (number of discoveries out of 10).

	Maximum depth					Depth of 20 m				
	f	Oxyg.	COD	Tot. P	Org. sed.	*f*	Oxyg.	COD	Tot. P	Org. sed.
Abundances of species:										
Lumbriculidae										
Stylodrilus heringianus	–	–	–	–	–	2	–	–	–	–
Naididae										
Chaetogaster langi	9	–	–	–	–	10	–	–	negat.	–
									42%	
C. setosus	2	– 0.04	–	– 0.04	–	4	–	–	–	–
				55%						
Amphichaeta leydigii	2	–	–	–	–	5	–	+ 0.04	+ 0.05	+ 0.05
								84%	42%	76%
Specaria josinae	4	–	–	–	–	5	– 0.04	+ 0.03	–	+ 0.05
Uncinais uncinata	4	–	– 0.03	–	–	3	+ 0.03	–	–	–
Vejdovskiella comata	2	–	–	–	–	5	+ 0.03	+ 0.03	+ 0.04	+ 0.03
Piguetilla blanci	4	+ 0.02	– 0.02	– 0.03	–	6	–	–	–	–
Dero digitata	3	–	–	–	–	3	– 0.02	–	–	+ 0.02
Tubificidae										
Tubifex tubifex	9	–	–	–	–	6	– 0.02	+ 0.03	+ 0.04	+ 0.03
							75%	78%	56%	74%
Limnodrilus hoffmeisteri	1	–	–	–	–	2	–	–	–	–
Spirosperma ferox	3	–	–	–	–	2	–	– 0.04	–	– 0.04
Aulodrilus pluriseta	5	–	–	–	–	0	–	–	–	–
Aeolosomatidae										
Aeolosoma hemprichi	4	+ 0.01	– 0.01	– 0.01	– 0.05	7	+ 0.05	–	–	– 0.04
A. quaternarium	4	+ 0.02	– 0.02	– 0.02	– 0.04	7	+ 0.04	–	–	–
Rheomorpha neiswestnovae	2	–	–	–	–	7	–	– 0.04	–	negat.
								45%		38%
Other Oligochaeta variables										
Oligochaeta, total abundance	10	–	–	–	–	10	–	–	–	–
Oligochaeta, total biomass	10	–	–	–	–	10	negat.	–	–	–
							81%	79%	73%	45%
Biomass of Naididae, without Ch.	10	–	–	–	–	9	–	–	–	–
Biomass of Tubificidae	10	–	–	–	–	8	– 0.04	–	–	–
							87%	82%	76%	45%
Biomass of Aeolosomatidae	4	+ 0.02	– 0.02	– 0.02	– 0.04	8	+ 0.05	–	–	–
Diversity of Oligochaeta	10	+ 0.02	– 0.02	– 0.02	–	10	–	–	–	–
		60%			39%					
Number of species of Oligoch.	10	+ 0.02	– 0.04	– 0.03	–	10	–	–	–	–
		56%	49%	46%	51%					

4. the group must not dominate any locality by virtue of an excessive number of species, and
5. it must be identifiable at the species level, easily sampled, and responsive to environmental perturbation.

Organic loading and trophic state, and possibly also organochlorine contamination, are the most important environmental parameters to be considered in these data. Oligochaetes fulfil most of the requirements presented by Cranston (1990).

In marine macrobenthic communities increased organic loading will reduce species richness and increase the total number of individuals of a few opportunistic species. There also tends to be a general reduction in biomass, although there may be an increase in the biomass of the opportunists, and a decrease in body size of the average species or individuals (Weston, 1990). These data show species richness to decrease as industrial loading increases (Table 3). At maximum depths, both diversity (Shannon, l_n based) and the number of species of oligochaetes are negatively correlated with total phosphorus and COD, which measures the organic loading, and thus species richness decreases as industrial loading increases. This is not seen in the upper part of the profundal zone (20 m) where the conditions are not as critical as at the greatest depths. An increase in numbers of individuals with increasing loading can be seen only in the 20 m zone for *Amphichaeta leydigii, Specaria josinae, Vejdovskyella comata, Dero digitata* and *Tubifex tubifex*. Therefore, these can be considered opportunistic species. On the other hand, certain other species seem to be indicators of oligotrophy or undisturbed conditions and prefer a high oxygen content. These are *Piguetiella blanci* at maximum depths and two species of *Aeolosoma* in both depth categories. *Rheomorpha neiswestnovae* also has a tendency to prefer undisturbed environments rather than organic loading.

No general reduction in Oligochaeta biomass can be seen. Biomass is positively correlated with variables indicating anthropogenic loading in the 20 m zone, although this is due to the tubificid biomass which is largely dependent on the opportunistic species *T. tubifex*. A reduction in biomass brought about by the loading can be seen only in the biomass of Aeolosomatidae at the greatest depths.

The Oligochaeta:Chironomidae ratio presented in Table 1, which has been used to rank trophic or pollution state (Wiederholm, 1980), should be highest at stations closest to the discharge. This ratio demonstrated little difference at the maximum depths, probably because of low numbers, particularly the small numbers of chironomid larvae contained in the small samples. The regional distribution in the 20 m zone corresponds better with the expected result.

Individual size

There has been some contradiction whether increased stress from anthropogenic loading and increased organic matter will affect body size (Thiel, 1975; Warwick *et al.*, 1986, 1987; Weston, 1990; Rapport, 1990)

The average individual biomass at maximum depths were not significantly correlated with the

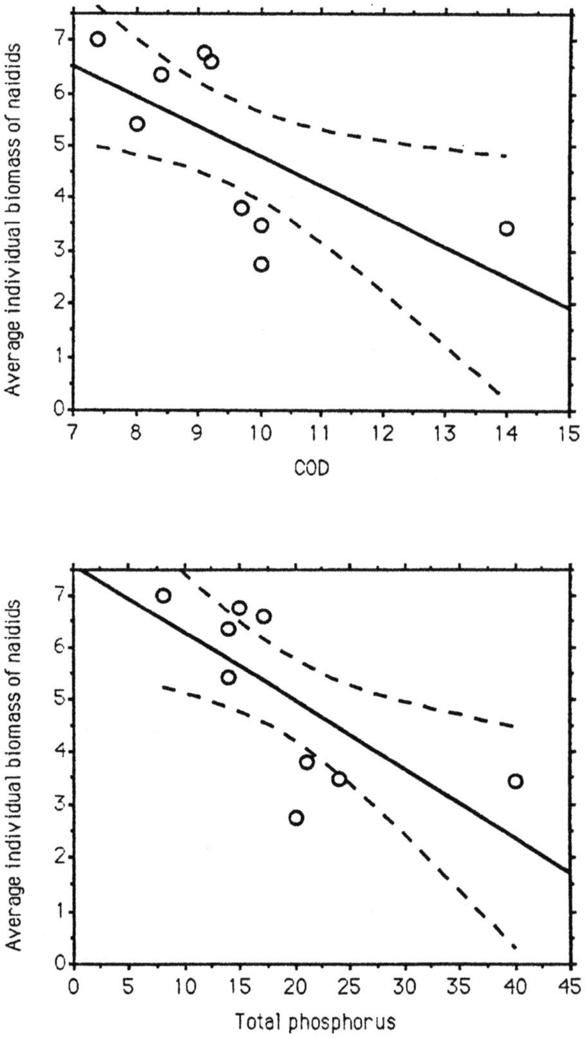

Fig. 2. Dependence between the average individual size of non-predatory naidids (μg dry w.), at the 20 m depth zone, and chemical oxygen demand (COD, measured as $KMnO_4$ consumption, mg O_2l^{-1}) and total phosphorus concentration (mg m^{-3}), with 90% confidence bands. Equation for COD: $y = -0.575x + 10.55$, $r = -0.649$, $r^2 = 0.421$, $p = 0.059$, for total P: $y = -0.13x + 7.563$, $r = -0.702$, $r^2 = 0.492$, $p = 0.031$.

Table 4. Influence of organic loading on marine macrobenthic (Weston 1990) and lacustrine meiobenthic oligochaete communities. * = naidids without *Chaetogaster langi.*

	Marine macrobenthos	Lacustrine meiobenthic oligochaetes	
		At max. depths	At 20 m depth
Decrease in species richness	yes	yes	no
Increase in numbers of opportunistic species	yes	no	yes for tubificids
Decrease in biomass	yes	yes for aeolosomatids	no
Increase in biomass of opportunistic species	yes	no	yes for tubificids
Decrease in body size	yes	no	yes for naidids*

organic matter content of the water (COD), organic sedimentation measured from the surface of the bottom at the deepest sites or the total phosphorus content of the water. An almost significant negative correlation ($p = 0.059$) was found between the average individual biomass of naidids (excluding the carnivore *C. langi*) and COD at a depth of 20 m (Fig. 2), and a significant ($p = 0.035$) negative correlation between this biomass and total phosphorus. Thus organic loading, and still more clearly eutrophication, reduce the body size of detrital and bacterial feeding naidids, just as in the marine macrobenthos (Weston, 1990). The correlations for all naidids were not significant at 20 m depth and those for all oligochaetes were even positive and significant, but this was caused by tubificids, which are abundant in polluted areas.

The effects of organic loading on marine macrobenthos (Weston, 1990) and lacustrine meiobenthic oligochaetes at maximum depths and at 20 m depth are summarized in Table 4. The decrease in body size in at least some of the oligochaetes in eutrophic and organically loaded environments emphasizes the importance of including meiobenthic oligochaetes in evaluations of the regional extent of pollution.

Correlations between Oligochaeta and environment

Table 5 presents the environmental preferences of the oligochaetes studied here based on their correlations with environmental variables. The preferences are similar in the upper and lower parts of the profundal zone. Meiobenthic species, especially *C. langi, Uncinais uncinata, P. blanci* and all three aeolosomatid species, seem to prefer oligotrophic and undisturbed conditions. *Specaria josinae* in both depth categories and *V. comata* and *D. digitata* at 20 m prefer disturbed conditions, as does the known macrobenthic pollution tolerant species *T. tubifex*. The total numbers of oligochaetes and the numbers of the various families, both expressed in terms of biomass and abundance, are on average higher in disturbed environments than in undisturbed ones, except for the numbers of Aeolosomatidae. The diversity and numbers of species are reduced by anthropogenic loading (Table 3). Communities in semilotic environments, as at station 1, more or less resemble those of oligotrophic environments, even though this station has a somewhat disturbed water quality.

Bathymetric differences

Some species of Naididae and Aeolosomatidae seem to prefer the maximum depths (Table 5) to the 20 m zone (Wilcoxon, matched-pairs signed-ranks). In the meiobenthic oligochaetes, this preference is shown by *C. langi, A. leydigii, U. uncinata, D. digitata* and *Aeolosoma quaternarium*, although the difference is significant only for *C. langi. Vejdovskyella comata, A. hemprichi* and *R.*

Table 5. Environmental and bathymetric preferences of species of Oligochaeta and other oligochaete variables. O = preferring oligotrophy and undisturbed conditions, E = preferring euthrophy and/or organic loading, M = mesotrophic or intermediary preference, SL = occurring especially in a semilotic environment. p values are for the Wilcoxon test between depth categories. NS = not significant. Biomass of Naididae excluding *C. langi.*

	Environmental preference		Bathymetrical preference	
	At max. depth	At 20 m depth	Preferred depth zone	Value of *p*
Chaetogaster langi	O	O	Max. d.	0.05
C. setosus	–	O, M	20 m	NS
Amphichaeta leydigii	–	E	Max. d.	NS
Specaria josinae	E	E	20 m	NS
Uncinais uncinata	O	O	Max. d.	NS
Vejdovskyella comata	–	E	20 m	0.05
Piguetilla blanci	O	O	20 m	NS
Dero digitata	?	E	Max. d.	NS
Tubifex tubifex	E	E	Max. d.	NS
Aulodrilus pluriseta	O	–	Max. d.	NS
Aeolosoma hemprichi	O	O, SL	20 m	0.06
A. quaternarium	O	O	Max. d.	NS
Rheomorpha neiswestnovae	O	O	20 m	0.03
Oligochaeta, total abundance	M	E	Max. d.	NS
Oligochaeta, total biomass	?	E	Max. d.	NS
Biomass of Naididae	O, SL	E	20 m	0.09
Biomass of Tubificidae	E	E	Max. d.	0.06
Biomass of Aeolosomatidae	O	O, SL	20 m	0.07
Diversity of Oligochaeta	O	?	20 m	NS
Number of species of Oligoch.	O, SL	?	20 m	NS

neiswestnovae are significantly more abundant at 20 m. Diversity and number of species are greater at 20 m, as expected, particularly in view of the less severe oxygen conditions, although it could also be expected that the more stable conditions of greater depths would increase diversity.

Multivariate analyses

Relationships between oligochaete numbers and biomass between stations were examined using cluster analysis. Figure 3A shows a relationship between *T. tubifex, S. josinae*, total phosphorus

Fig. 3. Dendrograms showing connections between ten species of Oligochaeta (abundances) and three environmental variables (oxygen saturation, total phosphorus and sedimentation of organic mater) (A and B), between other Oligochaeta variables and environmental variables (C and D), between sampling stations in terms of Oligochaeta abundances (E and F) and between sampling stations with regard to the above three environmental variables (G and H). The left side (A, C, E and G) applies to maximum depths, and the right side (B, D, F and H) to 20 m depth. Ol. tot bm = total biomass of oligochaetes, Ol. tot abund. = total number of individuals of oligochaetes, Olig. div. = diversity of oligochaetes, Ol. spec n = number of oligochaete species, Tub. bm, Naid. bm and Aeol. bm = biomass of Tubificidae, Naididae and Aeolosomatidae.

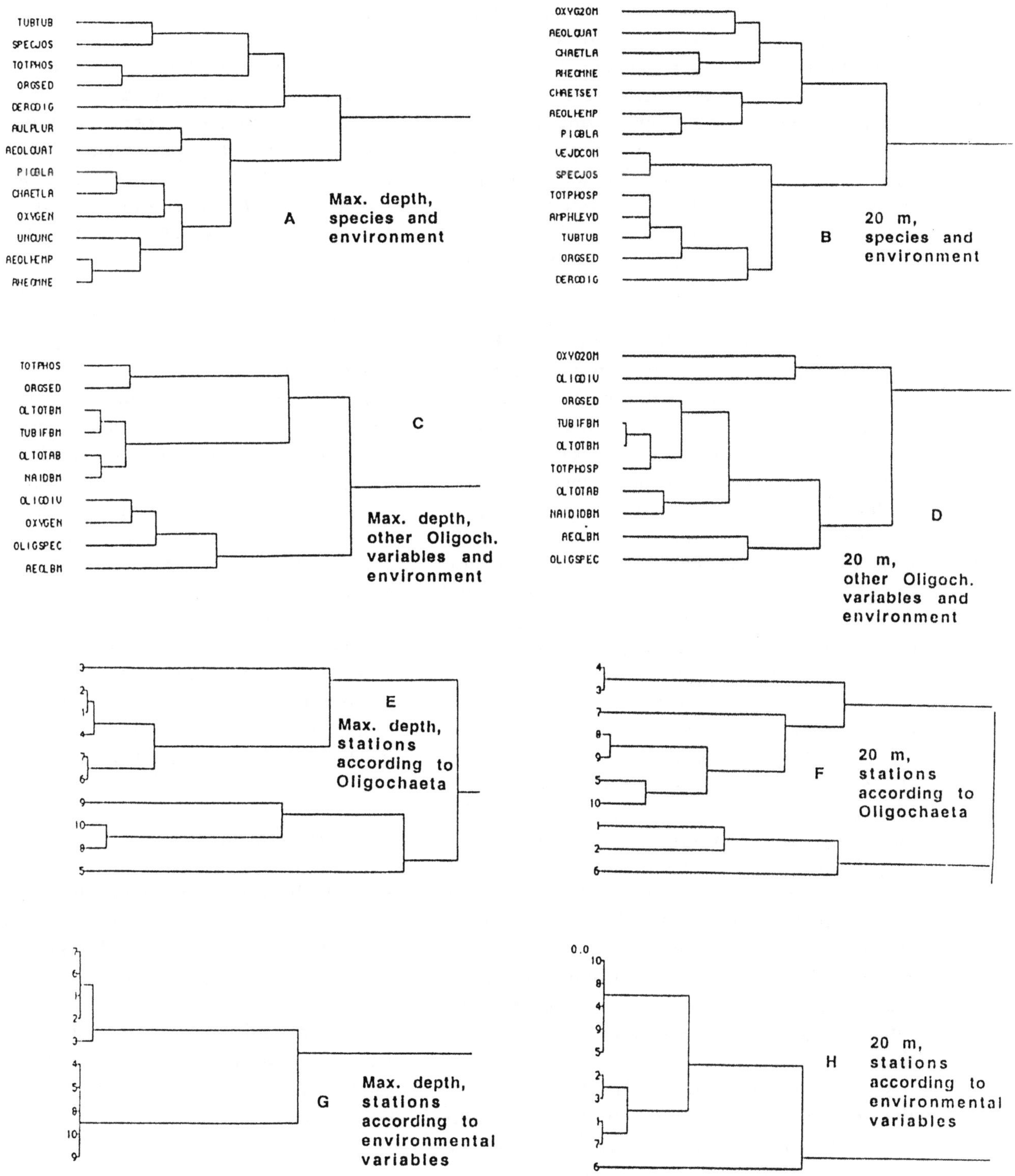
TUBTUB
SPECJOS
TOTPHOS
ORGSED
DERODIG
AULPLUR
AEOLQUAT
PIGBLA
CHAETLA
OXYGEN
UNCUNC
AEOLHEMP
RHECMNE
A
Max. depth, species and environment
OXYG20M
AEOLQUAT
CHAETLA
RHECMNE
CHAETSET
AEOLHEMP
PIGBLA
VEJDCOM
SPECJOS
TOTPHOSP
AMPHLEYD
TUBTUB
ORGSED
DERODIG
B
20 m, species and environment
TOTPHOS
ORGSED
OLTOTBM
TUBIFBM
OLTOTAB
NAIDBM
OLIGDIV
OXYGEN
OLIGSPEC
AEOLBM
C
Max. depth, other Oligoch. variables and environment
OXYG20M
OLIGDIV
ORGSED
TUBIFBM
OLTOTBM
TOTPHOSP
OLTOTAB
NAIDIDBM
AEOLBM
OLIGSPEC
D
20 m, other Oligoch. variables and environment
E
Max. depth, stations according to Oligochaeta
F
20 m, stations according to Oligochaeta
G
Max. depth, stations according to environmental variables
H
20 m, stations according to environmental variables

and organic sedimentation at maximum depths, suggesting that these species prefer an enriched environment. Other species were associated with high oxygen content; *A. hemprichi*, for example, is seen to prefer more oligotrophic conditions than *A. quaternarium*. The corresponding dendrogram for 20 m (Fig. 3B) is somewhat different. *Amphichaeta leydigii*, for instance, is associated with water quality variables indicative of anthropogenic loading, while *C. langi* has a stronger association with high oxygen content than does *Chaetogaster setosus*.

Variables describing the whole oligochaete community (Figs 3C & 3D) show that diversity and number of species at the maximum depths (Fig. 3C) are associated with high oxygen content, as is aeolosomatid biomass. On the other hand, the total biomass of oligochaetes and the naidid biomass are poor environmental indicators. The oxygen content is not as important at 20 m depth (Fig. 3D) as in the deeper profundal and has a strong association only with diversity of the oligochaetes. Aeolosomatid biomass is high when the number of species is high, *viz.* under undisturbed conditions, and the other biomass variables are high when sedimentation and total phosphorus are high.

The dendrograms for the sampling stations are similar whether they are based on the abundances of oligochaete species (Figs 3E, 3F) or on the most important environmental values (Figs 3G, 3H). Oligochaeta at maximum depths (Fig. 3E) show a group of clean stations (stations 5, 8, 9 and 10), polluted stations (6, 7) and intermediate stations (1, 2, 4). The deep station (3) stands out as a separate entity. Using environmental values (oxygen, total phosphorus and organic sedimentation) (Fig. 3G), five clean and five polluted stations can be discerned.

Clustering oligochaetes at 20 m (Fig. 3F) shows the clean stations 5, 8, 9 and 10 to form a group and the polluted stations 1, 2 and 6 form another group. While environmental variables (Fig. 3H) point to a clear group of clean stations (4, 5, 8, 9 and 10) and another of polluted stations group (1, 2, 3 and 7), the most polluted station (6) is still distinct from the rest.

Conclusions

Meiobenthic oligochaetes alone, can be used as indicators of the environment. The family Aeolosomatidae is an indicator of oligotrophy, *A. hemprichi* prefers cleaner environments than *A. quaternarium*. Of the naidids, the clearest oligotrophy indicators are *P. blanci* and *C. langi*. More opportunistic species which indicate organic loading or eutrophication are *A. leydigii, S. josinae, V. comata* and *D. digitata*. Small individuals of *T. tubifex* are very abundant in eutrophic environments. Total oligochaete abundance or biomass values are poor indicators.

The influence of environmental factors on oligochaete communities is similar at the maximum depths and in the uppermost part of the profundal zone, at 20 m. Oxygen may have a somewhat stronger effect on the oligochaetes than does total phosphorus or the organic matter content and the amount of organic sedimentation has the weakest effect.

The sampling areas clustered in a similar manner using oligochaetes and environmental variables. The clustering of the species together with environmental factors show groups that have internal affiliation concerning their relation to trophic state and organic loading.

Acknowledgements

I thank the Research Council for the Environmental Sciences (Academy of Finland), for the support for this work, and Mr Malcolm Hicks, M.A., for checking the English, Kathryn A. Coates and Trefor Reynoldson for editing, David Rosenberg and an anonymous referee for constructive criticism, and Tarmo Timm for organizing the congress.

References

Brinkhurst, R.O., 1974. The benthos of lakes. The Macmillan Press Ltd, London and Basingstoke, 190 pp.

Cranston, P.S., 1990. Biomonitoring and Invertebrate Taxonomy. Envir. Monit. Assess. 14: 265–273.

Maatela, P., J. Paasivirta, J. Särkkä & R. Paukku, 1990. Organic chlorine compounds in lake sediments. II. Organically bound chlorine. Chemosphere 21: 1343–1354.

Milbrink, G., 1980. Oligochaete communities in pollution biology: the European situation with special reference to lakes in Scandinavia. In R.O. Brinkhurst & D.G. Cook (eds), Aquatic Oligochaete Biology. Plenum Press, New York, 433 pp.

Rapport, D.J., 1990. Challenges in the detection and diagnosis of pathological change in aquatic ecosystems. J. Great Lakes Res. 16: 609–618.

Särkkä, J., 1989. Meiobenthic naidid and aeolosomatid oligochaetes from the profundal zone, and relations of species to eutrophication. Hydrobiologia 180/(Dev. Hydrobiol 51): 185–190.

Särkkä, J., 1992. Lacustrine profundal meiobenthos as an environmental indicator. Hydrobiologia 243/244 (Dev. Hydrobiol. 79): 333–340.

Thiel, H., 1975. The size structure of the deep-sea benthos. Int. Revue ges. Hydrobiol. 60: 575–606.

Warwick, R.M., N.R. Collins, J.M. Gee & C.L. George, 1986. Species size distributions of benthic and pelagic Metazoa: evidence for interaction? Mar. Ecol. Progr. Ser. 34: 63–68.

Warwick, R.M., T.H. Pearson & Ruswahyuni, 1987. Detection of pollution effects on marine macrobenthos: further evaluation of the species abundance/biomass method. Mar. Biol. 95: 193–200.

Weston, D.P., 1990. Quantitative examination of macrobenthic community changes along an organic enrichment gradient. Mar. Ecol. Prog. Ser. 61: 233–244.

Wiederholm, T., 1980. Use of benthos in lake monitoring. J. Wat. Pollut. Cont. Fed. 62: 537–547.

Hydrobiologia **278**: 243–249, 1994.
T.B. Reynoldson & K.A. Coates (eds), Aquatic Oligochaete Biology V.

Oligochaete response to changes in water flow in the Dokka Delta, Lake Randsfjorden (Norway), caused by hydroelectric power development

Svein-Erik Sloreid
Norwegian Institute for Nature Research, P.O. Box 1037, Blindern, N-0315 Oslo, Norway

Key words: hydroelectric power development, substrate disturbance, oligochaetes, abundance, community development

Abstract

The results reported here are part of a joint study on possible ecological effects in the delta area of Lake Randsfjorden, Southern Norway, caused by a reduction in water flow as a result of hydroelectric power development.

From 1988 to 1990 the mean daily water flow decreased sharply and the total load of suspended matter was greatly reduced. During the same period there was an increase in oligochaete abundance from 6,096 ind. m^{-2} in 1988 to 100,051 ind. m^{-2} in 1989 and 154,345 ind. m^{-2} in 1990.

Piguetiella blanci was the dominant species in 1988 making up 48.5% of the total community. In 1989, *Limnodrilus hoffmeisteri, Nais variabilis/Nais communis* and enchytraeids were the most abundant oligochaetes, making up 29.3%, 26.1% and 28.2%, respectively of the total number. Reduced water flow from 1988 to 1989 and 1990 caused reduced disturbance of the bottom sediments and is considered the main reason for the change in species composition and abundances.

Introduction

The results presented here are part of a joint study on the possible ecological effects in the delta area of Lake Randsfjorden, Southern Norway, caused by hydroelectric power development. A 45% reduction in water flow through the delta as a result of diversion of water was estimated. The project investigated benthic macroinvertebrates, planktonic and littoral crustaceans, aquatic macrophyte vegetation and hydrological processes. The investigations were started in 1988, one year prior to the development and continued until 1991.

It has been observed in other Norwegian lakes that water flow patterns and sedimentation patterns in delta areas are dependent upon the velocity of the inflowing water and the water stage in the lake. These variations lead to changes in the amount and size of the settling particles at a given location (Bogen, 1983, 1987). Hence, the reduction in discharge affected the flow regime in the Dokka delta in Lake Randsfjorden. This in turn would imply changed erosion and sedimentation patterns. It is thus likely that the high water discharge prior to power development caused repeated disturbance of the bottom sediments at the study site, and that the disturbances were greatly reduced after the diversion of water (Bogen, pers. comm.).

A disturbance can be defined as an event that causes changes in an existing community and initiates a process that allows the establishment of new individuals (Sousa, 1984). In its early successional stages a community is often dominated by a few species and these often occur in great numbers. Typically, early colonizers are small, are primarily found in the sediment-water interface, have high reproductive rates and grow rapidly (Soster & McCall, 1990). Voshell & Simmons (1984) found that naidids were the first oligochaetes to colonize a new reservoir, and Barnes (1983) showed that naidid oligochaetes dominated lakes younger than three years. Late colonizers grow more slowly, mature late in life and are usually deeper dwelling, head-down feeders. They tend to increase gradually in number and will eventually dominate the communities. Tubificid oligochaetes are typical late colonizers (Lopez, 1988, Soster & McCall, 1990). Between these two extremes intermediate communities will occur (Soster & McCall, 1990).

The study site was highly influenced by the discharge of the tributary river Dokka. The object of this study was (1) to characterize the oligochaete community during natural high water flow conditions and repeated disturbances of the bottom sediments and (2) to investigate how the oligochaete community varies with the reduction in water flow and subsequent reduced disturbance of the bottom sediments during the years following hydroelectric power development.

Study site

The oligotrophic Lake Randsfjorden is situated 132 m above sea level, it is 75 km long and its greatest width is 4.5 km. The surface area is 137 km^2 and the greatest depth is 120.5 m (Holtan, 1970). The lake is usually ice-covered from November to April. Lake Randsfjorden has been regulated since 1916. The water level fluctuates 2.8 m, with minimum and maximum depths occurring in April and late May, respectively. During summer and autumn there are minor water level fluctuations, but from December the water level decreases until the spring flood in April.

The study site is located in the inlet of Lake Randsfjorden delta where two branches of the River Dokka run into the lake (60° 50′ N, 10° 05′ E) (Fig. 1). The depth at the study site fluctuated between 1.2 m and 4.0 m. There was no submerged macrophyte vegetation and the surficial sediment consisted of 95% sand, 4% silt and 1% clay (October 1989). There was no thermal stratification or oxygen depletion. The study site was marked with a permanently anchored buoy. The area sampled was within 100 m^2 of the buoy.

From 1989 a part of the River Dokka was developed for hydroelectric power production. The dam construction and the diversion of water resulted in reduced water flow through the delta. In 1988 mean daily water flow (from April to November) was 9.0×10^6 m^3. In 1989 and 1990 mean daily water flow were 3.2×10^6 m^3 and 2.2×10^6 m^3 respectively (Bogen, 1989, 1990, 1991). Over the period the mean organic matter in suspension was reduced from 19.6×10^3 kg day^{-1} in 1988 to 5.7×10^3 kg day^{-1} in 1989 and 4.2×10^3 kg day^{-1} in 1990 (Bogen, op. cit.). The total load of suspended matter was 87.5×10^3 kg day^{-1}, 19.9×10^3 kg day^{-1} and 17.4×10^3 kg day^{-1} in 1988, 1989 and 1990, respectively (Bogen, op. cit.).

Materials and methods

Collections of oligochaetes were made using a core sampler of 38.5 cm^2 (d = 7 cm). Five replicates were taken monthly from April/May to October/November, 6 visits in 1988 and 1990 and 7 visits in 1989. The upper 10 cm of sediment was washed through a 250 μm sieve, preserved in 70% alcohol and the organisms were sorted in the laboratory under a dissecting microscope. The study site was the same all years and was randomly sampled each visit.

The oligochaetes were identified to species level. Immature tubificids without capilliform setae were considered to be *Limnodrilus hoffmeisteri* Claparede since this was the only mature tubificid found without capilliforms. The enchytraeids have not been identified to species level but their small size, mainly six setae per bundle and thin, relatively long sigmoid setae im-

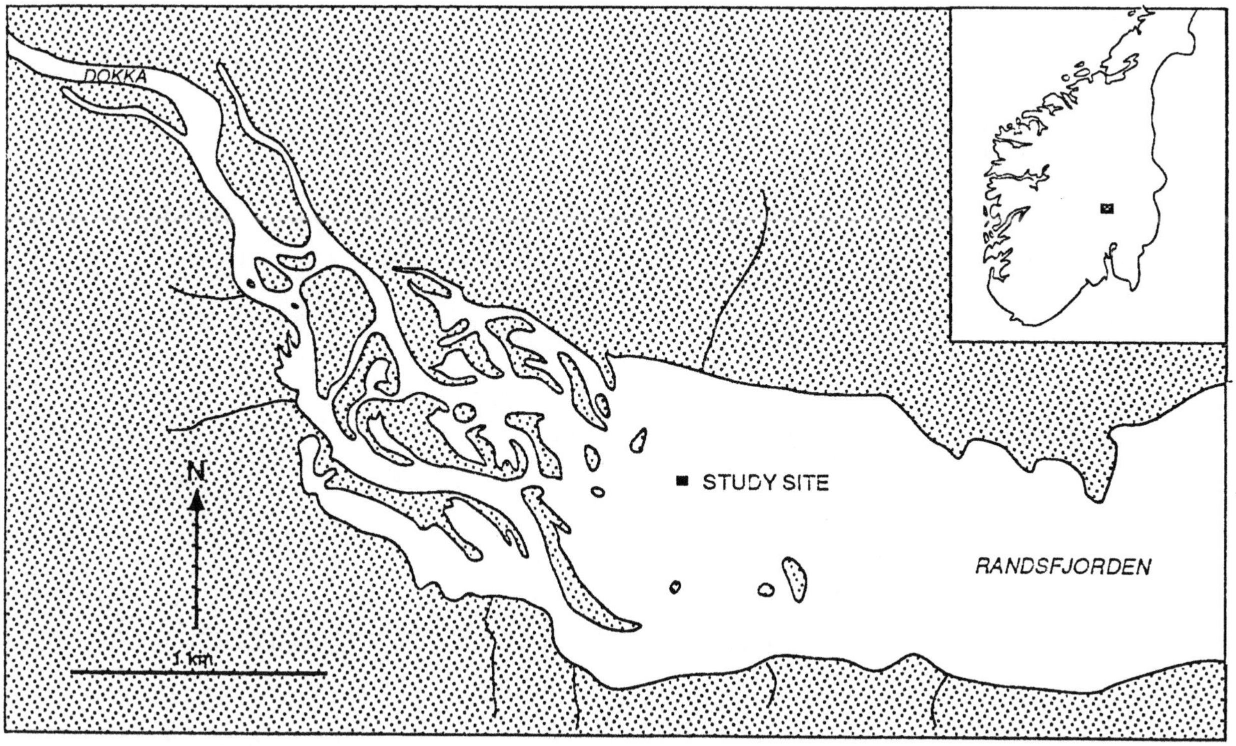

Fig. 1. Map of western part of lake Randsfjorden and tributary river Dokka, (60° 50′ N, 10° 05′ E), Southern Norway.

plies the genus *Cernosvitoviella*. *Nais variabilis* Piguet and *Nais communis* Piguet were combined because they are often difficult to separate on the basis of the shape of the setae as intermediate teeth lengths seem to occur.

At present only the oligochaete material from 1988 and 1989 has been identified to species level, but total oligochaete abundance from 1990 is available. All five replicates have not been fully processed to species, and full statistical treatment of the material must be postponed. The results are therefore preliminary.

Results

The preliminary analyses show a great increase in mean abundance of oligochaetes, especially from 1988 to 1989 and a somewhat lesser increase from 1989 to 1990 (Fig. 2). The mean abundances for all years were significantly different ($p<0.005$, Mann-Whitney U-test). In 1988 the mean number of oligochaetes was 6,110 ind. m^{-2} (S.E. 615). In 1989 there was an increase in abundance during the summer months stabilizing towards autumn. Mean number of oligochaetes in 1989 was 100,051 ind. m^{-2} (S.E. 12,368). In 1990 abundance during the study period was more stable than in previous years and the mean number increased to 154,345 ind. m^{-2} (S.E. 12,052). There was a tendency each year for oligochaete abundance to decline during or immediately after the spring flood (Fig. 2). In 1988 a decline in abundance was also found after a heavy rainfall in August, but there was no reduction in density after a similar event in June/July.

Figure 2 also illustrates the differences in discharge between these years. The decline in water flow in the spring of 1988 from around 600 m^3 s^{-1} to 200 m^3 s^{-1} in 1989 and 1990 demonstrates the impact of power development. Under normal conditions discharge is especially high in spring (e.g. 1988), but filling of the reservoir greatly reduces the peaks in spring flood (1989, 1990).

There was a profound change in species com-

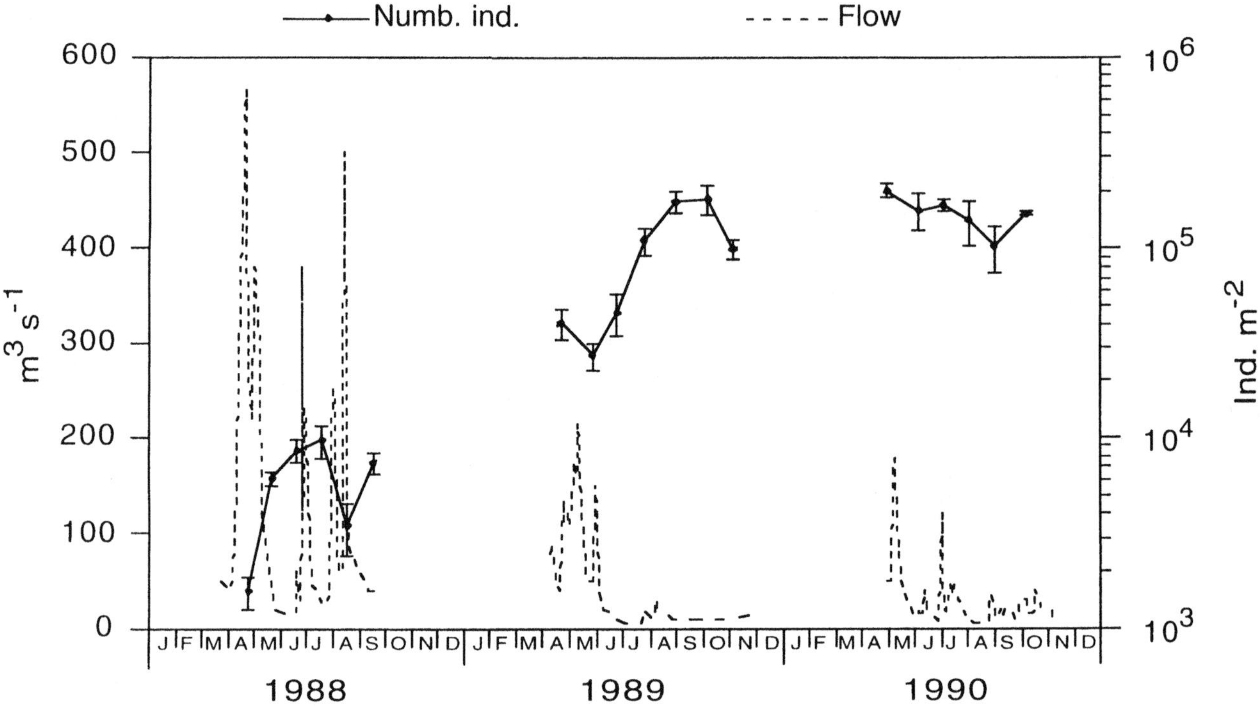

Fig. 2. Mean number of oligochaetes m^{-2} and mean daily water flow $m^3\ s^{-1}$ at the study site in lake Randsfjorden during the sampling season in 1988, 1989 and 1990. Vertical bars represent one standard error ($\pm$ 1 S.E.). Water flow data from Bogen (1989, 1990, 1991).

position from 1988 to 1989 (Table 1). The dominant species in 1988, *Piguetiella blanci* (Piguet) declined from 48.5% to 2.3% of the community in 1989, yet mean density decreased only slightly from 2,929 ind. m^{-2} in 1988 to 2,293 ind. m^{-2} in 1989. *Specaria josinae* (Vejdovsky) declined from 8.8% in 1988 to 5.0% in 1989, but its density increased from 529 ind. m^{-2} to 4,965 ind. m^{-2}. In addition, *Nais alpina* Sperber constituted 5.9% in 1988, but decreased to 1.8% of the oligochaetes in 1989. This represents 355 ind. m^{-2} and 1,796 ind. m^{-2}, respectively.

In 1988 only single specimens of *N. variabilis*/*N. communis* were found, but in 1989 they were the most abundant naidids with 26,152 ind. m^{-2}, comprising 26.1% of mean total number. At the same time, there was a great increase in the abundance of *L. hoffmeisteri*, from 650 ind. m^{-2} to 29,357 ind. m^{-2}, and Enchytraeidae from 641 ind. m^{-2} to 28,172 ind. m^{-2}. This represents an increase from 10.8% to 29.3% for *L. hoffmeisteri* and from 10.6% to 28.2% for the enchytraeids of the total mean number.

The naidid portion of the total community was reduced from 65.2% in 1988 to 38.4% in 1989, whereas the tubificids and enchytraeids increased from 18.6% to 30.8% and from 10.6% to 28.2% respectively, from 1988 to 1989.

Other benthic invertebrate groups have not been identified to species at this stage, but there was an increase in the density of small mussels, mainly *Pisidium* spp., from only single specimens in 1988 to 208 ind. m^{-2} in 1989 and 506 ind. m^{-2} in 1990. The most dominant macrobenthos in abundance were Oligochaeta, Chironomidae and Nematoda, their proportions varied each sampling date between 70–90%, 6–25% and 3–7% of the total benthic community, respectively.

Discussion and conclusions

It is reasonable to suggest that the high and variable water flow together with the high content of suspended particles, caused a great disturbance to the bottom sediments during the 1988 season.

Table 1. Mean numbere (ind. m^{-2}) and per cent of the different oligochaete species and taxons in 1988 and 1989. P = less than 10 ind. m^{-2} and S.E. = one standard error.

Taxon	1988			1989		
	%	Mean	S.E.	%	Mean	S.E.
Lumbriculus variegatus (Müller)	1.0	61	16	0.1	127	40
Stylodrilus heringianus Claparede	4.0	243	25	1.3	1,310	167
Tubifex ignotus (Stolc)	2.3	139	34	0.2	184	39
Tubifex tubifex (Müller)	4.3	260	49	1.0	986	133
Limnodrilus hoffmeisteri Claparede	10.8	650	122	29.3	29,357	3,194
Spirosperma ferox (Eisen)	1.0	61	10	0.1	134	40
Ryacodrilus coccineus (Vejdovsky)				0.1	72	27
Aulodrilus pluriseta (Piguet)				0.0	28	10
Slavina appendiculata d'Udekem		P		0.5	530	79
Ripistes parasita (Schmidt)		P				
Stylaria lacustris (Linnaeus)		P				
Piguetiella blanci (Piguet)	48.5	2,929	272	2.3	2,293	337
Nais pardalis Piguet		P		1.0	961	161
Nais elinguis Müller		P		0.1	92	35
Nais variabilis Piguet/*Nais communis* Piguet	1.3	78	18	26.1	26,152	4,006
Nais alpina Sperber	5.9	355	57	1.8	1,796	318
Nais bretscheri Michaelsen		P		0.2	176	39
Specaria josinae (Vejdovsky)	8.8	529	78	5.0	4,965	572
Pristina spp.	0.3	17	7	1.4	1,356	290
Naididae (Other)		P		0.1	107	41
Tubificidae (Other)	0.6	35	10	0.1	53	20
Enchytraeidae	10.6	641	50	28.2	28,172	1,622
Indet	0.7	43	7	1.2	1,173	225
Aelosomatidae		P		0.0	27	10
Total		6,041	514		100,051	9,710

Only a few species were able to maintain certain densities, and the dominance of the Naididae in 1988 suggests a community in its early stage of succession. The ability of naidids to reproduce asexually enables them to maintain densities despite repeated disturbances such as those in 1988. *P. blanci* was the most abundant oligochaete species that year. Together with *S. josinae* and *Nais alpina* Sperber they constituted most of the naidids. The dominance of naidids was probably characteristic for this disturbed site before power production development, and the species composition found in 1988 could be typical for this site during high water flow conditions.

There is little information in the literature on the ecological demands of *P. blanci*. Särkkä (1987) describes it as a mesotrophic species occurring both in the littoral and profundal zones of lakes and in lotic waters, but never in great numbers. The species seems to disappear under eutrophic or organically enriched situations. In colonization studies by McCall & Soster (1990) and Soster & McCall (1990), *S. josinae* is characterized as an intermediate species that appears as opportunists decline. In Finland, Särkkä (1987) has found *S. josinae* from oligotrophic to eutrophic conditions and in organically enriched localities. Griffiths (1991) found *S. josinae* associated with bottom sediments of gravel and sand in erosional and/or polluted conditions. *Nais alpina* seems to be relatively intolerant to organic enrichment (Learner *et al.*, 1978) and Wachs (1967) found it associated with stony substrates in the upper reaches of rivers.

L. hoffmeisteri and *Tubifex tubifex* (Müller) were present in low numbers in 1988, but increased in numbers in 1989. Particularly *L. hoffmeisteri* has a great ability to adapt its life history to local conditions, and has been found to breed throughout the year (Kennedy, 1966). Despite the disturbed conditions, *L. hoffmeisteri* may be able to maintain its abundance through continuous hatching of cocoons. It is well known that both species may be abundant in organically enriched situations, but *T. tubifex* tends to be associated with the profundal of lakes and *L. hoffmeisteri* is often abundant in shallow habitats. Soster & McCall (1990) and McCall & Soster (1990) describe *Limnodrilus* spp., *Ilyodrilus templetoni* (Southern) and to some extent *Aulodrilus pigueti* Kowalewski as late successional species. They tend to increase gradually in numbers and eventually dominate the communities, and *Limnodrilus* spp. seem to be particularly characteristic in late successional stages. The occurrence of *Rhyacodrilus coccineus* (Vejdovsky) and *Aulodrilus pluriseta* (Piguet) in 1989 could indicate more suitable conditions for tubificids that year although only single specimens were found.

Learner *et al.* (1978) state that both *N. communis* and *N. variabilis* appear to benefit from organic enrichment which can lead to a considerable increase in population densities. The two species seem to be more successful in the less disturbed conditions of 1989.

The enchytraeids were also abundant in 1989. They too are sediment dwellers and reproduce sexually, suggesting that they might have been vulnerable to the repeated disturbances in 1988, while the more stable situation in 1989 allowed them to breed successfully.

The reduced disturbance from 1988 to 1989 and to 1990 is considered the major controlling factor in the development of oligochaete abundance and the change in species composition. Tevesz (1985) found that disturbance was an important controlling factor in freshwater environments, and McCall & Soster (1990) found that distribution of macrobenthos was correlated with disturbance levels in western Lake Erie. The increase in bivalve abundance during the three years supports the idea that sediment disturbance plays an important role in structuring the benthic community, as they are considered to be late colonizers (Soster & McCall, 1990).

In addition, development of plant communities, availability of organic detritus, physical and chemical modification of the substrate and biotic interactions such as predation and competition are other environmental factors affecting macrobenthic succession (Tevesz, 1985).

Although there was a reduction in total organic load from 1988 to 1989 and to 1990 (Bogen, 1989, 1990, 1991), there was a large increase in oligochaete abundance and the development of a community with species that prefer organic enrichment. Unfortunately, data on the amount of sediments that actually settled at the study site are not available, but it is reasonable to believe that the reduced discharge resulted in increased sedimentation. This may influence the amount and quality of the available food and change the texture of the sediment, thus, contributing to the observed changes in the community. In addition, the increase in tubificid abundance in itself with the deposition of faecal matter at the sediment surface, accelerates the development of a suitable substrate for tubificids (McCall & Tevesz, 1982).

The importance of other factors in the development of the benthic community is difficult to assess, but it is probable that biotic interactions become increasingly important if the community develops without serious disturbance. The development of the community can therefore be viewed in a sequence from non-equilibrium to equilibrium (Wiens, 1984) or as an r-K continuum (Pianka, 1970).

In 1990 the physical/chemical conditions were very similar to 1989, and there was a further increase in oligochaete abundance. Identification, to species level, of the oligochaete material collected in 1990, will determine if the trend towards a community dominated by species that grow slowly, reproduce late in life and are sediment dwellers, continues.

Acknowledgements

I would like to thank E. Framstad, G. Halvorsen and I. P. Muniz for valuable comments on the manuscript and for improving the English and B. Walseng for co-operation both in the field and in the laboratory. I am also grateful to J. Bogen for interpretations of the fluvial processes in the Delta. The investigation has been financed by The Norwegian Institute for Nature Research and The Norwegian Water Resources and Energy Administration.

References

Barnes, L. E., 1983. The colonization of ball-clay ponds by macroinvertebrates and macrophytes. Freshwat. Biol. 13: 561–578.

Bogen, J., 1983. Morphology and sedimentology of deltas in fjord and fjord valley lakes. Sediment. Geol. 36: 246–267.

Bogen, J., 1987. Deltaic depositional in a glacier-fed lake: a model for fluvial/lacustrine interface. Soc. Econ. Paliont. Min. Spes. Pub. 39: 121–131.

Bogen, J., 1989. Etna-Dokka sedimenttransport 1988. NVE. VHB-Notat nr. 20/89 (in Norwegian).

Bogen, J., 1990. Etna-Dokka sedimenttransport. NVE. VHB-Notat nr. 21/90 (in Norwegian).

Bogen, J., 1991. Suspensjonstransport i Etna-Dokka 1990. NVE-Notat (in Norwegian).

Griffiths, R. W., 1991. Environmental quality assessment of the St. Clair River as reflected by the distribution of benthic macroinvertebrates in 1985. Hydrobiologia 219: 143–164.

Holtan, H., 1970. Randsfjorden – en limnologisk undersøkelse 1967–1968. NIVA 0–15/64 (in Norwegian).

Kennedy, C. R., 1966. The life history of *Limnodrilus hoffmeisteri* Clap. (Oligochaeta: Tubificidae) and its adaptive significance. Oikos 17: 158–168.

Learner, M. A., G. Lochhead & B. D. Hughes, 1978. A review of the biology of British Naididae (Oligochaeta) with emphasis on the lotic environment. Freshwat. Biol. 8: 357–375.

Lopez, G. R., 1988. Comparative ecology of the macrofauna of freshwater and marine muds. Limnol. Oceanogr. 33: 946–962.

McCall, P. L. & F. M. Soster, 1990. Benthos response to disturbance in western Lake Erie: regional faunal surveys. Can. J. Fish. aquat. Sci. 47: 1996–2009.

McCall, P. L. & M. J. S. Tevesz, 1982. The effects of benthos on physical properties of freshwater sediments. In P. L. McCall & M. J. S. Tevesz (eds), Animal-sediment relations. The biogenic alteration of sediments. Plenum Press, New York: 105–176.

Pianka, E. R., 1970. On r- and K-selection. Am. Nat. 104: 592–597.

Särkkä, J., 1987. The occurrence of oligochaetes in lake chains receiving pulp mill waste and their relation to eutrophication on the trophic scale. Hydrobiologia 155: 259–266.

Soster, F. M. & P. L. McCall, 1990. Benthos response to disturbance in western Lake Erie: field experiments. Can. J. Fish. aquat. Sci. 47: 1970–1985.

Sousa, W. P., 1984. The role of disturbance in natural communities. Ann. Rev. Ecol. Syst. 15: 353–391.

Tevesz, M. J. S., 1985. Benthic colonization in fresh water. Kirtlandia 41: 3–14.

Voshell, Jr., J. R. & G. M. Simmons, Jr., 1984. Colonization and succession of benthic macroinvertebrates in a new reservoir. Hydrobiologia 112: 27–39.

Wachs, B., 1967. Die Oligochaeten – Fauna der Fliessgewässer unter besonderer Berücksichtigung der Beziehungen zwischen der Tubificide Bessiedlung und dem Substrat. Arch. Hydrobiol. 63: 310–386.

Wiens, J. A., 1984. On understanding a non-equilibrium world: myth and reality in community patterns and processes. In D. R. Strong, Jr., D. Simberlof, L. G. Abele & A. B. Thistle (eds), Ecological communities: conceptual issues and the evidence. Princeton University Press, Princeton: 439–457.

Hydrobiologia **278**: 251–266, 1994.
T.B. Reynoldson & K.A. Coates (eds), Aquatic Oligochaete Biology V.

An experimental manipulation of oligochaete communities in mesocosms treated with chlorpyrifos or nutrient additions: multivariate analyses with Monte Carlo permutation tests

Piet F.M. Verdonschot[1] & Cajo J.F. Ter Braak[1,2]
[1]*Institute for Forestry and Nature Research, Department of Aquatic Ecology, P.O. Box 23, 6700 AA Wageningen, The Netherlands;*
[2]*Agricultural Mathematics Group-DLO, P.O. Box 100, 6700 AC Wageningen, The Netherlands*

Key words: BACI-design, chlorpyrifos, ditch, eutrophication, insecticide, large-scale experiments, oligochaete, toxic stress

Abstract

Oligochaete communities were monitored under semi-natural conditions in experimental ditch mesocosms. Twelve ditches were used in a Before-Control-After-Impact (BACI) experiment to study the effect of the insecticide chlorpyrifos. Another eight ditches were used in a randomized experiment to study the effect of eutrophication. Oligochaete communities were sampled by deploying trays of substratum for colonization over a 20-week period. The experiments were analyzed by multivariate analysis using redundancy analysis and Monte Carlo permutation to assess statistical significance. These novel methods have the advantage over classical multivariate analysis of variance (MANOVA) of being distribution-free and of having no restrictive upper limit on the number of species that can be analyzed simultaneously. In the BACI-experiment no significant effect of chlorpyrifos on oligochaete communities was detected. Eutrophication effects were observed at the higher eutrophication levels in clay ditches. Oligochaete abundances decreased in those ditches. Considerable variation was attributed to stochastic factors given that the ditches were in an early developmental stage when the experiments were initiated.

Large-scale experiments such as the ones that we describe require time to develop and stabilize before parameters of community structure like population abundance, can be employed to detect changes associated with water quality manipulations.

Introduction

Numerous authors have used oligochaete communities to assess the health of the aquatic environment (e.g. Hynes, 1960; Aston, 1973; Wiederholm, 1980). For example, the presence and abundance of oligochaetes are often indicative of different levels of organic pollution and/or eutrophication. This relationship is based on observational studies that incorporate nuisance variation associated with the complex reality of nature. In aquatic environments, eutrophication or saprobic pollution is usually accompanied by chemical pollution; for example, the presence of toxicants in sewage effluents or of pesticides in the nutrient-enriched drainage water from agricultural areas. Laboratory studies have revealed tolerance levels for different oligochaete species that are pollutant-specific and that depend on the prevailing environmental conditions (e.g. Chapman *et al.*, 1982). Unfortunately, to study oligochaete populations under any possible environmental condition would be very time consuming or even impossible. Large-scale outdoor experiments provide the opportunity to conduct experiments under controlled conditions in complex functioning (semi-natural) ecosystems. It is generally accepted that such mesocosm experiments reduce the disadvantage of the complexity without becoming too simple (e.g. Bowling *et al.*, 1980).

Ditches make up an important part of the Dutch landscape with an estimated length of 300,000–400,000 km (Higler, 1989). These ditches are mainly situated in agricultural areas. Therefore, they are often contaminated and/or eutrophied.

A series of experimental ditches was constructed to study the effect of chemical and nutrient disturbances, separately. This study examines the response of oligochaete communities to the separate effects of eutrophication by nitrogen and phosphate, and toxic stress by the insecticide, chlorpyrifos.

Materials and methods

Facility description and experimental design

The experimental facility consisted of 20 uniform ditches situated near Renkum, in the centre of The Netherlands. A number of physical properties are listed in Table 1. The ditch bed consisted of clay or sand. The clay was taken from an uncontaminated lake, mixed and put in the ditches. This construction step also introduced a number of seeds, cocoons, and other biological material to the ditch community. For the sand-lined ditches, the sand was taken from a deep, sterile sand layer. All the ditches were filled with water in February 1988. Before the experiments were began, the clay-lined ditches were inter-connected to permit the exchange of water and species. No similar equilibration was undertaken in the sand-lined ditches.

Colonization pattern and rate were studied in one clay-lined ditch.

The experiments with the insecticide chlorpyrifos employed twelve ditches with a clay bed. The experimental design followed a modified Before-After-Control-Impact (BACI) design (Green, 1979; Stewart-Oaten *et al.*, 1986). The design differed from the standard BACI design in that four levels of impact were used instead of one. The four concentrations of chlorpyrifos (Table 1) were applied to two ditches each on 8 May 1990. Four ditches served as control. Sampling took place three times in the six months before the application of the insecticide and four times in the eight months thereafter.

The eutrophication experiment started on 23 May 1989 and had a completely randomized design. Eight ditches (four clay bottoms and four sand bottoms, including two contols) were treated with additions of phosphorus, or phosphorus and nitrogen (Table 1).

Sampling technique

Oligochaetes were collected by deploying trays of substrate and retrieving the trays after a standard time period. Each tray consisted of an inner cylinder (height 6 cm, diameter 10.4 cm) which was closed at the bottom and placed in an outer cylinder which fitted just around the first. The outer cylinder was set into the ditch bottom. From preliminary studies the minimum surface area of the cylinder required to collect at least 100 specimens of tubificids was estimated to be 80 cm^2. The inner cylinder was filled with the same material as was used for the ditch bed. During sampling only the inner cylinder was removed without disturbing the outer one or the surrounding sediment. Before sampling the inner cylinder was enclosed by a gauze cover to prevent organisms from escaping, the cylinder was then lifted and the contents were washed out into a bucket and taken to the laboratory. After sampling the cylinder was refilled and returned. Two sets of cylinders were set into the bottom of each ditch. Consequently, each cylinder was exposed for twenty weeks. Sampling was undertaken every ten weeks start-

Table 1. Description of the ditches and the experimental design.

DITCHES

Length	40 m
Width at the water surface	3.5 m
Width at the top of the bed	2 m
Water depth	0.5 m
Total water volume	55 m^3
Depth of the bed	0.25 m
Water source	deep groundwater
Bed	clay: uncontaminated lake sand: deep sand layer

TREATMENT

insecticide:	
Formulation	chlorpyrifos (Dursban 4E)
Bed of	clay (ditch number 1 to 12)
Control	4 ditches (number 2, 3, 5, 11)
0.1 µg chlorpyrifos	2 ditches (number 1, 8)
0.7 µg chlorpyrifos	2 ditches (number 4, 10)
5 µg chlorpyrifos	2 ditches (number 7, 12)
35 µg chlorpyrifos	2 ditches (number 6, 9)
nutrients:	
Formulation	phosphate: K_2HPO_4, nitrogen: NH_4NO_3
Bed of	clay (4 ditches; number 13 to 16) sand (4 ditches; number 17 to 20)
Control	2 ditches (number 16, 17)
49 grP/y	2 ditches (number 15, 18); supplied each half year
155 grP/y, 559 grN/y	2 ditches (number 14, 19); supplied each half year
1488 grP/y, 8928 grN/y	2 ditches (number 13, 20); supplied monthly

LAY-OUT OF THE EXPERIMENTAL DITCHES AND THEIR TREATMENT

Ditch 1 Treatment 0.1 µg chl.	Ditch 2 Treatment control	Ditch 3 Treatment control	Ditch 4 Treatment 0.7 µg chl.
Ditch 5 Treatment control	Ditch 6 Treatment 35 µg chl.	Ditch 7 Treatment 5 µg chl.	Ditch 8 Treatment 0.1 µg chl.
Ditch 9 Treatment 35 µg chl.	Ditch 10 Treatment 0.7 µg chl.	Ditch 11 Treatment control	Ditch 12 Treatment 5 µg chl.
Ditch 13 Treatment high P/N	Ditch 14 Treatment medium P/N	Ditch 15 Treatment low P	Ditch 16 Treatment control
Ditch 17 Treatment control	Ditch 18 Treatment low P	Ditch 19 Treatment medium P/N	Ditch 20 Treatment high P/N

ing in July 1989 for the eutrophication experiment and in November 1989 for the chlorpyrifos experiment. To study the colonization process, 48 more trays were placed into one ditch (no. 16) at 26 June 1990. Every two weeks two trays were sampled for a total of 24 weeks.

The samples were sieved (mesh sizes 1.0, 0.5, and 0.3 mm) and sorted by eye. The oligochaetes were preserved in formalin (4%), identified and counted.

Statistical analysis

The oligochaete abundances were transformed to a logarithmic scale and the data were ordinated by principal component analysis (PCA) and redundancy analysis (RDA) with the program CANOCO 3.1 (Ter Braak, 1988; 1990). The following account is based on Ter Braak and Prentice (1988) and Ter Braak (1990). Both PCA and RDA assume a linear model for the relationship between the response of each taxon and the ordination axes. A linear model is appropriate for our data because a preliminary detrended correspondence analysis showed short gradient lenghts (<3 SD). The data are thus quite homogeneous in species composition. RDA is the constrained form of PCA of taxon data, in which the components (axes) are constrained by a linear model, i.e. by linear combinations of environmental variables. With quantitative environmental variables RDA is very similar to canonical correlation analysis, which the standard linear method for relating two set of variables. For the analysis of experiments, RDA is very similar to a multivariate analysis of variance (MANOVA) followed by the derivation of canonical variates (Green, 1979). There are two important differences. The most important being that RDA can analyze any number of species whereas both canonical correlation and MANOVA impose a restrictive upper limit which depends on the number of samples. Like ours, most ecological experiments are small in sample number but large in species number. In contrast to MANOVA, RDA allows analysis of the full species composition. The second difference lies in the way statistical tests are carried out. In MANOVA, the standard test uses statistical tables which require the assumption that the data are multivariate normal, whereas in RDA (with CANOCO) the standard test uses the computer intensive method of Monte Carlo permutation which does not require multivariate normality (Manly, 1991). The test statistic used in CANOCO 3.1 is a partial F-ratio which differs from the univariate partial F-ratio in that the regression sum of squares and the residual sums of squares are accumulated across species. It is thus a stacked univariate statistic. The principle of the Monte Carlo test is as follows. Under the null hypothesis of no effect, the samples are exchangeable among the various treatment groups. Consequently, samples in the taxon data can be randomly permuted giving, for each permutation, a 'random data set' that is analyzed in the same way as the original data. To determine the significance level of the observed F-ratio, it is ranked among F-ratio's calculated from the random data sets. If the observed F-ratio is among the 5% highest, the null hypothesis is rejected at the 5% significance level. The test has an exact Monte Carlo significance level (Marriot, 1979). The power of the test increases somewhat with the number of permutations, but for a test at the 5% level little can be gained by generating more than 99 permutations (Hall & Titterington, 1989). In analogy with the guidelines for the construction of bootstrap significance test (Hall & Wilson, 1991), loss of power is avoided by permuting residuals of the taxon data instead of the raw data (Ter Braak, 1990; 1992). For experimental data the residuals are obtained as in MANOVA.

Partial RDA ordination was used to control for nuisance variation, in particular spatial and/or temporal variation, and to focus the ordination on effects of treatments or other explanatory variables (Ter Braak & Prentice, 1988; Borcard, Legendre & Drapeau, 1992). Partial ordination was carried out by specifying covariables in CANOCO which yields an ordination of the residual variation in the taxa data after the covariables are factored out by multivariate linear regression.

The advantages of (partial) RDA over MANOVA also apply to the analysis of the BACI-experiment to assess the impact of chorpyrifos. In a BACI-experiment the impact is inferred by comparing the difference in each taxon's abundance between Control and Impact plots Before treatment with chorpyrifos with the difference After treatment. For more levels of treatment, the comparison is made for each level and the comparisons are then combined in a meaningful way. The inference assumes that the differences between ditches remains constant over time without the impact, i.e. that there is no ditch vs. time interaction. This assumption is not unreasonable for our data on a logaritmic scale. In the original BACI-design of Green (1979) the Control and Impact areas are singular. Replications were locations within areas. This was criticized by Hurlbert (1984) as pseudoreplication. Subsequently, Stewart-Oaten *et al.* (1986) advocated the use of replication in time for valid statistical inference. Our BACI-design takes the best aspects of those in Green (1979: 165–181) and in Stewart-Oaten *et al.* (1986). Ditches are independent rather than being nested within areas (as in Green, 1979) and there are repeated measurements over time (as in Stewart-Oaten *et al.* (1986). Jagers op Akkerhuis and van der Voet (1992) used the same type of design. Our analysis is a multivariate form of theirs. The analysis is based on a linear model in which the log-abundance y of each species is decomposed as:

$$y = \text{ditch} + \text{time} + \text{chlorpyrifos} + \text{error}\ .$$

In this model, chlorpyrifos is assumed to have a constant effect over time after it is applied. The interaction term chlorpyrifos.time is assumed negligible. If the chlorpyrifos effect was short lived, our analysis would have difficulty picking up this short-term effect and an analysis with an interaction term would be more approriate. For maximum power, the effect of chlorpyrifos was assumed linear on the logarithmic scale. In the analysis chlorpyrifos was thus a quantitative variable. For comparison with Green's analysis we note that the chlorpyrifos effect lies in the interaction space of ditch x time. For each taxon the model was fitted by least-squares giving estimates of the effect, the associated regression sum of squares and the residuals sum of squares. The sums of squares were added across taxa to form a single F-ratio as described above. These calculations were carried out with CANOCO by requesting a partial RDA with 'ditch' and 'sampling date' as covariables and 'chlorpyrifos' as a single quantitative environmental variable. The Monte Carlo test requires further consideration. A test that is valid 'by design' uses only the permutations that are feasible under the randomization scheme of experimental design. In the BACI-experiment, ditches were randomized among treatment levels. We are thus allowed to permute ditches, not individual samples. Ditches are permuted by permuting a whole series of samples from each ditch. Unfortunately, because the number of ditches in the experiment is small, the test has low power. In the smallest BACI-experiment (with a single control and a single treatment ditch) no permutations are possible under this scheme. Therefore, we also carried out a test that uses unrestricted random permutations. This test is likely to be too liberal if residuals are correlated in time.

In the analysis of the clay-lined ditches in eutrophication experiment, the clay-lined control ditches of the chlorpyrifos experiment are taken as additional controls.

The ordination results are presented as correlation biplots of taxa, samples and environmental variables (CANOCO scaling option 2). In these plots the horizontal and vertical axes concerns 1 and 2, respectively. Instead of individual samples, centroids of groups of samples are displayed, e.g. samples taken from the same ditch or taken in the same month. The mean abundance of a taxon in such a group can be inferred from the biplot using the rules of the usual taxon-sample biplot. Simplified these rules are as follows. By projecting the taxa points onto the arrow for a group, the taxa can be divided into two categories. The taxa that fall on the same side of the origin as the arrow head, have higher than average abundance in that group, whereas the taxa that fall on the other side have a lower than average abundance. As, the deviation becomes larger the greater the length of the arrow and the distance between the projection point of the taxon and the origin. The deviations thus obtained are in multiples of the standard deviation of each taxon. Also, notional arrows can be drawn from the origin to each taxon point. Projection of group centroids and other taxon points onto such an arrow for a particular taxon gives a ranking of its mean abundance values in the groups and of its correlations with the other taxa, respectively. In the ranking, the origin marks the taxon's mean abundance and zero correlation, respectively. Further rules can found in Ter Braak (1987: 127–129). We also used the idea of adding points to an existing biplot by projection. Such points are termed passive or supplementary. The eigenvalue of an ordination axis in PCA and RDA is the proportion of the total variance explained by that axis and indicates within an analysis the relative importance of the axis (Verdonschot & Higler, 1989).

As in multiple regression, qualitative predictor variables (e.g. 'sampling date' and 'ditch') were defined by series of indicator or dummy variables, one dummy variable for each category. A dummy variable takes the value of one when the sample belongs to the corresponding category and the value zero otherwise. The variables 'nutrient supply' and 'chlorpyrifos load' were treated as quantitative variables with values 0 (control) to 4 (highest supply or load). Compared to the original values (Table 1), the new values are approximately on a logarithmic scale.

Table 2. Average abundance (number/tray) of oligochaetes in the colonization samples (A) and in the regular samples of ditch 16 (B). Colonization week number 2 refers to 26 June 1990. The regular samples of ditch 16 are taken at 05.06.90, 03.08.90, 20.10.90 and 08.01.91, respectively. (w.s. = with hair setae, wh.s. = without hair setae).

A. Colonization week number	2	4	6	8	10	12	14	16	18	20	22	24
Tubificidae												
Limnodrilus hoffmeisteri			1					5		8		5
Limnodrilus udekemianus							1			2		
Tubifex tubifex									2			2
Ilyodrilus templetoni	1	1	1	1			1					5
Tubificidae j.w.s.	4	4	2	4	6	8	7	8	22	35	1	7
Tubificidae j.wh.s.	2	1	1	2	34	69	33	32	37	70	8	29
Naididae												
Dero digitata	128	66	147	293	510	435	454	526	383	270	42	100
Stylaria lacustris	1		1	1			3	3	6	28	3	1
Lumbriculidae												
Lumbriculus variegatus	1										7	6

B. Regular sample corresponding to week	2	6	16	24
Tubificidae				
Limnodrilus hoffmeisteri	6	9	1	3
Limnodrilus udekemianus	6		1	
Limnodrilus claparedeianus				2
Tubifex tubifex				1
Ilyodrilus templetoni	6	2		3
Tubificidae j.w.s.	32	8		18
Tubificidae j.wh.s.	76	43	52	10
Naididae				
Dero digitata	11	82	190	22
Stylaria lacustris			7	1

Results and discussion

Colonization pattern

The average numbers of oligochaetes per taxon per sampling date in the colonization samples are given in Table 2A. In comparing the colonization samples with the corresponding regular samples from ditch 16 (Table 2B), it appeared that *Limnodrilus claparedeianus* only occurred in the regular and *Lumbriculus variegatus* only in the colonization samples. After 18 weeks all other taxa occurring in ditch 16 were collected in the colonization samples. *Dero digitata* is a rapid colonizer. In contrast, juvenile tubificids need about 10-to-12 weeks to reach the average abundance level in the ditch; the mature tubificids required 18-to-20 weeks. The results of the colonization experiment are influenced by sampling season. This effect is shown by the decrease in numbers of *Dero digitata* and *Stylaria lacustris* in colonization weeks 22 and 24. This pattern was also evident by adding the colonization samples to a PCA-ordination of regular samples (Fig. 1). The regular samples of ditch 16 show a seasonal difference between samples from spring and summer (top of Fig. 1) versus the autumn and winter samples (middle). Although taken in summer, the first colonization samples lie at the very bottom of Fig. 1 and thus resemble more the regular winter samples. But after ten weeks of colonization the samples approach the regular August and October samples. We conclude that after ten weeks the colonization samples are similar to the regular ones. The colonization period of the regular samples was set at 20 weeks which was thus sufficiently long to ensure that samples were representative of the oligochaete community of the ditch.

Control clay ditches

The control clay ditches of both the experiments with chlorpyrifos and nutrients were used to examine the natural spatial and seasonal variation

Table 3. Average abundance (number/tray) of oligochaetes before (B) and after (A) the treatment with chlorpyrifos. The sampling dates are 02.11.89, 09.01.90 and 20.03.90 before treatment and 05.06.90, 03.08.90, 20.10.90 and 08.01.91 after treatment.

Chlorpyrifos concentration		0.1		0.7		5		35		Control	
Ditch number		1, 8		4, 10		7, 12		6, 9		2, 3, 5, 11, 16	
Treatment Before/After		B	A	B	A	B	A	B	A	B	A
Taxon	Abbreviation										
Tubificidae											
Limnodrilus hoffmeisteri	LIDRHOFF	11	7	9	13	13	14	26	12	14	9
Limnodrilus udekemianus	LIDRUDEK	3	1	1	2	1	5	1	3	1	4
Limnodrilus claparedeianus	LIDRCLAP	6	1	5	5	11	4	13	5	10	3
Tubifex tubifex	TUFETUBI	5		3	2	1	3	2	1	4	2
Ilyodrilus templetoni	ILDRTEMP	19	7	23	23	25	6	37	1	41	9
Aulodrilus pigueti	AUDRPIGU	1	1		3		4				6
Tubificidae j.w.s.	TUFIAEJM	25	21	60	57	37	20	41	15	72	65
Tubificidae j.wh.s.	TUFIAEJZ	16	61	123	114	141	229	208	113	152	114
Naididae											
Dero digitata	DERODIGI	62	114	47	61	141	520	46	7	44	170
Stylaria lacustris	STLALACU	1	6	3	3	2	64	4	49	3	5
Nais simplex	NAISSIMP		1						1		
Uncinais uncinata	UNCIUNCI	1	1								
Slavina appendiculata	SLAVAPPE		18		2						1
Nais variabilis	NAISVARI						1				
Total number of Oligochaeta		302	260	274	285	372	870	378	207	341	388

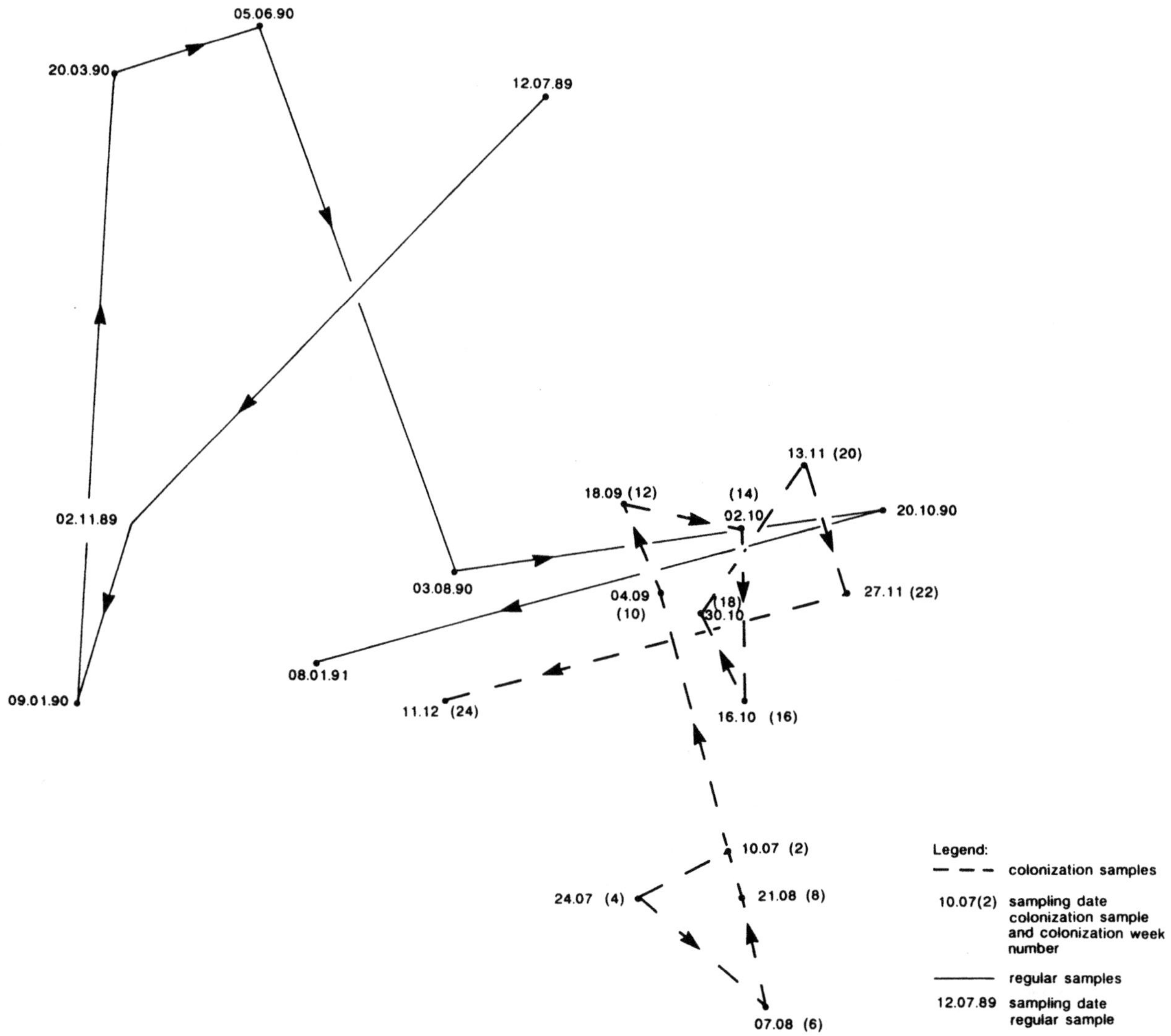

Fig. 1. Ordination (PCA) diagram of the regular oligochaete samples of ditch 16 showing their seasonal variation (straight line) with colonization samples (dotted line) added as supplementary points.

in oligochaete composition. The ordination (Fig. 2) illustrates the importance of both the variation among sampling dates and among ditches. Sampling date splits the diagram approximately into three parts; winter (January and March) 1990 in the left upper corner when *Limnodrilus claparedeianus, L. hoffmeisteri* and *Ilyodrilus templetoni* are more abundant, autumn (October) 1990 in the left lower corner when *Dero digitata* is highly abundant, and winter (January) 1991 to the right when numbers and diversity are low. The winter of 1990–1991 was much colder than the previous winter (Fig. 3). Cold periods, especially when ice formation causes oxygen levels to decrease, usually reduces invertebrate abundances.

The seasonal variation in abundance comprises about 35% of the total variance and makes the interpretation of variation among ditches more difficult. To highlight the variation among ditches, a partial ordination was carried out in which the variation attributed to sampling date was factored out (Fig. 4). The permutation test showed that the control ditches differed significantly ($P < 0.01$) in their oligochaete composition. *Aulodrilus pigueti*

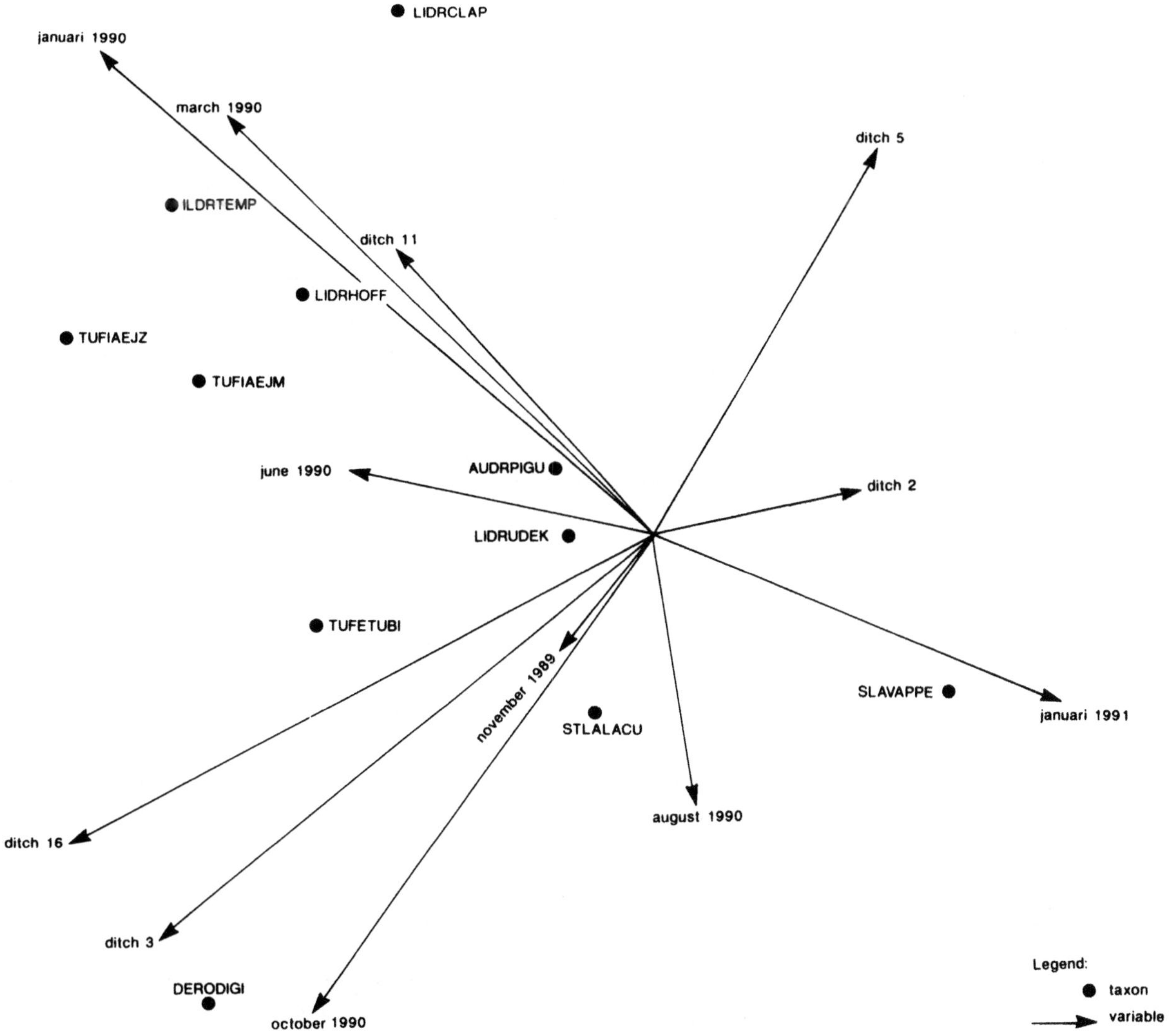

Fig. 2. Ordination (RDA) diagram showing the variation in the abundances of taxa (dots) among control ditches (arrows) and sampling dates (arrows). Abbreviations are explained in Table 3. The diagram explains 47% of the total variance in oligochaete abundance, the first two eigenvalues being 0.30, 0.17. The third and fourth eigenvalues are 0.07 and 0.04, respectively.

and *Slavina appendiculata* occur only in ditch 11 and 2, respectively. Ditch 11 is dominated by juvenile tubificids and *Limnodrilus hoffmeisteri*, ditches 3 and 16 are dominated by juvenile tubificids and *Dero digitata*. Ditches 2 and 5 have lower numbers of all oligochaetes.

In conclusion, over the three-year period the experimental ditches have developed differently. The ditches are dominated by opportunistic taxa which suggests that the ditches are still in an early developmental stage. This also explains the divergence in the development of the oligochaete communities in the different ditches.

Treatment with chlorpyrifos

Preliminary tests had shown that *Limnodrilus hoffmeisteri* had an LC_{50} (48h) of > 50 μg/l chlorpyrifos (Kersting, 1990). By contrast several other studies showed no direct effects on oligochaetes (Macek *et al.*, 1969; Nelson & Evans, 1973).

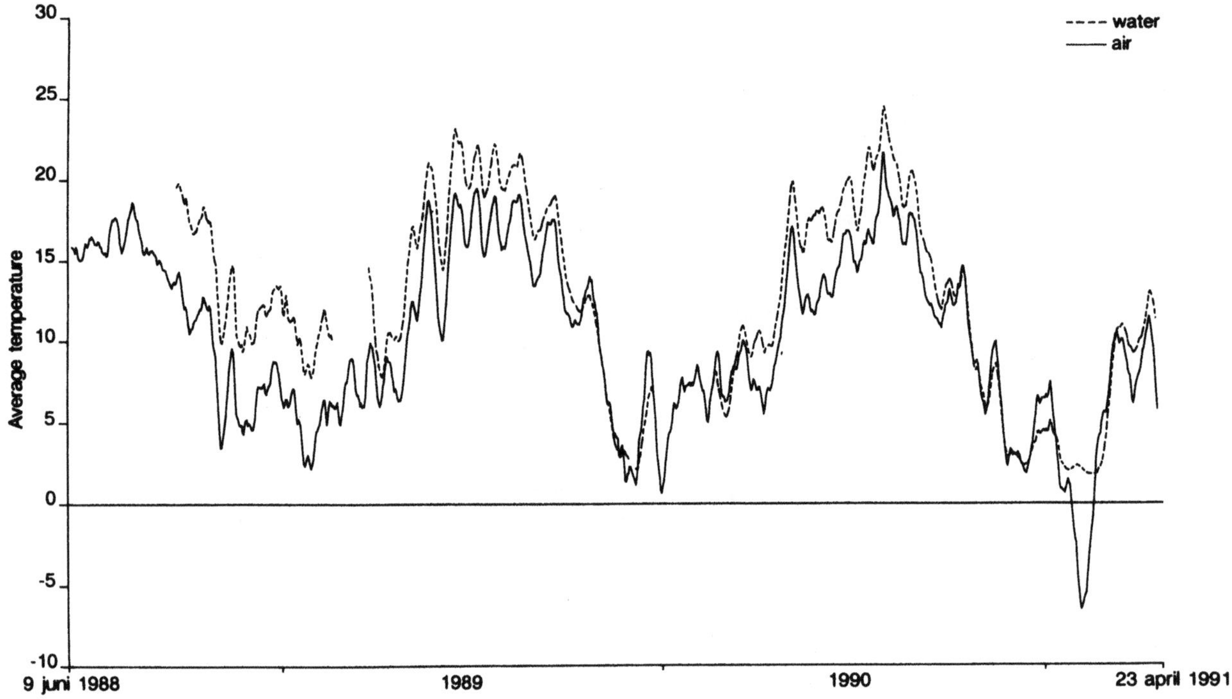

Fig. 3. Air and water temperature of ditch 2.

Although secondary effects, between 14 and 56 days after treatment, are suggested by van Wijngaarden & Leeuwangh (1989) who reported that oligochaetes decreased in numbers or even vanished. Therefore, the monitoring program of oligochaetes focused on long term changes (i.e. up to six months).

After the chlorpyrifos addition, there was an immediate mixing with the water in each ditch (Fig. 5). Fifty-six days later the concentrations in the water were strongly reduced. Soon after the treatment a conspicuous mortality of insects and crustaceans was observed (van Wijngaarden, pers. comm.). The average abundances before and after the chlorpyrifos treatment at the two highest concentration levels indicate a decrease in abundance of several taxa (e.g. *Ilyodrilus templetoni*, *Dero digitata* and tubificid juveniles with hair setae) although abundances in the control ditches also decreased (Table 3). Furthermore, the number of naidids and tubificids decreased to zero by 3 August 1990 whereupon the naidids remained at zero and the tubificids increased in the following two sampling periods in ditch 6 which received the highest chlorpyrifos treatment. In contrast, ditch 9 that also received the highest chlorpyrifos treatment, showed little response. Through secondary processes one ditch may be affected differently within a treatment, especially when the oligochaete communities are in an early developmental stage. The destabilization caused by the treatment sets the system back in its development. As is evident from the previous subsection, individual ditches can differ considerably.

Ordination (RDA with model 'ditch + chlorpyrifos load' and covariable 'sampling date') suggested a chlorpyrifos effect on the first axis based on the t-ratio associated with the canonical coefficient for chlorpyrifos (estimate/(apparent s.e.) = 2.2). The apparent standard error is, however, an underestimate. Therefore, the permutation test is required to validate this inference.

Ordination (RDA with model 'chlorpyrifos load' and covariables 'ditch' and 'sampling date') showed that the chlorpyrifos effect, if it exists, is small (explaining 1% of the total variance) compared to ditch and time differences (each account-

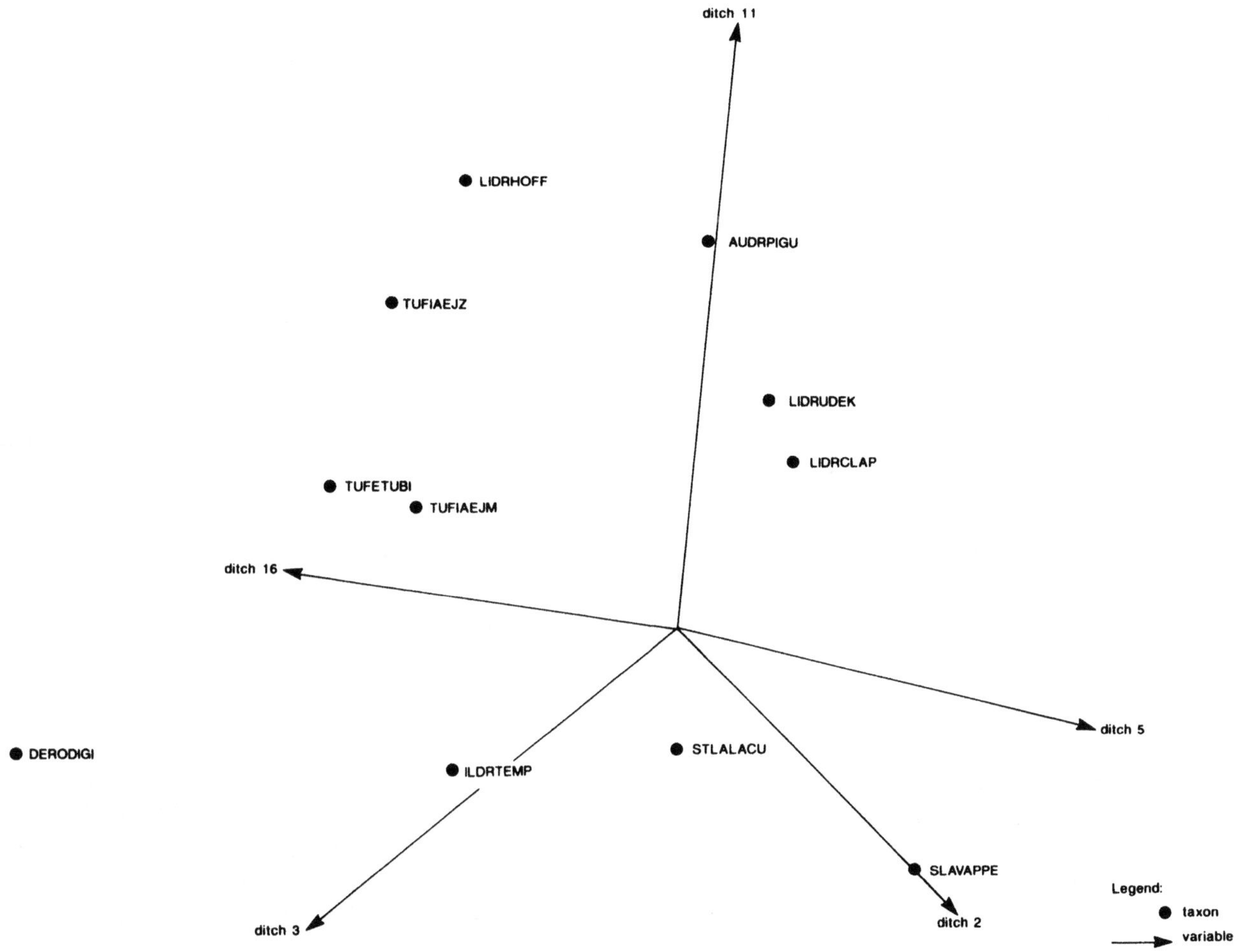

Fig. 4. Ordination (partial RDA) diagram showing the variation in the abundances of taxa (dots) between control ditches (arrows) after adjustment for seasonal variation. Abbreviations are explained in Table 3. The diagram explains 92% of the variance among ditches, the first two eigenvalues being 0.19, 0.06. The third and fourth eigenvalues are both c. 0.01.

ing for ca. 25% of the variance). Both the Monte Carlo test that is valid 'by design' and the more liberal unrestricted permutation test failed to demonstrate the existence of a chlorpyrifos effect ($P = 0.18$ and 0.14, respectively).

In conclusion no significant effect of chlorpyrifos on the oligochaete populations was detected. Although, we cannot exclude the possibility that through secondary processes, especially for oligochaete communities in ditches which are in an early developmental stage, taxa and abundances are influenced in a more stochastic way (e.g. compare ditch 6 and ditch 9).

Eutrophication of clay ditches

The eutrophication of both clay-lined and sand-lined ditches did not result in higher phosphate and nitrogen levels in the water column after the first one-and-a-half years of treatment, except at the highest levels of nutrient application (Table 4). It is more likely that the nutrients are incorporated into organic material and sediment.

Ditch 13 and 14 provide typical examples of the effects of nutrient additions to the oligochaetes in the clay ditches, (i.e. the ones with the two highest levels of nutrient additions; Table 5). Most

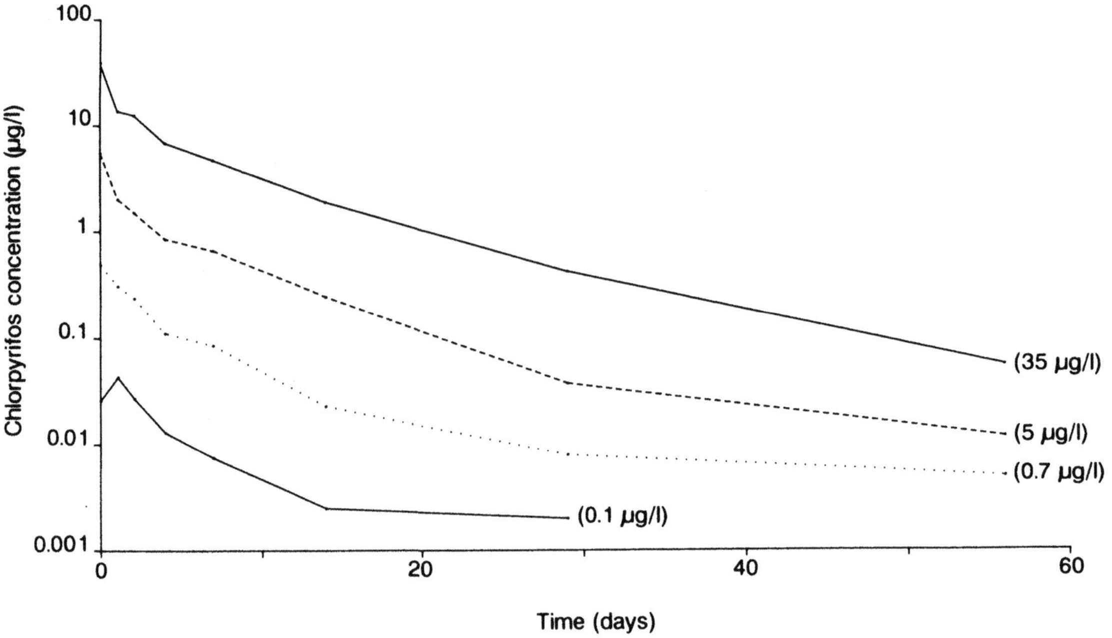

Fig. 5. The decline of average chlorpyrifos concentrations in time for the four different treatments.

likely, the eutrophication effects are mediated through changes in either the quantity or the quality of food available to the oligochaete populations and possibly through changes in the oxygen regime. In both ditches *Limnodrilus udekemianus* was absent and the abundances of *Ilyodrilus templetoni*, *Dero digitata* (also less abundant in ditch 15) and juvenile tubificids, especially those with hair setae, were low. In ditch 13 *Limnodrilus claparedeianus* was abundant.

Table 4. Chemical parameters (in mg/l) of the eutrophication ditches.

Ditch	Date	NH_4-N	Kj-N	NO_3-N	P-tot	PO_3-P*	Cl
13	14.11.89	0.04	2.0	1.9	0.56	0.42	7.9
13	19.11.90	2.5	4.6	0.12	3.2	1.6	9.6
14	14.11.89	0.04	1.8	0.00	0.10	<0.01	9.5
14	19.11.90	0.00	1.6	0.04	0.05	0.01	9.7
15	14.11.89	0.09	1.1	0.00	0.07	<0.01	8.0
15	19.11.90	0.01	1.7	0.04	0.03	0.01	9.4
16	14.11.89	0.05	1.3	0.00	0.06	<0.01	8.7
16	19.11.90	0.00	0.90	0.09	0.05	<0.01	7.9
17	14.11.89	0.13	1.4	0.05	0.04	<0.01	9.2
17	19.11.90	0.12	0.70	0.21	0.10	0.02	9.0
18	14.11.89	0.11	1.4	0.07	0.06	<0.01	9.2
18	19.11.90	0.03	2.6	0.14	0.01	<0.01	8.4
19	14.11.89	0.13	2.5	0.05	0.08	<0.01	9.3
19	19.11.90	0.20	1.3	0.12	0.02	0.01	8.8
20	14.11.89	0.33	3.2	3.8	1.6	0.76	8.5
20	19.11.90	5.7	6.0	0.21	4.3	3.8	8.4

* All samples under 14.11.89 refer to 10.07.89.

Table 5. Average abundance (number/tray) of oligochaetes in the eutrophication experiments over 10 sampling periods. The control ditches for clay include the control ditches of the chlorpyrifos experiment; ditch numbers are 2, 3, 5, 11 and 16. The control sand ditch is number 17.

	Clay ditches				Sand ditches			
Ditch number	13	14	15	Control	20	19	18	Control
Tubificidae								
Limnodrilus hoffmeisteri	43	11	38	24	23	9	1	8
Limnodrilus udekemianus			1	6		1		
Limnodrilus claparedeianus	53	17	6	12				
Tubifex tubifex		1		6				1
Ilyodrilus templetoni	4	13	36	50				
Aulodrilus pigueti				6	2			
Tubificidae j.w.s.	12	17	120	137		2	2	2
Tubificidae j.wh.s.	189	129	372	279	79	20	2	48
Naididae								
Dero digitata	46	28	21	214	17	226		
Stylaria lacustris		4		8	32			2
Nais simplex		1			1	109	5	7
Uncinais uncinata					568	3	14	11
Slavina appendiculata				1		100		
Nais variabilis					43	138	2	259
Lumbriculidae								
Lumbriculus variegatus					5			
Enchytraeidae	1	1						
Total number of Oligochaeta	348	205	594	743	770	608	26	338

Figure 6 illustrates the differences between the eutrophicated clay ditches (ordination based on RDA with model 'ditch' and covariable 'sampling date'). The first axis is highly correlated with the variable 'nutrient supply' ($r = -0.53$) which is added to Fig. 6 as a passive environmental variable. RDA with model 'nutrient supply' and a random permutation of the ditches (using permutations among ditches of a whole series of observations) confirms the significant effect of the variable 'nutrient supply' (P-value 0.01).

The significantly reduced abundances of tubificids and, especially, naidids in the more eutrophicated clay ditches contrast with the common opinion that both groups tend to increase when waters become eutrophied. The eutrophication process lasts already for almost two years now. The chemical analysis of the water does not yet show an increase in phosphate and nitrogen. This finding suggests that a lot of the nutrients that were added have been incorporated into the system by algae and macrophytes. However, it does not imply that this organic material is directly available as a food source. Furthermore, the higher primary production can imply greater fluctuations in, for example, oxygen. This combination of circumstances (poor food quality and low oxygen conditions) may explain the lower number of oligochaetes.

Eutrophication of sand ditches

The effect of the eutrophication experiment in the sand ditches was less obvious (Table 5). The higher abundances of *Limnodrilus hoffmeisteri*, *Uncinais uncinata*, *Stylaria lacustris*, and juvenile tubificids without hair setae indicated greater nutrient availability in ditch 20. Several naidids were abundant in ditch 19 (*Nais variabilis*, *N. simplex*, *Dero digitata*, *Slavina appendiculata*). Ditch 18 was depauperate in oligochaetes but ditch 17 was

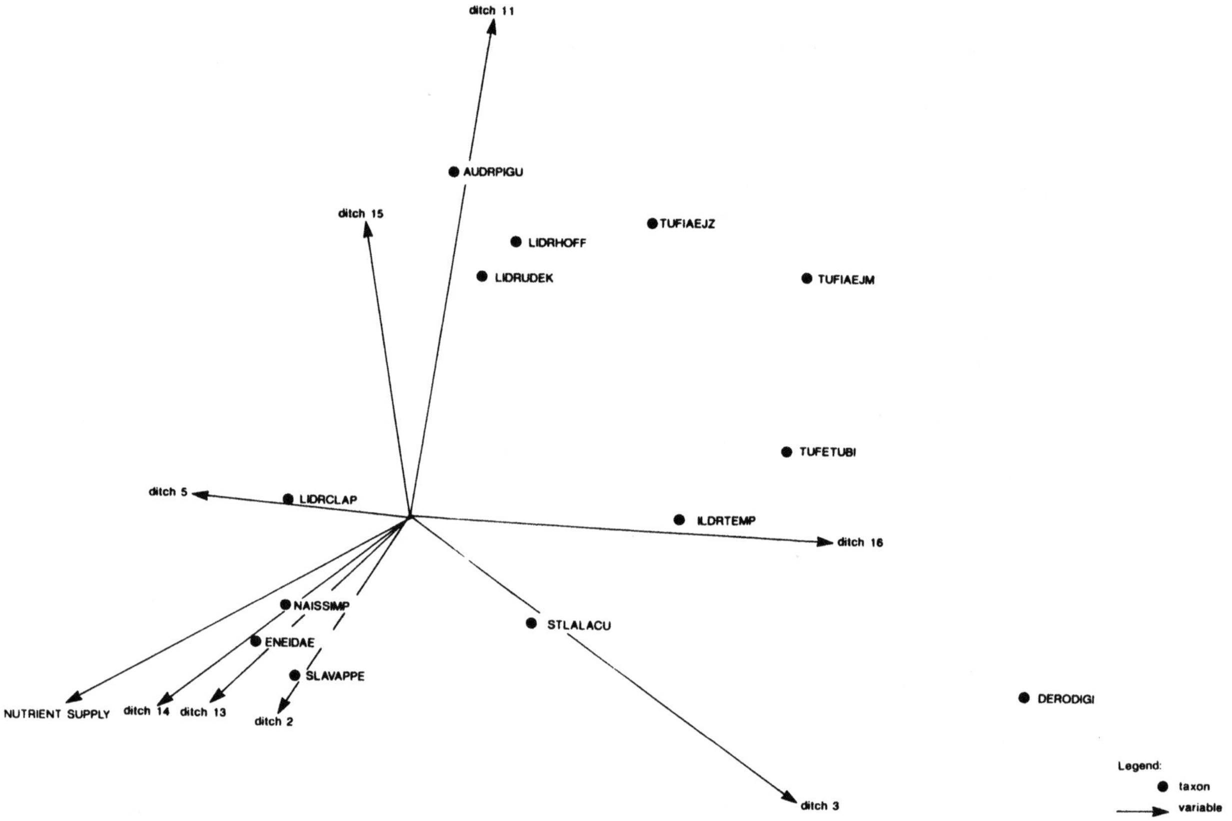

Fig. 6. Ordination (partial RDA) diagram showing the variation in the abundances of taxa (dots) between the control clay-lined ditches (arrows; ditch 2, 3, 5, 11, and 16), the eutrophied clay-lined ditches (arrows; ditch 13, 14, and 15) and nutrient supply (arrow). Abbreviations are explained in Table 3, except for ENEIDAE = Enchytraeidae.

richer in tubificids and some naidids, especially *Nais variabilis*, relative to ditch 19.

Figure 7 shows the differences between the eutrophic sand ditches (ordination based on RDA with model 'ditch' and covariable 'sampling date'). The second axis is highly correlated with the variable 'nutrient supply' ($r = -0.81$) which is included in the ordination as a passive environmental variable. RDA with model 'nutrient supply' and a random permutation of the ditches shows that the effect of eutrophication approaches significance (*P*-value 0.11).

The difference in individual development of these four ditches is very pronounced. The low nutrient status of the sandy bed at the beginning of the experiment apparently slows down the rate of colonization and eutrophication.

Odum (1971) described ecological succession as "an orderly process of community development that involves changes in species structure and community processes with time; it is reasonably directional and, therefore, predictable". At the start of the ditch experiments colonization had just started (see the development of the individual control ditches); the ditches were only one-to-two years old. A secondary succession took place in the clay ditches possible due to the higher nutrient content of the clay bed, a flow-through system of water transfer which assisted species exchange between the clay ditches before the experiments started and to the origin of the ditch bed (it was already inoculated with cocoons, eggs, seeds, and so on). The nutrient-poor sand with no initial inoculum led to a primary succession to evolve in the sand ditches. Consequently all experiments were initiated while the ditch sys-

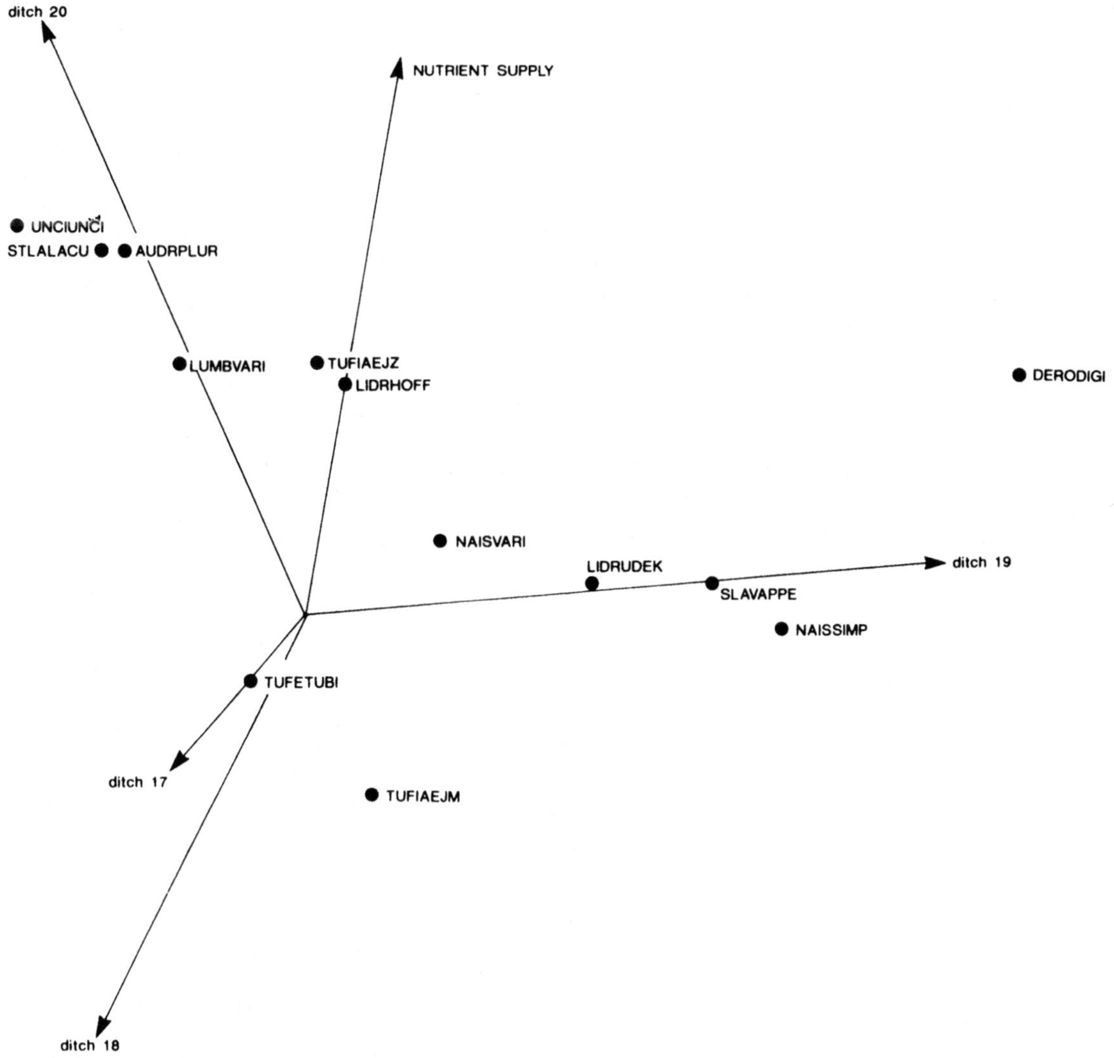

Fig. 7. Ordination (RDA) diagram showing the variation in the abundances of taxa (dots) between the control sand-lined ditch (arrow; ditch 17), the eutrophied sand-lined ditches (arrows; ditch 18, 19 and 20) and nutrient supply (arrow) are indicated. Abbreviations are explained in Table 3, except for AUDRPLUR = Aulodrilus pluriseta and LUMBVARI = Lumbriculus variegatus.

tems were still at an early developmental stage when opportunistic species tend to dominate the community. Furthermore, in this stage the turnover rate of species and population abundances are variable (Pianka, 1978). The temporal changes in the oligochaete populations in the control clay ditches, and, especially, in the sand ditches showed an individualistic and possible a stochastic development. These developmental patterns were unpredictable. The individual ditches show a divergence in time (Fig. 8). This stresses the importance of the colonization time or stabilization factor in large-scale experiments. Large-scale experiments or mesocosms need a certain time to develop and stabilize before one can use parameters of community structure like population abundance, to detect changes. After a certain period of time a convergence between the ditches is expected (Fig. 8). The amount of time necessary to reach this convergence in the oligochaete community structure requires further study.

Conclusions

1. A period of 20 weeks is enough to complete the colonization process in the trays.
2. The oligochaete 'communities' in colonization trays in the control ditches have diverged during the three years of existence of the ditches.
3. No significant chlorpyrifos treatment effects were revealed by the oligochaete populations.
4. A significant effect of nutrient enrichment on oligochaetes in clay ditches was observed.
5. The effect of nutrient enrichment on oligochaetes in sand ditches was not significant.
6. Large-scale experiments or mesocosms need a certain time to develop and stabilize before using community structure parameters like population abundance, to detect changes in oligochaete assemblages.

Acknowledgements

The authors thank Joke Schot, Nienke Korteweg and Henk van Rheede for assistance in the field and laboratory. Jan Drent and Rob Portielje provided the chemical data. Further thanks go to several colleagues of the Institute for their contributions. We thank R. K. M. Somers and R. C. Bailey for their constructive reviews.

References

Aston, R. J., 1973. Tubificids and water quality: a review. Environ. Pollut. 5: 1–10.

Bowling, J. W., J. P. Giesy, H. J. Kania & R. L. Knight, 1980. Large-scale microcosms for assessing fates and effects of trace contaminants. In J. P. Giesy (ed.), Microcosms in ecological research. DOE Symp. Ser. 52: 224–247.

Borcard D., P. Legendre & P. Drapeau, 1992. Partialling out the spatial component of ecological variation. Ecology 73: 1045–1055.

Chapman, P. M., M. A. Farrell & R. O. Brinkhurst, 1982. Relative tolerances of selected aquatic oligochaetes to individual pollutants and environmental factors. Aquat. Toxicol. 2: 47–78.

Green, R. H., 1979. Sampling design and statistical methods for environmental biologists. Wiley, New York.

Hall P. & D. M. Titterington, 1989. The effect of simulation order on level accuracy and power of Monte Carlo tests. J. R. Statist. Soc. B 51: 459–467.

Hall P. & S. R. Wilson, 1991. Two guidelines for bootstrap hypothesis testing. Biometrics 47: 757–762.

Higler, L. W. G., 1989. Hydrobiological research in peat polder ditches. Hydrobiol. Bull. 23: 105–109.

Hurlbert S. H., 1984. Pseudoreplication and the design of ecological field experiments. Ecological Monographs 54: 187–211.

Hynes, H. B. N., 1960. Biology of polluted waters. Liverpool Univ. Press. 202 pp.

Jagers op Akkerhuis G. A. J. M. & H. van der Voet, 1992. A dose-effect relationship for the effect of deltamethrin on a Linyphiid spider population in winter wheat. Arch. Environ. Contam. Toxicol. 22: 114–121.

Kersting, K., 1990. Effect of chlorpyrifos on oxygen and pH regime of experimental draining ditches. SETAC 1990, 11th Ann. Meeting, November 11–15, 1990, Arlington, Virginia.

Macek, K. J., D. F. Walsh, J. W. Hogen & D. D. Holz, 1969. Toxicity of the insecticide Dursban to fish and aquatic invertebrates in ponds. Trans. Am. Fish. Soc. 101: 420–427.

Manly B. F. J., 1990. Randomization and Monte Carlo methods in biology. Chapman and Hall, London.

Marriott F. H. C., 1979. Barnard's Monte Carlo tests: how many simulations? Applied Statistics 28: 75–77.

Nelson, J. H. & E. S. Evans Jr., 1973. Field evaluations of the larvicidal effectiveness. Effects on non-target species and environmental residues of a slow-release polymer formulation of chlorpyrifos. U.S. Army Environ. Hygiene Agency Rep. no. 44–022–73/75. 15 pp.

Odum, E. P., 1971. Fundamentals of ecology. Saunders, Philadelphia. 574 pp.

Pianka, E. R., 1978. Evolutionary ecology. Harper & Row, New York. 397 pp.

Stewart-Oaten, A., W. W. Murdoch & K. R. Parker, 1986. Environmental impact assessment: 'pseudoreplication' in time? Ecology 67: 929–940.

Ter Braak, C. J. F., 1987. Ordination. In R. H. G. Jongman, C. J. F. Ter Braak & O. F. R. Van Tongeren (eds), Data analysis in community and landscape ecology. Pudoc, Wageningen.: 91–173.

Ter Braak, C. J. F., 1988. CANOCO – A FORTRAN program for canonical community ordination by [partial] [detrended] [canonical] correspondence analysis, principal component analysis and redundancy analysis (version 2.1). Report LWA-88-02. Agricultural Mathematics Group, Wageningen.

Ter Braak, C. J. F. & Prentice, I. C., 1988. A theory of gradient analysis. Advances in ecological research 18: 271–317.

Ter Braak, C. J. F., 1990. Update notes: Canoco version 3.10. Agricultural Mathematics Group, Wageningen, The Netherlands.

Ter Braak, C. J. F., 1992. Permutation versus bootstrap significance tests in multiple regression and ANOVA. In K. H. Jökel *et al.* (eds.), Bootstrapping and related techniques. Springer-Verlag, Berlin.

Verdonschot, P. F. M. & Higler, L. W. G., 1989. Macroinvertebrates in Dutch ditches: a typological characterization and the status of the Demmerik ditches. Hydrobiol. Bull. 23: 135–142.

Wiederholm, T., 1980. Use of benthos in lake monitoring. J. Wat. Pollut. Cont. Fed. 52: 537–547.

Wijngaarden, R. van & P. Leeuwangh, 1989. Relation between toxicity in laboratory and pond: an ecotoxicological study with chlorpyrifos. Med. Fac. Landbouww. Rijksuniv. Gent 54: 1061–1069.

Hydrobiologia **278**: 267–274, 1994.
T.B. Reynoldson & K.A. Coates (eds), Aquatic Oligochaete Biology V.

Quantitative observations on the population ecology of *Branchiura sowerbyi* (Oligochaeta, Tubificidae)

C. Bonacina [1], A. Pasteris [2], G. Bonomi [2] & D. Marzuoli [2]
[1] *CNR-Istituto Italiano di Idrobiologia, Largo Tonolli, 50, I 28048 Pallanza NO, Italy;* [2] *Dipartimento di Biologia evoluzionistica sperimentale, Universita di Bologna, Via Belmeloro, 8, I 40126, Bologna BO, Italy*

Key words: Branchiura sowerbyi, population ecology, cohort cultures

Abstract

Cohorts of *Branchiura sowerbyi* were reared at different temperatures and initial population densities in order to obtain data suitable for the interpretation of population dynamics in field populations. Percent hatching from cocoons reaches its maximum at 25 °C and decreases towards lower and higher temperatures. Embryonic development time, T_E, was measured and the relative threshold temperature, 10 °C, calculated by extrapolation. The degree day requirement for embryo development is 195 °C d. The time of first cocoon laying (T_{gm}) was observed and the ratio T_E/T_{gm} was seen to fit with that of other tubificid species cultured so far. Embryo mortality is rather high, while worm mortality is low or very low. Fecundity increases from 15 to 20 °C but decreases at 25 °C. A mathematical model for the simulation of population densities with four size-stage compartments is suggested. It could be used for the optimization of worm uptake (simulated as stage specific mortality) in mass cultures reared for the production of *Branchiura*, to be used as food for fish fingerlings.

Introduction

Over the last few years we have devoted part of our activity to the study of reservoirs, which provides good opportunities for profundal benthic studies.

These studies are normally funded by water- or electric power-supply agencies, requiring information on the state of the water bodies or a forecast of their future trophic state after some hydrological manipulation, but we see these occasions mainly as opportunities for studying oligochaete communities with simple taxonomic structures. Indeed, most of our previous papers deal with populations from reservoirs (Adreani & Bonomi, 1980; Bonomi & Adreani, 1978) and with cohort and mass cultures (Adreani *et al.*, 1984; Bonacina *et al.*, 1987, 1989a, 1989b), which we studied in order to obtain quantitative data on the life cycle of some species (mainly *Tubifex tubifex* and *Limnodrilus hoffmeisteri*) and to describe their population dynamics in the field.

Recently we studied a pumped storage system, the upper reservoir of which (Lake Brasimone) supported a profundal community dominated by *T. tubifex*, *L. hoffmeisteri* and *Branchiura sowerbyi.*

Lake Brasimone is located at an altitude of 845 m a.s.l., has a maximum depth of 25 m, a volume of 6.6 10^6 m^3 and an estimated water renewal time of 3–4 days. The lake waters are therefore fully mixed throughout the year, oxygenation is very good in the profundal and temperature displays an annual range of 4–20 °C (Salmoiraghi, in press). This is probably the main reason for the presence, in this reservoir, of *B. sowerbyi* a species rarely found in the profundal of reservoirs (Margalef *et al.*, 1976). The *Branchiura* population of Lake Brasimone was studied in detail (Bonacina *et al.*, in press) in the laboratory to obtain quantitative data suitable for developing a population dynamics model and evaluating the field population parameters.

Branchiura sowerbyi appears to prefer warm waters and in temperate countries is often found in artificially heated habitats, though it is not restricted to them (Aston, 1968; Brinkhurst & Jamieson, 1971; Casellato, 1984). Cocoons are large and transparent and the embryos show early development of posterior gill filaments, a peculiar feature of this species. These distinctive characters make field populations of *Branchiura* suitable for population studies.

Our laboratory investigations of *Branchiura* were also supported by the Italian Electric Power Agency (ENEL). This agency uses discarded warm water from electric power plants for fish farming and is also interested in using this warm water for the mass culture of Oligochaeta for fish (fingerlings) food.

Materials and methods

Large numbers of *B. sowerbyi* were collected from storage ponds at the electric power station of Castel S. Giovanni (Piacenza, N. Italy) near the River Po. All the individuals collected were kept in large aquaria at room temperature or in glass jars in a thermostat at 17 °C. The larger individuals were then transferred into cultures kept in thermostats (temperature fluctuation: ±0.1 °C) at 12, 15, 17, 20, 25, 30, 32.5 and 35 °C using glass jars (11 cm in diameter) half-filled with sand covered with dechlorinated tap water. After about a month, cocoons were collected each day from these initial cultures and used to start cocoon cohort cultures at the same temperatures. Cocoon cultures were inspected every second day and newly hatched worms counted and removed.

Furthermore, cocoons from the 15, 17, 20, 25 °C initial cultures were collected weekly and kept in culture (9.5 diameter jars) at the same temperatures in order to obtain cohort cultures of worms. These cocoon cohort cultures were inspected weekly, and the newly hatched worms counted. The same type of sand and water was used as for the initial cultures. At every weekly inspection, the sand was washed, food (deep frozen spinach) added, the individuals counted according to the biological stage reached, and weighed monthly, as soon as it was possible to do so without damaging them. Each individual was dried on filter paper and then weighed to the nearest 0.1 mg using a Mettler AE 260 balance.

Results

A – Cocoon cultures

Egg and embryo mortality rates are higher in *B. sowerbyi* than in other tubificid species in similar culture conditions (Adreani *et al.*, 1984; Bonacina *et al.*, 1987, 1989a; Bonomi & Di Cola, 1980); percent hatching of worms from cocoons (Fig. 1) appears to reach the maximum at 25 °C and decreases towards lower and higher temperatures. No worms were hatched in cultures at 12 and 35 °C.

There is, on the other hand, a lower mortality of the hatched worms during the experiments (Table 1 and Bonacina *et al.*, 1987).

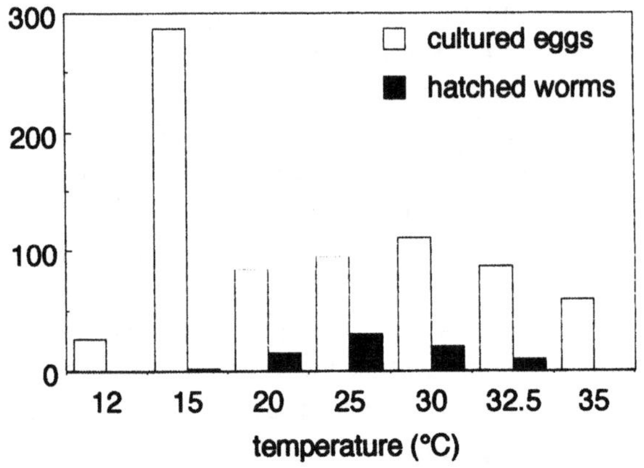

Fig. 1. Cultured eggs and hatched worms (ordinate) in the range 12.0–35.0 °C.

Table 1. Long term cohorts reared at different temperatures (a), number of cocoons (b), eggs (c) and newborn (e) used to start the experiment. Clutch size (d), i.e. mean number of eggs/cocoon is also given.

T (°C)	(a)	(b)	(c)	(d)	(e)
15	4	445	719	1.6	46
17	5	229	429	1.9	53
20	7	212	457	2.2	194
25	7	212	377	1.8	166

The relationship between temperature and the time required for embryonic development (time spent inside the cocoon) is shown in Fig. 2a. The threshold temperature for embryonic development was calculated as 10 °C (Fig. 2b). As with other tubificid species studied (Bonomi & Di Cola, 1980; Adreani *et al.*, 1984), *B. sowerbyi* displays high variability in development time. The results for development time agree with those of Casellato (1984) who cultured *Branchiura* from a

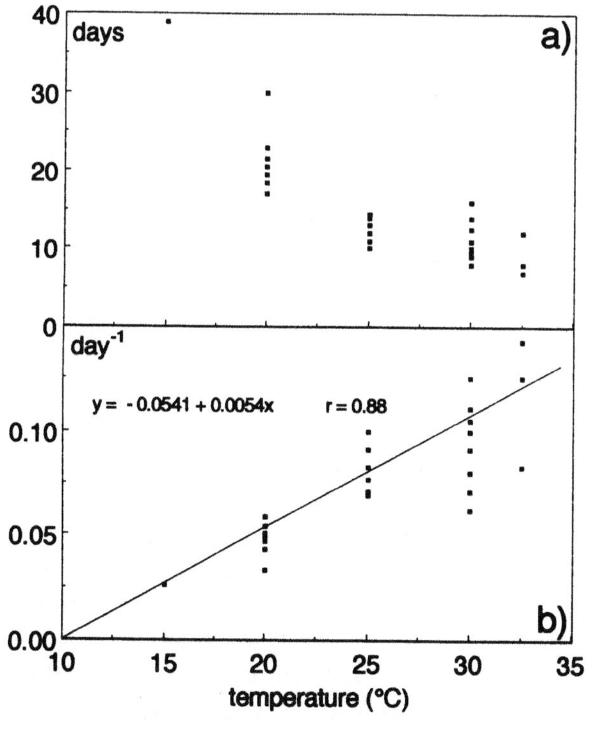

Fig. 2. Relationship between temperature and (a) the time spent inside the cocoon, (b) the embryos development rate. The intercept gives a threshold temperature of 10 °C.

pond in the Botanical Gardens of the University of Padua. The clutch size (1.8 eggs/cocoon) of the cocoons laid by *Branchiura* (Casellato, 1984) is close to that of our cultures (Table 1). Aston *et al.* (1982) report for their cultures a clutch size of ca. 2.5 eggs/cocoon. They do not observe any cocoon laying at 13 and 17 °C but only at higher temperatures, presumably because of their relatively short observation period.

B – Long term cohort cultures

Cohorts with low initial density

The aim of these cohort cultures was that of obtaining a measure of growth rates and the duration of the different life stages. A total of 20 cohorts were studied and the numbers of young + mature and those of ovigerous individuals as well as those of cocoons and eggs produced were counted weekly. Hatching rates, as already mentioned, were much higher at higher temperatures (in the range 15–25 °C); as a result, the initial number of young individuals was higher at 20–25 than at 15 °C. Rearing temperatures were 12, 15, 17, 20, 25 °C. At 12 °C, cocoons are laid but they do not develop; consequently, no long term cohort cultures were obtained at this temperature (Table 1).

To minimize the effects of density on the development processes (Bonacina *et al.*, 1989a), a first analysis of the data was made using only cohorts with N_0 ranging from 4 to 15. The four cohorts reared at 15 °C (N_0 = 15, 12, 10, 9) show remarkable homogeneity; the time of first cocoon laying lies between 230 and 260 days and their growth rates are comparable (Fig. 3).

Of the two cohorts reared at 17 °C (N_0 = 12 and 8), the first reached first deposition at 240 days (some individuals showed a very slow initial growth but attained the same final weight) (Fig. 4a, 4b), the second grew much faster and attained first deposition at 135 days (Fig. 4c, 4d). It is likely that we operated an unintentional selection, forming the first cohort with 'slow growing' individuals and the second one with 'fast growing' individuals (Adreani *et al.*, 1981).

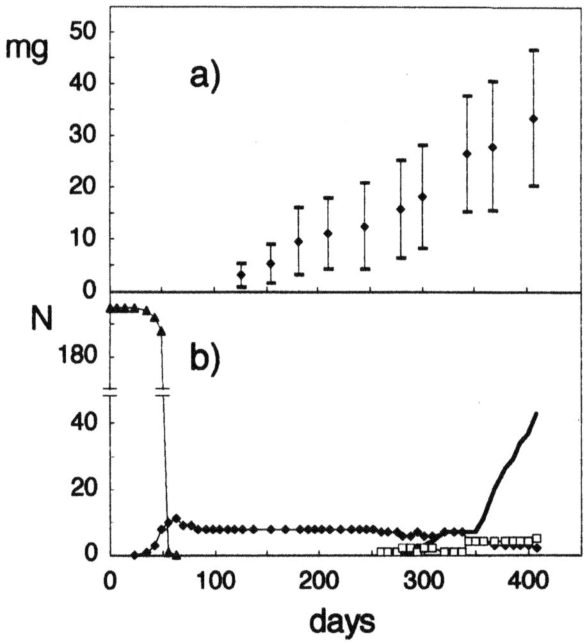

Fig. 3. Example of a cohort cultured at 15 °C. (a) mean weight and standard deviation, (b) ▲ = eggs + embryos; ◆ = young + mature; □ = ovigerous; – = cumulative plot of eggs production.

Of the two cohorts reared at 20 °C ($N_0 = 4$ and 8), the first attained first deposition at 105 days (Fig. 5); the second did not survive beyond 119 days.

Two cohorts were also reared at 25 °C ($N_0 = 12$ and 13); they showed a first deposition time of 90–110 days and a homogeneous growth rate (Fig. 6).

The results seem to indicate that at low N_0 and low temperatures minimum deposition time is ca. 240 days, at higher temperatures (20 and 25 °C) minimum deposition time ranges from 90–110 days.

Cohorts with 'high' initial density

The only 'high' initial density cohort that we were able to culture at 17 °C (Fig. 7, $N_0 = 25$) displayed a long phase with insignificant growth rate before 230 days and a first cocoon laying at 250 days.

Growth at 20 and 25 °C is rather limited in all 'high' density cultures.

At 20 °C, two cohorts ($N_0 = 40$, $N_0 = 31$) all died at about 300 days, but produced the first cocoons after 240–260 days. The first had a lower growth rate and only 1 in 34 became ovigerous. A better growth rate was seen in the second cohort, in which ovigeration rate reached 6 in 30. Two other cohorts ($N_0 = 30$, $N_0 = 31$) grew comparatively better and produced cocoons at 180–220 days. There was a long lag phase in growth and a wide range of weight (12–39 mg; 3–28 mg respectively at 350 days) characterized the cohorts.

The temperature of 25 °C seems to be rather unfavourable for the growth of *Branchiura* cohorts cultured in the conditions we adopted for our experiments. Indeed, the worms did not appear to feed directly on the chlorophyll parenchima of the spinach leaves, and tended to become long and very slender (thread-like). Aston (1966; quoted in Brinkhurst & Jamieson, 1971: 139) claims that at 20 °C with an oxygen saturation below 22% defecation ceases in *B. sowerbyi*; he suggests that feeding activity also ceases in these conditions. This observation could explain our results, but unfortunately we have no data about the oxygen saturation in our cultures.

Nevertheless, mortality is low and cocoon laying, if any, occurs at 260–320 days, the delay being clearly attributable to the limited growth, so that an individual threshold weight for ovigeration and cocoon laying is attained only exceptionally.

Concluding remarks

The data obtained on the hatching rates of young *Branchiura* from the cocoons at different temperatures (Fig. 1) seem to show clearly that at least the embryonic development of this species is favoured by high water temperature. No hatching was observed below 15 °C and the maximum hatching success occurred at 25 °C. This feature is confirmed by the relationship between the time required by intra-cocoon development (T_E) and the

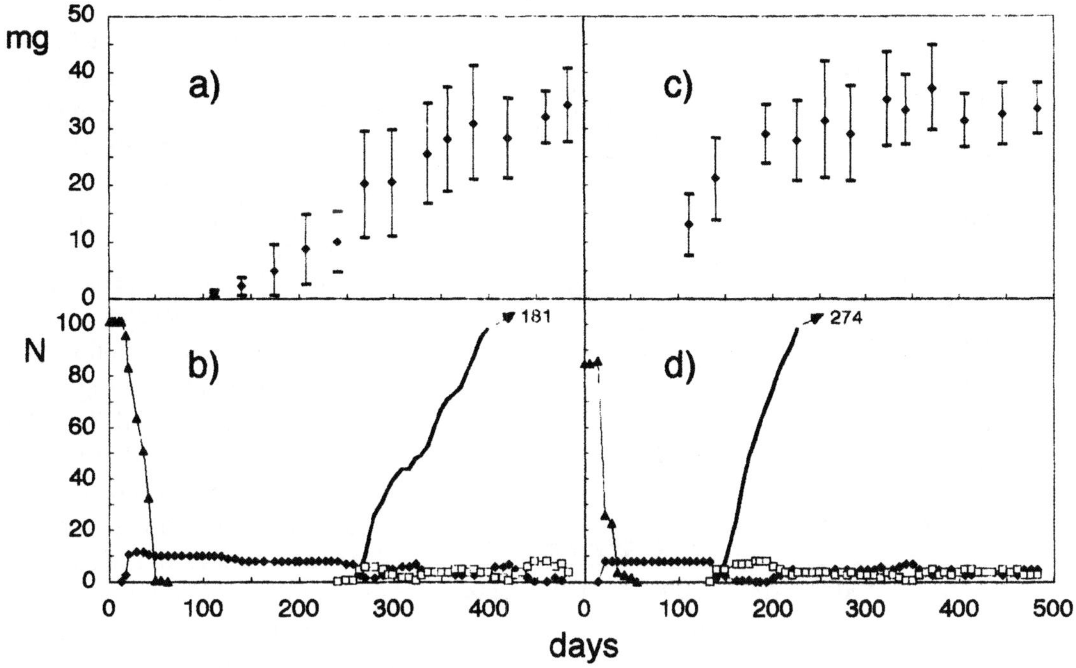

Fig. 4. Growth (a,c) and development (b,d) of the two cohorts cultured at 17 °C. Symbols as in Fig. 3.

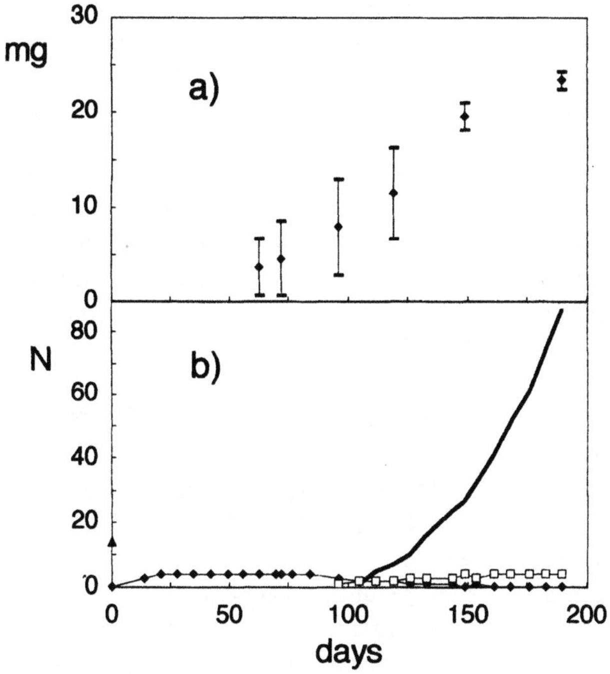

Fig. 5. Cohort cultured at 20 °C. Symbols as in Fig. 3.

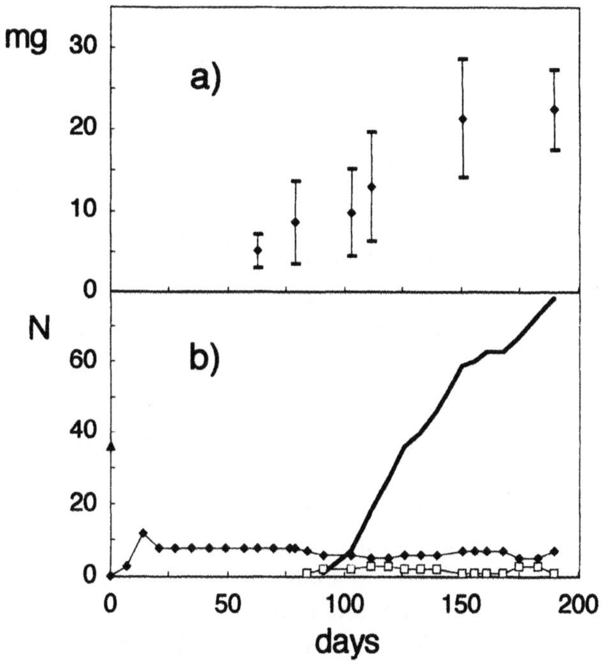

Fig. 6. One of the two cohorts cultured at 25 °C. Symbols as in Fig. 3.

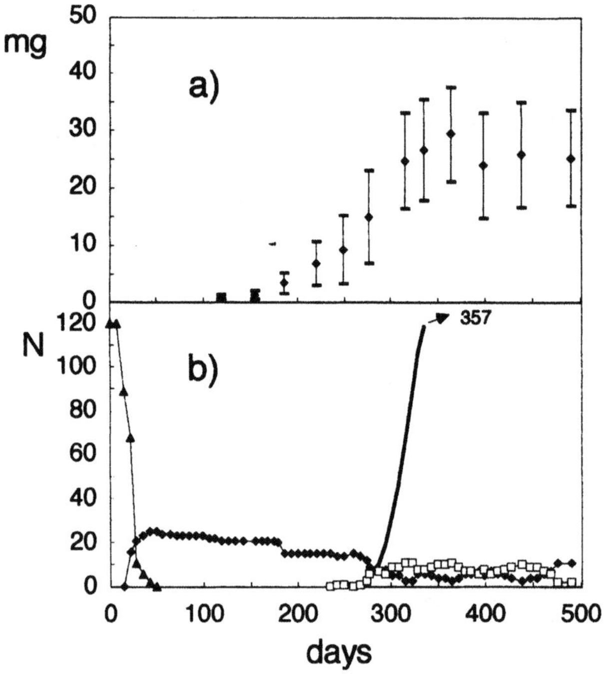

Fig. 7. Cohort cultured at 17 °C ($N_0 = 25$). Symbols as in Fig. 3.

water temperature (Fig. 2), which makes it possible to determine, by extrapolation, a threshold temperature of 10 °C. This is the highest threshold temperature we have measured in our tubificid cultures (Bonacina *et al.*, 1987): previously, it ranged from 0 °C (*T. tubifex* from Lake Pertusillo) to 4 °C (*L. hoffmeisteri*, again from Lake Pertusillo). But it must be remembered that all our previous biological material came from the profundal zone of lakes (Lake Maggiore, Lake Mergozzo, Lake Orta, Lake Garda) or reservoirs (Lake Pertusillo). *Branchiura* is typically found in the littoral or sub-littoral zone of lakes (*e.g.* Lake Garda, Bonomi, 1974; Lake Bolsena, Bonomi, 1968; Lake Nemi, Mastrantuono, 1983) or in small water bodies or streams, i.e. environments in which the temperature attains rather high seasonal values.

If we calculate the T_E of our *Branchiura* taking into account that $T_0 = 10$ °C, *i.e.* if we express T_E in °C days $[(T - T_0)\ d = K]$, we obtain $K = 190$ °C d. A comparison of this value with those given by Bonacina *et al.* (1987, Fig. 1) reveals that the T_E of *Branchiura* is the lowest measured so far. However, when making this comparison, we have to bear in mind that the values of T_E for the 11 species given by the above authors basically ranked the species in the range eutrophic to oligotrophic, while it is evident that the 190 °C d of *Branchiura* is a threshold temperature-dominated value.

The minimum generation time (T_{gm} as used by Bonacina *et al.*, 1987), *i.e.* the time spent inside the cocoon + the time before the laying of the first cocoon, is indicated in Fig. 8 and may be roughly calculated as 2400 °C d. The plot T_E/T_{gm} is added in Fig. 8, and shows that the data for *Branchiura* fit well with those of the other six tubificid species investigated from this perspective.

As the ultimate aim of our investigation is to set up a mathematical model suitable for estimating the main demographic parameters in field populations of *Branchiura*, we may now consider what progress our present results allow in this direction. As knowledge of the population ecology of this species of tubificid is still relatively incomplete, in particular as regards the time spent in the different life stages at different temperatures, a first step could be to consider only a few biological compartments and set up a simulation model, to investigate through 'experiments' with the computer and find out more suitable formulations. A provisional approach might be to consider only four (size & stage) classes: E (eggs + embryos); Y (worms under 5.5 mg); M (worms between 5.5 and 20.0 mg); O (worms over 20.0 mg). We might then describe the dynamics of the population by the following set of differential equations, involving parameters based on the lab observations at 17 °C.

$$dN_E/dt = bN_O - (m_E - 1/T_E)\ N_E$$
$$dN_Y/dt = N_E/T_E - (m_y + 1/T_Y)\ N_Y$$
$$dN_M/dt = N_Y/T_Y - (m_M + 1/T_M)\ N_M$$
$$dN_O/dt = f(N_O, N_M)\ N_M/T_M - m_O N_O$$

where:

b: fecundity rate = 0.25 d^{-1}

T_E: residence time in stage E = 25 d

T_Y: residence time in stage y = 110 d

T_M: residence time in stage M = 110 d

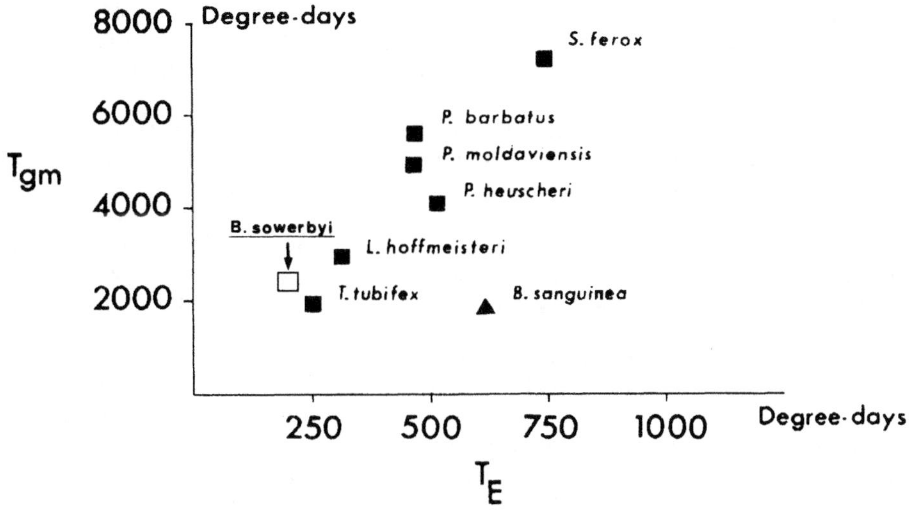

Fig. 8. Plot of the degree days required by the egg + embryo stage, T_E, versus those corresponding to minimum generation time, T_{gm} (from Bonacina *et al.*, 1987, modified).

m_E: mortality rate during stage E = 0.092 d^{-1}
m_Y: mortality rate during stage Y = 0.002 d^{-1}
m_M: mortality rate during stage M = 0.000 d^{-1}
m_O: mortality rate during stage O = 0.000 d^{-1}

A description of the feed-back function f (N_O, N_M) and a suitable way of taking into account the observed great differences in individual growth rates will be the next steps.

Our laboratory results may also supply useful suggestions on the most favourable conditions to be adopted for mass cultures. Considering that fecundity, as well as hatching success, is maximum at 20–25 °C (if population density is relatively low; e.g. 1000–1500 ind m^{-2}) and remembering that the generation time is minimum at 20 °C, a water temperature of 20 °C may well be that suggested for a good worm production (indoor tanks). Using coarse sand as substrate, supplying decaying waste vegetables from food manufacturing as food for worms and maintaining at least 30% 0_2 saturation by means of slow water flow (Aston, 1966) one should guarantee good culture conditions.

The above proposed simulation model might be used to optimize the product uptake (mortality) by the utilizers.

Acknowledgements

We are greatly indebted to Prof. Sandra Spence, who kindly revised the manuscript. Thanks are due to Prof. Giulio Di Cola, Dept. of Mathematics, University of Parma for the formulation of the mathematical model.

References

Adreani, L., C. Bonacina, G. Bonomi & C. Monti, 1981. Production and population dynamics in profundal lacustrine oligochaeta. Verh. int. Ver. Limnol. 21: 967–974.

Adreani, L., C. Bonacina, G. Bonomi & C. Monti, 1984. Cohort cultures of *Psammoryctides barbatus* (Grube) and *Spirosperma ferox* Eisen: a tool for a better understanding of demographic strategies in Tubificids. Hydrobiologia 115: 113–119.

Adreani, L. & G. Bonomi, 1980. Elementi per una dinamica di popolazione dei Tubificidi (Annellida, Oligochaeta). In Atti 3° Congresso Associazione Italiana di Oceanologia e Limnologia, Sorrento, 18–20 Dicembre 1978: 355–366.

Aston, R. J. 1984. The culture of *Branchiura sowerbyi* (Tubificidae, Oligochaeta) using cellulose substrate. Aquaculture 40: 89–94.

Aston, R. J. 1966. Temperature relations, respiration and burrowing in *Branchiura sowerbyi* Beddard (Tubificidae, Oligochaeta). Unpublished Ph. D. Thesis, Univ. of Reading. In R. O. Brinkhurst & B. G. M. Jamieson, 1971. Aquatic

Oligochaeta of the World. Oliver & Boyd, Edinburgh: 139.

Aston, R. J. 1968. The effect of temperature on the life cycle, growth and fecundity of *Branchiura sowerbyi* (Oligochaeta, Tubificidae). J. Zool. 154: 29–40.

Aston, R. J., K. Slader & A. G. P. Milner, 1982. The effect of temperature on the culture of *Branchiura sowerbyi* (Oligochaeta, Tubificidae) on activated sludge. Aquaculture 29: 137–145.

Bonacina, C., G. Bonomi, S. Gazzera & A. Pasteris, 1992. Struttura, dinamica e distribuzione dei popolamenti macrobentonici in un sistema lago artificiale-lago serbatoio. Riv. Idrobiol. (in press.)

Bonacina, C., G. Bonomi & C. Monti, 1987. Progress in cohort cultures of aquatic Oligochaeta. Hydrobiologia 155: 163–169.

Bonacina, C., G. Bonomi & C. Monti, 1989a. Density dependent processes in cohort of *Tubifex tubifex*, with special emphasis on the control of fecundity. Hydrobiologia 180: 135–142.

Bonacina, C., G. Bonomi & C. Monti, 1989b. Population analysis in mass culture of *Tubifex tubifex*. Hydrobiologia 180: 127–134.

Bonomi, G. 1974. Benton profondo. In Indagini sul Lago di Garda. Quaderni dell Insituto di Ricerca Sulle Acque 18: 211–223.

Bonomi, G. & L. Adreani, 1978. Significato adattativo della struttura comunitaria e della dinamica di poplazione nel macrobenton profondo di un lago artificiale. In Il lago di Pietra del Pertusillo: definizione delle sue caratteristiche limno-ecologiche. Ed. Ist. ital. Idrobiologia, Pallanza: 133–201.

Bonomi, G. & G. Di Cola, 1980. Population dynamics of *Tubifex tubifex* studied by means of a new model. In R. O. Brinkhurst, D. G. Cook (eds), Aquatic Oligochaete Biology. Plenum Press, New York: 185–203.

Bonomi, G. & D. Ruggiu, 1968. Lago di Bolsena: composizione, distribuzione e biomassa del benton profondo. Mem. Ist. ital. Idrobiol. 23: 209–244.

Brinkhurst, R. O. & B. G. M. Jamieson, 1971. Aquatic Oligochaeta of the World. Oliver & Boyd, Edinburgh, 860 pp.

Casellato, S., 1984. Life cycle and cariology of *Branchiura sowerbyi* Beddard (Oligochaeta, Tubificidae). Hydrobiologia 115: 65–69.

Margalef, R., D. Planas, J. Armengol, A. Vidal, N. Prat, A. Guiset, J. Toja & M. Estrada, 1976. Limnologia del los emblases espanoles. Direccion General de Obras Hidraulicas, Ministerio de Obras Publicas-Madrid: 422 pp.

Mastrantuono, L., 1983. Struttura del macrobenthos litorale del lago di Nemi (Italia centrale) in rapporto ad un marcato fenomeno di inquinamento organico. Riv. Idrobiol. 22: 229–242.

Salmoiraghi, G. Il sistema limnico Suviana-Brasimone: caratteristiche idrologiche, termiche e ottiche delle acque invasate. Riv. Idrobiol. (in press.)

Hydrobiologia **278**: 275–279, 1994.
T.B. Reynoldson & K.A. Coates (eds), Aquatic Oligochaete Biology V.

Biology of *Limnodrilus claparedeanus* Ratzel (Oligochaeta, Tubificidae) in the Tsimlyansk Reservoir

Tatyana M. Dolidze
Regional Nature Protection Committee, Volgograd, Russia, 400001, Volgograd 1. Profsoyznaya, 30

Key words: life cycle, maturation, breeding, fertility

Abstract

Limnodrilus claparedeanus with its high productivity is the most numerous tubificid species in the Tsimlyansk Reservoir. As a result of frequent reproduction, young worms are abundant in the population. Individuals mature in a short time and are able to lay cocoons nine weeks after hatching. The number of embryos in the cocoons varies from one to twenty-six, thirteen being most frequent. The young worms mature rapidly and early reproduction results in an increase in the population in autumn.

Introduction

The Tsimlyansk Reservoir is situated in the south-east of european Russia and it was established in 1952 by the construction of a dam in the lower Don River which is part of the Black Sea drainage system. It is situated in a region with an arid continental climate and long, hot summers and has a naturallly high productivity. The reservoir is a typical plains-and-river waterbody, and is characterized by low flow (water exchange factor 0.9) a warm summer temperature (>20 °C), and a shallow depth (average –8.8 m). Nutrient levels in the water are high and stimulate plankton growth, a large part of which sinks after dying off, which has resulted in an organic silt. Enrichment of the sediment with organic substances creates favourable conditions for benthic organisms. The diversity of the benthic fauna is limited by low oxygen levels in some parts of the benthic zone, and by fluctuations in the water level. These conditions tend to favour large oligochaete populations. This group is most tolerant of the poor oxygen conditions and prefers the silty sediments which are found in the deeper waters. At present oligochaetes are the most numerous group of animals and, excluding molluscs, account for 91% by number and 56% by biomass of the benthic community.

Material and methods

The study of *Limnodrilus claparedeanus* Ratzel was undertaken at a permanent station with grey silt sediment at a depth of 11 m. Currents are only observed at this station in high water conditions. The average organic content of the sediment is 2.05%. Samples were collected 1–4 times a month from May–October 1982, and from April–October 1983. In 1984, samples were taken monthly from May till October at 56 stations of different sediments and depths. A Peterson grab with a sampling area of 0.025 m^2 was used. Sediments were sieved through a U.S. sieve No. 32, and fixed in 8% neutral formalin. Sorting of the

Table 1. Quantitative charateristics of benthos and its main groups in the Tsimlyansk Reservoir in 1984.

	ind. m^{-2}	g m^{-2}	
The whole benthos	17,989	580.701	
'Soft' benthos	16,210	18.382	
Oligochaetes	14,653	10.153	
L. claparedeanus	7,697	4.121	
0–3 m	325	0.975	6%
3–6 m	2,340	3.190	26%
>6 m	13,000	5.525	100%
sand (>6 m)	0	0	–
silt (>6 m)	>20,000	>10.000	–

oligochaetes was done under a microscope at low magnification. The worms were divided into 2.5 mm size groups: as well as individuals with a developing reproductive system; mature individuals with reproductive organs and a clitellum, and; individuals with a resorbed reproductive system. Mature worms were measured and selectively dissected to determine the level of development of the reproductive system according to descriptions of Poddubnaya (1971). Cocoons were measured, weighed and dissected to count embryos and determine their level of development.

Additional laboratory observations were made. Worms were kept at room temperature in aquaria with silt washed through gas sieve No. 35 to separate them from foreign material and remove organisms. Silt was used as both food and a substratum, and changed 2–4 times a month. Two densities were established in the aquaria, 714 Ind. m^{-2} and 2532 Ind. m^{-2}.

Results and discussion

Limnodrilus claparedeanus is the most common species of tubificid in the Tsimlyansk Reservoir. It comprises an average of 53% by number and 41% by biomass of the oligochaete community and is most abundant (Table 1) in the deep water areas (Table 1) and prefers silty sediments independent of depth.

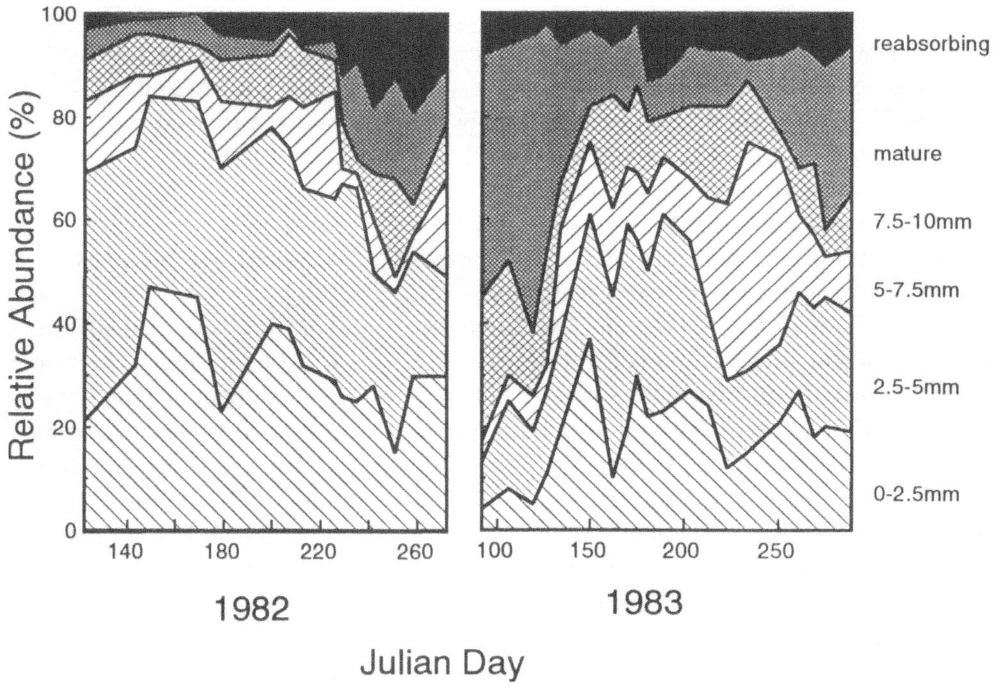

Fig. 1. Change of the relative abundance of the worms at different stages of maturity. 1–4-young worms: 1 – up to 2.5 mm; 2 – 2.5–5 mm; 3 – 5–7.5 mm; 4 – 7.5–10 mm long; 5 – mature with a clitellum; 6 – with a developing or resorbing reproductive system.

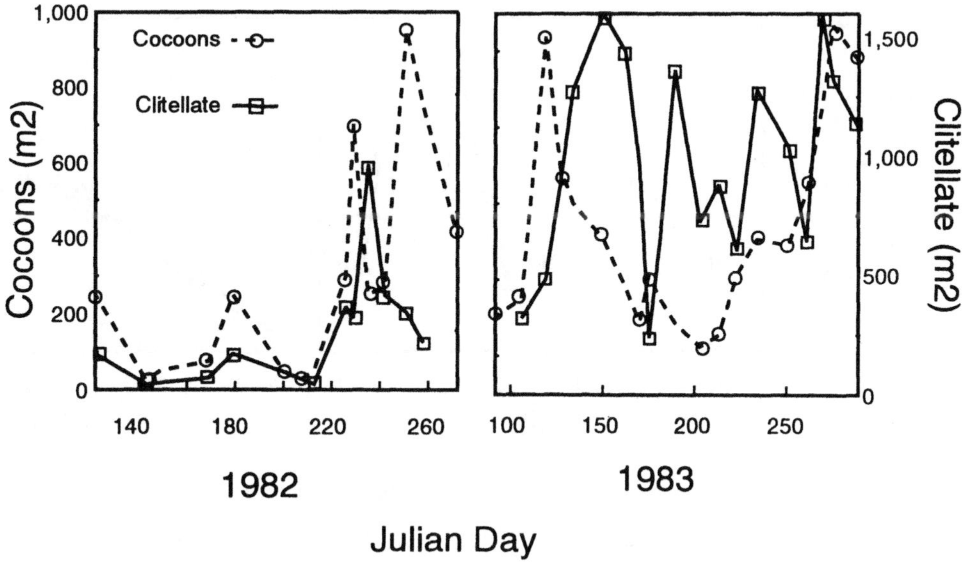

Fig. 2. Dynamics of cocoons and clitellate individuals in 1982 and in 1983.

Limnodrilus claparedeanus matures very quickly. Under laboratory conditions it can lay cocoons after 56 days, but more often at 64–70 days. Worms of the same age kept under the same conditions do not develop simultaneously. Some individuals do not begin laying cocoons until after 78 days. Early maturity is an important feature of *L. claparedeanus* which determines the seasonal dynamics of the populations and their distribution in the Tsimlyansk Reservoir.

Limnodrilus claparedeanus begins reproducing in spring, at which time the population consists of individuals of different ages and stages of maturity. The age was approximated from their age-and-size characteristics and from the state of the reproductive system. In the Tsimlyansk Reservoir the worms live for a minimum of 2 years. Older worms, that have already reproduced, are large and their reproductive system begins to develop earlier than in younger individuals. At the end of March and beginning of April they have eggs in egg sacs and spermatozoa in sperm sacs. The reproductive system of younger individuals is at the same time only developing yolk. Other age-and-size groups in the population are not significant at this time (Fig. 1).

Cocoons first appear in early April when the bottom water temperature is 4–6 °C. The numbers gradually increase as more individuals, particularly small animals, begin reproducing (Fig. 1 and 2).

Hatching is observed in late April and leads to changes in the age-and-size structure of the *L. claparedeanus* population. The size and number of clitellate individuals become smaller. Individuals with a resorbing reproductive system appear although their number is not large as many of the older worms die after reproduction. In May the population of *L. claparedeanus* consists mainly of young individuals not more than 5 mm long. The importance of other age-and size groups is insignificant.

The summer months are characterized by warm water temperature. At 22 metres the water temperature can reach 22 °C, at 11 metres the temperature is 26°. Summer is a period of repeated breeding and intensive hatching and growth of the young under the favourable temperature and trophic conditions. The absolute and the relative number of clitellate individuals is insignificant at the beginning of summer. At this time some of the old individuals as well as one-year-olds and two-years-olds that have redeveloped their reproductive system breed repeatedly. High summer water temperatures reduce the embryonic development time to 10 days. In summer the popula-

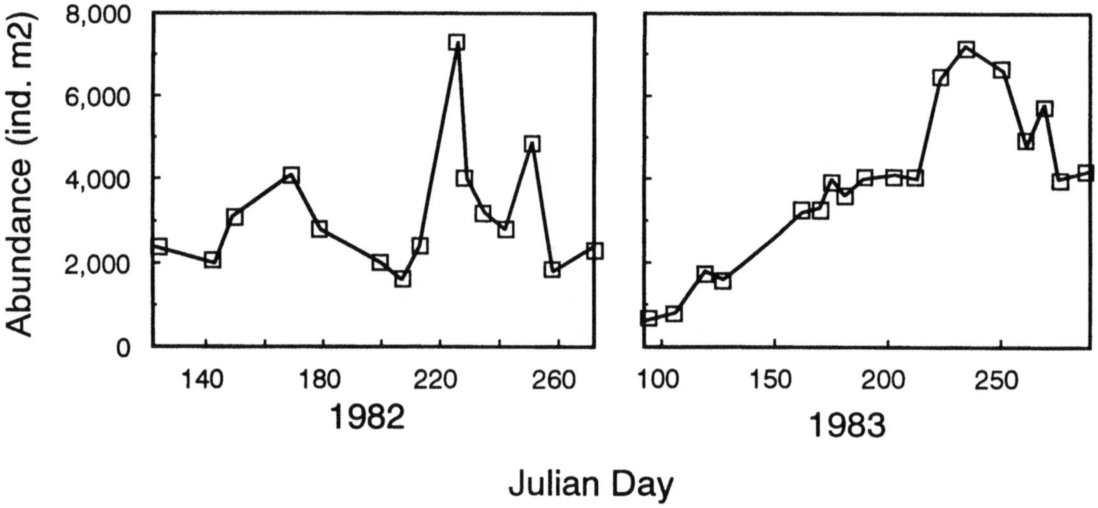

Fig. 3. Dynamics of total abundance of *L. claparedeanus* in 1982 and in 1983.

tion of *L. claparedeanus* is characterized by dominance of the new-born and established young worms.

In the autumn reproduction occurs again. The number of clitellate and post-reproductive worms reaches its maximum at this time (Fig. 2). The number of worms with resorbing or developing reproductive systems is noticeably higher than in spring-summer. This is caused by the decline in the number of the oldest individuals, which die after reproduction, and the prevalence of the younger post-reproductive specimens. The young still constitute more than half of the population.

Rapid development of the spring and summer generations leads to constant replenishment of the reproducing members of the population. As a result the number of clitellate individuals and number of cocoons increases in autumn.

Cocoons of *L. claparedeanus* are cigar-shaped and covered with small detritus particles. The embryos are not visible through the casing. The number of embryos in the cocoons from the reservoir varies from 1 to 26, but usually it is about 12. Under the laboratory conditions the worms maintain their rhythm of reproduction but their individual fertility and the duration of the reproductive period increase. Winter reproduction and an increase in the number of embryos up to as many as 70 in a single cocoon can be observed when cultivating worms at 18–24 °C. The mean number of embryos observed in the laboratory is 17.3 (S.D. 7.6, $n = 328$) in the 18–24 °C range. Over 169 days at a population density of 2532 specimens/m^2 the newly matured worms laid an average of 15 cocoons containing a total of 240 embryos.

The conditions of the Tsimlyansk Reservoir appear to favour high reproduction in *L. claparedeanus*. The abundance and biomass of the population peaked in the autumn (Fig. 3). This pattern was described by Poddubnaya (1973), in the Ivankovskoye Reservoir in a warm water zone, where the number and biomass of *Limnodrilus hoffmeisteri* Claparede in autumn increased many times.

In both reservoirs the phenomenon is explained by quick maturation of young and their early reproduction.

This type of seasonal pattern is more obvious under optimum conditions. The increase in number and biomass of the worms is not always observed in the shallow zone because of decomposing algal masses or on sand because of the lower trophic values.

Early maturation, a long reproductive period and high fertility of *L. claparedeanus* result in its wide distribution in the Tsimlyansk Reservoir. Before the reservoir was created, *L. claparedeanus* was found in silty and warm waterbodies in the flood-plain of the Don River. At present the species is the most numerous representative of Tubificidae in this habitat.

References

Poddubnaya, T. L., 1971. Resorption and regeneration of the reproductive system of Tubificidae on the example of *Isochaetides newaensis* Mich. (Oligochaeta, Tubificidae). In Biology and Physiology of Freshwater Organisms. Leningrad: 81–90 (in Russian).

Poddubnaya, T. L., 1973. Tubificidae production in the Ivankovo Reservoir in the zone of warm waters effluent of the Konakovo State Regional Power Station. Hydrobiol. J. 9: 52–57 (in Russian).

Hydrobiologia **278**: 281–286, 1994.
T.B. Reynoldson & K.A. Coates (eds), Aquatic Oligochaete Biology V.

Population dynamics of *Potamothrix alatus paravanicus* Poddubnaya & Pataridze (Tubificidae) in different areas of Lake Sevan

K. Jenderedjian
Institute of Hydroecology and Ichthyology of the Academy of Sciences of Armenia, Kirov 186, 378610 Sevan, Armenia

Key words: Potamothrix alatus paravanicus, population dynamics, littoral, sublittoral, profundal

Abstract

Potamothrix alatus paravanicus Poddubnaya & Pataridze (Tubificidae) is the most abundant species of benthic invertebrate in Lake Sevan and the only species in the deep profundal. Differences in annual temperature, length of the stratification period (low oxygen content) and bottom sediment characteristics influence the population of *P. a. paravanicus* in different depth zones of Lake Sevan. From shallow to deep water a regular increase of the following indices was observed in 1984–1987: annual biomass from 1.2–2.0 to 8.2–17.0 g m^{-2} wet weight; mean individual wet weight of clitellate specimens from 2.0–3.8 to 8.3–16.6 mg; total life span from one and a half to more than 20 years; the length of breeding season from 2–3 to 12 months.

An important ecological distinction between *P. a. paravanicus* and allied *Potamothrix hammoniensis* (Michaelsen) is the much lower fecundity of the former species.

Introduction

Earlier investigations (Arnoldi, 1929; Fridman, 1950; Markosian, 1959; Meshkova, 1976; Nikolaev, 1985; Ostrovski, 1985) showed the most important benthic animal group of Lake Sevan to be the family Tubificidae. Of these, *Potamothrix alatus paravanicus* Poddubnaya & Pataridze, is found in the greatest numbers (95%) and biomass (98%); the other three important species are *Limnodrilus claparedeanus* Ratzel, *Tubifex tubifex* (Müller) and *Limnodrilus hoffmeisteri* Claparede (Jenderedjian & Poddubnaya, 1987).

P. a. paravanicus is the only species which inhabits all types of bottom sediment from 0.2 m to the maximum depth (80 m) and it is the only macrobenthic species in the deep profundal. *Potamothrix a. paravanicus* is only absent from the heavily polluted areas near river mouths, where it is replaced by *L. hoffmeisteri* and *T. tubifex* (Jenderedjian, 1990a). However, only certain aspects of the population dynamics have been examined (Jenderedjian, 1989, 1990b), this paper summarizes the work done to date.

The species has been previously recorded in Lake Sevan as *Hyodrilus hammoniensis* (Fridman, 1950), *Euilyodrius hammoniensis* (Churakova, 1972), *Ilyodrilus hammoniensis* (Chekanovskaya, 1981), *Potamothrix hammoniensis* (Jenderedjian, 1990a). It was recently described as a separate species *Potamothrix paravanicus* (Poddubnaya & Pataridze, 1989) but most recently was described by Finogenova (Finogenova & Poddubnaya, 1990) as *Potamothrix alatus paravanicus.*

The study area

Lake Sevan is located in Armenia at an altitude of 1898 m a.s.l. and consists of the deeper Minor Sevan (average depth 39 m, surface 328 km^2) and the comparatively shallow Major Sevan (average depth 24 m, surface 916 km^2) (Fig. 1). Using temparature and oxygen conditions and sediment characteristics five main zones are distinguished: (1) littoral (0–8 m) and (2) sublittoral (8–27 m), in both Major and Minor Sevan, (3) a transitional zone (27–35 m) from sublittoral to profundal in Minor Sevan, (4) profundal of Minor Sevan (35–80 m) and (5) profundal of Major Sevan (27–34 m). Some morphometric and limnological characteristics of these areas are given in Table 1.

Materials and methods

A Petersen dredge enclosing 0.025 m^2 was used for sampling (two grabs per sampling event). The samples were taken along transect 16 in Minor Sevan and transect 2 in Major Sevan (Fig. 1) from one to five times per month during 1984–87 at 2, 4, 7, 10, 15, 20, 25, 30, 40, 50, and 60 m depths. The material includes approximately 68,000 specimens of *P. a. paravanicus* from 1320 samples.

Samples were washed on a sieve with a mesh size of 0.16 mm and preserved in 4% formalin. The worms, cocoons and embryos (in cocoons) were sorted and counted under 16 times magnification. Individuals of *Potamothrix a. paravanicus* were identified as either: (1) immature (four size groups), (2) mature without clitellum, (3) clitellate individuals without or (4) with spermatozeugmata in spermatheca (copulated specimens).

Worms were weighed in groups to within 0.1 mg wet weight. Intact mature specimens were measured (body length) to within 1 mm.

Results

At depths of 2, 4 (Fig. 2a) and 7 m in both transects specimens with spermatozeugmata occurred from June to October and cocoons from June to August. Mature worms were not found after October. The seasonal biomass was high in April–June (1.5–4.1 g m^{-2} wet weight) and low from August until the end of the year (0.1–0.5 g m^{-2}). In Major Sevan the pattern was similar but the values were lower.

At depths of 10 (Fig. 2b), 15, 20 and 25 m in

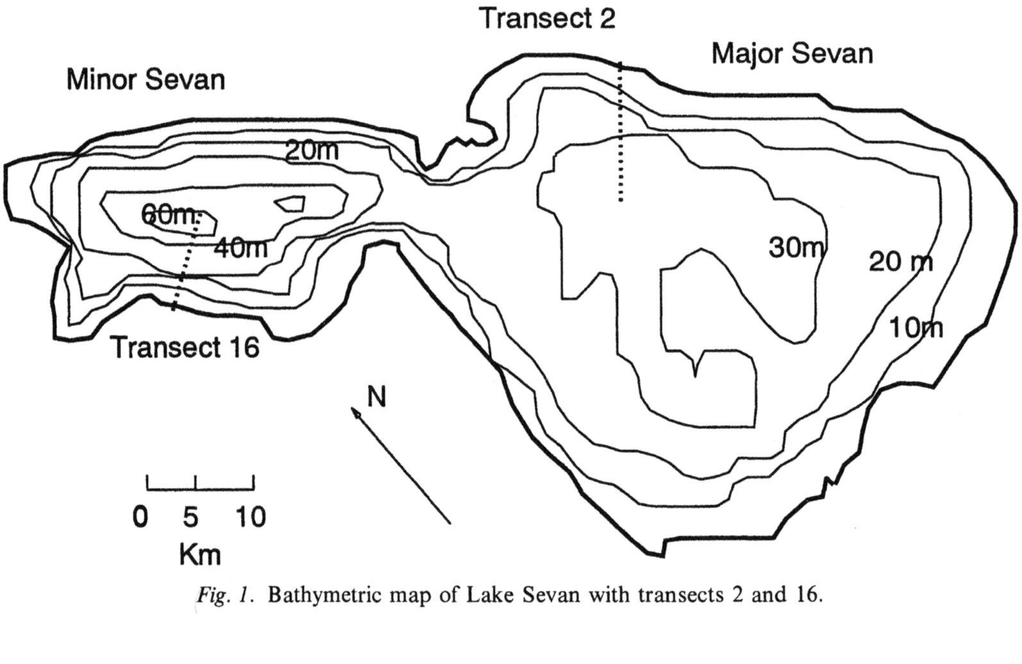

Fig. 1. Bathymetric map of Lake Sevan with transects 2 and 16.

Table 1. Some morphometric and limnological characteristics of five zones in Lake Sevan (O_2 after Parparova, 1985).

Character	1	2	3	4	5
Depth	0–8	8–27	27–35	35–80	27–34
Surface Area (km^2)	100	560	34	220	330
Min. Ann. Temp. (°C)	0–1	1–2	2	2	2
Max. Ann. Temp. (°C)	20–22	18–20	16	5–8	12
Ann. Degree Days (No.)	3200–3700	3000–3500	2400–2900	1300–2300	2100–2600
Length of anoxia (d)	0	0	0–30	60–120	30–60
Sediment type	sand	yellow silt	brown silt	black silt	black silt

Minor Sevan specimens with spermatozeugmata occurred from May to August, cocoons from the end of May (temp. 10–11 °C) to August. Clitellate worms occurred year round but in autumn their number was small. Once again the same pattern was observed in Major Sevan. The seasonal biomass was high in May–June (2.5–4.8 g m^{-2} in Minor Sevan and 8.0–22.4 g m^{-2} in Major Sevan) and low in August–October (1.6–3.2 and 3.5–5.7 g m^{-2} wet weight respectively). In November–December a second increase in biomass occurred.

In the transitional zone in Minor Sevan at a depth of 30 m (Fig. 2c) specimens with sperma-

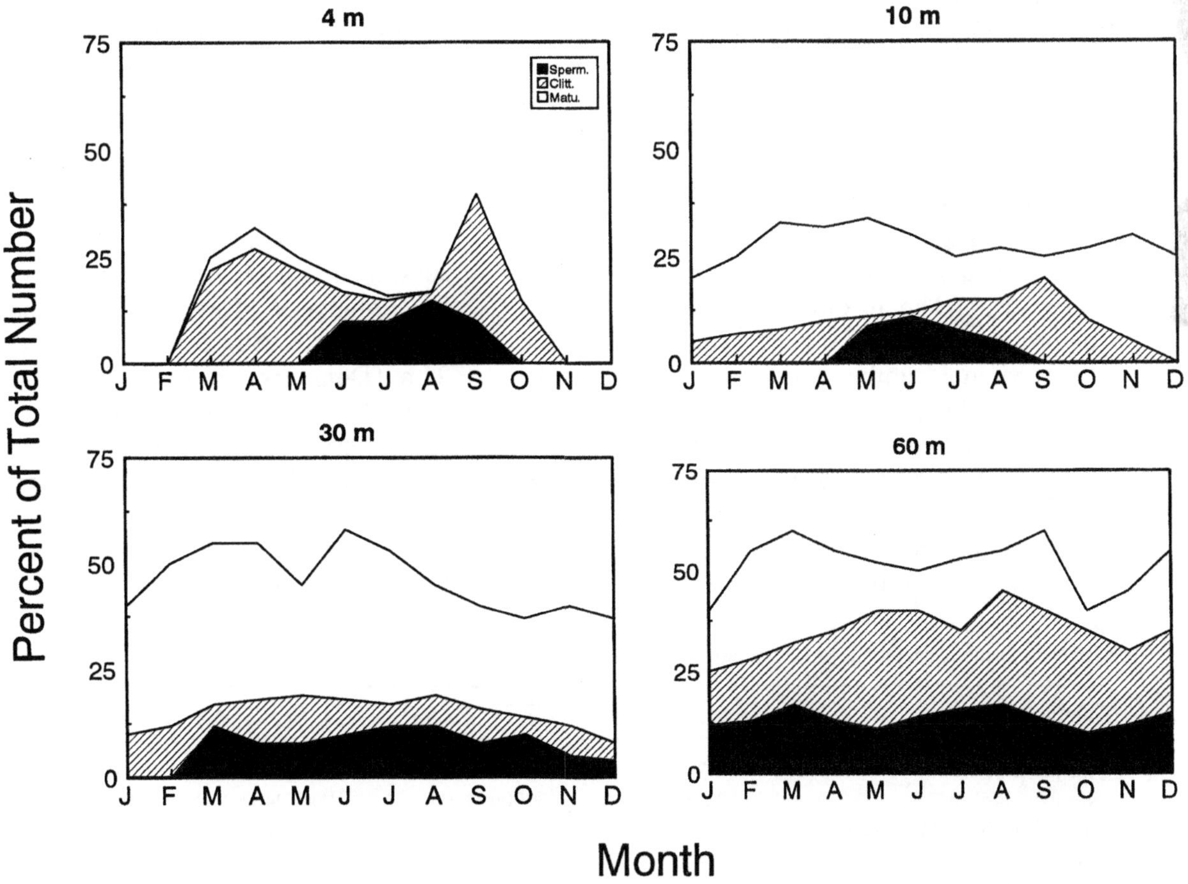

Fig. 2. Average seasonal dynamics (1984–87) of mature specimens of *P. a. paravanicus*, expressed as percent of total numbers: (open) without clitellum, (hatched) with clitellum, and (solid) with spermatozeugmata; at depths of 4, 10, 30 and 60 m.

tozeugmata occurred from May to December, cocoons from June (temp. 6–8 °C) to October. In 1985 the breeding period began in February at a temperature of 2 °C. Clitellate worms occurred all year round with a maximum number in June-August and a minimum in November-December. The seasonal biomass was usually higher in May-October (6.6–9.9 g m^{-2} wet weight) and lower in February-March (2.9–4.0 g m^{-2}).

In the profundal of Minor Sevan at depths of 40, 50 and 60 m (Fig. 2d) specimens with spermatozeugmata as well as other mature individuals and cocoons occurred the whole year round at temperatures of 2–8 °C at 40 m depth and 2–5 °C at 50–60 m. The biomass varied between 3.2–5.8 and 13.6–14.2 g m^{-2} wet weight and there was no annual pattern in the biomass.

In the profundal of Major Sevan at a depth of 30 m specimens with spermatozeugmata occurred from May to November, cocoons from June (6–8 °C) to October. Clitellate worms occurred all year round with a maximum number in June-August and a minimum in December. The seasonal biomass varied between 5.2–6.8 and 17.0–26.5 g m^{-2} wet weight. There was no apparent seasonal pattern in biomass.

Table 2 summarizes the population characteristics of *P. a. paravanicus* in the different zones of Lake Sevan. The maximum biomass occurred in the profundal (8.2–17.0 g m^{-2} wet weight) while the maximum number was observed in the sublittoral (2,000–3,500 ind. m^{-2}). The mean weight of mature individuals increased from 2.0–3.8 mg wet weight in the littoral to 8.3–16.6 in the profundal, i.e. 4–5 times. The maximum number of embryos per cocoon increased from 3 in shallow water to 8 in the profundal zone. The length of breeding period varied from 2–3 months in the littoral to all year in the deeper profundal.

Discussion

In all areas, except the profundal of Minor Sevan, cocoons were first deposited at the end of May or in the first few days of June when temperature is 6–11 °C. In the profundal of Minor Sevan where temperature is uniformally low coccon deposition took place the whole year round. The effects of annual temperature variations on the population dynamics of *P. a. paravanicus* are similar to those for other tubificids; sexual organs develop in the winter, reproduction occurs at the beginning of summer followed by resorption and then a second period of development of sexual organs (Poddubnaya, 1980; Timm, 1987).

The length of the breeding period for *P. a. paravanicus* seems to be controlled primarily by temperature: the smaller the annual temperature range, the longer the breeding season. The duration of the breeding season is 2–3 months at 2 to 25 m (16–22 °C range), 5–7 months at 30 m (10–14 °C range) and 12 months at 40 to 60 m (3–6 °C range). The length of the breeding season of, *e.g. P. hammoniensis* is 2–3 months in Rybinskoe Reservoir (Russia) at 10 m (Arkhipova, 1980), 5 months in Lake Esrom at 22 m (Denmark) (Jónasson & Thorhauge, 1972) and 12 months in the profundal of deep-water Lake Ohrid (Macedonia) (Šapkarev, 1959).

The fecundity of *P. a. paravanicus* is much lower than in *P. hammoniensis*. Mean numbers of embryos per cocoon of *P. a. paravanicus* varied between 1.9 and 3.1 with the maximum number

Table 2. Some chracteristics of P. a. parvanicus in 1984–87 in different zones of Lake Sevan.

Character	1	2	3	4	5
Annual Biomass	1.2–2.0	4.5–8.0	5.7–7.1	8.2–12.2	13.0–17.0
Number (ind. m^{-2})	1100–1900	2000–3500	1900–2400	1800–2700	2100–2800
Mean Weight of Matures (mg)	2.0–3.8	4.4–6.7	4.9–10.0	8.3–12.5	13.5–16.6
Embryo.coccon (mean, SD)	1.9 ± 0.6	2.4 ± 1.0	2.6 ± 1.4	2.8 ± 1.4	3.1 ± 1.2
Age of Clitellate ind. (yr)	1.1–1.3	1.6–2.0	2.4–2.7	4.8–15.2	5.2–7.0
Life Span (yr)	1.6	1.9–2.3	3.6–4.1	7.2–22.6	6.7–8.5

of embryos per cocoon being 8. Mean numbers of embryos per cocoon of *P. hammoniensis* varied between 4.0–10.5 (maximum 19) in Lake Vôrtsjärv (Estonia) (Timm, 1987), 3.0–11.0 (maximum 20) in Rybinskoe Reservoir (Arkhipova, 1980) and 5.5–10.5 (maximum 28) in Lake Esrom (Jónasson & Thorhauge, 1976).

Direct determination of the age of oligochaetes is impossible (Poddubnaya, 1963). However, the age of *P. a. paravanicus* was estimated from weight, temperature and oxygen (see Tables 1 and 2) based on linear correlations between growth rate L(t), mm, water temperature (T, °C) and time (t, day) (Jenderedjian, 1989):

$$L(t) = 5.23\ 10^{-3}(T - 1.02)t \quad r = 0.95 \qquad (1)$$

and coefficient of conditions of *P. a. paravanicus* (Jenderedjian & Hounanian, 1987):

$$\lg W = -2.105 + 1.997 \lg L \quad r = 0.93 \qquad (2)$$

where W – wet weight, mg, L – length, mm. From equation (2) equation (1) may be rewritten:

$$Y = 2167 W^{0.5} S^{-1}$$

where Y – age, year, W – wet weight, mg, S – sum of effective degree-days per year, °C. In equation (3): it is more accurate to use $W(t) = W - W_o$ instead of W, but in this case the error is minimal. From equation (1), the base growth temperature of *P. a. paravanicus* is roughly 1 °C. Minimum base growth temperature of *P. hammoniensis* varied between 0.5 and 2 °C (Jónasson & Thorhauge, 1972; Arkhipova, 1980), of *T. tubifex* 0 and 1.6 °C (Kosiorek, 1974; Adreani *et al.*, 1981; Bonomi and Di Cola, 1980; Finogenova & Lobasheva, 1987), of *L. hoffmeisteri* 0 and 4 °C (Adreani *et al.*, 1981; Bonacina *et al.*, 1987). Jónasson and Thorhauge (1972) observed in the profundal of Lake Esrom a small decrease in weight of *P. hammoniensis* during summer stagnation, therefore it was assumed that when oxygen content near the bottom was less than 1 mg l^{-1} the growth rate of *P. a. paravanicus* will be zero.

Estimation shows that the total life span of *P. a. paravanicus* increases from one and a half years in the littoral to roughly 20 years in the deepest profundal. The mean age of clitellate specimens varies from 1 year in the littoral to 15 years in the deepest profundal. Estimated total life spans and mean ages of mature specimens are consistent with population dynamics at different depths (see Fig. 2). Timm (1983, 1984, 1987) stated the value of potential life span of tubificids 5–10–15 years. Generally at low temperature and long periods of anoxia life span is longer than in high temperature and favourable oxygen conditions. At depths of 60 m the constant low temperature (2–5 °C) and 4 months of continuous oxygen deficiency it is suggested that a life span of 20 years is realistic.

Conclusion

The effects of annual temperature variations on the population dynamics of *P. a. paravanicus* are similar to those described in other tubificids: sexual organs develop in the winter, reproduction occurs at the beginning of the summer followed by resorption and then a regeneration of sexual organs.

The duration of the breeding season of *P. a. paravanicus* depends on temperature: the smaller the temperature fluctuations, the longer the breeding period.

An important ecological distinction between *P. a. paravanicus* and allied *P. hammoniensis* is the much lower fecundity of the former species.

Potamothrix a. paravanicus reached maximum biomass in the profundal where there is: a high quantity of planktonogenic organic material in the sediments; constant low temperature; a long period of oxygen deficiency and; an absence of competitors.

The life span of *P. a. paravanicus* increased from one and a half years in the littoral to 20 years in the deeper profundal as a result of the temperature and oxygen regime.

Acknowledgements

My thanks are due to Dr T. Timm for benevolent criticism, to Mrs I. Aidinian for help during the preparation of the manuscript.

References

Adreani, L., C. Bonacina & G. Bonomi, 1981. Production and population dynamics in profundal lacustrine Oligochaeta. Verh. Ver. Limnol. 21: 967–974.

Arkhipova, N.R., 1980. Objectives of biology of *Potamothrix hammoniensis* (Mich.) (Oligochaeta, Tubificidae) of Rybinskoe Reservoir. Transact. Inst. Biol. Inland Waters 44: 14–27. [in Russian]

Arnoldi, L.V., 1929. Materials on the bottom productivity studies of Lake Sevan. Transact. Sevan Hydrobiol. Station 2: 5–96. [in Russian]

Bonacina, C., G. Bonomi & C. Monti, 1987. Progress in cohort cultures of aquatic Oligochaeta. Hydrobiologia 155: 163–169.

Bonomi, G. & G. Di Cola, 1980. Population dynamics of *Tubifex tubifex*, studied by means of a new model. In R.O. Brinkhurst & D.G. Cook (eds), Aquatic Oligochaete Biology, Plenum Press, N.Y.: 185–203.

Chekanovskaya, O.V., 1981. Aquatic Oligochaeta of the USSR. Amerind Publishing Co. Pvt. Ltd., New Delhi. 513 pp.

Churakova, K.P., 1972. Objectives of taxonomy and distribution of Oligochaeta of Lake Sevan. In Aquatic Oligochaeta. Nauka, Moscow: 75–82. [in Russian]

Finogenova, N.P. & T.M. Lobasheva, 1987. Growth of *Tubifex tubifex* (Müller) (Oligochaeta, Tubificidae) under various trophic conditions. Int. Revue ges. Hydrobiol. 72: 709–726.

Finogenova, N.P. & T.L. Poddubnaya, 1990. One more revision of the genus *Potamothrix* Vejdovsky et Mrázek 1902 (Oligochaeta, Tubificidae). Zool. Jb. Syst. 117: 55–83.

Fridman, G.M., 1950. Bottom fauna of the Lake Sevan. Transact. Sevan Hydrobiol. Station 11: 7–92. [in Russian]

Glushchenko, L.O., 1989. Particulate organic matter and its role in the formation of water quality in Lake Sevan (Armenia). Arch. Hydrobiol. Beih. 33: 265–271.

Jenderedjian, K., 1989. Peculiarities of biology and the production of Oligochaeta of Lake Sevan. Cand. biol. sci. thesis, Leningrad, 24 pp. [in Russian]

Jenderedjian, K., 1990a. Distribution of *Potamothrix hammoniensis* in Lake Sevan. In Species in area: biology, ecology and productivity of water invertebrates. Navuka i Tekhnika, Minsk: 214–217. [in Russian]

Jenderedjian, K., 1990b. Seasonal dynamics of *Potamothrix hammoniensis* along the model semi-section of Lake Sevan. In Species in area: biology, ecology and productivity of water invertebrates. navuka i Tekhnika, Minsk: 218–221. [in Russian]

Jenderedjian, K. & A. Hounanian, 1987. Connection between weight and some linear parameters of tubificids on the example of *Potamothrix hammoniensis* (Michaelsen). In Aquatic Oligochaeta. Riga: 64–66. [in Russian]

Jenderedjian, K. & T.L. Poddubnaya, 1987. Species composition and distribution of Oligochata in Lake Sevan. Biol. J. Armenia 40: 36–42. [in Russian]

Jónasson, P.M. & F. Thorhauge, 1972. Life cycle of *Potamothrix hammoniensis* (Tubificidae) in the profundal of a eutrophic lake. Oikos 23: 151–158.

Jónasson, P.M. & F. Thorhauge, 1976. Population dynamics of *Potamothrix hammoniensis* in the profundal of Lake Esrom with special reference to environmental and competitive factors. Oikos 27: 193–203.

Kosiorek, D., 1974. Development cycle of *Tubifex tubifex* (Müll.) in experimental culture. Pol. Arch. Hydrobiol. 21: 411–422.

Markosian, A.K., 1959. Benthos productivity in Lake Sevan. In Transact. 6th Conf. Probl. Inland Waters Biol., Leningrad: 139–145. [in Russian]

Meshkova, T.M., 1976. Eutrophication of Lake Sevan. Biol. J. Armenia 29: 14–22. [in Russian]

Nikolaev, S.G., 1985. Structure of zoobenthos communities in Lake Sevan in years of sharp increase of trophic status and its changes in comparison to an oligotrophic period. Cand. biol. sci. thesis, Moscow, 24 pp. [in Russian]

Ostrovsky, I.S., 1985. Zoobenthos of Lake Sevan and its dynamics, Transact. Sevan Hydrobiol. Station 20: 132–187. [in Russian]

Parparova, R.M., 1985. Peculiarities of the cycle of phosphorus in Lake Sevan on a background of the changes of its hydrochemical regime in relation with anthropogenic influence. Cand. chem. sci. thesis, Rostov-na-Donu, 24 pp. [in Russian]

Poddubnaya, T.L., 1963. Life cycle and growth rate of *Limnodrilus newaensis* Mich. (Oligochaeta, Tubificidae). Transact. Inst. Biol. Reservoirs 5: 46–56. [in Russian]

Poddubnaya, T.L., 1980. Life cycle of the mass species of Tubificidae. In R.O. Brinkhurst & D.G. Cook (eds.), Aquatic Oligochaete Biology, Plenum Press, N.Y.: 175–184.

Poddubnaya, T.L. & A.I. Pataridze, 1989. A new species of the genus *Potamothrix* (Oligochaeta, Tubificidae) from the high mountain lakes of the Caucasus. Zool. J. 68: 153–156. [in Russian]

Šapkarev, J., 1959. *Ilyodrilus hammoniensis* Mich. (Oligochaeta) in the large lakes of the Macedonia (Ohrid, Prespa and Dojran). Sect. Sci. nat. Univ. Skopje, Station Hydrobiol. Ohrid, Rec. trav. 7 40: 1–12.

Timm, T., 1983. On the potential age of Oligochaeta. In Aquatic Oligochaeta. Tbilisi: 113–117. [in Russian]

Timm, T., 1984. Potential age of aquatic Oligochaeta. Hydrobiologia 115: 101–104.

Timm, T., 1987. Aquatic Oligochaeta of the northwestern part of the USSR. Valgus, Tallinn, 299 pp. [in Russian].

Hydrobiologia **278**: 287–290, 1994.
T.B. Reynoldson & K.A. Coates (eds), Aquatic Oligochaete Biology V.

Influence of environmental factors on the production of *Potamothrix alatus paravanicus* Poddubnaya & Pataridze (Tubificidae) in different areas of Lake Sevan

K. Jenderedjian
Institute of Hydroecology and Ichthyology of the Academy of Sciences of Armenia, Kirov 186, 378610 Sevan, Armenia

Key words: Potamothrix alatus paravanicus, P/B ratio, depth, temperature, low oxygen content

Abstract

A modification of the Hynes' method was used to estimate production of *Potamothrix alatus paravanicus* Poddubnaya & Pataridze (Tubificidae), the dominant species of benthic invertebrate in Lake Sevan. Maximum production (5–24 g m^{-2} wet weight) occurred in the sublittoral, while maximum biomass was found (8–17 g m^{-2} wet weight) in the profundal. The turnover ratio of the 'average cohort' varied between 2.4 and 5.0. The annual turnover ratio (*P/B*) decreased from 2.1–3.1 in the littoral to 0.1–0.4 in the profundal zone.

A logarithmic correlation was found between *P/B* and temperature and oxygen regime, and depth and average size of clitellate specimens.

Introduction

Potamothrix alatus paravanicus Poddubnaya & Pataridze (Tubificidae) is the most important bottom animal of Lake Sevan and represents 98% of tubificids and two-thirds of the biomass of the zoobenthos (Jenderedjian & Poddubnaya, 1987; Jenderedjian, 1989). The productivity of *P. alatus paravanicus* is of special interest because of the wide distribution of this species in shallow and deep waters. The varying temperature and oxygen regimes to which this species is exposed in Lake Sevan have resulted in differences in the life cycles and population dynamics of *P. alatus paravanicus* (Jenderedjian, 1994) this paper examines the impact on production dynamics.

Material and methods

The study area, sampling techniques, sieving, preservation and sorting methods were described in the preceding paper (Jenderedjian, this volume). Quantitative data collected over the 1984–1986 period were used for estimating the productivity of *P. alatus paravanicus* at different depths in Lake Sevan.

Hynes' method (Hynes & Coleman, 1968; Hamilton, 1969; Benke, 1979; Menzie, 1980) was modified to estimate productivity. This method is preferred as it does not require separate cohorts to be identified (Timm, 1990b). Using the Hynes' method production of the 'average cohort' (*P*) can be calculated:

$$P = \sum_{j=1}^{i} (N_j - N_{j+1})(W_j \cdot W_{j+1})^{0.5}, \quad (1)$$

where: j – size group, i – number of size groups, N_j, N_{j+1} and W_j, W_{j+1} – numbers and average weights of specimens in size groups j and $j+1$. When $j = i$, $N_{j+1} = 0$, $W_{j+1} = W_{\max}$, i.e., the oldest cohort ceases to exist when the final individual dies.

Furthermore,

$$N_j = n_j(P_e \cdot P_a^{-1})_j^{-1} \text{CPI}, \quad (2)$$

where: n_j – average number of specimens per annum in size group j, $(P_e P_a^{-1})$ – part of the life span spent in size group j, CPI – cohort production interval (in this case CPI is equal to the length of the total life span (Jenderedjian, 1994).

The biomass of the 'average cohort' (B_c) may be calculated by equation:

$$B_c = \sum_{j=1}^{i} b_j(P_e \cdot P_a^{-1})_j, \quad (3)$$

where: b_j – biomass of the 'average cohort' when the number of specimens is N_j with an average wet weight W_j.

It is evident that the average biomass per annum (B_y) is equal:

$$B_y = B_c \,\text{CPI}. \quad (4)$$

The smallest immature worms were not retained by the sieves used, but it is possible to determine their number on the basis of the number of cocoons m^{-2} and embryos per cocoon and it was assumed that the death rate of a cohort decreased geometrically:

$$N_{j-1} = N_j^2 \cdot N_{j+1}^{-1}, \quad (5)$$

where: N_{j-1} renewed number of specimens in size group $j-1$, N_j and N_{j+1} – real number of specimens in size group j and $j+1$.

Waters (1979) has shown the failure to obtain young specimens during sampling usually reduces the production rate by 5–10%.

Because the increase in length of *P. alatus paravanicus* in constant temperature is linear (Jenderedjian, 1994; equation 1):

$$(P_e \cdot P_a^{-1})_j = 1_j(L_{max} - L_0)^{-1}, \quad (6)$$

where $1:_j$ – the length of size group j, $L_{\max}$ – maximal length of the biggest individual in studied area L_0 – mean length of new-born specimens.

Results and discussion

As shown in Table 1, the maximum production of *P.a. paravanicus* occured at depths of 10–30 m in Minor Sevan (6–8 $g \cdot m^{-2}$ wet weight) and at 15–25 m in Major Sevan (10–24 $g \cdot m^{-2}$), while biomass genarally increased with depth in both basins. The maximum biomass occurred in the profundal of Minor Sevan at 40–60 m and in the deep sublittoral and profundal of Major Sevan at 25–30 m. The productivity of *P.a. paravanicus* in Minor Sevan (3.7–5.5 $g \cdot m^{-2}$) was half that of Major Sevan (8.3–11.9 $g \cdot m^{-2}$). This is attributed to differences in morphometry, temperature, oxygen and the organic content of the sediment (Jenderidjian, 1994, Table 1; Gluschenko, 1989).

Turnover ratios of the 'average cohorts' (P/B_c) had a tendency to decrease from shallow to deepwater: 3.8 ± 0.8 in the littoral, 3.5 ± 0.2 in the sublittoral, 2.9 ± 0.3 in the profundal of Minor Sevan and 2.4 ± 0.3 in the profundal of Major Sevan. The difference between P/B_c ratios seems to be a result of the different population structure and dynamics in the littoral, sublittoral and profundal (Jenderedjian, 1994). Winberg (1971) showed that the growth rate of old specimens is usually lower than younger animals. In the littoral mature specimens made up an average of 28% of the biomass and 31% of the production of the population, in the sublittoral 64 and 49% and in the profundal 93 and 80%, respectively.

Estimated annual turnover ratios (P/B_y) of *P. alatus paravanicus* varied between: 2.0–3.1 in the littoral (2–7 m), 1.5–1.9 in the sublittoral (10–25 m), 0.9–1.1 in the transitional zone (30 m) from sublittoral to profundal and 9.1–0.4 in the profundal. Turnover ratios of the related species *Potamothrix hammoniensis* (Michaelsen, 1901) varied between 0.8 and 6.2 (Borutski, 1939; Popchenko, 1973; Lukyanova, 1974; Poddub-

Table 1. Annual biomass (B, P, g · m^{-2} wet weight) and turnover ratios (P/B_y) of *P.a. paravanicus* calculated using Hynes' method and from equation 13 (text).

Location	Depth	1984			1985			1986		
		B	P	P/B_y	B	P	P/B_y	B	P	P/B_y
Minor Sevan	2	1.1	3.4	3.1	0.8	2.3	2.9*	0.5	1.5	3.0*
	4	1.5	3.1	2.1	0.9	2.1	2.3*	1.2	3.1	2.5*
	7	0.6	1.3	2.4	0.3	0.7	2.3*	0.7	1.5	2.3*
	10	3.5	5.8	1.7	3.4	5.4	1.6	3.0	5.0	1.6
	15	3.3	5.7	1.7	3.0	4.4	1.5*	1.8	2.9	1.6*
	20	3.5	6.6	1.9	2.4	3.6	1.5*	2.9	4.4	1.5*
	25	3.4	5.0	1.5	3.7	4.8	1.3*	1.7	2.1	1.2*
	30	5.7	5.5	1.0	5.4	4.9	0.9	5.7	6.1	1.1
	40	7.8	3.2	0.4	7.6	3.8	0.5*	5.8	2.9	0.5*
	50	7.6	2.1	0.3	7.7	1.8	0.2	5.1	1.4	0.3
	60	8.8	1.6	0.2	6.9	0.8	0.1	5.2	0.7	0.1
MajorSevan	2	0.1	0.2	3.3*	0.1	0.2	2.9*	0.1	0.3	3.4*
	4	0.1	0.2	3.1*	0.1	0.3	3.0*	0.1	0.3	3.0*
	7	1.7	4.7	2.8*	0.6	1.1	1.9*	0.9	2.4	2.7*
	10	3.2	6.2	2.0	1.3	2.3	1.8*	1.8	3.3	1.8
	15	5.1	9.0	1.8	2.8	4.2	1.5*	2.5	3.8	1.5*
	20	7.7	13.1	1.7	7.3	9.6	1.3*	7.0	9.8	1.4*
	25	15.1	22.8	1.5	13.8	18.0	1.3*	10.0	13.0	1.3*
	30	17.0	5.6	0.3	13.9	3.5	0.2	1.4	4.4	0.4

naya, 1975; Jonasson & Thorhauge, 1976; Pidgaiko *et al.*, 1972; Slepukhina, 1977; Gavrilov, 1985; Timm, 1990a; Karataeva, 1990). The difference between turnover ratios of *P. alatus paravanicus* and *P. hammoniensis* can be explained by differences in fecundity, life span, environment and methods (Timm, 1990b).

Correlation analysis ($n = 27$) showed a relationship between P/B_y ratio of *P. alatus paravanicus* and the sum of degree-days per year (S, °C), the length of the anoxic period (D, number of days per year, when oxygen content near the bottom is below 1 mg l^{-1}), depth (H, m) and the size of clitellate specimens (W, mg wet weight):

$$\ln P/B_y = -14.646 - 0.668 \ln W + 1.998 \ln s \quad r = 0.99, \qquad (7)$$

$$\ln P/B_y = -23.968 - 0.701 \ln W + r.332 \ln D \quad r = 0.99, \qquad (8)$$

$$\ln P/B_y = -26.316 - 0.370 \ln H + 4.699 \ln D \quad r = 0.97, \qquad (9)$$

$$\ln P/B_y = -18.259 - 0.210 \ln H + 2.379 \ln S \quad r = 0.96, \qquad (10)$$

$$\ln P/B_y = -29.201 + 1.621 \ln S + 2.806 \ln D \quad r = 0.96, \qquad (11)$$

$$\ln P/B_y = -2.979 - 1.031 \ln W - 0.379 \ln H \quad r = 0.90. \qquad (12)$$

The integral form of equations (7–12) is:

$$\ln P/B_y = -18.235 + \ln S + 1.973 \ln D - 0.168 \ln H - 0.402 \ln W. \qquad (13)$$

The difference between P/B_y ratios, calculated from the Hynes method and by equation (13) was less than 20% (Table 1).

Conclusions

Maximum production of *P. alatus paravanicus* (5.4–23.8 g m^{-2} wet weight) was observed in the sublittoral zone, at the 10–30 m depth in Minor Sevan and 15–25 m in Major Sevan.

Turnover ratios of the 'average cohorts' had a tendency to decrease from shallow to deep-water. The difference between values of P/B_c seem to be a result of population structure: the smaller the number of mature specimens in a population, the greater the value of the P/B_c ratio.

Annual turnover ratios of *P. alatus paravanicus* varied between 2.0–3.1 in the littoral and 0.1–0.4 in the profundal. The value of the P/B_y ratio depends on temperature and oxygen regimes, depth and size of mature specimens.

Acknowledgements

This paper is dedicated to my teacher, Dr T.L. Poddubnaya. My thanks are due to Dr T. Timm for valuable criticism and to Mrs I. Aidinian for her help during the preparation of the manuscript.

References

Benke, A., 1979. A modification of the Hynes' method for estimating secondary production with particular significance for multivoltine populations. Limnol. and Oceanogr. 24: 168–171.

Borutsky, E.V., 1939. Dynamics of the total biomass of the benthos in the profundal of Lake Beloye. Proc. Limnol. Station Kosino 22: 196–218 [in Russian].

Gavrilov, S.I., 1985. Macrozoobenthos. In Ecological systems of the Naroch Lakes. Navuka i Tekhnika, Minsk: 182–184 [in Russian].

Gluschenko, L.O., 1989. Particulate organic matter and its role in the formation of water quality in Lake Sevan (Armenia). Arch. Hydrobiol. Beih. 33: 265–271.

Hamilton, A.L., 1969. On estimating annual production. Limnol. and Oceanogr. 14: 771–782.

Hynes, H.B.N. & M.J. Coleman, 1968. A simple method of assessing the annual production of stream benthos. Limnol. and Oceanogr. 13: 569–573.

Jenderedjian, K., 1989. Peculiarities of biology and the production of Oligochaeta of Lake Sevan. Cand. biol. sci. thesis, Leningrad, 24 pp. [in Russian].

Jenderedjian, K., 1994. Population dynamics of *Potamothrix alatus paravanicus* Poddubnaya & Pataridze (Tubificidae) in different areas of Lake Sevan. Hydrobiologia, 278 (Dev. Hydrobiol. 95): 281–286.

Jenderedjian, K. & T.L. Poddubnaya, 1987. Species composition and distribution of Oligochaeta in Lake Sevan. Biol. J. Armenia 40: 36–42 [in Russian].

Jonasson, P.M. & F. Thorhauge, 1976. Production *Potamothrix hammoniensis* in the profundal of eutrophic Lake Esrom. Oikos 27: 204–209.

Karataeva, I.V., 1990. The rate of growth and production of populations of *Potamothrix hammoniensis* (Mich.) in electric power station cooling reservoir. In Species in area: biology, ecology and productivity of water invertebrates. Navuka i Tekhnika, Minsk: 203–207 [in Russian].

Lukyanova, V.P., 1974. Benthos of the lake Ilmen, its production and utilization by fishes. Cand. biol. sci. thesis, Leningrad, 19 pp. [in Russian].

Menzie, C.A., 1980. A note on the Hynes' method of estimating secondary production. Limnol. and Oceanogr. 25: 770–773.

Pidgaiko, M.L., V.S. Grin, L.A. Kititsina, L.G. Lenchina, M.F. Polivannaya, O.A. Sergeeva & T.A. Vinogradskaya, 1972. Biological productivity of Kurakhov's Power Station cooling reservoir. In Productivity problems of freshwaters. Warszawa: 477–491.

Poddubnaya, T.L., 1975. Estimation of production of the tubificids. In Methods of studies of the biogeocoenosis of internal waters. Nauka, Moscow: 192–199 [in Russian].

Popchenko, V.I., 1973. Productivity of Oligochaeta in the small humified lakes of Karelia. In Biological studies on internal waters of Baltic countries. Minsk: 84–88 [in Russian].

Slepukhina, T.D., 1977. Zoobenthos and phytophilous fauna of Lake Kubenskoye. In Lake Kubenskoye. Leningrad: 51–87 [in Russian].

Timm, H., 1990a. Production of *Potamothrix hammoniensis* at 1984–1986 in the produndal of Lake Chudskoye. In Species in area: biology, ecology and productivity of water invertebrates. Navuka i Tekhnika, Minsk: 208–210 [in Russian].

Timm, H., 1990b. Zoobenthos production of Lake Chudskoye. Cand. biol. sci. thesis, Moscow, 16 pp. [in Russian].

Waters, T.F., 1979. Influence of benthos life history upon the estimation of secondary production. J. Fish. Res. Bd. Can. 36: 1425–1430.

Winberg, G.G., 1971. Methods for the estimation of production of aquatic animals. Academic Press, London.

Hydrobiologia **278**: 291–301, 1994.
T.B. Reynoldson & K.A. Coates (eds), Aquatic Oligochaete Biology V.

Population dynamics and energy budget of *Marionina southerni* (Cernosvitov) (Enchytraeidae, Oligochaeta) in the shallow littoral of Lake Esrom, Denmark

Claus Lindegaard, Kirsten Hamburger & Peter C. Dall
Freshwater Biological Laboratory, University of Copenhagen, 51 Helsingørsgade, DK–3400 Hillerød, Denmark

Key words: Enchytraeidae, population dynamics, production, respiration, energy budget, Lake Esrom

Abstract

Marionina southerni (Cernosvitov) was numerically the dominant oligochaete in a shallow, strongly exposed surf zone locality (0–1 m depth) in the mesotrophic Lake Esrom. It comprised 21,500 ind m^{-2} or nearly 50% of the total oligochaete community, which otherwise was dominated by *Nais* spp. and the lumbriculid *Stylodrilus heringianus* Claparède. *M. southerni* appears to have a 1–y life cycle. Individual biomass ranged from 3 to 48 μg ash free dry weight. Annual net production at the site was 5.1 kcal m^{-2} (size-frequency method) with a $P/\bar{B}$ ratio of 2.5. This is about 0.5% of the estimated mean zoobenthic production in the littoral zone from 0 to 2 m depth in Lake Esrom. At 5 °C *M. southerni* showed a constant oxygen uptake down to 25% oxygen saturation, but practically no regulatory respiration was found at 20 °C. In the field *M. southerni* was never found in sediment with less than 60% oxygen saturation. The annual community respiration of *M. southerni* was 14.2 kcal m^{-2}, and annual assimilation thus made up 19.3 kcal m^{-2} with a net production efficiency of 26.5%.

Introduction

An investigation of the invertebrate communities in the exposed littoral zone of Lake Esrom was initiated by the Lake Esrom Littoral Research Project in 1979 (Dall *et al.*, 1984a). Abundance and distribution of the oligochaetes were outlined by Lindegaard & Dall (1988). It was shown that the enchytraeid *Marionina southerni* (Cernosvitov) was found exclusively at the most exposed eastern 0.4 m site together with *Nais barbata* (Müller), *Rhyacodrilus coccineus* (Vejdovsky) and *Stylodrilus heringianus* Claparède in a community characteristic of oligotrophic conditions. The aim of the littoral project is to quantify the energy flow of the most important organisms in the littoral zone. Population dynamics, secondary production and community respiration are essential components in studies of energy flow and have gained much attention in recent decades (e. g., Benke, 1984).

Due to continual breeding and difficulties in determining the life span of oligochaetes, only a few studies on these themes have been carried out (cf. Jónasson & Thorhauge, 1976; Brinkhurst,

Lake Esrom Littoral Research Project, Publ. No. 31. Publication No. 536 from the Freshwater Biological Laboratory, University of Copenhagen, 51 Helsingørsgade, DK–3400 Hillerød, Denmark.

1980; Lazim & Learner, 1986a, 1986b). The size-frequency method of production estimates (e. g., Hamilton, 1969; Benke, 1984) and the experimental life span studies of oligochaetes by Adreani *et al.* (1984) and Bonacina *et al.* (1987) made it possible to estimate realistic production values of *S. heringianus* in Lake Esrom (Lindegaard *et al.*, 1990). However, life span studies of aquatic Enchytraeidae are few and concern only populations in filter beds of waste water purification plants (e. g., Reynoldson, 1948; Williams et al., 1969). This paper describes the energy flow of *M. southerni* in Lake Esrom based on "traditional" population data.

Study area

Lake Esrom is situated in the northern part of Zealand, Denmark (56° N, 12° E). Descriptions are given by Berg (1938), Jónasson (1972, 1977, 1984) and Jónasson *et al.* (1974). Sampling was made at two depths (0.4 and 2 m) on a western and eastern site of the lake shore. Detailed descriptions are given by Dall et al. (1984a) and Lindegaard & Dall (1988). *Marionina southerni* occurred primarily at the eastern 0.4 m site (Lindegaard & Dall, 1988).

Methods

Quantitative sampling was carried out from May 1979 to June 1980 at monthly intervals in summer and 2–3 months intervals in winter. A special grab (Dall, 1981) for sampling stones and the underlying matrix, enclosing 350 cm^2, was used. Three to five samples were taken at each date. All samples were sieved with a mesh size of 100 μm. This should ensure retention of small oligochaetes. However, as sorting was done manually without magnification, the sorting efficiency of very small oligochaetes was low. To obtain reliable values also for the meiofauna, subsamples of each sample were sorted under a stereo microscope (12.5 × magnification). The sampling programme and methods are described in detail by Dall *et al.* (1984a) and Lindegaard & Dall (1988).

Size classes of *M. southerni* were determined by measuring the length of the worms to the nearest 0.1 mm under a stereo microscope using a drawing device and a map distance reader. Worm biomass was based on the relationship between the length and the ash free dry weight (AFDW) of individual fresh worms (killed in 4% formalin) weighed on an automatic microelectro balance (Sartorius S4) with an accuracy of ± 0.1 μg. AFDW was found by subtracting ash content after ignition (1 h, 550 °C) from dry weight (24 h, 60 °C). Calorific values were calculated from AFDW by assuming that 1 g AFDW equals 5.5 kcal (Salonen *et al.*, 1976; Waters, 1977). In order to compare the data with results from the literature, we used the following conversion factors: 6 g wet weight = 1 g dry weight = 0.9 g AFDW = 0.134 g N (Nielsen, 1961; Waters, 1977).

Respiration was measured in the laboratory in closed bottles (pycnometers) using a Clark oxygen electrode (Radiometer, Denmark). As suspension medium we used autoclaved, 0.45 μm millipore-filtered tapwater, initially adjusted by bubbling with N_2 to between 15 and 100% oxygen saturation. A suitable number of animals (30–50) was placed in each bottle (10 ml) so that the oxygen tension during the experiment (7–20 h) was reduced by 10–20 mm Hg. The oxygen concentration was measured at the start and the end of the experiment in 0.5 ml samples withdrawn from the bottles. Two bottles without animals served as controls. Respiration experiments were carried out at 2, 5, 10, 15 and 20 °C with animals collected at approximately the same lake temperatures.

Annual community respiration was calculated from the monthly average population biomass (g AFDW m^{-2}) and equations (cf. Table 1) relating the oxygen consumption to temperature and oxygen concentration in the sediment.

Monthly average population biomass (g AFDW m^{-2}) was calculated from the abundance (cf. Fig. 1), the size frequency histograms (cf. Fig. 2) and the relationship between length (mm) and AFDW (mg) (cf. Fig. 3). Measurements in the sediment (5 and 10 cm depth) have shown actual variations in the range 6.1–11.0 mg O_2 l^{-1}

Table 1. Equations for regression lines relating the weight-specific oxygen uptake [$\mu l\ O_2$ mg AFDW^{-1} h^{-1}(y)] to oxygen concentration [mg O_2 l^{-1} (x)] in the form y = a + bx for *Marionina southerni* in Lake Esrom at different temperatures and oxygen levels. a, b = coefficients, P = probability, r^2 = coefficient of determination, n = number of experiments.

Temperature °C	Oxygen range mg O_2 l^{-1}	a	b	*P*	r^2	*n*
2	2–14	0.0116	0.0245	<0.001	0.46	21
5	1– 3	0.0088	0.1101	<0.0005	0.70	14
5	3–12	0.3793	0.0075	<0.01	0.20	27
10	1– 5	0.0017	0.1177	<0.005	0.42	16
10	5–11	0.6570	0.0026	≫0.5	0.004	18
15	1– 5	0.2494	0.2120	≪0.0001	0.84	18
15	5–10	1.4296	0.0134	≫0.5	0.006	10
20	1– 6	−0.2356	0.3663	≪0.001	0.79	15
20	6–10	1.6986	0.0269	≫0.5	0.002	16

which, correlated to field temperatures, gave saturations from 45 to 85% (Dall *et al.*, 1990). In the present calculation we assumed that the worms lived in sediment with saturations above 60%.

Results

The abundance of *M. southerni* varied without significant trends between 8,000 and 43,000 ind m^{-2} during the sampling period with an average density of 21,500 ind m^{-2} (Fig. 1).

The size classes of *M. southerni* found in the samples are shown in Figure 2. The smallest and largest size classes were 1.9 and 8.3 mm (length of body), respectively. Small, but obviously not newly hatched, individuals (2–3 mm) were present throughout the year with a maximum from June to October.

Mature individuals with a clitellum were found in all size classes from 3 mm upwards and were present throughout the year, though they were very scarce from August to January. From January to July mature individuals made up from 45–95% of the population. The number of mature worms declined rapidly from July to August and was almost zero during autumn (Fig. 1). There was a small decline in size of mature worms from winter to summer. This coincided with a general decline in numbers of large animals, which suggests that most worms died after breeding (Fig. 2). Cocoons were not sampled quantitatively. The *M. southerni* data suggest a 1-y life cycle in Lake Esrom with an overwintering generation starting to mature in January and with reproduction continuing until July.

The relationship between the length and the AFDW of newly killed *M. southerni* is shown in Figure 3. The regression line was used in the estimation of biomass in the size classes.

Using the size-frequency method (Hamilton, 1969; Benke, 1984) supplemented by the improvements of Krueger & Martin (1980) and Morin *et al.* (1987), the annual net production from June 1979 to June 1980 at the eastern 0.4 m depth site was estimated at 923 mg AFDW m^{-2}. The 95% confidence limit was estimated to be 637–1221 mg AFDW m^{-2} using the bootstrap method with 500 iterations. This equals 5.1 kcal m^{-2} and gave an annual P/$\bar{B}$ ratio of 2.5. It was presumed that (1) the cohort production interval (CPI) was 365 days and (2) the time spent in each of the 30 size classes was equal (12 days). *Marionina southerni* is restricted to the shallow exposed eastern shore and thus contributes only 1.3 kcal m^{-2} yr^{-1} or

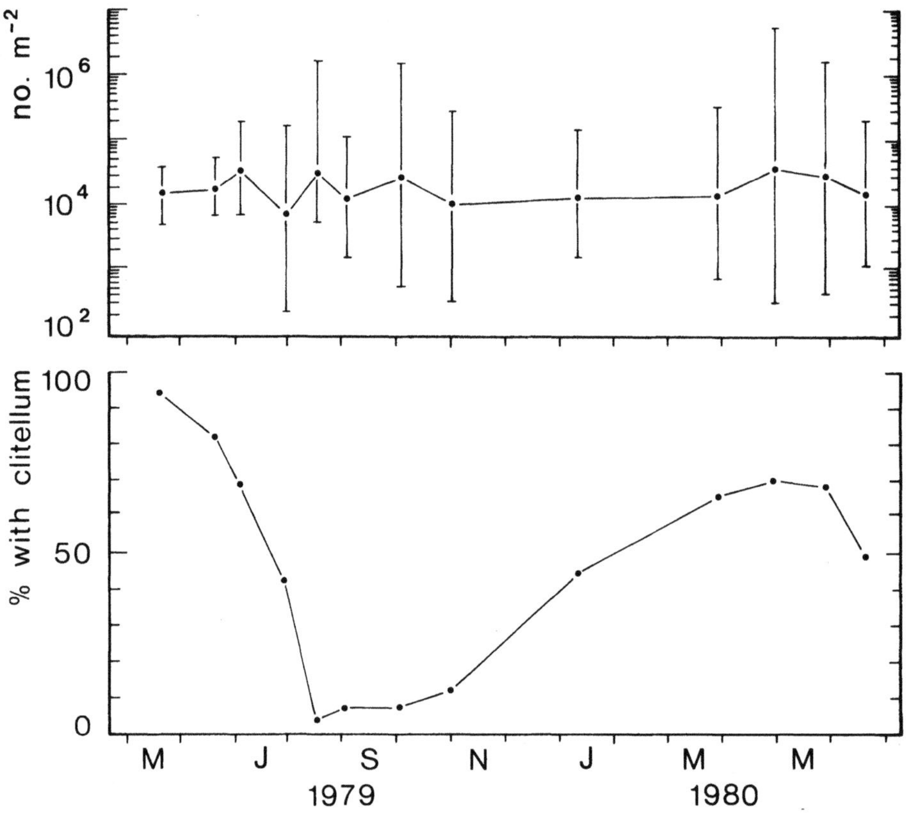

Fig. 1. Abundance and percentage of mature *Marionina southerni* at the exposed eastern littoral zone of Lake Esrom during 1979–80. The vertical bars show 95% confidence limits.

0.5% of the estimated mean zoobenthic production in the littoral zone from 0–2 m depth.

Oxygen consumption of *M. southerni* at varying oxygen concentrations and temperatures is shown in Figure 4. The results suggest that the capability of *M. southerni* for regulating oxygen consumption is rather limited. We have arbitrarily chosen the critical points below which regulation fails at 3, 5.5, 6 and 5.5 mg $O_2\,l^{-1}$ at 5, 10, 15 and 20 °C, respectively, and fitted two regressions for each set of observations (Table 1). Below the critical points the rate of oxygen consumption declined almost in proportion to the oxygen concentration; above the oxygen consumption only changed slowly. The decline from saturation down to the critical point equalled 13, 3, 4 and 8% of maximum oxygen consumption at the four temperatures indicated above. At 2 °C the number and range of observations did not allow distinction of a critical point, but the single fitted regression indicates a steady decline amounting to about 50% as the oxygen concentration changed from saturation (13 mg $O_2\,l^{-1}$) to 2 mg $O_2\,l^{-1}$. No significant differences in oxygen uptake were found between animals taken from spring and autumn populations (Fig. 4: 5 and 10 °C) although only the spring animals were in the reproductive state (Fig. 2).

Annual community respiration was calculated to be 14.2 kcal m^{-2}. The annual population production was 5.1 kcal m^{-2} and the assimilation consequently 19.3 kcal m^{-2} with a net production efficiency of about 26.5%.

Discussion

According to Nielsen & Christensen (1959) *M. southerni* is a brackish water species commonly found in Danish coastal areas. Giere (1976) men-

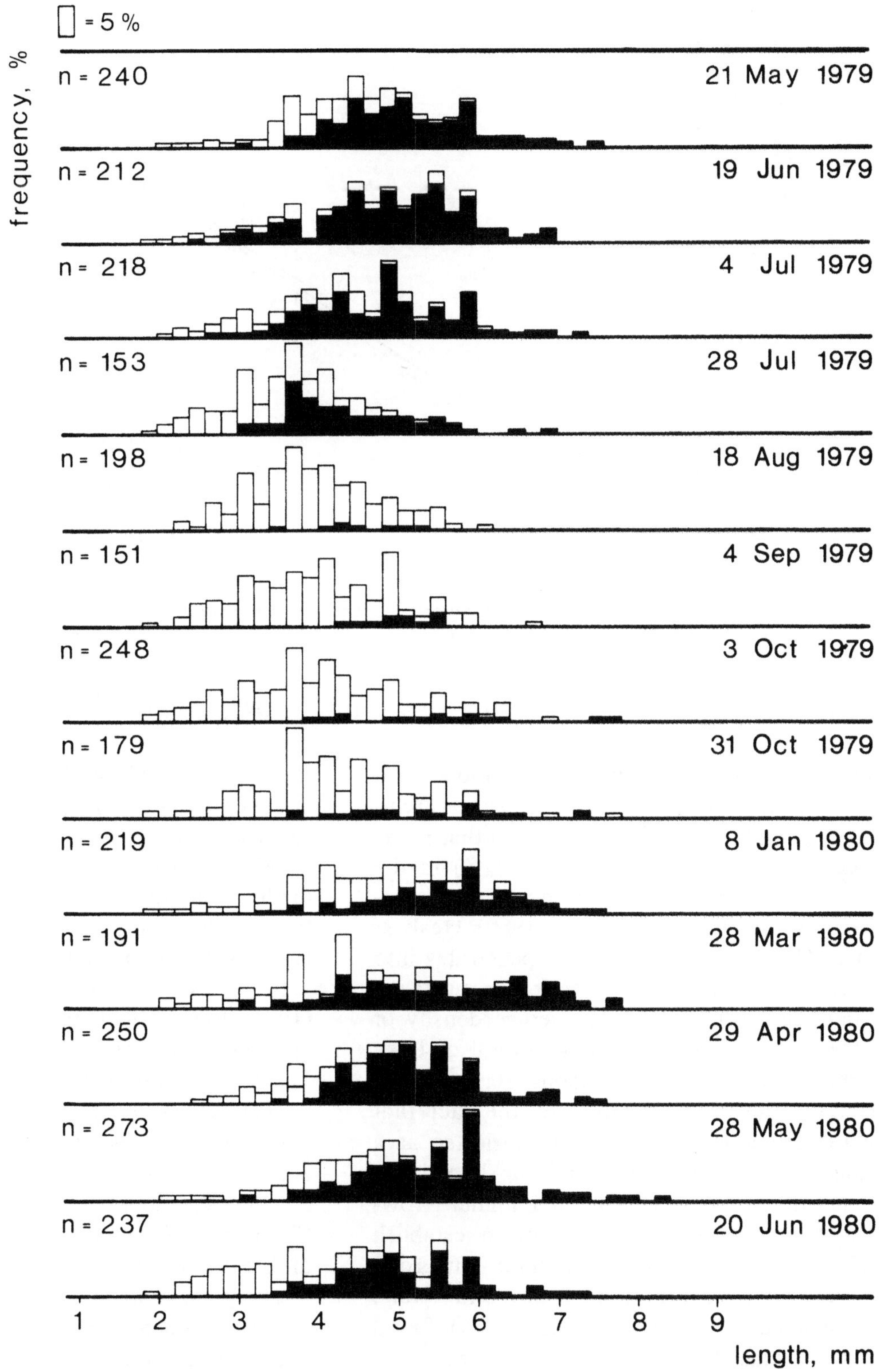

Fig. 2. Size frequency histograms for *Marionina southerni* based on the total length (mm). Black areas indicate individuals with clitellum. n = numbers measured.

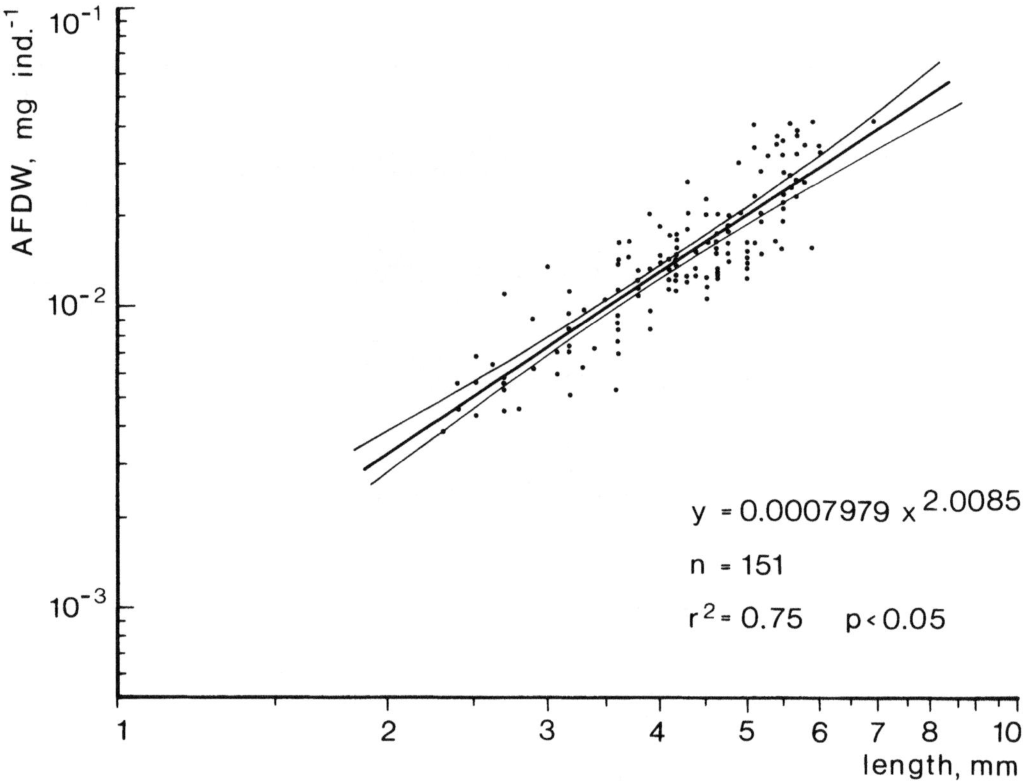

Fig. 3. Relationship between *Marionina southerni* length (mm) and AFDW (mg). Curved lines show 95% confidence limits.

tioned it as a dominant oligochaete in moist sand above the waterlevel in the southwestern archipelago of Finland. He further showed that it has a preference for low salinity (2–7‰ S) in the western part of the Baltic Sea, though it also occurred in pure freshwater (Giere, 1980). Healy & Bolger (1984) mentioned that it penetrates into freshwater in Ireland. However, it was surprising to find this species with an average density of 21,500 ind m^{-2} in the shallow littoral of Lake Esrom, which is located about 10 km from the sea. This might be due to the fact that Enchytraeidae have been largely neglected as a limnectic group and therefore our knowledge of enchytraeids in freshwater is minimal. However, a number of records seem to establish enchytraeids as a numerically important group of the zoobenthos in both lakes and rivers. For instance, Lindegaard *et al.* (1975) found Enchytraeidae at a density of 2,600 ind m^{-2} in a moss carpet of a Danish helocrene spring, and Whiteside & Lindegaard (1982) recorded Enchytraeidae with densities up to 6,000 ind m^{-2} in the upper littoral zone of the oligotrophic Danish lake Grane Langsø. Further, Lindegaard (1992) found that five species of Enchytraeidae made up about 14% of the average invertebrate fauna of the oligotrophic Icelandic Lake Thingvallavatn with the maximum densities occurring at 4 to 6 m depth. Also Maitland and Hudspith (1974) found Enchytraeidae in high numbers (11,000 ind m^{-2}) in the sandy littoral of Loch Leven, Scotland. This, together with Kasprzak's (1977) records of 12 species of Enchytraeidae regularly occurring in the eulittoral and littoral zones of Polish lakes and the occurrence of 27 Enchytraeidae species in German streams (Schwank, 1981) indicate that this group can be an important element in freshwater communities.

The present size-frequency data did not allow an accurate determination of the life cycle, however we suggest, based on the data we did obtain and the results of other studies, that *M. southerni* is univoltine in Lake Esrom. Reynoldson (1943)

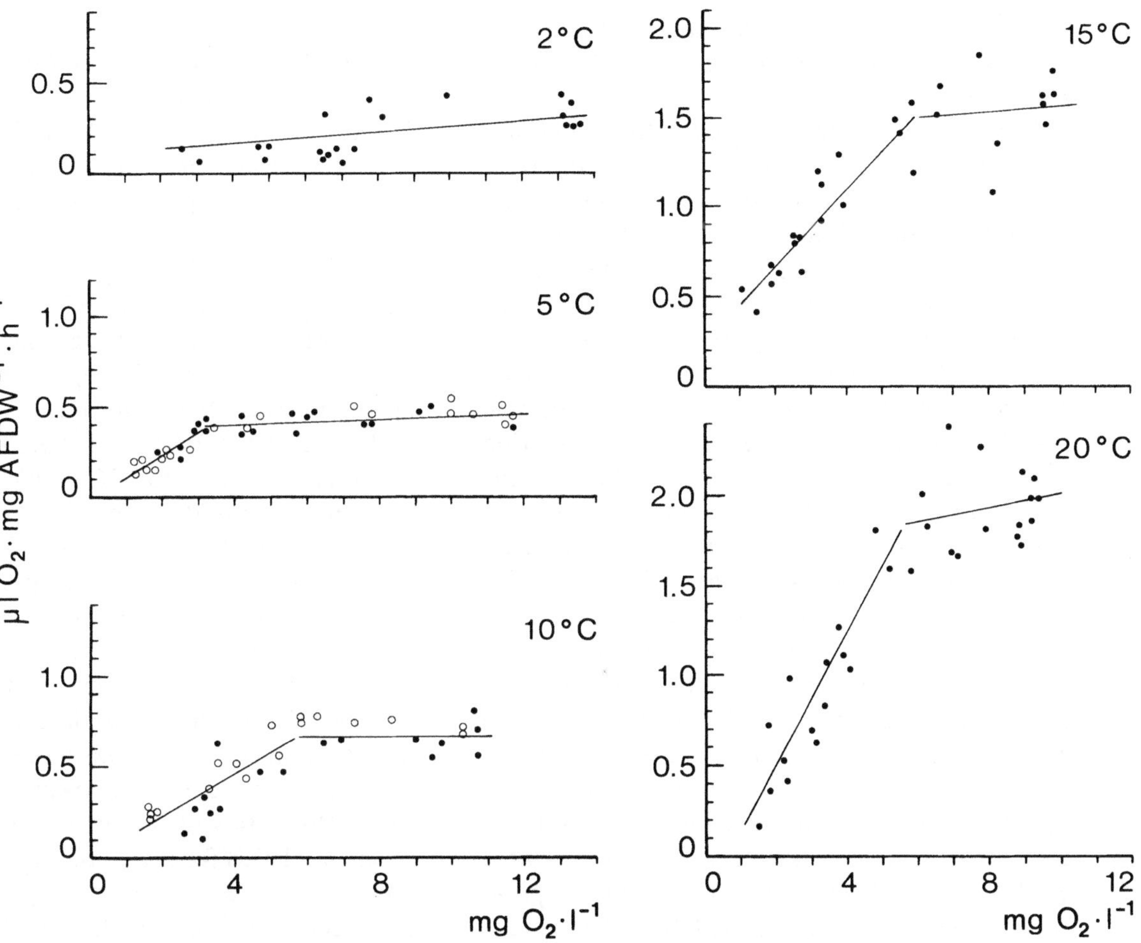

Fig. 4. The weight-specific oxygen uptake [μl O_2 mg AFDW^{-1} h^{-1} (y)] shown against the oxygen concentration [mg O_2 l^{-1} (x)] at 2, 5, 10, 15 and 20 °C. Regression lines were fitted for concentrations both above and below the critical point except at 2 °C, where only one regression was calculated. Filled and open circles at 5 and 10 °C denote experiments carried out at decreasing temperatures during autumn and increasing temperatures during spring, respectively.

found, in laboratory studies, that the total life cycle of *Lumbricillus rivalis* (Levinsen) and *Enchytraeus coronatus* (Nielsen & Christensen) (originally identified as *Lumbricillus lineatus* Müller and *Enchytraeus albidus* Henle) from bacteria beds varied from about 30 to 200 days, depending on temperature. In terms of degree days (dd), the life cycle of *L. rivalis* and *E. coronatus* averaged about 1100 and 665 dd, respectively. Reynoldson (1943) suggested that these results were applicable to field conditions. Springett (1970) studied a number of moorland Enchytraeidae and found, experimentally, life cycles varying from 56 to 321 days at 10 °C. Her field estimates were a little lower, ranging from 30 to 300 days. In particular, she found that *Marionina clavata* Nielsen & Christensen had two different types of life cycle: a fast one taking 560 dd and a slow one lasting 2230 dd. Learner (1972) studied four Enchytraeidae species from sewage percolating filters and found life cycles averaging 700 dd with a range from 405 to 1140 dd. Especially, he found that *L. rivalis* only took 675 dd to reach maturity, which was nearly half the number of dd found by Reynoldson (1943). Learner (1972) suggested methodological differences to account for this discrepancy. Recently, Bouguenec & Giani (1989) found, in laboratory mass cultures, that the terrestrial *Enchytraeus variatus* Bouguenec & Giani had a generation period of 26 days at 18–22 °C ($\approx$500 dd).

Field studies on life cycles of Enchytraeidae are few. O'Connor (1958) studied the seasonal trend in a number of enchytraeids in a coniferous forest soil in North Wales. He concluded that cocoons were laid during winter and hatched in spring. Williams *et al.* (1969) studied *L. rivalis* and *E. coronatus* in a sewage percolating filter and concluded that maximum depositions of cocoons occurred during February-May, but that the size distribution data were too variable to describe growth rates precisely. Springett (1970) concluded from both laboratory experiments and field data on five species of moorland enchytraeids, that two species were univoltine and three species had life cycles longer than one year.

Terrestrial enchytraeids appear to have a longer life cycle than aquatic, with drought periods during summer being responsible for minimizing the growth period in terrestrial habitats (cf. Nielsen, 1961). Life cycles of about 400 to 1200 dd found in laboratory experiments allow hypothetically 2–8 generations per year in Lake Esrom, where 3,200 dd are available (or 2,000 dd with temperatures above 4 °C). However, laboratory experiments are difficult to apply to field conditions as the influence of a number of factors – especially available food and inter- and intraspecific competition – are minimized in laboratory experiments. Lindegaard & Mortensen (1988) concluded that life cycles based on field studies of Chironomidae in temperate streams are much longer than those obtained in laboratory experiments (Mackey, 1977), the latter allowed up to 30–40 generations per year for small species (cf. Benke *et al.*, 1984). Therefore, with the distinct decline in mature worms during late summer and autumn and with the modest number of 2–8 hypothetical generations, it seems reasonable to suggest a univoltine life cycle for *M. southerni* in Lake Esrom.

Total zoobenthic production in the upper littoral zone of Lake Esrom was estimated to about 200 kcal m^{-2} y^{-1} (Jónasson *et al.*, 1990). At the 0.4 m station on the eastern shore the trichopteran *Tinodes waeneri* (Linnaeus) produced 96 kcal m^{-2} y^{-1} (Dall *et al.*, 1984b), *Stylodrilus heringianus* 20 kcal m^{-2} y^{-1} (Lindegaard *et al.*, 1990), the Hirudinea 2 kcal m^{-2} y^{-1} (Dall, 1987), the coleopterans *Oulimnius* spp. 1 kcal m^{-2} y^{-1} (Dall & Lindegaard, in prep.) and *Marionina southerni* 5.1 kcal m^{-2} y^{-1}. These taxa together account for about 125 kcal m^{-2} y^{-1}, but additional important groups are expected to contribute with high production values (e. g., the remaining oligochaetes, chironomids, ephemeropterans, the zebra mussel, *Dreissena polymorpha* Pallas: cf. Dall *et al.*, 1984a). The calculated production of 5.1 kcal m^{-2} y^{-1} for *M. southerni* seems low compared to the 20 kcal m^{-2} y^{-1} produced by *S. heringianus*. As the $P/\bar{B}$ ratio is 2.5 and 3.0 for *M. southerni* and *S. heringianus*, respectively, the production efficiency of the two species is approximately the same. The maximum biomass of *M. southerni* is 48 μg and according to Banse & Mosher (1980) the theoretical $P/\bar{B}$ ratio of an animal of this size should be about 12. The present $P/\bar{B}$ ratio of 2.5 is thus remarkably low. However, lower $P/\bar{B}$ ratios of 0.8 to 1.2 were recorded by Standen (1973) for a moorland enchytraeid, *Cognettia sphagnetorum* (Vejdovsky) reproducing by fragmentation. Further, she found a $P/\bar{B}$ ratio of 2.6 for the sexually reproducing *C. glandulosa* (Michaelsen), which corresponded to values by Lindegaard (1992) for an Enchytraeidae population in the Icelandic lake Thingvallavatn. These values for Enchytraeidae are generally low, but within the range given for univoltine species (Waters, 1977).

The respiration curves in Figure 4 show that *M. southerni* is able to regulate its oxygen consumption a little, but the critical point – especially at high temperatures – is high (60% oxygen saturation at 10–20 °C). Other species of the same genus are less sensitive to low oxygen concentrations. For example *Marionina achaeta*, also a brackish water species, is able to regulate oxygen uptake down to 1 ml O_2 l^{-1} ($\approx$ 15% saturation) at 20 °C, and survives up to 3 days at 0.5 ml O_2 l^{-1} (Lasserre, 1976). A high demand for oxygen might explain why *M. southerni* is limited to habitats with high degrees of exposure and consequently high oxygen concentrations in sediment interstitial water (cf. Dall *et al.*, 1990). However, the lumbriculid worm *S. heringianus* shows the same restricted distribution as *M. southerni* in Lake Esrom (Lindegaard & Dall, 1988) although it is able to regulate its oxygen uptake down to 17% oxygen saturation at 20 °C (Lindegaard *et al.*, 1990). Therefore, other factors (e.g. H_2S) probably in combination with oxygen concentration determine the distribution of oligochaetes in Lake Esrom. High densities of decaying weeds produce redox discontinuity layers which give rise to anoxic conditions and to toxic compounds such as hydrogen sulphide (H_2S), which is even at low concentrations toxic to most meiofauna species including *M. achaeta* (Lasserre, 1976).

Respiration rate of *M. southerni* varied between 0.2 and 2.0 μl O_2 mg AFDW^{-1} h^{-1} in winter and summer, respectively (Fig. 4). This is a high consumption rate compared to *S. heringianus*, which in summer (20 °C) respired only 0.7 μl O_2 mg AFDW^{-1} h^{-1} (Lindegaard *et al.*, 1990). However, considering the difference in size (AFDW of *S. heringianus* ranges from 0.05–2 mg ind^{-1}), the respiration rate of *M. southerni* agrees well with this and also other oligochaete species in the littoral zone of Lake Esrom (Hamburger & Dall, 1990).

Marionina southerni is a relatively small enchytraeid worm (individual weight ranges from 3 to 48 μg AFDW) and specimens used in respiration experiments were in the range of 11 to 26 μg AFDW ind^{-1}. Williams *et al.* (1969) found a considerably higher oxygen consumption of 2.5 and 4.5 μl O_2 mg AFDW^{-1} h^{-1} for the small *Lumbricillus rivalis* (18 μg AFDW ind^{-1}) at 10 and 20 °C, respectively, whereas the corresponding values for large individuals (180 μg AFDW ind^{-1}) were 1.6 and 2.5 μl O_2 mg AFDW^{-1} h^{-1}. Also, O'Connor (1963) found an increasing respiration rate with decreasing size of three enchytraeid species from a coniferous forest soil. For the small species *Marionina cambrensis* (O'Connor) and *Achaeta eiseni* (Vejdovsky), with maximum individual weights about 36 μg AFDW, he found an oxygen uptake about 2.6 μl O_2 mg AFDW^{-1} h^{-1}, whereas it was 5.0 μl O_2 mg AFDW^{-1} h^{-1} in the larger *Cognettia cognettii* (Issel) which had a maximum individual weight of about 60 μg AFDW. Both results were for individuals weighing 18 μg AFDW and at 20 °C. Further, Nielsen (1961) found a constant oxygen consumption of about 1.9 μl O_2 mg AFDW^{-1} h^{-1} for a number of enchytraeid species weighing 70–500 μg AFDW ind^{-1} at 20 °C, whereas smaller individuals generally showed a higher respiration rate (up to about 7–8 μl O_2 mg AFDW^{-1} h^{-1}). However, two of his experiments were with *M. southerni* with individual weights ranging from 12 to 22 μg AFDW. Their oxygen consumption was 2.1 and 2.5 μl O_2 mg AFDW^{-1} h^{-1}, which is well within the range found in our experiments at 20 °C (Fig. 4).

An average net production efficiency (NPE) of 26.5%, as found for *M. southerni*, is low compared to the 40% found for *Stylodrilus heringianus* at the same habitat in Lake Esrom (Lindegaard *et al.*, 1990), but higher than the 14–17% found by Standen (1973) for the terrestrial *C. sphagnetorum*. Standen (1980) recalculated production in four other enchytraeid populations using a $P/\bar{B}$ ratio of 2.5. She found NPE values between 15 and 20% in three cases and only in one case a NPE of nearly 50%. However, our NPE of 26.5% is right in the middle of the range of NPE from 22 to 30% for zoobenthic populations as calculated by Alimov (1983) by reviewing a large number of studies.

Acknowledgements

The authors wish to thank T. B. Reynoldson, P. M. Chapman and P. F. M. Verdonschot for valuable suggestions and linguistic correction of the manuscript. We further thank C. Overgaard Nielsen for the identification of *Marionina southerni*, and Ms L. Eplov, Ms H. Sørensen and Mr F. Petersen for technical assistance. The investigation was supported by grants from the Danish Natural Science Research Council to C. Lindegaard and P. C. Dall, and from the Carlsberg Foundation to K. Hamburger and C. Lindegaard.

References

Adreani, L., C. Bonacina, G. Bonomi & C. Monti, 1984. Cohort cultures of *Psammoryctides barbatus* (Grube) and *Spirosperma ferox* Eisen: a tool for a better understanding of demographic strategies in Tubificidae. Hydrobiologia 115: 113–119.

Alimov, A. F., 1983. Energy flows in populations and communities of aquatic animals. Int. Revue ges. Hydrobiol. 68: 1–12.

Banse, K. & S. Mosher, 1980. Adult body mass and annual production/biomass relationships of field populations. Ecol. Monogr. 50: 355–379.

Benke, A. C., 1984. Secondary production of aquatic insects. In V. H. Resh & D. M. Rosenberg (eds), The Ecology of Aquatic Insects. Praeger, New York: 289–322.

Benke, A. C., T. C. Van Arsdall Jr. & D. M. Gillespie, 1984. Invertebrate productivity in a subtropical blackwater river: The importance of habitat and life history. Ecol. Monogr. 54: 25–63.

Berg, K., 1938. Studies on the bottom animals of Esrom Lake. K. Danske Vidensk. Selsk. Skr. Nat. Mat. Afd. 9 (8): 1–255.

Bonacina, C., G. Bonomi & C. Monti, 1987. Progress in cohort cultures of aquatic Oligochaeta. Hydrobiologia 155: 163–169.

Brinkhurst, R. O., 1980. The production biology of the Tubificidae (Oligochaeta). In R. O. Brinkhurst & D. G. Cook (eds), Aquatic Oligochaete Biology. Plenum Press, New York: 205–209.

Bouguenec, V. & N. Giani, 1989. Biological studies upon *Enchytraeus variatus* Bouguenec & Giani 1987 in breeding cultures. Hydrobiologia 180: 151–165.

Dall, P. C., 1981. A new grab for the sampling of zoobenthos in the upper stony littoral zone. Arch. Hydrobiol. 92: 396–405.

Dall, P. C., 1987. The ecology of the littoral leech fauna (Hirudinea) in Lake Esrom, Denmark. Arch. Hydrobiol. Suppl. 76: 256–313.

Dall, P. C. & C. Lindegaard, in prep. Population dynamics and energy budget of *Oulimnius troglodytes* and *O. tuberculatus* (Coleoptera, Elmidae) in Lake Esrom, Denmark.

Dall, P. C., C. Lindegaard, E. Jónsson, G. Jónsson & P. M. Jónasson, 1984a. Invertebrate communities and their environment in the exposed littoral zone of Lake Esrom, Denmark. Arch. Hydrobiol. Suppl. 69: 477–524.

Dall, P. C., H. Heegaard & A. F. Fullerton, 1984b. Life-history strategies and production of *Tinodes waeneri* (L.) (Trichoptera) in Lake Esrom, Denmark. Hydrobiologia 112: 93–104.

Dall, P. C., C. Lindegaard & P. M. Jónasson, 1990. In-lake variations in the compositions of zoobenthos in the littoral of Lake Esrom, Denmark. Verh. int. Ver. Limnol. 24: 613–620.

Giere, O., 1976. Zur Kenntnis der litoralen Oligochaetenfauna Südfinlands. Ann. Zoo. Fenn. 13: 156–160.

Giere, O., 1980. Tolerance and preference reactions of marine Oligochaeta in relation to their distribution. In R. O. Brinkhurst & D. G. Cook (eds), Aquatic Oligochaete Biology. Plenum Press, New York: 385–409.

Hamburger, K. & P. C. Dall, 1990. The respiration of common benthic invertebrate species from the shallow littoral zone of Lake Esrom, Denmark. Hydrobiologia 199: 117–130.

Hamilton, A. L., 1969: On estimating annual production. Limnol. Oceanogr. 14: 771–782.

Healy, B. & T. Bolger, 1984. The occurrence of species of semi-aquatic Enchytraeidae (Oligochaeta) in Ireland. Hydrobiologia 115: 159–170.

Jónasson, P. M., 1972. Ecology and production of the profundal benthos in relation to phytoplankton in Lake Esrom. Oikos Suppl. 14: 1–148.

Jónasson, P. M., 1977. Lake Esrom Research 1867–1977. Fol. Limnol. Scand. 17: 67–89.

Jónasson, P. M., 1984. The ecosystem of eutrophic Lake Esrom. In F. B. Taub (ed.), Lakes and Reservoirs. Elsevier Science, Amsterdam: 177–204.

Jónasson, P. M. & F. Thorhauge, 1976. Production of *Potamothrix hammoniensis* in the profundal of eutrophic Lake Esrom. Oikos 27: 204–209.

Jónasson, P. M., E. Lastein & A. Rebsdorf, 1974. Production, insolation, and nutrient budget of eutrophic Lake Esrom. Oikos 25: 255–277.

Jónasson, P. M., C. Lindegaard & K. Hamburger, 1990. Energy budget of Lake Esrom, Denmark. Verh. int. Ver. Limnol. 24: 632–640.

Kasprzak, K., 1977. Oligochaeta. The bottom fauna of the heated Konin Lakes. Monografie Fauny Polsky 7: 49–146.

Krueger, C. C. & F. B. Martin, 1980. Computation of confidence intervals for the size-frequency (Hynes) method of estimating secondary production. Limnol. Oceanogr. 25: 773–777.

Lasserre, P., 1976. Metabolic activities of benthic microfauna and meiofauna. Recent advances and review of suitable methods of analysis. In I. N. McCave (ed.), The Benthic Boundary Layer. Plenum Press, New York: 95–142.

Lazim, M. N. & M. A. Learner, 1986a: The life-cycle and production of *Limnodrilus hoffmeisteri* and *L. udekemianus* (Tubificidae; Oligochaeta) in the organically enriched Moat-Feeder Stream, Cardiff, South Wales. Arch. Hydrobiol. Suppl. 74: 200–225.

Lazim, M. N. & M. A. Learner, 1986b. The life-cycle and productivity of *Tubifex tubifex* (Oligochaeta; Tubificidae) in the Moat-Feeder Stream, Cardiff, South Wales. Holarct. Ecol. 185–193.

Learner, M. A., 1972. Laboratory studies on the life-histories of four enchytraeid worms (Oligochaeta) which inhabit sewage percolating filters. Ann. Appl. Biol. 70: 251–266.

Lindegaard, C., 1992. Zoobenthos ecology of Thingvallavatn: vertical distribution, abundance, population dynamics and production. Oikos 64: 257–304.

Lindegaard, C. & E. Mortensen, 1988. Abundance, life history and production of Chironomidae (Diptera) in a Danish lowland stream. Arch. Hydrobiol. Suppl. 81: 563–587.

Lindegaard, C. & P. C. Dall, 1988. Abundance and distribution of Oligochaeta in the exposed littoral zone of Lake Esrom, Denmark. Arch. Hydrobiol. Suppl. 81: 533–562.

Lindegaard, C., J. Thorup & M. Bahn, 1975. The invertebrate fauna of the moss carpet in the danish spring Ravnkilde and its seasonal, vertical and horizontal distribution. Arch. Hydrobiol. 75: 109–139.

Lindegaard, C., K. Hamburger & P. C. Dall, 1990. Population dynamics and energy budget of *Stylodrilus heringianus* Clap. (Lumbriculidae, Oligochaeta) in the shallow littoral of Lake Esrom, Denmark. Verh. int. Ver. Limnol. 24: 626–631.

Mackey, A. P., 1977. Growth and development of larval Chironomidae. Oikos 28: 270–275.

Maitland, P. S. & P. M. G. Hudspith, 1974. The zoobenthos of Loch Leven, Kinross, and estimates of its production in the sandy littoral area during 1970 and 1971. Proc. R. Soc. Edinb. B 74: 219–239.

Morin, A., T. A. Mousseau & D. A. Roff, 1987. Accuracy and precision of secondary production estimates. Limnol. Oceanogr. 32: 1342–1352.

Nielsen, C. O., 1961. Respiratory metabolism of some populations of enchytraeid worms and freeliving nematodes. Oikos 12: 17–35.

Nielsen, C. O. & B. Christensen, 1959. The Enchytraeidae. Critical revision and taxonomy of European species. Natura Jutlandica 8–9: 1–160.

O'Connor, F. B., 1958. Age class composition and sexual maturity in the enchytraeid worm population of a coniferous forest soil. Oikos 9: 272–281.

O'Connor, F. B., 1963. Oxygen consumption and population metabolism of some populations of Enchytraeidae from North Wales. In J. Doeksen (ed.), Soil Organisms. North-Holland Publ. Co., Amsterdam: 32–48.

Reynoldson, T. B., 1943. A comparative account of the life cycles of *Lumbricillus lineatus* Mull. and *Enchytraeus albidus* Henle in relation to temperature. Ann. Appl. Biol. 30: 60–66.

Reynoldson, T. B., 1948. An ecological study of the enchytraeid worm population of sewage bacteria beds: synthesis of field and laboratory data. J. Anim. Ecol. 17: 27–38.

Salonen, K., J. Sarvala, I. Hakala & M.-L. Viljanen, 1976. The relation of energy and organic carbon in aquatic invertebrates. Limnol. Oceanogr. 21: 724–730.

Schwank, P., 1981. Turbellarien, Oligochaeten und Archianneliden des Breitenbachs und anderer oberhessischer Mittelgebirgsbäche. II. Die Systematik und Autökologie der einzelnen Arten. Arch. Hydrobiol. Suppl. 62: 86–147.

Springett, J. A., 1970. The distribution and life histories of some moorland Enchytraeidae (Oligochaeta). J. Anim. Ecol. 39: 725–737.

Standen, V., 1973. The production and respiration of an enchytraeid population in blanket bog. J. Anim. Ecol. 42: 219–245.

Standen, V., 1980. The production biology of terrestrial Enchytraeidae (Oligochaeta). In R. O. Brinkhurst & D. G. Cook (eds), Aquatic Oligochaete Biology. Plenum Press, New York: 211–223.

Waters, T. F., 1977. Secondary production in inland waters. Adv. Ecol. Res. 10: 91–104.

Whiteside, M. C. & C. Lindegaard, 1982. Summer distribution of zoobenthos in Grane Langsø, Denmark. Freshwat. Invertebr. Biol. 1: 2–16.

Williams, N. V., J. F. de L. G. Solbé & R. W. Edwards, 1969. Aspects of the distribution, life history and metabolism of the enchytraeid worms *Lumbricillus rivalis* Levinsen and *Enchytraeus coronatus* (N & C) in a percolating filter. J. Appl. Ecol. 6: 171–183.

Hydrobiologia **278**: 303–308, 1994.
T.B. Reynoldson & K.A. Coates (eds), Aquatic Oligochaete Biology V.

Population dynamics of *Narapa bonettoi* Righi and Varela, 1983 (Oligochaeta: Narapidae) from the main channel of the Middle Paraná River, Argentina

M. Marchese
National Institute of Limnology (CONICET), José Maciá 1933, (3016) Santo Tomê (Santa Fe), Argentina

Key words: Paraná River, Argentina, Oligochaeta, population dynamics, Narapa bonettoi

Abstract

The aim of the present study was to determine the life-cycle of *Narapa bonettoi* in the main channel of the Middle Paraná River. The average density varied between 17 and 624 ind. sample^{-1}. The spatial distribution was contagious with a higher degree of aggregation during the first months of the observed cycle, than during the last months. Mature specimens were more abundant in September 1985–86, resulting in greater breeding activity in early summer, when there was a large population increase. Two reproductive strategies were observed: asexual and sexual. After breeding, worms undergo a resorption process of their reproductive organs, returning to a non-sexual condition. *Narapa bonettoi* probably uses asexual reproduction as an adaptative strategy, because cocoon development will be affected by environmental factors such as the drift in the Paraná River which is highly erodable. Population growth was rapid during the first four months and then stabilized.

Introduction

Narapa bonettoi Righi and Varela, 1983 has been reported in lotic waters with high current velocity and discharge, sandy sediments with little organic matter and mobile sediments with dunes up to 8 meters high. This species is dominant and often the only benthic species present in the main channel of the Paraná River and secondary channels with similar characteristics (Fig. 1) (Ezcurra de Drago, 1980; Marchese, 1981, 1984, 1987; Marchese & Ezcurra de Drago, 1983, 1992; Varela *et al.*, 1983; Bertoldi de Pomar *et al.*, 1986).

The aim of the present study was to determine the life-cycle of *N. bonettoi* in the main channel of the Middle Paraná River.

Materials and methods

Site description

The study site was the main channel of the Middle Paraná River (31° 32′ S–60° 17′ W). The water level (Fig. 2a) corresponds to the common water regime of the Paraná River.

The average discharge was 16000 m^3 s^{-1} and the current velocity varied between 1 and 1.56 ms^{-1} at the low and high water stages, respectively. The minimum temperature of the water was 16 °C (June) and the maximum was 29.5 °C (February). The Secchi disc (Fig. 2c) varied between 8 cm (January 1986) and 45 cm (August 1985). The minimum and maximum values were at the low and high water levels, respectively. The

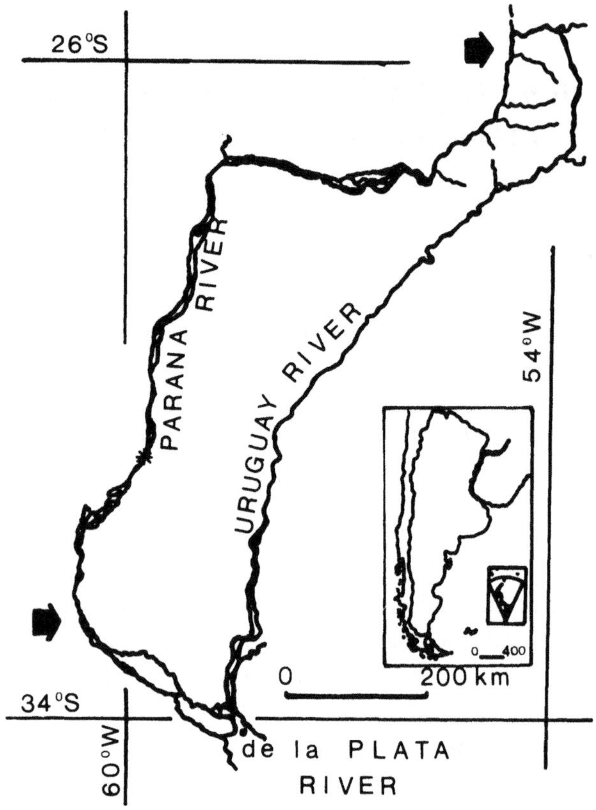

Fig. 1. Geographical distribution of *Narapa bonettoi*. The arrows show the reach of the Paraná River where the species was recorded. * Site samples.

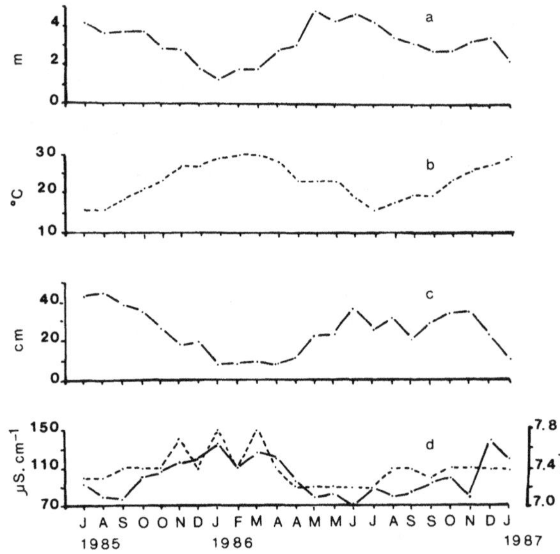

Fig. 2. Limnological characteristics of the Midde Paraná River: (a) water level at Parana city, Entre Ríos, (b) temperature in the main channel, (c) values for Secchi disc, (d) conductivity (solid) and pH (dash).

pH was between 7.0 and 7.8, reaching the highest values during the low water period (Fig. 2d).

The conductivity (Fig. 2d) varied between 70 (June) and 140 $\mu s\ cm^{-1}$ (December). The substratum was sandy (98–100% sand) with a very low organic matter content (0.004 to 0.007 gC%).

Sampling the worms

Samples were mostly collected at monthly, or occasionally fortnightly, intervals, from July 1985 to January 1987 (23 sampling events) using a 319 cm^2 Tamura grab. On each sampling occasion, 10 grabs (replicates) were obtained from one sampling point in the centre of the site, and were preserved in formalin in the field. The samples were taken to the laboratory, filtered through a 100 μ mesh sieve. The material retained with the sieve was kept in 70% ethanol. Then, the worms, that had been previously stained with eritrosyn were hand-picked from the sand. After each worm had been identified, the body length was measured using an eyepiece micrometer (6444 worms were measured).

Whenever the total number of worms found was too large to process, a sub-sample of approximately 500 worms was measured. Two hundred and thirty replicates were analyzed.

To determine differences between arithmetic means of abundance ANOVA was used. All samples; summer, spring, autumn and winter, high and low density period and hydrologic cycle were used to establish this difference.

The spatial distribution was analyzed using Morisita's Index of Dispersion (Elliott, 1977). Also, a length frequency distribution diagram from each sample event was constructed. The minimum number of samples was calculated according to Wetzel and Likens (1980).

$$N_{\min} = \frac{s^2}{\bar{x}(0.1)}.$$

The abundance data were logarithmically transformed, $\log_{10}(x+1)$, before 95% confidence intervals (CI) were calculated. The population growth was estimated using Bertalanffy's equation.

The linear correlation coefficients between the recorded density, physical and chemical parameters were estimated.

Results

Relationship between the number of individuals and physical and chemical characteristics of the environment

The correlation between the number of individuals and the values of the physical and chemical parameters (water regime, pH, current velocity, temperature, conductivity and transparency of the water) were only significant with the water regime $r = 0.46$ ($p < 0.05$) and pH, $r = 0.43$ ($p < 0.05$).

Population density and spatial distribution

The application of Morisita's Index of Dispersion to the abundance data, showed *N. bonettoi* to be contagiously distributed throughout the study period (Fig. 3). During August, October and February the greatest aggregation coincided with the breeding period.

Seasonal patterns in the abundance of the worms were observed (Fig. 4). They were more abundant during spring ($\bar{x} = 297$ ind. sample^{-1}) and summer ($\bar{x} = 314$) than during autumn ($\bar{x} = 62$ ind. sample^{-1}) and winter ($\bar{x} = 210$). The application of ANOVA to the abundance data showed that the differences between the means within both summer and autumn were not significant, $F = 2.17$ ($p > 0.05$) and $F = 0.90$ ($p > 0.01$), respectively. However, the differences between the means within both spring and winter were significant. The difference between the means within both periods, from February to July 1986 (with the lowest density) and from August 1986 to January 1987 (with the highest density) were not significant, with $F = 0.55$ ($p > 0.05$) and $F = 7.03$ ($p > 0.01$) respectively.

The calculation of the minimum sample number showed that with 10 replicates per sampling event the average error of the mean was 30%.

Life cycle

The size frequency data obtained from the immature and mature classes are show in Fig. 5. Mature specimens were more abundant in September 1985–86, resulting in greater breeding activity in early summer, when there was a large population increase. Another peak of breeding activity was observed in February, but it may be due to another generation. The decline in numbers in autumn appears to be directly related to the high water level regime.

From the summarized date in Fig. 5, it appears that *Narapa bonettoi* has a univoltine life cycle

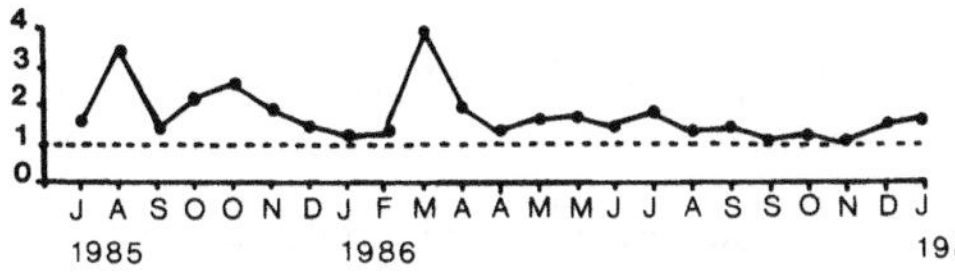

Fig. 3. Temporal changes in the dispersion pattern of *N. bonettoi* at the study site as measured by Morisita's Index of Dispersion. An Index of, 1.0 (--) indicates a random distribution, > 1.0 the distribution is contagious and < 1.0 regular.

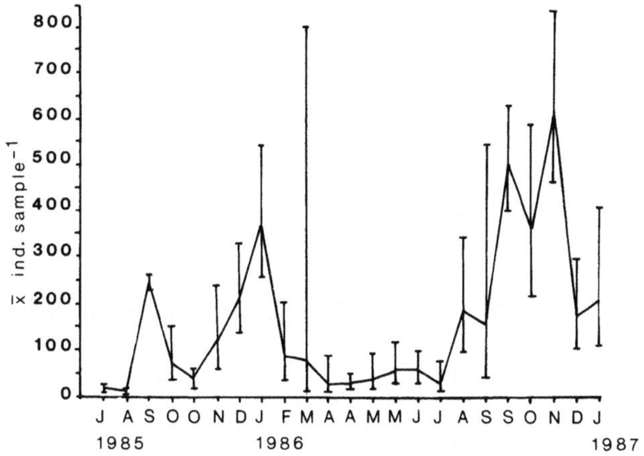

Fig. 4. Temporal changes in the average population density at the study site. The 95% confidence intervals shown are derived from $\log_{10}(x + 1)$ transformed data.

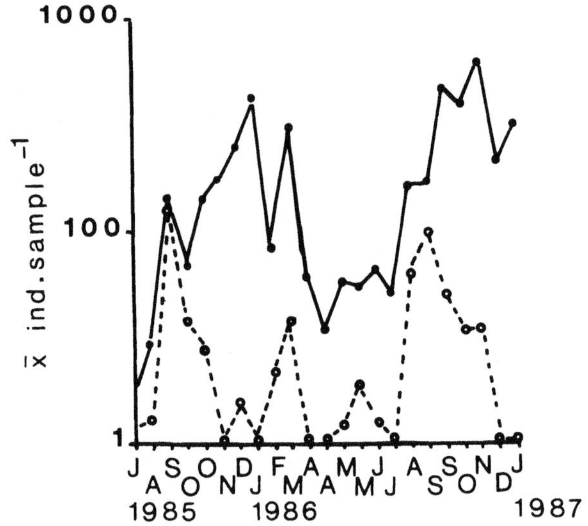

Fig. 5. Average number of ind. sample^{-1}, immature (solid) and mature (dash).

with a well defined breeding period in the spring months. A low proportion of the population breeding asexually, by fission, was also observed.

The smallest and largest *N. bonettoi* found were 0.97 mm (December 1985) and 4.90 mm (September 1985) (Fig. 6). Immature worms occurred in size classes from 0.97 mm up to 4.40 mm, and mature worms occurred from 1.8 up to 4.9 mm.

The smallest worms were observed in July, October, November and December 1985, and February, July, September, November and December 1986. These sampling periods are coincident with the breeding periods.

Maximum length was obtained using the Ford-Walford method and taking into account the average lengths for each sample. The calculated equation was $y = 1.645 + 0.328x$ that inter-

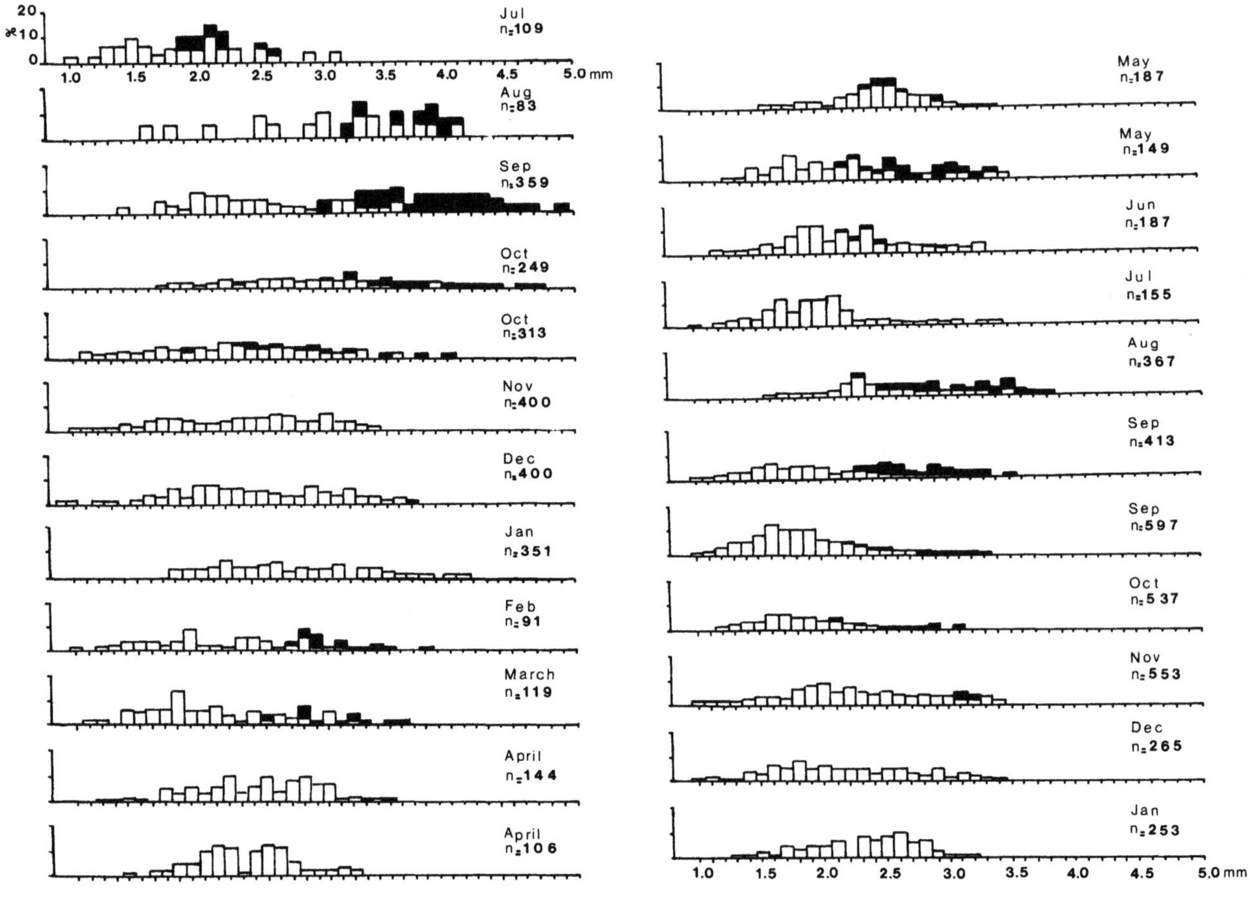

Fig. 6. Size frequency histograms based on the length (mm) of *N. bonettoi*. n = number of individuals measured. Immature (□), mature (■).

sects the quadrant's bisect xy for standard length = 2.46, and is a good approximation at $l\infty$. The growth curve (Fig. 7) obtained by Bertalanffy's equation showed rapid growth during the first month of life, then it is gradual until the fourth month, and from then on it was stable. The equation estimated was $L_t = 2.44\,(1 - e^{-1.115\,(t-0.603)})$.

Discussion and conclusions

The population density was high during the study period, except in autumn, when the water level increased. The spatial distribution was contagious with the greatest aggregation during the first months of the observed cycles, reaching a maximum value during periods of intense breeding activity.

Population growth was rapid during the first four months, after which it stabilized. This appears to be a strategy adapted to the rigorous habitat, with high current velocity and discharge, mobile sediments and sand beds. It is probable that most of the *N. bonettoi* reach maturity in approximately four months, some of them breeding more than once over a life-span. Therefore, there is considerable overlap between generations.

Two reproductive strategies were observed:

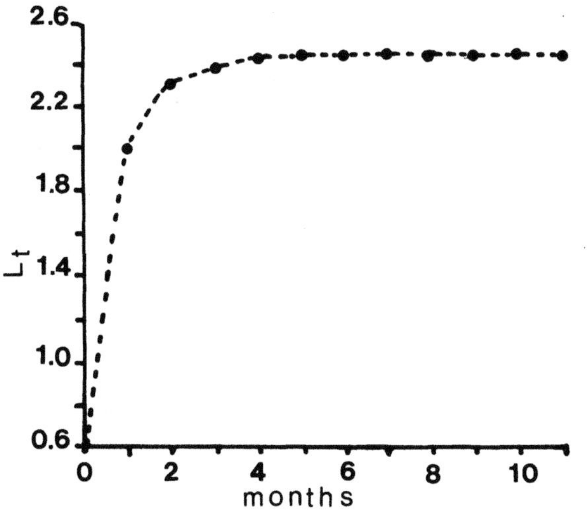

Fig. 7. Growth of *N. bonettoi* according to von Bertalanffy's equation.

asexual and sexual. It is noteworthy that this is the first observation of asexual reproduction in this species. Worms with asexual and sexual reproduction were found in the same period. This is not common among the oligochaetes. In Naididae, for example during favourable environmental conditions asexual reproduction is common, but when the conditions are unfavourable sexual reproduction takes place, mainly in high water and environments with variable temperature (Loden, 1981).

Cocoons may allow the population to survive environmental stress. Mehra (1920) refers to cocoon production as a strategy used when the conditions are adverse and Learner *et al.* (1978) has made a similar observation. *Narapa bonettoi* uses asexual reproduction as an adaptative strategy to assure its survival, because cocoon development would be more affected by environmental factors, such as the drifting processes of the mobile bed. After breeding, worms reabsorb their reproductive organs, returning to an immature condition.

In live specimens, it was observed that sand grains were attached to the body. It may be assumed that this is a stabilizing strategy to resist disturbance.

Acknowledgements

I am particulary grateful to Prof. I. Ezcurra de Drago for her valuable suggestions and critical reading of the manuscript.

References

Bertolid de Pomer, H., C. Copes, I. Ezcurra de Drago & M. Marchese, 1986. Caracteristicas limnologicas del rio Parana y sus principales tributarios en el tramo Goya-Diamante. Los sedimentos y su fauna. Rev. Asoc. Cienc. Nat. Litoral 17: 79–97.

Elliott, J. M., 1977. Some methods for statistical analyses of samples of benthic invertebrates. (3nd edn), Sci. Publ. Freshwat. Biol. Assoc., Ambleside 25, 160 pp.

Ezcurra de Drago, I., 1980. Campana limnologica 'Keratella I' en el rio Parana Medio: Complejo bentonico del rio y ambientes leniticos asociados. Ecologiá 4: 89–101.

Learner, M. A., G. Lochhead & B. D. Hughes, 1978. A review of the biology of British Naididae (Oligochaeta) with emphasis on the lotic environments. Freshwat. Biol. 8: 357–375.

Loden, M., 1981. Reproductive ecology of Naididae (Oligochaeta). Hydrobiologia 83: 115–123.

Marchese, M., 1981. Contribución al conocimiento del complejo bentónico del río Paraná medio. Ecología 6: 55–65.

Marchese, M., 1984. Estudios limnologicos en una seccíon transversal del tramo medio del río Paraná. XI: Zoobentos. Rev. Asoc. Cienc. Nat. Litoral 15: 157–174.

Marchese, M., 1987. The ecology of some benthic Oligochaeta from Paraná River, Argentina. Hydrobiologia 155: 209–214.

Marchese, M. & I. Ezcurra de Drago, 1983. Zoobentos de los principales tributarios del río Paraná medio en el tramo Goya-Diamante. Su relación con el cauce principal y cauces secundarios. Rev. Asoc. Cienc. Nat. Litoral 14: 97–107.

Marchese, M. & I. Ezcurra de Drago, 1992. Benthos of the lotic environments in the Middle Paraná River system.Transverse zonation. Hydrobiologia 237: 1–14.

Mehra, H. R., 1920. On the sexual phase in certain Indian Naididae. Proc. zool. Soc., London 31: 457–465.

Righi, G. & M. E. Varela, 1983. *Narapa bonettoi*, gen. nov., sp. nov. (Oligochaeta, Narapidae, Fam. nov.) de agua doce da Argentina. Rev. Asoc. Cienc. Nat. Litoral 14: 7–15.

Wetzel, R. G. & G. E. Likens, 1991. Limnological Analyses. 2nd Edition. Springer-Verlag, New York, 391 pp.

Hydrobiologia **278**: 309–313, 1994.
T.B. Reynoldson & K.A. Coates (eds), Aquatic Oligochaete Biology V.

Production of some oligochaete species in large Estonian lakes

Henn Timm
Institute of Zoology and Botany of the Estonian Academy of Sciences, Vôrtsjärv Limnological Station, EE2454, Rannu, Tartumaa, Estonia

Key words: Oligochaeta, lakes, production

Abstract

The production of *Potamothrix hammoniensis*, *Lamprodrilus isoporus*, and *Uncinais uncinata* was estimated in the eutrophic lakes Peipsi-Pihkva and Vôrtsjärv. Growth characteristics of these species are briefly described. Calculated production values were 6–32 (*P. hammoniensis*), 19–79 (*L. isoporus*), and 1–4 kJ m^2 (*U. uncinata*), respectively. The annual P/B ratios for the same species were, 0.8–1.7, 2.5–4.8, and 4.4–5.4, respectively.

Introduction

Lake Peipsi-Pihkva on the boundary of Estonia and Russia, and Lake Vôrtsjärv are the biggest inland water-bodies of Estonia (Fig. 1). Their biocenoses are best examined in comparison with other lakes. Studies of macrozoobenthos production in Estonia have only been carried out to date in these two lakes (Kangur, 1977; Antipova, 1982; Timm, 1990). Both lakes are shallow and eutrophic (Table 1).

Material and methods

The macrozoobenthos was sampled using the Borutski and Zabolotski grab bottom samplers (225 cm^2). Samples were taken once a week in summer, fortnightly in spring and autumn, and one or two times in winter. Four sampling sites were chosen in each lake (Fig. 1). The substrate was a brownish-grey mud at the sites 1 and 2 in Lake Peipsi-Pihkva (depth 9–10 m), a fine sand with *Dreissena polymorpha* at site 3 (6–7 m), and sand at site 4 (2–3 m). Insufficient data were available for calculation of the production of any oligochaete species at site 3. When calculating production, all sites from Vôrtsjärv (greyish mud, mud on sand or marl, depth 2.5–5 m) were treated together. A total of 826 samples were taken (1984–1986) from Lake Peipsi-Pihkva, and 408 samples (1984–1988) from Lake Vôrtsjärv.

The samples were washed through a 300 μm sieve. To estimate the possible numerical loss of

Table 1. Some data about the lakes (after Mäemets 1977, and Simm 1975).

	Vôrtsjärv	Peipsi-Pihkva
Area (km^2)	270	3555
Maximal depth (m)	6	15.3
Volume (km^3)	0.8	25
pH	7.1–8.8	7.6–8.6
HCO_3^- (mg/l)	171–227	118–146
Limnological type	Eutrophic, non-stratified, hard-water, with hyper-trophic features	Eutrophic, slightly stratified, with mesotrophic features in northern part

animals, a portion of the samples from the sandy bottom of Lake Peipsi-Pihkva were washed through a 100 μm sieve in 1984. A significant difference between 300 μm (usual) and 100 μm (control) samples was observed in the number of *Uncinais uncinata* in summer. Therefore, the standing stock and production of this species will be underestimated.

Live animals were sorted in the laboratory, fixed and preserved in 70% ethanol. To determine the effect of fixation on the wet weight of the animals, a part of the specimens were first weighed alive and then after weight stabilization in ethanol (3 months). The ratio of the fixed to live weight of oligochaetes from the sandy bottom of Lake Peipsi-Pihkva (mainly *Lamprodrilus isoporus*) was 0.58 and from mud sites (*Potamothrix hammoniensis*) 0.43. These ratios were used to calculate production in energy units.

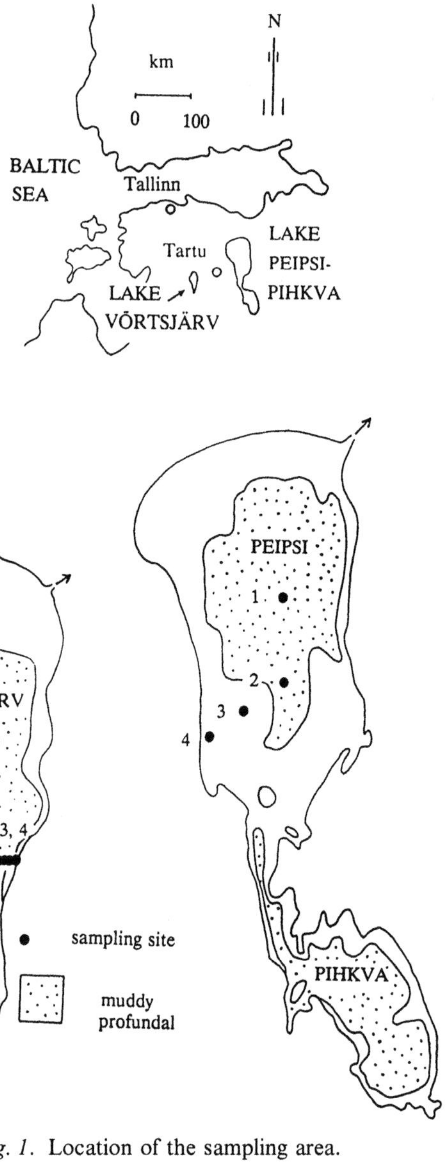

Fig. 1. Location of the sampling area.

For estimating the oligochaete production seven different, mainly indirect methods, were compared by the author using *L. isoporus* (Timm, 1987a). The instantaneous growth rate method (Chapman, 1978) was chosen as producing the most authentic production values. Cohorts were distinguished from the stage of sexual maturity and the width of the eighth segment.

Results and discussion

Production estimates of *P. hammoniensis*, *L. isoporus*, and *U. uncinata* are presented in Table 2. Due to the low density, and insufficient information about the cohort distribution, it was not possible to calculate production of the other oligochaete species (mainly tubificids). The error in the instantaneous growth rate production of a cohort of *L. isoporus* was calculated by the nonparametric bootstrap method (Efron & Tibshirani, 1986). The relative error of the cohort production was 7% and the error of P/B ratio was 28%.

The life cycle of *L. isoporus* in Lake Peipsi-Pihkva is from 1 to 1.5 years. Cocoons in sandy envelopes appear in winter. *Lamprodrilus isoporus* dominates both the standing stock and production of the macrozoobenthos on sandy substrates, at a depth of 2–3 m. The role of *U. uncinata* in the same biotope is small with respect to these parameters. Regardless of this, it had the highest P/B ratio (Table 2), although the methods of sampling and weighing did not include the smallest, asexually breeding specimens. Conventionally, *U. uncinata* is supposed to have one 'generation' per year.

In both lakes Peipsi-Pihkva and Lake Võrtsjärv, *P. hammoniensis* often has a 3 years life cycle. Cocoons usually appear in spring or at the beginning of summer. In the profundal, *P. ham-*

Table 2. Production of oligochaetes in Lakes Peipsi-Pihkva and Vôrtsjärv. P, B, and B_{max} – annual production, mean and maximum standing stock (all g m^2, live weight). In brackets production in kJ m^2, considering 5 g live weight equal to 1 g dry weight (T. Timm, personal information), and 1 g dry weight to 22.4 kJ (Brey *et al.*, 1988).

Lake, species & years	Sampling site	Number of samples	*P*	P/B	P/B_{max}
Peipsi-Pihkva					
P. hammoniensis					
1984–1985	1 & 2	176	3.3 (14.8)	1.2	0.6
1985–1986	1 & 2	210	3.1 (13.9)	0.8	0.3
L. isoporus					
1984–1985	4	145	4.3 (19.3)	2.5	1.6
1985–1986	4	105	17.6 (78.8)	4.8	2.2
U. uncinata					
1984–1985	4	145	0.2 (0.9)	5.4	0.5
1985–1986	4	105	0.8 (3.6)	4.5	0.2
Vôrtsjärv					
P. hammoniensis					
1987	1 & 2 & 3 & 4	288	7.2 (32.3)	1.7	0.5
1988	1 & 2 & 3 & 4	120	1.4 (6.3)	0.9	0.4

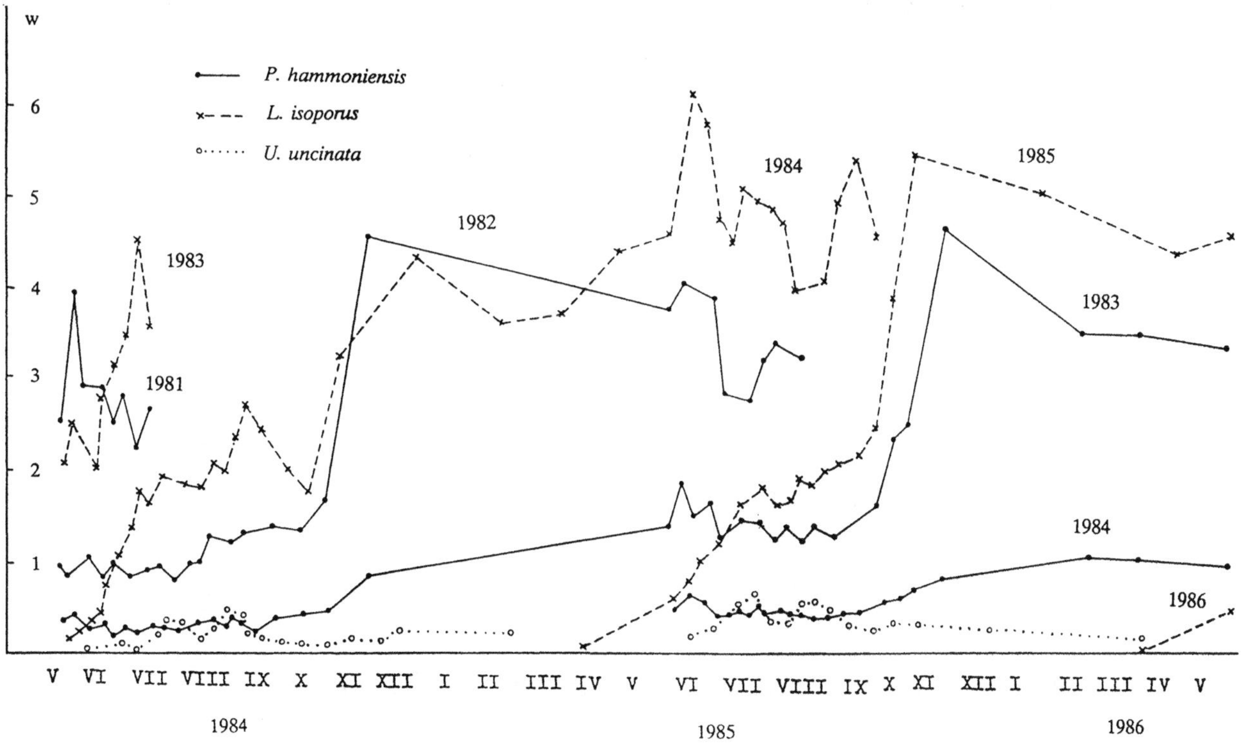

Fig. 2. Growth curves of some oligochaete species in Lake Peipsi-Pihkva. w – mean individual fixed weight (mg).

moniensis is usually the dominant oligochaete in terms of production but the midge *Chironomus plumosus* is much more important. For example, the production of this chironomid exceeded that of *Potamothrix* by 40 times at Lake Peipsi-Pihkva in 1984 (Timm, 1987b). This was due to the low growth rate and a relatively long age; the P/B of *P. hammoniensis* is smaller than in the case of other oligochaetes.

In Lake Peipsi-Pihkva, *P. hammoniensis* and *L. isoporus* grow intensively (at maximum production) in late autumn during total water circulation but not in summer during the period of maximum water temperature (Fig. 2). On the contrary, *U. uncinata* has the greatest abundance and highest production in summer. In Lake Vôrtsjärv, the growth of *P. hammoniensis* is almost constant throughout the year (Fig. 3).

The data from Estonian lakes compared with the literature (mean values) shows similar values in the production of *Potamothrix* in lakes (3.8 versus 3.5 g m^2 in wet weight respectively) (Gavrilov, 1985; Jenderedjian, 1989; Jonasson & Thorhauge, 1976; Popchenko, 1973), and *U. uncinata* versus naidids in lakes (both have production <1 g m^2, P/B 4.9 and 3.8, respectively) (Kamlyuk & Koval'chuk, 1972; Popchenko, 1973). The production and P/B of *L. isoporus* in the eutrophic Lake Peipsi-Pihkva exceeded slightly the corresponding values in the oligotrophic Lake Ladoga (Kuzmenko, 1987). The P/B of *Potamothrix*, as well as all P/B_{max} values proved to be lower in Estonian lakes when compared to the literature data.

Acknowledgements

I am greatly indebted to all my colleagues from the Vôrtsjärv Limnological Station who were engaged in collecting and sorting the production samples, as well as to the crew of the research vessel of the Tartu University 'Prometheus' from carrying out the expeditions on Lake Peipsi-Pihkva. I am especially grateful to Dr Tarmo Timm for his helpful discussion and criticism of the work, and for fitting the *P. hammoniensis* population from Lake Vôrtsjärv into cohorts. I would like to thank Dr Tônu Oja (Laboratory of Ecosystems of Tartu University) who composed

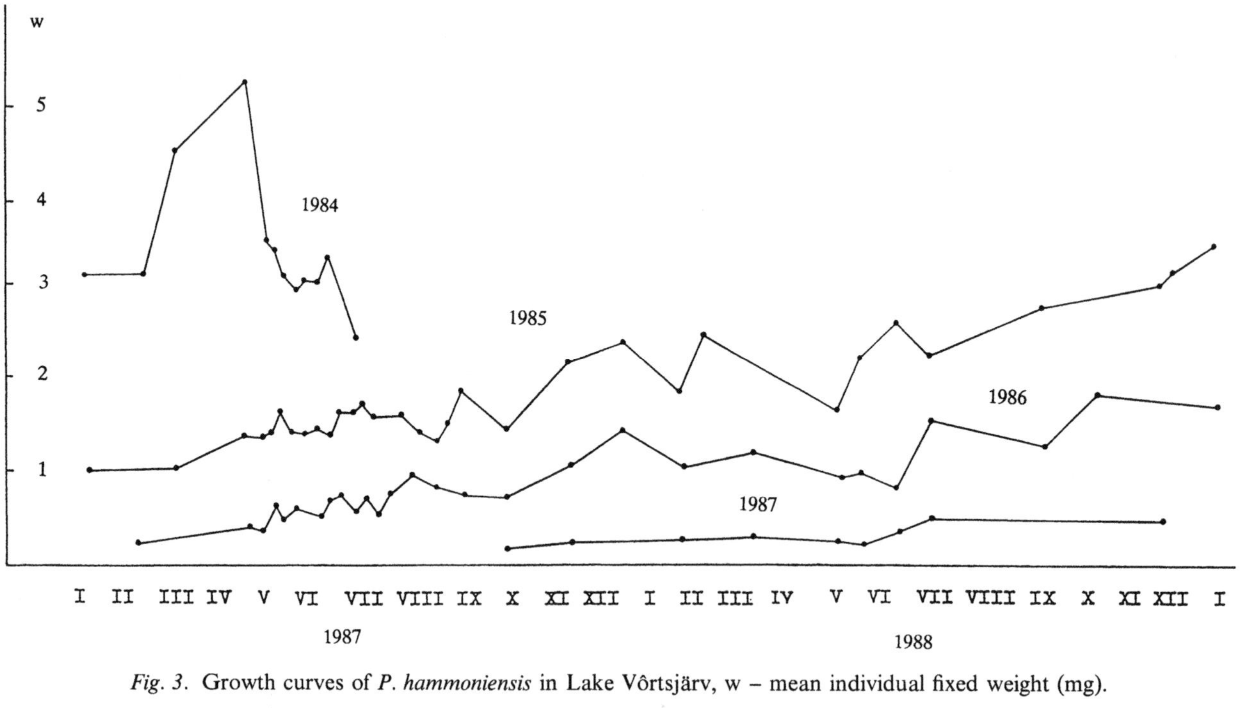

Fig. 3. Growth curves of *P. hammoniensis* in Lake Vôrtsjärv, w – mean individual fixed weight (mg).

a special program for estimating the accuracy of production values. Mrs Ester Jaigma is sincerely thanked for help with the English manuscript.

References

Antipova, L. F., 1982. Food basis and the efficiency of its utilization by main benthophagous fish species in Pskovsk-Chudskoyes lake system. Cand. Biol. Sc. Thesis. Leningrad (in Russian).

Brey, T., H. Rumohr & S. Ankar, 1988. Energy content of macrobenthic invertebrates: general conversion factors from weight to energy. J. Exp. Mar. Biol. Ecol. 117: 271–278.

Chapman, W. D., 1978. Production. In Methods for the assessment of fish production in fresh waters. T. Bagenall, Oxford: 202–217.

Efron, B. & R. Tibshirani, 1986. Bootstrap methods for standard errors, confidence intervals, and other measures of statistical accuracy. Statistical Science 1: 54–77.

Gavrilov, S. I., 1985. Macrozoobenthos. In The ecological system of the Naroch lakes. Minsk: 182–194 (in Russian).

Jenderedjian, K., 1989. Peculiarities in the biology and production of oligochaetes in Lake Sevan. Cand. Biol. Sc. thesis. Leningrad (in Russian).

Jonasson, P. M. & F. Thorhauge, 1976. Production of *Potamothrix hammoniensis* in the profundal of eutrophic Lake Esrom. Oikos 27: 204–209.

Kamlyuk, L. V. & M. M. Koval'chuk, 1972. Some data about the abundance, growth and production of *Stylaria lacustris* L. (Naididae, Oligochaeta) in the littoral of Lake Naroch. In Aquatic Oligochaeta. Materials of the Second All-Union Symp. held in Borok, June 27–30, 1972, Yaroslavl: 148–151 (in Russian).

Kangur, K. E., 1977. Ecology and productivity of *Chironomus plumosus* L. in Lake Vôrtsjärv. Cand. Biol. Sc. thesis. Tartu (in Russian).

Kuzmenko, K. N., 1987. Production characteristics of benthic invertebrates. In Contemporary state of Lake Ladoga. Leningrad: 137–139 (in Russian).

Popchenko, V. P., 1973. Productivity of oligochaetes in small humified waterbodies in Karelia. In Biological investigations on Baltic inland waterbodies. Minsk: 84–88 (in Russian).

Timm, H., 1987a. A comparison of different methods for calculating the production of aquatic oligochaetes. In Aquatic Oligochaeta. Materials of the 6th All-Union Symp. held in Salaspils, April 27–30, 1987, Riga: 135–139 (in Russian).

Timm, H., 1987b. Production of *Potamothrix hammoniensis* (Oligochaeta, Tubificidae) in the profundal of Lake Peipsi. Proc. Acad. Sci Estonian SSR, Biol. 36: 162–166.

Timm, H., 1990. Production of macrozoobenthos in Lake Peipsi. Proc. Estonian Acad. Sci. Biol. 39: 121–128.

Hydrobiologia **278**: 315–320, 1994.
T.B. Reynoldson & K.A. Coates (eds), Aquatic Oligochaete Biology V.

Observations on cohorts of *Tubifex tubifex* cultured at different food levels, using cellulose substrate

Andrea Pasteris [1], Carla Bonacina [2] & Giuliano Bonomi [1]
[1] *Dipartimento di Biologia Evoluzionistica Sperimentale, Università degli Studi di Bologna, Via S. Giacomo 9, I-40126 Bologna, Italy;* [2] *CNR – Istituto Italiano di Idrobiologia, Largo Tonolli 50, I 28048 Pallanza, Italy*

Key words: Oligochaeta, *Tubifex tubifex*, cohort cultures, population dynamics, food availability, cellulose

Abstract

In order to clarify the relationship between food availability and population dynamics in aquatic oligochaetes, short term cultures and cohort cultures of *Tubifex tubifex* were reared using substrates with different food concentrations, obtained by mixing sand with different amounts of cellulose powder.

In short term experiments, *T. tubifex* seemed capable of utilizing cellulose substrates for growth or egg production. Yet, in the cohort experiment, newly hatched worms could not grow on cellulose substrates. Survivorship, however, appears to be influenced by cellulose concentration giving an indication that this material, although inadequate, is somehow utilized.

Introduction

Food availability is supposed to play a major role in the distribution and abundance of tubificids (Brinkhurst & Kennedy, 1965; Brinkhurst, 1967). However, when the correlation of abundance of tubificids with total organic matter in sediments has been investigated, the results have been contradictory. Sometimes positive correlation has been observed (Brinkhurst, 1970), but in many instances tubificid abundance was not correlated or was even correlated negatively with the organic carbon content of sediments (McMurty *et al.*, 1983; Robbins *et al.*, 1989).

Poor correlation can occur because total organic carbon includes not only valuable food sources such as bacteria or recent algal remains but also substances resistant to decomposition and presumably unavailable as a food source (e.g., lignin and humic materials). For this reason total organic carbon is an unsatisfactory measure of the nutrients available to the tubificids.

Indeed, in the late sixties Brinkhurst (1967) was already suggesting that some more critical evaluation of the type of food material available in the sediments might be required to establish habitat preferences for these worms. We think that a suitable measure of food availability and an adequate knowledge of the trophic requirements of different species are still far from being achieved.

We believe that laboratory experiments in which demographic parameters and individual growth rates are measured using organisms fed on artificial sediments of known composition may play a significant role in arriving at an understanding of the value of different sediment fractions as a food source for different tubificid species. Such experiments may also help to improve our estimates of population production. In fact, these estimates require the classification by size of individuals and knowledge of the residence time in each size class. Tubificid worms do not show recognizable cohorts in nature, thus size class durations must be estimated from laboratory cultures. Currently, production estimates rely on the assumption, often implicit, that only temperature influences size class duration. This seems to be because there is little laboratory data on the effect on production of other factors, including food availability.

Compared to the importance attached to trophic conditions in the ecology of tubificid worms, little experimental work has been done in the laboratory with the aim of investigating the relationship between food availability and the population dynamics of these organisms.

Kosiorek (1974) compared growth and fecundity among groups of *Tubifex tubifex* (Müller, 1774) fed with yeast, bacteria, tainted lettuce and rat food. Aston & Milner (1981) reared *Branchiura sowerbyi* Beddard, 1832 on activated sludge mixed in different proportions with sand, silt, clay or river mud. Aston (1984) cultured *B. sowerbyi* on substrates in which the basic energy source was cellulose at different concentrations. Finogenova & Lobasheva (1987) compared the growth and fecundity of *T. tubifex* on activated sludge and natural silt.

In our opinion, the contribution these works make to a general view of the influence of trophic factors on the population ecology of tubificids is diminished by one or both of two experimental characteristics: the use of poorly standardized and measurable foods, and the short duration of experiments preventing the observation of an adequate portion of the life cycle. The present work is an attempt to remedy these limitations.

Cellulose, a major component of the organic fraction of sediments, is commercially available in preparations of constant quality and Aston (1984) reported that *B. sowerbyi* is able to grow and lay eggs when cultured on cellulose substrata. Thus cellulose seemed to be a suitable material for our experiments. Rather than the filter paper used by Aston (*op. cit.*), we used cellulose powder because we found it easier to mix this homogeneously with sand.

Materials and methods

Worms (*T. tubifex*) were collected from Lake Orta, N. Italy, and kept for some months in a tank, with water temperature varying from 7 °C to 15 °C, before they were used for the experiments. During all the experiments the worms were maintained at 20 °C in glass jars (diameter: 8 cm). The jars were half filled with a mixture of calcinated sand, diatomic powder and cellulose powder (Aldrich, particle size: 20 μm). A thin layer of sand was added on top, then the vessels were filled with mineral water (nitrate content in the region of 12 mg l^{-1}).

Short term experiments

Two short term experiments were carried out to verify whether it was possible to culture *T. tubifex* using cellulose powder as the food source and to identify a range of cellulose concentrations for the cohort experiment.

Worms had an initial weight between 2 and 6 mg, which was determined after 24 hours in water without sediment, allowing clearance of the gut contents and drying with filter paper. Then they were randomly divided into small groups and reared on substrates with different cellulose contents. At each substrate composition two cultures were done. Survivors and deposited eggs were counted and substrate renewed every week. After roughly one month the worms were weighed again using the same procedure.

The first experiment lasted 32 days. The initial number of worms was 6 per culture. Substrate composition is given in Table 1. The second experiment lasted 33 days. The initial number of worms was 8 per culture. Substrate composition is given in Table 2.

Table 1. First short term experiment: substrate composition, average weight of each culture at the end of the experiment (± standard error) and significance level of the t-Test between the weight of all the individuals before the experiment (mean: 40 mg, standard error: 0.1 mg) and the weight of the individuals in each culture at the end of the experiment.

Substrate composition				Average weight at the end of the experiment (mg) ± standard error	Significance level
Bottom layer			Upper layer		
Cellulose		Sand	Sand		
(g)	(%)	(g)	(g)		
0.00	0.00	60.00	40.00	2.6 ± 0.2 2.8 ± 0.3	0.001 0.003
1.00	1.67	59.00	40.00	5.4 ± 0.6 6.1 ± 0.8	0.001 0.001
3.00	5.00	57.00	40.00	5.1 ± 0.2 5.4 ± 0.5	0.017 0.001
9.00	15.00	51.00	40.00	5.9 ± 0.7 6.2 ± 0.9	0.021 0.001

Cohort experiment

The cohorts were started from cocoons laid by a "mother culture" of about 150 individuals kept for two months at 20 °C and fed with frozen spinach. Each culture was started with 39–41 eggs (10–11 cocoons) laid over a four-day time span. Cocoons were buried in the sand using a pipette. Five levels of cellulose were used (substrate compositions in Table 3). Two complete sets of five cultures, named A and B, were prepared. The vessels were submerged in two different water tanks, one for each set. The water in the vessels was free to mix with the common water of the tanks which was gently aerated. Every fortnight, the substrate was renewed, the cohorts were examined and survivors counted and placed again in the vessels. Cocoons containing viable embryos were reburied in the bottom sediments.

Results

Short term experiments

During the first short term experiment no mortality occurred in the cultures except in one with 5.00% cellulose and one with 15.0%, in which 2 and 1 individuals died, respectively. Cocoons

Table 2. Second short term experiment: substrate composition and observed survivorship, fecundity (average number of eggs per individual) and percent variation in average weight over a period of 33 days.

Substrate composition					Survivorship	Fecundity (eggs/ind)	Percent variation in average weight
Bottom layer				Upper layer			
Cellulose		Diatomic powder	Sand	Sand			
(g)	(%)	(g)	(g)	(g)			
0.00	0.00	5.00	75.00	20.00	0.38 0.38	2.1 0.0	-41 -35
0.50	0.63	4.50	75.00	20.00	1.00 0.88	7.6 10.8	-36 -15
1.00	1.25	4.00	75.00	20.00	0.88 0.88	13.6 11.3	-16 -24
2.00	2.50	3.00	75.00	20.00	0.13 1.00	7.5 10.9	-31 -27
4.00	5.00	1.00	75.00	20.00	0.75 1.00	15.0 22.1	3 -5

Table 3. Cohort experiment: substrate composition.

Bottom layer				Upper layer
Cellulose		Diatomic powder	Sand	Sand
(g)	(%)	(g)	(g)	(g)
0.16	0.20	6.09	73.75	20.00
0.40	0.50	5.85	73.75	20.00
1.00	1.25	5.25	73.75	20.00
2.50	3.13	3.75	73.75	20.00
6.25	7.81	0.00	73.75	20.00

were found only at the final count (32 or 33 days) ten and 26 eggs were found in the cultures with 1.67% cellulose.

The average individual weight of the worms before the experiment was 4.0 mg (standard error: 0.1 mg). The average weight of each culture at the end of the experiment is shown in Table 1. Average weights of worms increased in all cultures with cellulose and decreased in two with no cellulose. Student's t-Test between the weight of all the individuals before the experiment and the weight of the individuals in each culture at the end of the experiment is always significant and the significance level is reported in Table 1. While the difference between cultures with and without cellulose is evident, no difference is detectable among substrates in which cellulose was present in different proportions.

The survivorship at the end of the second experiment (Table 2) does not appear to be influenced by the proportion of cellulose in the substrate. The average number of eggs laid by one individual during the experimental period is shown in Table 2. The effect of cellulose concentration on egg production results is significant if tested by means of one way ANOVA. Average weight at the end of the experiment is lower than at the beginning in all cultures except the ones at 5% cellulose, the highest concentration used in the second experiment, but the weight loss is statistically significant only in some cases. However, the effect of cellulose concentration on percent variation of average weight (Table 2) is significant if tested by means of one way ANOVA ($p<0.05$). Egg production and weight variation seem to be affected not only by the presence or absence of cellulose but also by the proportion of cellulose in the substrates.

Cohort experiment

Cohorts were reared for a period from 180 to 197 days. During this period worms failed to grow, they seemed to get no bigger than newly hatched individuals.

In spite of their apparent starvation hatched worms survived for quite a long time. The survivorship curves are shown in Fig. 1; the 0.20% culture of the set B was accidentally lost at day 42, so the corresponding survivorship curve is not given. The numbers of embryos should be considered only rough estimates as only embryos in clearly degenerated cocoons were counted as dead (difficulty in identifying viable individuals). The survivorship of the cohorts at the end of the hatching period is rather low: from 32 to 50%. A period of low mortality lasts until day 120 when, in several cultures, a considerable decline in the number of survivors begins.

The lower limit of the survivorship span shown in Fig. 2, 60 days, was chosen because after this date cocoons were no longer present in the cultures and so embryonic mortality, likely unaffected by food availability, is excluded from the analysis. The upper limit, 180 days, is the minimum number of days a cohort was reared. Survivorship is higher at the intermediate cellulose concentrations, with cohorts of set B showing higher survivorship than corresponding cohorts of set A. Differences among cellulose concentrations and between the two sets of cultures are both significant if tested by means of two way ANOVA after arcsine transformation (Sokal & Rohlf, 1981: 427); the single 0.2% culture was grouped with the 0.5% cultures.

Discussion and conclusions

As stated above, the short term experiments were aimed at obtaining preliminary information for

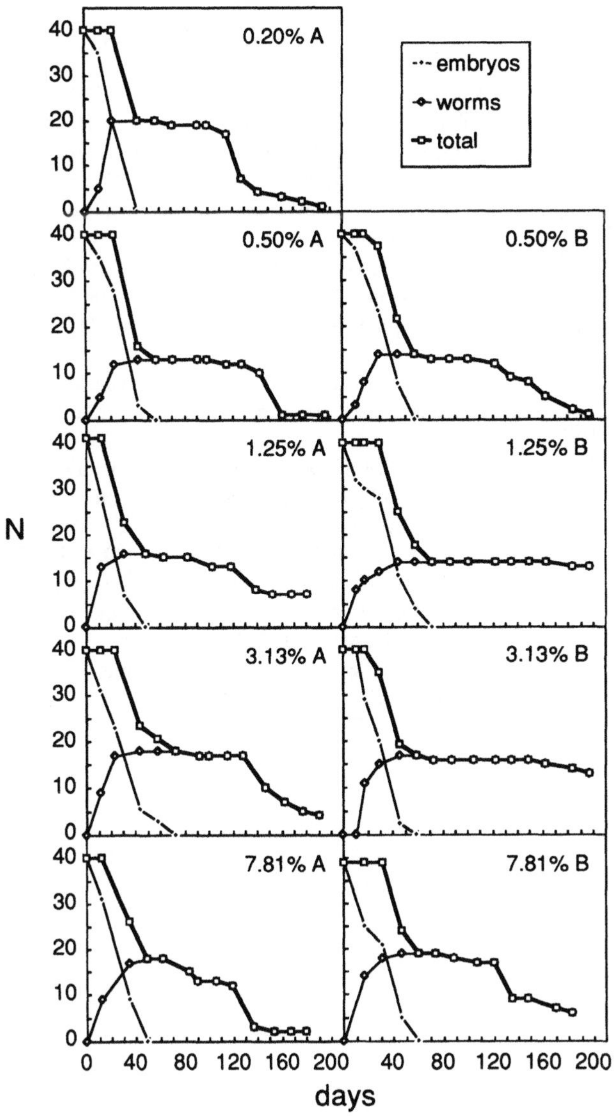

Fig. 1. Number of embryos or worms per culture and total survivorship in cohorts cultured on substrates containing different proportions of cellulose.

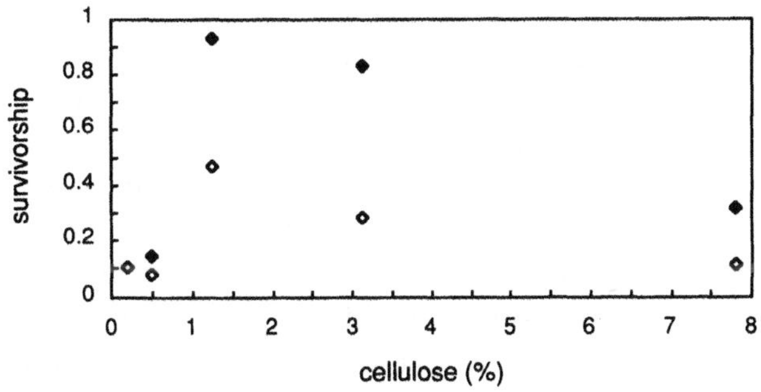

Fig. 2. Survivorship between 60 and 180 days (i.e. the proportion of individuals alive at day 60 and that survive until day 180) in cohorts cultured on substrates containing different proportions of cellulose. White diamonds: cohort A; black diamonds cohort B.

the design of the long and laborious cohort experiment. The outcomes of the two short term experiments were quite different: while during the first experiment worms laid few eggs but increased their weight, during the second, in contrast, worms laid several eggs but did not grow and in most cultures even lost weight. As the ranges of cellulose concentrations largely overlapped, it seems unlikely that such dissimilarities were due to the differences in the substrates used in the two experiments, primarily the presence or absence of diatomic powder. The cause of the observed differences might be the different water temperature of the stock culture when individuals for the first and the second experiment where collected, 14 and 8 °C, respectively. The greater temperature difference between the stock culture and the experimental cultures (20 °C) in the second experiment might have acted as a stimulus to reproductive activity biasing resource allocation towards eggs production rather than growth. However, despite these differences, *T. tubifex* seemed capable of utilizing cellulose substrates. During the first experiment worms increased in mean weight by 20 to 60% when cellulose was present, while they lost weight, by 30%, when cellulose was absent. In the second experiment cellulose proportion in the substrates is positively correlated with the number of eggs laid and negatively correlated with weight loss.

The short term experiments, and similar experiments of Aston (1984) on *B. sowerbyi,* suggested that it was possible to culture tubificids using cellulose powder as the food source. Yet, in the cohort experiment, newly hatched worms could not grow on substrates in which the basic energy source was cellulose. However, survivorship seems to be influenced by cellulose concentration, giving an indication that this material, although inadequate, is somehow utilized. In the culture with the highest cellulose content, microorganism colonies formed a thick tangle in which worms were apparently trapped. This could explain the low survivorship at the highest cellulose concentration.

There is no obvious explanation for the observed differences between set A and set B. Certainly they reveal that standardization of culture conditions, one of the goals of our experimental design, was not achieved. These experiments should be regarded merely as a first attempt to deal with the effects of food availability on the population dynamics of tubificids by means of laboratory cultures. A major drawback is that microbiological aspects have been completely neglected.

References

Aston, R.J., 1984. The culture of *Branchiura sowerbyi* (Tubificidae, Oligochaeta) using cellulose substrate. Aquaculture 40: 89–94.

Aston, R.J. & A.G.P. Milner, 1981. Conditions required for the culture of *Branchiura sowerbyi* (Oligochaeta:Tubificidae) in activated sludge. Aquaculture 26: 155–160.

Brinkhurst, R.O., 1967. The distribution of aquatic oligochaetes in Saginaw Bay, Lake Huron. Limnol. Oceanogr. 12: 137–143.

Brinkhurst, R.O., 1970. Distribution and abundance of tubificid (Oligochaeta) species in Toronto Harbour, Lake Ontario. J. Fish. Res. Bd Can. 27: 1961–1969.

Brinkhurst, R.O. & C.R. Kennedy, 1965. Studies on the biology of the Tubificidae (Annelida; Oligochaeta) in a polluted stream. J. anim. Ecol. 34: 429–443.

Finogenova, N.P. & T.M. Lobasheva, 1987. Growth of *Tubifex tubifex* Müller (Oligochaeta, Tubificidae) under various trophic conditions. Int. Revue ges. Hydrobiol. 72: 709–726.

Kosiorek, D., 1974. Development cycle of *Tubifex tubifex* Müller in experimental culture. Pol. Arch. Idrobiol. 78: 71–85.

McMurty, M.J., D.J. Rapport & K.E. Chua, 1983. Substrate selection by tubificid oligochaetes. Can. J. Fish. aquat. Sci. 40: 1639–1646.

Robbins, J.A., T. Keilty, D.S. White & D.N. Eddington, 1989. Relationship among tubificid abundances, sediment composition and accumulation rates in Lake Erie. Can. J. Fish. aquat. Sci. 46: 223–231.

Sokal, R.R. & F.J. Rohlf, 1981. Biometry. Freeman & Co., New York, 859 pp.

GPSR Compliance

The European Union's (EU) General Product Safety Regulation (GPSR) is a set of rules that requires consumer products to be safe and our obligations to ensure this.

If you have any concerns about our products, you can contact us on ProductSafety@springernature.com

In case Publisher is established outside the EU, the EU authorized representative is:

Springer Nature Customer Service Center GmbH
Europaplatz 3
69115 Heidelberg, Germany

Batch number: 10800763

Printed by Printforce, the Netherlands